ACCLAIM FOR EVAN EISENBERG's

The Ecology of Eden

"Eisenberg proves to be one of those all-too-rare literary creatures: a serious environmental thinker who is also a sprightly entertainer and a born raconteur. He winningly riffs his way across centuries of history and broad swaths of science, tracing how our notions of 'a time or place of perfect harmony between humans and nature' have both inspired hopeful nostalgia and collided with reality." —*Outside*

"Delightfully written . . . wonderfully original and provocative." —*The Boston Globe*

"A blend of information and visionary knowledge . . . around which a new cultural moment might form." —*The Reader's Catalog*

"A broadly rooted yet gorgeously precise narrative that ranges from biblical interpretation to musings on wilderness, agriculture, bacteria, chaos theory, and music. . . . A breathtakingly versatile and sagacious performance." —*Booklist*

"A huge, audacious book overflowing with ideas. . . . Delightful, passionate, thought-provoking." —*The Kansas City Star*

"Far-reaching . . . a rewarding and fascinating endeavor." —*The Plain Dealer*

EVAN EISENBERG

The Ecology of Eden

Evan Eisenberg's first book, *The Recording Angel*, a pathbreaking study of the cultural impact of recorded music, has been translated into French, German, and Italian. His writings on nature, culture, and technology have appeared in *The Atlantic*, *The New Republic*, *The Village Voice*, *Natural History*, and other periodicals. He has been a music columnist for *The Nation*, a synagogue cantor, and a gardener for the New York City Parks Department. Born in New York City, Eisenberg studied philosophy and classics at Harvard and Princeton, and biology at the University of Massachusetts at Amherst. He lives in Manhattan with his wife, an urban planner, and their daughter.

ALSO

BY

EVAN EISENBERG

The

Recording

Angel

EVAN EISENBERG

THE
ECOLOGY
OF EDEN

VINTAGE BOOKS
A DIVISION OF RANDOM HOUSE, INC.
NEW YORK

The Library of Congress has cataloged the Knopf edition as follows:
Eisenberg, Evan.
The ecology of Eden / Evan Eisenberg.
p.   cm.
Includes index.

1. Environmentalism.   2. Nature—Effect of human beings on.
3. Human ecology.   I. Title.
GE195.E38 1998
304.2—dc21   97-29329   CIP

ISBN-13: 978-0-375-70560-1

ISBN-10: 0-375-70560-0

www.vintagebooks.com

FOR FREDA

A thicket teeming with slender leaves grew in our courtyard
Sprung from the trunk of a full-grown olive, thick as a pillar
—Odyssey

CONTENTS

ACKNOWLEDGMENTS

MANY SUBJECTS ARE TOUCHED on in this book, and I am not an accredited expert on any of them. I have followed the trail of my inquiry wherever it led, tramping from one field to another and perhaps leaving some well-plowed furrows in disarray. To the scholars and scientists from whom I have learned so much—a debt my notes and bibliography can only sketch—I offer my apologies and my thanks.

To make this book as accurate as the present state of knowledge admits, an army of research assistants, fact-checkers, graduate students, expert readers, and benevolent professors would have been required. Failing that, a small troop of learned friends and acquaintances have read the book, or parts of it, at various stages in its growth and have offered suggestions, corrections, kicks in the pants, and shots in the arm. Ellen Bernstein, Mary Bernstein, Walter Dubler, Jeff Eiseman, Joel Kaminsky, Ann Marlowe, John Martin, Bill McKibben, Adam Narva, Paul Sanford, and Eilon Schwartz offered detailed comments; Gerald Cohen, Lenny Groopman, Yochanan Muffs, David Orr, David Rothenberg, Michael Singer, Caroline Stern, Burt Visotzky, and Lotte Weisz offered general comments and kind words, which were also prized. Many of these people have also shared ideas with me in conversation or correspondence—as have Jon Ashen, Larissa Brown, Shoshana Brown, John Carlucci, Gershom Gorenberg, Art Green, Josh Gutoff, Charles McKinney, David Seidenberg, Ira Sharkey, Steve Shaw, Elisabeth Sifton, Gordon Tucker, the members of Havurat Ha-Emek in western Massachusetts and Minyan Ma'at in Manhattan, and many others.

A few readers deserve special mention. Jon Beckmann provided a fresh, keen eye and a level head when both were badly needed—and, for good measure, an idyll in his own Arcadia, otherwise known as Sonoma. If I may borrow his own metaphor, he helped make the sprawling delta of the latter part of the book into something like a navigable channel.

Bill MacLeish and Dick Todd read the book when its spirits were drooping and gave it a well-timed boost. Mitch Thomashow found a solution to a problem that other readers had detected but could not resolve. Dan

Hillel read the book and offered his blessing; I am indebted, as well, to his writings, his conversation, and his example of interdisciplinary vigor.

Herb Bernstein read most of the manuscript and offered detailed comments, sometimes doing the math I had fudged. Also, he kept up a constant flow of humor, warmth, and mental stimulation, the last supplied both by the seminars of his Institute for Science and Interdisciplinary Studies and by his own polymathic perversity. I am grateful to ISIS, the University of Massachusetts at Amherst, and Harvard's Center for the Study of World Religions for allowing me to present some of my ideas to scholarly and responsive audiences; and to the libraries of the University of Massachusetts for their splendid views and liberal loan policies.

Lynn Margulis generously read several chapters with her usual acumen, making detailed and very useful comments (not all of which I was able to incorporate). No doubt there remain many things in this book with which she will disagree; but I hope she will not regret the inspiration I have drawn from her writing, from her teaching, and from the model she offers of a scientist interested in everything and cowed by no one.

No one cows Wes Jackson, either, and to him I am grateful for reading and commenting on the book, as well as for his friendship and intellectual fire. Readers who know their work will recognize Wes, Lynn, and Wendell Berry (whom I have had the pleasure of talking with a few times, and of reading many times) as the guiding spirits of this book.

The errors that remain are my own (some of them, no doubt, so deeply my own that no one else could have made them). Readers who bring them to my attention will win my undying thanks, especially if they do it nicely. With their benevolent weeding, this book may at length become a garden—or a wilderness—not wholly unworthy of its title.

THAT SOUNDS LIKE A PERORATION, but I have other people to thank. Besides showing patience above and beyond the call of a publisher's duty, Sonny Mehta kindly entertained an author's unschooled thoughts about design; so did Knopf's gifted art director, Carol Devine Carson. The production staff at Knopf—in particular, the designer, M. Kristen Bearse; the production manager, Tracy Cabanis; and the copy editor, Terry Zaroff-Evans—did a superb job under harrowing conditions, most of them of my making. I am very grateful to them. Though they may never want to read another sentence of mine, I hope they will read these three.

Sophie Cottrell gets thanks in advance for draping my name across every bookstore vitrine in the land. Claudine O'Hearn was the intermediary between delinquent author and impatient production staff, a thankless task if ever there was one, for which she has my heartfelt thanks. So has my editor, Dan Frank, who mixed patience with perspicuity and rigor, insisting from first to last that this book not be less than it might be.

To my literary agent, Joe Spieler, I am grateful for a host of ill-assorted things, ranging from diplomacy worthy of Talleyrand to Zen koans, Yiddish anecdotes, and bicycles. Above all, I am grateful to him for standing by his author through thick and thin. And I am grateful to those colleagues and European counterparts of his I have had some contact with, among them Abner Stein, Suzy Violette, Paul Marsh, Ada Muellner, and John Thornton.

For their hospitality in Jerusalem, Cheveuoz, and London, I am grateful to the Fishman and Thayer families, and to Ramsey Margolis.

Since blood is thicker than ink (or is it? Too late to check this fact), my greatest debts are to kith and kin. Many in those two categories have helped this project along, if only by not asking how it was going. In particular, my uncles Victor Shargai and John Aniello read a draft and offered comments and encouragement. My grandfather and grandmother, Harry and Millie Scharaga (who, by the way, celebrated their sixty-fifth wedding anniversary this year) were a one-man clipping service and a one-woman cheerleading squad. My parents, Arlene and Howard Eisenberg, shared the wisdom gained in two lifetimes of plying the writer's trade; they gave sterling editorial advice, sturdy last-minute research assistance, and constant moral support, assuring me I was on the right track even when I was blinded by approaching headlights.

In a twist on the story of Penelope and Odysseus, my wife Freda bravely made her way in the world while I sat weaving, unweaving, and reweaving my tapestry. For her, may the long-delayed completion of this book be a homecoming. Though it ill becomes a writer to quote himself, I am reminded of something I wrote in the acknowledgments to my first book (the context being that there are some performances, in life as in music, to which no record can do justice): "May the love, courage, and grace of my wife Freda live on only as legend." But in fact those things have lived on, not as legends but as realities for which I am grateful every day of my life.

Last and littlest, but hardly least, comes my daughter, Sara Xing. She has filled the bleakest days with laughter, deftly putting this book—and everything else in the universe—in its proper place.

Leverett, Massachusetts
March 1998

INTRODUCTION

FOR THOUSANDS OF YEARS we have dreamed of going back. To what, exactly?

Eden is a home from which we have been evicted. We stand on the sidewalk on tiptoe and find that the window is just a few inches too high to peer in. Exactly what evicted us? History? Society? Sheer numbers? The need for food? The retreat of the glaciers? Adulthood? Birth?

This is a book about humankind's place in nature, real and imagined. There are several reasons why such a book should find its center in images of paradise. First of all, those images show our muddled feelings about our place in nature: our guilty pride, our snug discomfort. They reveal our sense that something has gone wrong somewhere down the line. By imagining a time or place of perfect harmony between humans and nature, they indict the discord we feel here and now. Whether the harmony they imagine is, or was, real is a good question, but in a way beside the point. The discord is real enough.

Besides, if you insert a probe into any body of environmental thought, you will find, somewhere near its heart, a firm if amorphous idea about Eden. Consider (as we shall have occasion to do later) two schools of thought conspicuous in the present debate: I will call them the Planet Managers and the Planet Fetishers. The Fetishers dream of returning to Eden, restoring a state of harmony in which wilderness reclaims the planet and man is lost in the foliage, a smart but self-effacing ape. The Managers dream of a man-made paradise, an earth managed by wise humans in its own best interest and, by happy chance, humankind's as well. The Fetishers want to get past the fiery sword that guards Eden by crawling humbly under; the Managers, by vaulting over.

Both these dreams grow from confusions about our role in nature. Mainly, they grow from a failure to see that wildness and civilization have, and must have, their own domains. The health of the planet depends on the health of both. Though they are always shifting and always mingling, it is crucial that each maintain a distinct center where it is free to follow its own laws. The human condition is (among other things) the state of trying to keep one's balance on the shifting soil between these two domains.

That balancing act is what this book is about. In order to stand in a right relation to wilderness, one must stand in a right relation to civilization. It is tricky, for by temperament or argument each of us leans toward one or the other. Since ancient times, some peoples have felt that the Mountain— the wild place, the source of water and life—must be the center of the world. Others have claimed that role for the Tower, the human construction from whose heights nature is scanned and controlled. Still others have sought a middle realm where wildness and civilization are mingled in just the right proportions—a place known to the ancients as Arcadia, to us as suburbia.

In these pages I will look at various ideas or images of our role in nature and compare them with the facts (in other words, with *my* ideas, formed by what I know of science, history, and daily life). I will try to see where certain ideas have done good service and where they have led us astray. I will try to revive old ideas that might yet be useful, patch them up, and put them back to work. I will put forward some new ideas and ask you to try them on for size, holding up a mirror for your convenience.

But this is not a fashion show. Ideas matter. Ideas that are elegantly made, but do not fit the human body or the human soul, may cramp us or trip us up. Ideas unsuited to the outside world may leave us parched, drenched, or frozen. The ideas now vying for our custom—Planet Management, Deep Ecology, bioregionalism, Gaia—may or may not fit. Either way, they will, if we adopt them, change us: by causing us to act in accordance with their dictates, or by causing us to think one way and act another.

A Jewish curse goes, "May he inherit a hotel of a hundred rooms, and be found dead in every one of them." For the heirs of Western civilization, the curse seems to be coming true. We live in a first-class hotel, a place of remarkable beauty and comfort, full of marble busts and leather-bound books; yet in one room the tap water tastes of indeterminate chemicals, in a second insects are dying in midair and plummeting to the floor, in a third the temperature rises ominously and there is no way to open the window. The fact that it is our nonhuman guests who are dying first—first them, and then the non-Western humans in the cheaper rooms—does not offer much reassurance.

The search for Eden and its meaning is not just a sentimental journey; it is a matter of life and death. I will dwell on our present ecological ills only where they illustrate a point; but the newspaper headlines that shout of doom (and if one out of ten is accurate, that is doom enough to go around) must be understood as the context of the book, part of the concrete meaning of exile from Eden.

How do you strike a balance between nature and culture while being knocked about by waves of change? How do you find a paradise that does not depend on the misuse of nature or your fellow humans? How do you live in harmony with nature when nature itself is a caterwauling brawl?

To answer these questions, it is not enough to examine the way things stand at present. Freud once said that the chunks of dream you remember on waking are like the toadstool or visible part of a mushroom, beneath which a tangle of fibers leads you deeper and deeper into the moist darkness of the unconscious. In the case of the dream of paradise, those fibers lead us back into history and prehistory, in whose varying darknesses we find not only the roots of the dream but the pulpy stuff of conflict and decay on which it feeds.

But this makes the way before us seem narrower and muckier than it really is. The journey inward is also a journey outward, traversing great spans of time and space. It takes us through scenes of profound beauty, high irony, and low comedy: a little song, a little dance, a little seltzer down your pants (the last in a more literal sense than you might imagine).

The book has four parts. The first describes, in biological terms, the waves of change that have swept the planet at narrowing intervals over the past few million years. The second recounts some of the myths people have shored up against those waves, mainly in the ancient Near East, and asks what ecological facts the myths reflect, cover up, or help create. Part Three chases this inquiry across the whole length of Western history and a good stretch of our mental landscape, from pastoral verse to housing developments and from gardens to the Gaia hypothesis. Part Four, finally, applies the lessons of this journey to the problems of the present hour.

Let me see if I can present, in a few pages, a sort of overture to the drama ahead, sounding some of the themes that will be featured in each act.

Start with Part One. Viewed through a biological lens—either the close-up lens that focuses on the molecular commerce of matter and energy, or the wide-angle lens that takes in great arcs of space and time—the human remaking of the planet may not look either as villainous or as heroic as we tend to think. Have we really conquered the world single-handedly, or have we had help from other species? If we have had help, have our own alliances been the first to march across the globe, driving out other species and trampling ecosystems into the dust? Or might such conquests have been going on for millions of years before we stumbled onto the scene?

Suppose it turns out that our remaking of the world is in some sense "natural." Does that mean it is nothing to lose sleep over? Before we jump to this restful conclusion, we need to consider the particular effects of our conquests on the ecological systems we depend on. A difference of degree can become a difference of kind; a shift in tempo can turn a jaunty allegro into a dance of death.

In Part Two we turn to myth: in particular, the myths of the ancient Near East, which pose questions with which the West still wrestles today. For instance: What is the true center of the world? Is it the city, with its power and wealth? Or is it wilderness, which eludes human power but is the primal source of human wealth and well-being? Is the axis about which the world turns the Mountain, or the Tower?

For six thousand years we have careened betwixt these two great masses like baffled planets of a binary star, and we are not done yet. Placed in this context, some familiar stories take on new meaning. What stage or state of being does Eden murmur of? Is it childhood, or animalhood, or hunting and gathering, or even gardening of some sort? Is Eden a garden in the literal sense, irrigated and planted in rows? Or is it God's garden: the wild place at the heart of the world where all life bubbles up?

It is tempting to think that, if we fathom Eden truly, we can go back after all. Perhaps the Fall was the fortuitous result of a bad menu choice, and we need only recall the waiter and make good our mistake. If this were possible, would it be wise? For we must ask whether Eden is really such a great place—for people, that is—whether, indeed, it is a place for people at all. Is the blade that whirls at the gate of Eden, preventing re-entry, really our enemy? Or does it protect Eden and, in so doing, protect us?

More questions bubble up. Once expelled from Eden, do we lose all hope of enjoying its bounty? Or can the rivers that flow from Eden reach us, and nourish us, even in our fallen world of steel and glass?

To these questions raised by a childish fable, the answers are of the greatest practical importance. The flow of nutrients, water, energy, and genetic information; the maintenance of climate; the recipe of oceans and air—all these depend, in a sense I hope to make clear, on Eden and our relation to Eden.

The exile from Eden is the bedrock fact about humankind, but it is not a fact with which we rest easy. It doesn't feel like bedrock: it feels like a chasm. The Mountain pulls us one way, the Tower the other: the world is fallen—fallen away—and we are left hanging. To close this fissure in the world and in ourselves, all kinds of effort have been expended: mythic, literary, architectural, horticultural, even scientific. In Part Three, we will recast Western history as a somewhat muddled epic whose heroes—Virgil,

Nero, St. Augustine, Lorenzo the Magnificent, Louis XIV, Alexander Pope, Henry Ford, Frank Lloyd Wright, Lynn Margulis, and legions of others— struggle to regain a paradise lost. How close have they come, and at what cost?

In Part Four, the scene shifts firmly to the present. We start by looking at two sprawling but distinct camps in the present environmental debate: the Planet Managers and the Planet Fetishers. According to the Fetishers, humans have no right to a more prominent role in nature than bears. Having lengthily chewed up the scenery, we should now have the good grace to retire from center stage. In strict practice, this would mean reverting to hunting and gathering, and letting most of the world revert to wilderness. The Planet Managers, by contrast, take for granted that we will keep changing the world; they just think we should do a better job of it. Instead of acting blindly upon the planet, we must use all the resources of science to manage it: all of it, settled and wild alike.

Are these modern cults mere relics of those ancient ones, the cults of the Mountain and of the Tower? If not, how do they differ? Has either learned the lessons the ancient myths have to teach? If there is no escaping the need to manage nature, how can we minimize that need? How can we help wild nature regain the elbow-room and elbow-grease it needs to manage itself? And is there some way of working with nature less meddlesome than management: something more on the order of collaboration?

At the heart of this book is a Hebrew koan. Humans, we are told, were stationed in Eden "to work it and protect it." You might say this was a cruel joke, for it must have been well known in heavenly circles that humans would promptly move outward in waves of "work," changing Eden into something else. But the command can also be seen as a kind of riddle that we urgently need to solve. How do we protect nature from our work, and so keep from fouling the source of our own life? And how do we work with nature in a way that leaves both nature and human nature undiminished?

A FEW WORDS about words, and about method.

This book is less a series of syllogisms than a series of images. For images (whether myths, or scientific models, or metaphors, or just resonant words) are far more determinative than pure logic of how we see and act upon the world. To address matters of value with logic alone is to miss the heart of the matter, and our hearts as well.

For the most part, I use ordinary words in ordinary ways. Unfortunately, this means their reference may slide. The two stars of this semantic icecapade—skating about each other with maddening elusiveness—are "nature" and "we." Depending on context, "nature" may mean all living things on the planet Earth; or all living things except for humans; or the genetic tendencies of living things, as opposed to their cultural expression; or everything in the universe. "We" may mean humans; or Neolithic-and-later humans; or humans of the modern West.

If this last telescoping series seems ethnocentric, that is because it is. Western civilization is the one I know best; for better or worse, it is also the one most responsible for the present state of the planet. Returning to Freud and his toadstool, we might say that one task of this book is to put Western civilization on the couch. If we (and we know who we are) can get to the bottom of our bad conscience about nature, maybe we can see things a bit more clearly and act a bit less foolishly.

IF WE WANT TO DREAM of Eden, we had better go back to sleep. But if we want to understand the dream, we had better learn something of the past from which it sprang. Only then can we hope to divine its meaning for the hard daylight hours ahead.

PROLOGUE:
PERSONS FROM PORLOCK

IN THE SADDEST PREFATORY note I know of, Coleridge says that under the influence of "an anodyne" he dreamed a poem four or five times as long as the "Kubla Khan" we now have. Awakening, he hurried to write it down, but managed only fifty-four lines before "he was unfortunately called out by a person on business from Porlock." On regaining his room, he was dismayed to find that, "with the exception of eight or ten scattered lines and images, all the rest had passed away like the images on the surface of a stream into which a stone has been cast, but, alas! without the after restoration of the latter!"

To speak of "the dream of paradise" seems hackneyed or else derogatory, as if it were just a daydream or a pipe dream. But the sense of a dream broken off, a nest torn to shreds, which you struggle to piece together in all its brittle mystery—this means something. You want to sleep again, dream again, pick up exactly where you left off.

This is the way we dream of Eden. We cannot help feeling that it still exists, but just beyond our reach—beyond the wall of dark fire kept roaring by the cherubim of the brain.

The Dreamtime of the Aborigines, when the totemic ancestors walked the world and sang it into being, can be partly regained by those whose lungs and feet are up to the job. With the aid of shamans or mushrooms, other indigenous peoples are sometimes able to call back the age when heaven and earth were not yet sundered and humans talked companionably with animals and gods. But such moments are shot through with danger, and quick to evaporate. By and large, "primitive" peoples seem to feel, as we do, that something has wrenched them out of the dream.

The near universality of Golden Age and Fall stories gives the lie to the notion that "primitive" peoples live in complete harmony with nature, or even think they do. All seem to remember a time when, or a place where, the harmony was greater. All feel an estrangement. If they feel close to the animals, there is a dissonance between that closeness and the fact that they kill and eat them, and that some animals return the favor. No human can be without a sense of conflict with nature—nature which, broadly construed, includes the failings of human nature, of the flesh, of the spirit. The easiest way to explain Fall myths is to say that they are explanations of evil. But this is not to say they lack any nut of historical truth.

In the King James text God says to Adam, "Cursed is the ground for thy sake." With only a little stretching, the same Hebrew letters can be read to mean, "Cursed is the ground by thy passing over."

In other words, the earth is not cursed all at once, but by degrees—by human agency. Wherever humans go, the earth is despoiled. Though that is probably not what the Bible meant, it is largely a fact: and it casts a cold light on Milton's image of a chastened but hopeful first couple setting out from the gates of Eden, the world "all before them."

Eden, Arcadia, and the Golden Age can be seen as artifacts of the same process. As humankind—or farming, or civilization, or what have you—disperses from its point of origin and fills each new place, there is a sense of loss. The game thins out, the soil washes away, neighbors quarrel over wells and boundary stones. Armed men appear demanding rent and taxes. There are too many horses and not enough pasture, too many cars and not enough parking, too many restaurants and not enough shoemakers. We cannot help thinking that things were better once or are better elsewhere. And we ease our minds by telling a story.

Three stories are commonly told.

One is the story of Eden, the navel of the world, the umbilicus from which we were untimely severed and went our several ways. If the story is about the spread of civilization, then the navel becomes a city, such as Babel or Atlantis. More often it is a mountain, an island, or a garden.

As we spread outward over the earth, first "improving" and then degrading each new place, the place that is just enough but not too much improved is Arcadia. That is the second story. Depending on the level of civilization one likes, Arcadia can be frontier, or farmland, or suburbia, or the cheap bohemian quarter of a great city. Its classic setting is the hills in which herdsmen summer their flocks—in the classical world, the border between tilled land and wilderness. This setting is especially apt in that pastoralism is in the short run the easiest and pleasantest, in the long run the most destructive of ancient ways of life. Whatever form it takes, Arcadia contains in its ripeness the germ of its own spoiling.

For someone lost within the swelling circle—someone living in a place through which Arcadia has already passed—what is craved is either another place at this time, or this place at another time. The former, a place just inside the edge of the circle, is Arcadia; the latter is the Golden Age. And that is the third story. As those who tell it rarely have a clear sense of the process that has stabbed them in the back, they assume that this Golden Age held sway everywhere at once.

It is easy to see why even hunter-gatherers might tell such a story. The totemic ancestors of the Aboriginal Dreamtime "live in an environment infinitely superior to that in which the present-day native has to battle for his existence. The great sire and his horde of sons live under ideal conditions, in a land of natural riches and plenty." No doubt Australia was just such a place when the Aborigines got there. No doubt many other places were similarly bountiful when humans first arrived. It may be that the Golden Age myths of the hunters and gatherers preserve that memory. That might help to explain, too, the role of animals as ancestors. The newly arrived humans want to think of themselves as autochthonous, sprung from the soil of this place; knowing on some level that the animals are the real natives, they adopt them as foster parents.

Clearly the third story is temporal, the second geographical, and the first both at once. Eden is a certain place at a certain time. We like to think, however, that Eden is still Eden even if we can't get back in. That may be a mistake; if my image of mankind's ravaging march is at all correct, then the point of origin should by rights be the most ravaged place on earth. But a place that is damaged can sometimes repair itself, especially if it has been damaged so badly that the people go away.

A lot depends, too, on what kind of origin we are talking about. If human patterns evolve slowly enough, the place in which they evolve may have a chance to adapt to them. The Rift Valley of East Africa, where several million years of hominid evolution took place, remains (to some eyes at least) a paradise. Africa as a whole stood up well to the human presence until quite recently. The main reason is that its flora and fauna were able to keep up with each stumbling step humans took. They were rarely caught off guard. Parasites, especially, were adept at keeping human numbers in check.

But when the ratio between biology and culture shifts, human progress (if that is the word) goes into high gear. Then it is very hard for other species to keep up. Thus the "tree where man was born," though cleft and scorched, still stands—but the cradle of civilization is now mainly a sandbox.

BEFORE WE TRY TO FATHOM these stories of Eden, Arcadia, and the Golden Age, we had better get a sense of the waves of human advance themselves: what has impelled them, what their impacts have been. Only then can we begin to know to what degree the myths are myths in the bad sense—mystifications—and to what degree they are useful pictures of our place in nature. To understand the waves themselves, we turn to biology.

PART ONE

WAVES

1

THE MARRIAGE
OF GRASS AND MAN

THE FALL OF MAN is not a sheer drop. It is more like a scrambling outward and downward that must feel at times like a helpless sliding, as when your heels refuse to catch in a scree slope; at times like a giddy romp; at times, even, like a triumphant ascent. The waves of human-led change—our alliances with certain species, and roughshod crushing of others; our unhousing of the soil community; our stirring up of long-buried hydrocarbons—are a matter of concern to creatures on whom sermons about free will might well be lost. In various senses, we take a lot of nonhuman nature down with us.

NEARLY ALL CREATURES vie for a place in the sun. Like geese snapping at tossed crumbs, we snap at photons, or at packets of energy derived from photons. First-hand or fifth-hand, it all comes from the same place. It is true that some creatures, such as the bacteria at deep-sea vents, derive their energy from inorganic chemicals that owe nothing to the sun. But these are freaks. If photosynthesis had not been invented, nearly all life on earth would have run out of steam eons ago.

At the end of E. M. Thomas's prehistorical novel *Reindeer Moon*, the men at the Camps of the Dead spear the sun, cut it into strips, and eat it. ("Hot and crisp, mine burned my mouth but tasted good because, just as I thought, it was mostly fat. What else can make such a fire?") Far from being a conceit, this is an acknowledgment of the way the world works. What all hunters, all farmers, all plants and animals do is hunt the sun and eat it. The trick is to find the most efficient path between the sun and the mouth.

The flow of energy and nutrients in nature is not a relay race in which each runner waits patiently for the baton. It is not a contradance in which partners smile and curtsy and move seamlessly down the line. It is more like what would happen if you dumped a bale of hundred-dollar bills into Times Square.

Every strong chemical bond, whether forged by sunlight or found in the bubblings of the darkest deep, will be popped like a beer can and drained. The lower-energy bonds in the dregs or waste product will be seized on by some humbler creature and sucked dry. If no life form exists that is capable of using that energy, nature is likely to invent one. But since most of the chemicals plentiful on earth have been around for billions of years, the main ways of getting energy out of them have already been devised. The basic moves in the game of life—fermentation, photosynthesis, respiration, as well as protein synthesis and genetic transmission—were all invented by bacteria. Compared with these, the refinements added by more recent players are so much fancy footwork. That footwork is worth watching, though, especially when our own progress depends on it.

THE FIRST WAVE

What most of us call the food chain—the hierarchy of who eats whom—is better called the trophic pyramid, since it is broad at the base and tiny at the top. As you step up a level (the rule of thumb says) the biomass, or total amount of living matter, shrinks by a factor of ten. The reason is simple: in any conversion of energy, some gets wasted. Of course, "waste" is a subjective term. From the point of view of those at the top of the pyramid, those on the next level down waste a great deal of energy running away.

When, in the transition from apehood, our ancestors moved up a step on the trophic pyramid—or, rather, a half-step, from herbivory to partial carnivory—their numbers should by rights have shrunk. They were able to sidestep this law of diminishing returns by moving from the trophic pyramid of the rain forest to that of the savanna. Here the vast platform provided by grass, and above that by quadrupeds, gave them plenty of room to spread out. They were aided by fire, which limited the spread of thorny acacia trees, removed dry grass, and caused fresh grass to spring up and draw game.

Let us look at this first marriage of grass and man from grass's point of view. In effect, grass enlisted the aid of humans in its turf fight with trees for control of the African savanna. Its strategy (not a conscious strategy, of course, but a pattern favored by natural selection) was to make its leaves and stems plentiful and palatable to four-legged beasts. Humans, being fond of those beasts, would then set fires that would keep trees scarce, grass fresh and abundant. By dividing its energy between edible leaves and stems above ground and fireproof roots and runners below, grasses could thrive amid fires set by lightning and fires set by men.

The alliance of humans, perennial grasses, and grazing animals brought about the first great wave of human-led expansion. Glaciers, too, lent a hand: in effect, the Pleistocene Ice Age extended the African savanna over much of the dry surface of the globe. The temperate zone, where not intemperately iced over, took the shape of tundra or steppe. Trees fled from the prevailing cold, and grasses took their place. Humans joined tigers, lions, bears, and other carnivores in a movable feast such as the earth had never seen.

Every hundred thousand years or so, an interglacial period would crash this picnic and cramp the trophic pyramid of the grasslands. The planet would warm up and the glaciers would recede, tugging the steppe after them. Forests—far less hospitable to big game—would edge northward. Human numbers would plummet, then recover twenty or thirty thousand years later as the glaciers and grasses came back.

Tested perhaps twenty times over the course of two million years, humans finally found a way out. Whether that had to do with the direness of this particular challenge—to which, after all, such awful beasts as the cave bear and woolly mammoth succumbed—or with the fact that it came when a certain threshold of technical skill had been crossed, we do not know.

At first, their response to this interglacial was a more violent version of their usual response: to discourage trees and encourage perennial grasses. Indeed, the clearing of the forests of Europe was begun by Mesolithic peoples as a response to the postglacial expansion of forests. It was not an assault on the forest primeval; it was an act of self-defense against an invading host.

We are a savanna species. When the Ice Ages extended the savanna into Northern Europe and Asia, we followed it there. But when the savanna began its Holocene retreat, we refused to retreat with it. We stood our ground and mowed down the trees as they came at us. Humans cleared and burned the advancing forests, making them look more like open woodlands or savannas. By this means, numerous plants edible by man, or edible by animals edible by man, were saved from the devouring shadow of the woods.

But this was at best a holding action. The big meal tickets, such as the mammoths, were extinct, and in the long run man and grass would have lost ground. Their solution was to make their partnership more efficient by eliminating the four-legged middlemen. Now this awesome primary productivity (as ecologists call green stuff), instead of being run through four-footed converters that wasted so much of it trying to run from us, was running directly into our mouths.

By stepping down a level—returning to an apish diet, but with a difference—humans were able to increase their numbers by a factor far greater than ten: a factor that has now reached about five hundred, and whose full magnitude has yet to be reckoned. But they could do so only by altering the size and composition of the next level down, the green level.

THE SECOND WAVE

"Agriculture" is the name we give that alteration. It is often talked about, but rarely from the point of view of the plant.

Remember that grass's first strategy, tried and found true for half a million years, was to divide its energy between edible leaves and stems above ground, and fireproof, perennial roots and runners below ground. Humans would set fires to discourage trees and encourage fresh grass, which in turn would draw game. Later, humans clinched this arrangement by herding the quadrupeds instead of chasing them, but the principle remained the same.

The other strategy was bolder, and did not hit its stride until some ten thousand years ago; but the payoff, when it came, was spectacular.

There were two phases. The first was a matter of investing less energy in root systems and more in seeds: in other words, being an annual grass rather than a perennial. By this means a species could colonize stretches of ground that had been burned, trodden on, shat upon, and otherwise altered by the actions of human beings.

In the second phase, the grasses that were camp followers of man—that thrived in refuse and in disturbed soils, and so were technically weeds—had the daring notion of involving him more intimately in their sex life. He would make their bed and plant their seed, and they in return would feed him. The outcome of this unlikely ménage was an explosion in both populations. This second great wave of human-led expansion, based on the alliance of humans and annual grasses, has not yet spent its terrific force.

In the space of a few millennia, the balance of numbers has tipped decisively against the trees. In effect, the alliance of man and a handful of plants has turned back the geologic clock. But with a difference: in the short term, at least, we have made for ourselves the best of all possible worlds, one that is warm but mostly grassy. Such a world is abnormal. History, we might say, is the hectic dream of a feverish earth. It is the creature of an interglacial, a bout of fever that began only a dozen millennia ago. Humankind is a bacterium that was kept in check while the refrigerator was functioning, but took advantage of a power outage to run amok.

Standing at the shore of the grasslands, man has commanded the postglacial tide of trees to go back. It has gone back. But the seas which, as a side effect of this and other triumphs, have already begun to rise may prove less obedient. And as the trees that we have retained for their decorative or practical value succumb to heat or drought, we may find that it is too late to command the forests' return.

TREES ARE AMONG the few creatures on earth that really compete with us. Of course fire, the girdling cord, the ax, and now the chain saw have given us a distinct edge. By now we would have exterminated all vascular plants taller than ourselves had not some of us noticed that some of them can be our allies. Olives and almonds, shade and lemons, mast for pigs and masts for ships—all these boons were evident long before anyone knew about flood control or the absorption of greenhouse gases.

The habitat of the wild grains that would become tame grains was hilly woodland. The first people we know of who made full use of those wild grains, the Natufians of the Levant, made use of nuts and acorns as well. We might say that trees tried to shake man's allegiance to grass—tried to buy off this implacable and fiery enemy. Some trees had some success: olives soon covered large parts of the Mediterranean uplands.

But the best they could do was to temper our general dislike. Humans are fond of trees individually but fear them in the aggregate. Just as the soldier who kills a dozen of the enemy without flinching may, if the situation arises, nurse a single wounded enemy back to health, so the man who levels a forest will lavish fertilizer, water, and other attentions on a sapling he has planted or an old wounded tree whose spirit moves him.

As the reference to fruit and nut and oil trees shows, not all our allies have been grasses. We have also made common cause with broad-leaved plants that have edible foliage, roots, tubers, pods, or fleshy fruits. But with the exception of trees, most of our allies have been plants that flourish in places cleared of trees.

PIONEERS AND CONQUERORS

These, then, were the first two great waves of human conquest. In the first case our major allies were perennial grasses, in the second annual grasses.

The second alliance was perhaps more compromising, for many annual grasses are best described as weeds.

Weeds are pioneer species. They take advantage of readily available nutrients stirred up by some disturbance. Their aggression is nasty but brief, and serves the larger weal of the ecosystem. For it sets succession in motion, allowing the ecosystem (or some part of it) to rise from its own ashes.

Pioneer species, once they have leveled the playing field by their mere existence, lose their edge. They create conditions in which they themselves do not flourish, and so are outcompeted by their successors. To be truly expansionist you must have elbows. You must be able to knock entrenched species—even dusty climax communities—off the shelf. To do this you must be tougher and more aggressive than weeds, which are no more than sneaky opportunists.

It has been said (both in shame and in pride) that humans are weeds. But weeds do not start fires or plow fields. They must wait for their chance, and seize it. Either that, or they must team up with a tougher species.

Taken as a team, humans and their domesticates have elbows. We do not have to wait for disturbances; we create them. We spark a catastrophe that forces succession to start from scratch, and on our terms. Instead of letting any and all comers invade the soil we have cleared and disturbed, we sow the weeds we find useful, such as wheat, oats, or maize. And instead of letting succession take its course, in which the grasses might be followed by shrubs and pioneer trees, and these by the climax forest, we arrest its development at the grassy stage.

Perhaps because they must act quickly to seize the main chance, pioneers tend to be loners. By contrast, truly expansionist species—species that colonize wholly new environments or drive other species out of old ones—often work in teams. And when a weedy species such as an annual grass does manage to be truly expansionist, it is usually by teaming up with another species, such as man. A single species is rarely able to modify a new environment to suit its needs. It needs help. This it gets by joining with one or more other species to form a sort of portable microenvironment, or what we might call an "ecomodule"—in effect, a space ship or space station that colonizes some part of Earth as if it were another planet. In the next chapter we will look at some examples of this technique, which was practiced by other species for several billion years before humans got hold of it.

2

AXIS POWERS

I F O U R R E M A K I N G of the world is a sin, it is not an original sin. If it is a triumph, it is by no means a unique triumph. Not only did we have help from other, humbler species; the truth is that other, humbler species have done similar things quite without our help. Human generalship is clearly not required, for alliances of species were remaking the world long before we came along.

The world of the Cenozoic, which began some sixty-five million years ago, is a world of fur, feathers, and flowers. The dominant plants of the Mesozoic had been gymnosperms (conifers, for example), most of which relied on wind and dumb luck to scatter their pollen and seed. But the present age, the Cenozoic, is the age of angiosperms. These plants use all the methods of modern packaging and advertising to get consumers—insects, birds, and mammals—to do the work of pollination and propagation for them. Neon colors, junk food, sexy images (as when the *Ophrys* orchid mimics a female bee's behind): anything goes. Of the famous studies of mutualism in the ecological literature, the greatest number trace the gossamer, spun-steel threads connecting flowering plants, insects, birds, and mammals.

"Mutualism"—a relationship between two species that yields greater population growth for both—is a slippery term. If you look at all the threads connecting all the species of flowering plants, birds, mammals, and insects, most will involve predation or competition. Yet the whole tangle taken together does seem to form a commonwealth of interest. It has been a kind of steel wool scrubbing out the remnants of Mesozoic life. Though insect taxa do not seem to have expanded much in the early Cenozoic, the other partners in this grand alliance—birds, mammals, and flowering plants—expanded prodigiously. Of course, their conquest of the world seems a bit less heroic when one considers the likelihood that many of their rivals, such as the dinosaurs, were killed off by the smack of a giant meteor, which threw up enough dust to hide the sun for decades. But great victories often happen that way. You cannot always tell whether the rise of one set of creatures is the cause or the effect—or both—of the fall of another set.

If this grand Cenozoic alliance seems too broad to bear comparison with human-led leagues, we can look at some narrower ones. Among the flowering plants in the Cenozoic alliance were grasses. Obviously people were not the first mammals to ally themselves with grasses; in fact, the alliance of perennial grasses and two-legged mammals could take place only through the mediation of four-legged mammals. Grasses and grass-eating mammals coevolved; and though their relationship was, strictly speaking, that of eater and eaten, they nevertheless enjoyed a kind of alliance in which each spurred the spread of the other. Grazing, like fire, fans the growth of fresh shoots and the sidewise reaching out of roots and runners.

But even this tight transaction has its middlemen. All herbivorous mammals depend to some degree on gut microbes to digest cellulose and other tough stuff. Without methanogenic bacteria, the vast grasslands of the earth would have been closed to ruminants—antelope, bison, musk oxen—and so to meat-eating mammals as well. In fact, neither the grasslands, the grass-eating mammals, nor the meat-eating mammals would have evolved in anything like the forms we know or the profusion that allows us (with just a hint of jingoism) to call the Cenozoic the Age of Mammals.

But the Age of Reptiles, too, would have been a more modest affair had it not been for certain microbes. For mammals were by no means the first big creatures to use very small creatures as an aid to digestion. With cellulose-digesting bacteria under their belts, vegetarian dinosaurs were able to make themselves at home in the high-fiber forests of the Mesozoic, with their forbidding muesli of cycad, ginkgo, conifer, and fern. Beasts such as the apatosaurus were vast fermentation vats on legs, with perhaps five thousand kilograms of slurry sloshing around inside. Absent their tiny allies, "thunder lizards" would have been less thunderous and much fewer in number.

If any relationship can be called mutualistic, that between an herbivore and the cellulose-chomping microbes in its gut can surely be so called: the host providing shelter and a steady supply of groceries, the guest doing the cooking. (The shelter the host provides is above all shelter from oxygen, which is toxic to these bacteria as it is to all the most ancient forms of life.) But it can also be called by a less slippery name: symbiosis, which simply means that two species live in close and near-constant contiguity. Again and again in the course of geologic time, symbiosis has allowed the partners to colonize or conquer new environments. Giant tube worms used sulfur-metabolizing bacteria as their passport to the bubbling sulfur hot springs of the deep-sea floor. Deep-sea cold seeps were colonized by mussels playing host to bacteria powered by natural gas. Corals and giant clams were able to colonize the nutrient-poor but sun-rich waters of the seashore by offering homes to single-celled algae and taking sun-made sugars as rent. Each of these arrangements is millions of years old but is still in force.

Symbiosis is just as common on land. By installing woody-hungry microbes in their guts, cockroaches and termites gained a kingdom in the heart of the Carboniferous forest, and later in the heart of human settlements. To make sure that each has a full complement of these vital microbes, termites engage in coprophagy: that is, they eat one another's stool. The hindgut of a modern termite is a zoo, many of whose denizens have amazing symbioses among themselves. For instance, a lozenge-shaped microbe called *Mixotricha paradoxa* is propelled through the fluids of the gut by half a million spirochetes that cling to its side, beating in synchrony like galley slaves. They are paid with nutrients secreted by the host ship.

As this last example hints, some symbioses are so intimate that you can hardly tell where one organism leaves off and the other begins. A lichen, for instance, is not really an organism at all, but the union of a fungus and an alga. With the alga's photosynthetic skill and the fungus's tenacity, lichens are among the hardiest entities on earth. You can dip them in liquid nitrogen (at −196 degrees C.) and they will bounce right back, photosynthesizing at full strength a few hours later. They colonize deserts, alpine peaks, and polar ice sheets. By darkening the color of an arctic land surface, they can lower its albedo (reflection of sunlight), thus raising its temperature and tempting other species to venture in. They have been proposed as the ideal first colonists of another planet.

The success of lichens has been impressive, given that they seem to have arisen only fifteen million years ago. But an older symbiosis of a very similar kind has had a far more spectacular success, one that you can probably see from where you sit right now. Vascular plants first colonized the dry land some 450 million years ago. Examination of the first land plants in the fossil record suggests that they were inside-out lichens—in other words, a symbiosis of fungus and alga in which the alga was the dominant partner. The fungus contributed a fledgling root system, so that the alga could draw water and minerals from dry land. The alga contributed the ability to draw energy from sunlight. The third great requirement for life on land—a waxy cuticle, so that the plant would not quickly lose its water to the dry air—

seems to have been the product of chemical teamwork between the two partners. In relatively short order, this ecosystem in a suitcase established itself all over the world, displacing cyanobacteria in some places and bringing life to other places for the very first time.

THE COMMUNITY OF THE CELL

The oldest alliance is the grandest of all, and the smallest as well. It has conquered nearly the whole surface of the planet, yet its forces are marshaled in each cell of your body. In fact, they are marshaled in the cells of all living things except bacteria—an exception that is easily explained, since the forces themselves are, or were, bacteria.

Every cell of every "higher" organism is an alliance of bacterial species. Before we can make sense of this statement, we must make some distinctions. Of all distinctions that can be made between living things, the one that cuts deepest is that between bacteria and everything else. Bacteria are prokaryotes—that is, cells without a true nucleus. All other creatures are eukaryotes: either they are cells with a true, membrane-bounded nucleus (amebae, paramecia) or they are made up of such cells (snails, mushrooms, redwoods, people). Cells with a true nucleus also have other organelles that bacteria lack, such as mitochondria, which burn sugar to release energy, and—in the case of green plants—chloroplasts, which use sunlight to make sugar.

Recent studies have found that the DNA of these organelles is different from that of the cell as a whole, which is stored in the nucleus. This and other evidence has led biologists to accept a theory they once thought harebrained: that the eukaryotic cell arose from a series of intimate partnerships between species of bacteria. Although no one can be sure about things that happened billions of years ago, the sequence seems to have gone something like this. A very tough bacterium that lived in hot, acidic water (which was much more common in the earth's hot, volcanic youth) was nibbled on by a restless bacterium much like a modern spirochete. This may have happened millions of times before the two species worked out an arrangement—perhaps an arrangement something like the one between the microbe boat and its oarsmen in the termite's hindgut. One thing led to another, and in time a new species arose in which the spirochete became a whiplike oar. The new species was both tough and mobile, and covered a greater range than either of its parents had done.

Millions of years later, descendants of this mongrel species were invaded by voracious oxygen-breathing bacteria that tried to devour them from the inside out. At length the host species and its guests learned to coexist, reproducing at the same rate and sharing the energy the guests so furiously produced. The oxygen-breathers became mitochondria—power plants for the cell. The new species combined toughness, mobility, and oxygen power and could range freely on the surface of the earth's shallow seas, breathing up a storm.

More millions of years later, some grandchild of this second-order Creole, inured to the habit of eating small photosynthetic bacteria, figured out that, instead of making a single meal of them, it could get numerous meals out of them by keeping them alive inside itself, busily photosynthesizing. In time, they became chloroplasts. As this pattern repeated itself, new kinds of species, much like modern phytoplankton, filled the upper layer of the offshore waters, supplanting the older photosynthetic bacteria.

If all these symbioses seem far-fetched, it should be noted that very similar ones have been seen under the microscope in extant species. In one case, the birth of a new, chimerical species was actually observed.

More important than the success of any of these immediate mongrels is the fact that eukaryotic cells, in their various forms and aggregations—microbes, fungi, plants, and animals—have taken over the world. Though bacteria remain widespread and vital to the health of the planet, it is clear that they have lost market share.

Between this alliance and that of humans, grains, and grazing animals there are many glaring differences. This alliance involved symbioses in the strict sense, where two species are in near-constant physical contact; and those symbioses evolved into actual mergers of two species into one. Nothing so intimate has occurred in the case of people and corn and cows. Yet the basic principle is the same. A close functional association of two or more species is able to do more in the way of energy capture, energy use, and the conquest of new environments than either could have done alone.

As far as function goes, the eukaryotic cell is not unlike a farm. The chloroplasts are the crops, which capture energy from the sun. The mitochondria are the livestock, which convert energy into forms humans can use. The nucleus is the farmhouse, from which most of the work is directed.

Of course this is just an analogy, and a strained one at that. Can any alliance in nonhuman nature really be compared to farming? Don't the tilling, planting, and harvesting that humans do put them in a class apart—a technological class to which no other creatures need apply? Before we rush to answer this question, we had better consider the ways of the ant.

ANTS AS FARMERS AND RANCHERS

Leafcutter ants are the dominant ants of the New World tropics. They are also the most hated by human beings, thanks to their habit of stripping a whole garden bare overnight. In a year, a single colony of leafcutters can cut two thousand kilograms of greenery. In Neotropical forests, leafcutter ants consume about one-fifth of all leaf production, which would make them the prime herbivores of the region if in fact they ate the leaves they cut. Instead, they use them as the substrate for fungus gardens.

"Garden" is not quite the right word, though: what the ants do is more like factory farming. They have raised the assembly line to a plane hardly touched by the most up-to-the-minute human agribusiness. The largest workers cut leaves, shoots, and flowers from a wide variety of plants and pass them on to slightly smaller workers, who cut them into strips. Smaller ants chew the strips into lumps, and still smaller ones add the lumps to piles of pulp in cramped chambers of the nest. The smallest ants plant the fungus in the pulp, fertilize it with droplets of their own feces, and tend it as it grows. The crop consists of a single species—a monoculture, like so much of our own farming. It is kept pure by a number of methods: weeding out other fungi; manuring with nutrients, enzymes, and growth hormones that favor the favored fungus; and applying pesticides that suppress bacteria and the spores of other fungi. The factories that produce these chemicals are the bodies of the ants themselves.

The fungus is a domesticated species—more fully domesticated, in fact, than the mushrooms that people cultivate. It is incapable of forming spores, and relies entirely on the ants for propagation. In return, it forms swellings at the ends of its fibers which their German discoverer called "heads of kohlrabi," but which are more like edible balloons filled with a solution of sugars and other nutrients. The workers pluck these and drink their juice, or feed them whole to the larvae of the nest. While the juice is merely a high-powered supplement to the workers' main diet of sap (absorbed directly from the leaves they chew), nectar, and other insects, the balloons are breakfast, lunch, and supper for the larvae.

Why don't leafcutters just eat the leaves they cut, instead of going to all the trouble of cultivating a fungus? The catholicity of their cutting points to the answer. They could not possibly eat such a wide range of plants, because they could not possibly produce in their own bodies the wide range of enzymes necessary to break down all the terpenoids, alkaloids, and other chemicals plants use for self-defense. So they hire a fungus—whose mission in life is breaking things down—to do the job for them. In fact, their dependence on the fungus has caused them to lose some of the digestive enzymes they had in the first place. But the ants' physical breaking down of the leaves makes life much easier for the fungus, which would otherwise have to wait for the leaves to fall from the plants and be degraded by a succession of organisms before it could have a crack at them.

Although fungi are neither animal nor plant, but a kingdom unto themselves, people who grow mushrooms are usually thought to be raising a crop. But the leafcutter fungus is less like our crops than like our cattle, which convert raw leafage into meat and milk. The maceration of greenery by the ants is comparable to (though harder on the jaws than) the process by which we make silage for our livestock from cornstalks and other roughage.

Other ants do raise genuine crops. In Central and South America, one frequently sees something that looks like a florist's arrangement stuck in the crotch of a sapling or a clinging vine. It is a ball of fibrous stuff that looks something like what florists stick stems into. Sticking out of it are various kinds of epiphytic plants. The ball is an ant nest, made of the ant-manufactured composite material known as carton. The plants are cultivated plants—cultivated by ants—most of which do not grow wild. The ants plant their seeds in the carton, which is laced with ant manure and other good things, and harvest fruit pulp, nectar, and the food bodies of seeds. As a bonus, the roots of the plants help hold the nest together. The whole ensemble is called an ant garden—a name that is not fanciful, but exact.

The prevalent form of husbandry among ants, though, is animal husbandry. A vast range of ant species, from the African weaver ant to the common wood ant of Europe, are dairy farmers. Their cows are aphids—those tiny insects, despised by human gardeners, that look like pale-green teardrops and live by sucking the sap out of plants. Even more than the ancestral cow, aphids beg to be domesticated. For whereas the ancestral cow gave milk for its calves, aphids shit a highly nutritious syrup—a solution of sugars, amino acids, proteins, vitamins, and minerals—for no apparent reason. Some aphids can excrete 133 percent of their body weight in an hour. Their strangely wasteful metabolism can only be explained as a product of glut—of the prodigal food source they have tapped into—or maybe, in some cases, as a product of selection by ants.

The treacly syrup that aphids and their cousins exude, known to us as honeydew, was known to the Israelites in the desert as *man*, or mana. It is still gathered by the Arabs, who call it *man*, and by Australian Aborigines, who call it "sugar-lerp." Aphids are too small to be herded or milked easily by humans, but ants began the process of domestication at least 30 million years ago.

Ants do not have a monopoly on aphid-raising; certain bees, wasps, beetles, and butterflies have gotten into the business, if not quite at the ants' level. Nor do aphids have a corner on the ants' attentions. Many ant species also tend lycaenid-butterfly larvae, which reward them with a sugar solution secreted from a special gland. In fact there are hundreds of species, both large and small, with which ants have formed alliances that border on the agricultural. No wonder ants are among the most successful of all life forms, making up from 10 to 15 percent of the total animal biomass in most land ecosystems. Not only have ant species led many alliances that conquered large parts of the earth, but they have done so in a manner almost eerily manlike. They have shown a genius for well-oiled cooperation both with other creatures and among themselves. In other words, like most human societies, they are both urban and (in many cases) agricultural.

MUTUALISM IS NOT NICE

Some of the more adorable instances of mutualism uncovered in recent years have been made much of in the popular press. The cockles of our hearts are easily warmed by stories of ants that protect acacia trees, or luminescent bacteria that act as lightbulbs for "flashlight fish." Nature red in tooth and claw is strong stuff, and we are glad to have it diluted with a little niceness. Surely such niceness must be what makes nature run smoothly.

But mutualism is not always as nice as it seems. When two parties benefit, it is generally at the direct expense of some third party (for instance, the insects from which ants protect acacias) and the indirect expense of many others. And while predation tends to stabilize ecosystems, mutualism can sometimes destabilize them. Predator and prey control each others' numbers in a rough, fluctuating balance of power; mutualism, on the other hand, can give a group of species an edge on those around them—an edge that slices through ecosystems. To return to the image that gives this chapter its title, it is clear that Germany, Italy, and Japan together did more to destabilize the world than Germany could have done by itself.

Invasions by single species are common enough, of course. But there is a difference between an invasion in this weak sense and an outright conquest. In the first case, a species merely fills an empty or half-filled niche, or displaces a few other species. In the second case, a whole ecosystem—even a whole planet—is demolished and remade. This had never happened until we came along, or so we supposed. In fact it had happened before, but only when two or more kinds of creatures joined forces. (To this rule the one great exception is not humankind but, rather, the cyanobacteria—blue-green algae—who, some two billion years ago, single-handedly poisoned almost all their contemporaries: not with cyanide, but with oxygen.)

Lone species like to infiltrate mature ecosystems in which, for whatever reason, their niche is unfilled or is filled by a weaker species. But they are rarely up to the task of colonizing a new or skimpily used environment, or conquering an ecosystem and driving many of the natives out. To do that, what is needed is a sort of ecosystem in a suitcase, able to start up or take over the hard metabolic work—the processing of nutrients and energy—that a mature ecosystem does so well. Eukaryotic cells; lichens; land plants; ruminants and gut microbes; ants and fungi—each of these alliances takes on a range of biological functions, achieving a kind of "vertical integration" (to use the economist's term for a classic type of monopoly). Compared with this, what a lone species does is just freeloading.

A LEAP OF FAITH

Talk of the "alliance" or "symbiosis" or "mutualism" of man and wheat may sound fanciful. What is in it for the wheat? It gets eaten, for God's sake. But this is to forget that annuals do not care a fig for length of life, as long as they get to set seed. True, most of the seed wheat sets gets eaten. Even so, far more of it germinates and sprouts than could ever happen without human help.

The life of wheat might seem empty and contemptible to a wild plant, if a wild plant could form an opinion—an endless round of servitude, year in and year out, with no point but its own proliferation. Apparently, that is how the life of the farmer looks to the hunter. And the life of the industrial proletarian, leashed to wheat more distantly but no more flexibly, does not look a whole lot better. Yet in terms of sheer reproductive success the arrangement is very satisfactory.

The development of the nonshattering rachis, or fruiting stalk—the trait that makes a grain refrain from dispersing its seeds to the wind, and instead wait for humans to harvest them—is a kind of evolutionary Rubicon. It marries human and plant, committing each to the other's tender mercies. Henceforward they will be almost as vital to each other's reproduction as are man and woman.

Mutualism may often demand such a leap of faith, especially if it is to arise between eater and eaten. Man and wheat (wild emmer and wild einkorn) had been coevolving for some time, but mainly in the measure-and-countermeasure, Tom and Jerry manner of prey and predator. In such a case, the first step toward reconciliation comes when the predator begins to manage his prey: by eliminating its competitors, or its other predators (his competitors), or by setting fires that stimulate its growth.

The path from foraging to agriculture is not straight. Pure foraging can cause negative selection (at least of plants whose seeds and tubers are completely digested), as the most desirable plants are eaten before they can reproduce. Positive selection is an about-face. That does not mean it must be the result of a shouted order or some kind of rational intent. According to the "dump-heap" theory of agricultural origins, seeds or tubers dropped by accident in the refuse pile or kitchen midden of a camp, or left undigested in the droppings of an open-air latrine, might profit from the disturbed and well-fertilized soil to germinate or root. The same might happen to grains or roots or fruits cached in the earth against a lean winter. In this way, a plant that had been "selected" by foragers for its size, promised flavor, or even a nonshattering rachis might at length colonize an area handy to foragers. The dump heap would become the first field.

OTHER ALLIES

Even those who see farming as a kind of mutualism rarely see just how broad a mutualism it is. Consider the common bean, a crop perhaps as old as wheat and so primal that it is the main thing Thoreau deigned to grow at Walden. Its great value lies in its ability to "fix" nitrogen: to draw that critical element from the soil air and convert it into a form usable by plants and animals. No plant or animal can do this by itself. Most nitrogen-fixing plants have root systems that invite colonization by bacteria called rhizobia. In exchange for a modest ration of sugar exuded by the roots, these bacteria form tumorlike nodules in which they set up shop and go about the business of nitrogen fixation. Without them, and certain free-living soil bacteria that do the same thing, not only beans but all plants and animals would be starved for nitrogen. Without nitrogen, they would be unable to make proteins—that is, to make themselves. (It is not by accident that beans and other leguminous, or pod-bearing, plants are rich in protein.) Moreover, beans, like most plants, would be starved for phosphorus and short of several other nutrients were it not for the fungi known as mycorrhizae that cling to their root hairs, greatly extending their reach and grasp.

What seemed to be an entente between animal and plant (man and bean) thus turns out to involve four of the five kingdoms into which biologists now divide the empire of life: animals, plants, fungi, and bacteria. On still closer inspection, the fifth kingdom—the protoctists, or single-celled (and colonial many-celled) organisms that have a membrane-bounded nucleus—proves to be involved as well. No plant, nitrogen-fixing or otherwise, could long survive without the host of creatures in the soil that turn cadavers and wastes of all kinds into usable nutrients. Indeed, to say that they are "in the soil" is to sell them short. They make the soil. To a large extent they are the soil. Among these creatures are subjects of all five kingdoms.

Or consider the cozy, almost filial link between humans and cows. That a third party, grass, is involved is plain enough, but in fact the ménage is larger. Just as we are unable to digest leaves of grass without the help of cows and other ruminants who first turn them into milk or meat, so (as I said earlier) those ruminants are unable to digest grass without the help of the anaerobic bacteria in their rumens. In exchange for this service the bacteria get a steady diet of cellulose as well as shelter from the air, which from their point of view has been lethally polluted—with oxygen—for the past two billion years. For that matter, kindred if somewhat more finicky anaerobes in the human colon allow us to process the more modest amounts of cellulose in the vegetable matter we do eat.

These examples suggest why some biologists refuse to use the term "mutualism." All of life on earth can be seen as one great tangle of backscratching, one great net in which each fiber depends on the sum of all the others. Equally, it can be seen as a great battle royal or free-for-all. One can draw a line linking two species and declare that their relationship is one of mutualism or competition or predation. But if one steps back for a moment, that line is quickly lost in the general tangle.

A RAGTAG ARMY

Among the species that a wave of humans and their allies displaces may be other humans (or hominids) and their allies. The genus *Homo* displaces the genus *Australopithecus*; modern humans displace Neanderthals; farmers displace hunter-gatherers; civilized peoples displace or conquer primitive farmers; nomadic pastoralists displace or conquer civilized peoples; industrialized peoples displace, or conquer, or economically subjugate whoever is left. In most of the cases we know about, the displaced humans are viewed by the displacers as "nature." Often their allies and the ecosystems they manage are viewed as "wilderness."

In the work of displacement, the new wave of humans often finds itself aided by an unlikely set of allies: its own parasites and pathogens.

When you make an alliance, you do not get a neat and uniformed army.

You get a mixed multitude, a ragtag collocation of camp followers. You get freeloaders, con men, cutpurses, and cutthroats: ticks, lice, fleas, tapeworms, hookworms, roundworms, bacteria, viruses. (In computer simulations of life, some of the first life forms to arise spontaneously are parasites. This is true in social evolution, too. Fleas must be included in any clear-eyed view of nature. Of the pastoral genre's many lies, the biggest is the absence of lice, fleas, and flies.)

As soon as domestication begins, disease increases. When humans live cheek by jowl with livestock, the pathogens of sheep and cows and fowl mutate to take advantage of this new host. As human settlement gets denser, the problem intensifies. It is not just that people in close quarters sneeze and cough and ejaculate on and in each other, though that is a large part of it. The quarters themselves become hosts.

Organisms that have evolved to live in and live off the human body do not, as a rule, last very long outside it, except in the moist tropics. Cold kills the pathogens; air and sunlight and hydrolytic enzymes break them down; other microbes eat them up. But dense settlement can make a whole town resemble a giant human body. We build houses that are warm enough to keep pathogens alive. We build cities whose streets and streams are bowels and ureters coursing with human and animal waste. Soil and water, over-taxed with effluvia, no longer purify themselves. Irrigated fields and paddies offer parasites a sheltered taxi stand where they can wait in comfort for their next ride to come along.

These irregular troops are a nuisance and sometimes much worse, but you learn to tolerate them; and when you are invading foreign territory they can be very useful, doing in their petty nastiness more damage than all your caparisoned host. For by now you and your pathogens have come to an arrangement. Having run through the available population, killing off a large part of it, a virulent pathogen has to moderate its demands on its hosts, or else find itself homeless. A strain that extinguishes whole villages may soon be extinct itself, while gentler strains flourish. Likewise, a strain that prostrates its victims, or kills them out of hand, will not spread as rapidly as one that lets them walk, talk, cough, shake hands, and make love. (There are exceptions to this pattern, especially among pathogens that use insects as their mode of transport: they can immobilize a host and still move on to new pastures.) The remaining hosts, meanwhile, have developed some degree of resistance. Instead of a deadly plague, what you have is one of the civilized diseases that everyone gets at an early age and most people survive with only a few scars.

But the people you are invading (by whatever means) have no such arrangement with your pathogens. For them, measles and chicken pox and influenza—even the common cold—can be deadly epidemics: a principle luridly illustrated when Europe invaded America. The Age of Exploration was also the Age of Infection. American Indians were particularly defenseless, as they had no domesticated animals apart from dogs and turkeys. While Europe exported dozens of diseases to America, America had virtually none to send back; its trade deficit was staggering. (The infamous exception, syphilis, has lately been thrown into question: studies of European skeletons suggest that the disease may have existed in Europe long before 1492.) In general, the Indians' allies, both regular and irregular, were no match for the Europeans', which had evolved in places more crowded and competitive.

What Europe did to America, Asia did to Europe. For what Europe was to America, Asia was to Europe: an older place, a place in which humans and animals had been living in closer proximity for a longer time. Throughout recorded history, war and trade have brought Asian diseases to Europe, some of which killed off as much as two-thirds of the population before making themselves at home. New strains of the Asian flu still come from East Asia, where poultry, cattle, pigs, sheep, goats, and humans live in great intimacy. Most of these viruses arise among migrating wild ducks, then are passed along to domestic fowl and finally to people. Pigs may interpose, serving as mixing drums for viral genes. Much as Bloomingdale's and Macy's send buyers to Paris and Milan in the spring to see the fashions that Americans will be wearing in the fall, so the Centers for Disease Control sends scouts to China each year to scope out the new strains of influenza and bring them home for study, in hopes of having vaccines and other countermeasures ready by the time the bugs hit these shores.

Perhaps this westward flow of disease began in the early days of farming, when the Near East invaded Europe. In any case, it is clear that expanding human groups, especially civilized ones, have often had the horsemen of pestilence as their vanguard. Zones of devastation and depopulation have appeared at their edges, furthering the illusion that the places they were invading were "wilderness."

WHEN YOU LOOK at humankind's conquest of the world through the lens of biology, you can see that it has not been single-handed, it has not been unique, it has not been "unnatural." That is not to say that it has been harmless. The dark side of the Eden myth—in our reading, that the earth is cursed by man's passing over—is only thrown into deeper shadow by the harsh light of science.

3

DIRT CHEAP

THE ALLIANCE OF grass and man has conquered the world about as thoroughly as any previous alliance, and in record time. The genus *Homo* had struggled for two million years to reach a population of five million; once the marriage with annual grasses was solemnized, it took only ten thousand years to reach the present level of almost six billion. Humans now control at least 25 percent of the primary productivity—that is, the green stuff—on the planet's land surface. Much of this consists of annual grasses.

But conquering the world is the easy part. The hard part is running it once you've got it.

When prior axis powers conquered the world, they either left intact enough of the previous biota to maintain the processes that all life depends on—the flow of energy, the cycling of nutrients, the regulation of climate and the atmosphere, the absorption and release of rainwater, and so on—or else learned to do some of these jobs themselves. The alliance of grass and man has not been so prudent. Many of the species that we have trampled underfoot did jobs that we have not been able to take over.

A major case in point is that of the soil community. These species recycled organic matter, retained moisture, fixed nitrogen, and made nutrients available to the roots of plants. They were allies we ought to have recruited, and treated with kid gloves. Instead, we treated them like dirt.

Our ancestors had the excuse of ignorance. After all, most of the soil community was invisible, and even its aggregate role was inkled only in the vaguest way. We have no such excuse. Knowing the role of the soil population, we persist in its casual massacre, or else keep it working like slave labor under gulag conditions. It may not hold out much longer.

To speak of "the soil community" may sound odd, but the fact is that dirt is alive—at least, healthy dirt is. An acre of good topsoil may house eleven tons of insects, worms, nematodes, fungi, and microbes, each with a major role in the great Passion play of death and resurrection. Even without counting roots, there is often more living biomass below the surface than above it. These decomposers are at once at the top of the food chain and at the bottom, taking nutrients from the loftiest organisms and passing them down to the lowest. They are the clasp of the food chain, without which it dangles limp and broken.

THE BOWELS OF THE EARTH

Soil is an extraordinary thing, unlike any thing on or in the earth. But it is such an occult thing—hidden from us even when we dig into it, slipping through our fingers in more ways than one—that we feel the need to compare it to something else. If I had to choose the two aboveground things that soil is most like, I would choose a great city and a mess of entrails.

The city part is easy enough. Soil is, to put it mildly, densely settled. One teaspoon of good grassland soil may contain 5 billion bacteria, 20 million fungi, and 1 million protoctists. Expand the census to a square meter and you will find, besides unthinkable numbers of the creatures already mentioned, perhaps 1,000 each of ants, spiders, wood lice, beetles and their larvae, and fly larvae; 2,000 each of earthworms and large myriapods (millipedes and centipedes); 8,000 slugs and snails; 20,000 pot worms, 40,000 springtails, 120,000 mites, and 12 million nematodes.

These citizens are often in motion, hurrying along the vast expressways made by moles, the boulevards of earthworms, the alleys between particles of sand or clay, the dank canals that these alleys often become. Certain districts and certain intersections—mainly close to the roots of plants—get especially busy. The citizens move in the dark, sniffing at chemical trails. They are constantly doing business with one another. They traffic in molecules: minerals, organic compounds, packets of energy. Their interactions are sometimes friendly, sometimes competitive, often predatory.

Though city-haters of all stripes will readily grant the likeness of cities to bowels, I had better explain why a city-lover such as myself should set them (transitively and approximately) equal. The main reason is that both systems are arranged, or arrange themselves, to maximize the exchange of energy, matter, and information. The maze of streets and the honeycomb of shops and offices in a city make possible millions of interactions between people each day. The arrangement of your guts similarly makes possible billions of interactions between molecules. Their compact, switchback windings, the villi or tiny projections that line them, the blood vessels that marble them: all these features effectively pack acreage into the space between your ribs and loins, so that you may extract as much energy and nutrition as possible from the food you eat.

Aristotle called earthworms the intestines of the earth, and he was not wrong. In fact, though, the soil itself is a tangled slaw of intestines, some as thick as your arm, some a few molecules wide. The business of the soil, like that of the gut, is digestion and assimilation, and the creatures of the soil are always enlarging the surface area where these can take place.

Fungi, ranging in size from the great mycelial clot beneath a toadstool to tiny threadlike microbes, send out long fingers called hyphae that penetrate dead tissue. Besides roughing it up physically, they exude enzymes that digest it chemically, so that the simplified nutrients can be absorbed into the fungus. If you could somehow unravel the fungi in a single ounce of rich forest soil, they might easily stretch two miles.

A like figure might be racked up by filamentous bacteria called actino-mycetes, which under the electron microscope look like something midway between antlers and bean sprouts. Among the most plentiful of all soil organisms, they give damp soil and damp basements their heavy, heady smell. They also secrete enzymes that break down dead tissue into nutri-ents, which are then absorbed along the surface of the branched, springy filaments—though, like all bacteria, they lack the fungus's ability to break tissue down by brute force.

Roots, which do not digest but merely absorb, are not quite so rich in surface area, but they come close. The root system of a single four-month-old rye plant was found to have a surface area of 639 square meters—130 times the surface area of the aboveground plant—all of this packed into about six liters of soil. Much of this expanse is accounted for by root hairs, which sprout like a scraggly beard near the growing tip of the root. If a mere stripling of an annual plant can so pack the soil with roots, rootlets, and root hairs, one can imagine what lies beneath the sod of a mature perennial grassland or the litter of a forest.

Most plants, though, are unsatisfied with the mileage they get out of rootlets and root hairs. As I mentioned in an earlier chapter, they enlist the help of fungi that invade or cling to the roots and branch out into the soil. These unions of roots and fungi, known as mycorrhizae, not only increase the surface area of the root network, but also give the plant access to cer-tain nutrients, such as phosphate, that it would be hard put to gather on its own. In return, the fungus gets a trickle of sugar.

Fungi maximize surface area—or, to be precise, the ratio of surface area to volume—by stretching out. Bacteria, their only peers in decomposition, achieve the same end by staying very small. As any object gets bigger, its volume increases faster than its surface area. (Volume increases with the cube of the radius, surface area with the square.) To put it even more crudely, the bigger something gets, the farther its insides get from the outside world. If you want to exchange energy and matter with the outside world as efficiently as possible, you had better stay small.

It is no accident that, apart from the soil, the other great habitat for bacteria on land is the guts of humans and other animals. Not only their small size, but their chemical virtuosity—the range of enzymes they can produce, the unlikely stuffs from which they can pluck energy—makes them as invaluable in your bowels as in the bowels of the earth.

SIX ACRES TO THE OUNCE

As small as they are, bacteria still need a place to live. And here we come to the way living things conspire to expand the surface area of the soil itself.

Surface area, I said, is needed for the exchange of matter and energy: not only for digestion, but for metabolism in general. In short, it is needed for life. For most organisms, it is useful not only to have a surface, but to be on a surface.

According to the great ecologist G. Evelyn Hutchinson, one of the things that make this planet hospitable to life is its interfaces between the solid, liquid, and gaseous states of matter: seashores, stream banks, the surface of the ocean, the surface of the earth. On closer inspection, each of these interfaces is infinitely involuted. Reference books tell us that the dry surface area of the earth is 57.5 million square miles, but the truth is that it can't be measured, any more than you can measure the surface area of a sponge. Solid ground is not as solid as it seems. A good loam is only half solid matter; the other half is space, filled in varying proportions with air and water.

At Rothamsted Experimental Station in England, where a good part of the foundation of modern soil science was laid, researchers tried to measure the surfaces of the particles in a single ounce of clay-rich soil. They came up with a total of six acres.

Like many of the conditions that make the earth hospitable to life, this interface has been made by life itself. The making of soil starts with the breaking down of rock. In this work, wind and water are joined by the roots of plants and the hyphae of fungi and lichens. The work still goes on thousands of years later, when the bedrock is covered by yards of rich soil. Roots probe the bedrock, opening fissures which are then pried wider by the freezing and thawing of water and the trickling of natural acids. Under the insistent fingertips of rootlets and hyphae, rocks are ground to pebbles,

pebbles to sand, sand to clay. Some of the finer grinding goes on in the guts of earthworms and other creatures.

In the topsoil proper, mineral particles range in size from coarse sand grains to invisible specks of clay a thousand times smaller. The smallest particles are the most vital. They are colloidal: small enough so that their surfaces bristle with molecular hooks and charges. As we all know from experience, clay holds much more water than sand does. The surface of a clay particle eagerly coats itself with a film of water. It also grapples to itself positive ions, or cations, such as those of calcium, magnesium, and potassium. Clay is thus a reservoir from which water and nutrients are slowly released to the roots of plants.

But a soil with too much clay in it becomes damp, heavy, and airless, a bane not only to farmers but to plant roots and air-breathing soil organisms. From the point of view of all these parties, a good soil is one in which particles of various sizes are judiciously mixed. The mixing is done by the soil organisms themselves. Ants, termites, bees, wasps, moles, shrews, and other excavators bring up coarse particles from the lower soil horizons and blend them with the finer particles above. And nearly all the soil biota takes part in adding the most crucial ingredient of all: organic matter.

In the 50 percent of a good soil that is solid matter, 45 percent is mineral and only 5 percent is organic. Yet that 5 percent is roughly the equal of the other 45 percent in its ability to hold water and nutrients. Bits of organic matter are colloids like clay, but even more chemically active. In addition, they bind with soil particles to form aggregates, or crumbs: the texture you feel when you sift good soil through your fingers. In other words, they are able to granulate soil, like granulated sugar, so that instead of a sticky mess you get a light, crumbly stuff with plenty of pores for air and water, and big enough pores so that excess water drains out. Both the slimy halfway products of decay and the drier, more durable remnant known as humus have this vital knack. Besides being bound together with organic matter, a soil crumb may be glued together by the mucus of earthworms, slugs, and snails. Almost certainly it is sewn together by the threads of fungi, rootlets, and filamentous bacteria, both alive and dead. As I said before, the soil organisms do not just make the soil; to a great extent, they are the soil.

In multiplying the surface area of the world, the soil creatures create a multitude of worlds. Each aggregate is a tiny planet, complete with oceans, soil (organic matter), and wildlife. The drier parts are settled by aerobic bacteria, the filmy oceans by anaerobic bacteria and by the great leviathans of the one-celled world—amebae, hairy ciliates, undulipodiates that move through the water by swishing their tails. In place of gravity, the planet has electrochemical forces: forces that, on this bejeweled scale, are far more potent than gravity. That is why the soil holds water after a rain, instead of letting it all drain away into the water table. In fact, so tightly is water bound that the most desperate plant rootlets cannot suck all of it away. Even during droughts that make plants wilt and die, a great quantity of water still clings to the soil particles—a good thing, since it means that the life of the soil goes on. The strength of the surface charges also means that mineral ions and other nutrients are not washed away with the rain but are let go slowly, like time-release vitamins, to meet the needs of plants.

Despite all the forces that would keep one at home, there is a lot of interplanetary travel. When the soil is not bone-dry, the smaller spaces between particles are filled with water. In these linked oceans swim small aquatic creatures such as nematodes, rotifers, and crustaceans, many of them pygmy cousins of the ones found in ponds. Water streams toward thirsty roots, bearing dissolved nutrients. Micro-organisms are carried along with the tide, or swim that way on purpose, tracking the chemicals that roots exude. After a heavy rain, water may fill the larger soil pores, too, driving out the soil air. For the aerobic bacteria, it is a cataclysm: they die off by the trillions. But their place is quickly taken by anaerobes, multiplying at the hair-raising pace that only bacteria can manage, or migrating upward from the water table. Protoctists pop out of the protective cysts in which they weather dry spells and make a holiday, feasting on the dying aerobes. Later, as the water drains away, there are new holocausts, new boom times, new migrations. Migrations happen every day, in any case, as organisms move upward toward the sun-warmed surface, then downward to escape the night's chill.

All soil life has a hand in creating the world, or worlds, of the soil. All soil creatures add their quota of organic matter to the soil; all excrete, all die, all play a role in the process of decay. Though fungi and bacteria do most of the work, they have help. Ants, mites, millipedes, springtails, slugs, snails, beetles, and other invertebrates feed on plant debris, breaking it down physically (which exposes a larger surface area to the action of microbes) and doing some simple and some not-so-simple digestion. Earthworms literally eat their way through the soil, which on its way through their lengthy guts is refined, enriched, and mixed with the proceeds of leaves that the worms have dragged into their burrows and eaten. Unlike cats, who like to bury their feces, earthworms go to some trouble to deposit theirs on the surface of the ground. Known as castings, these deposits are fine, dark, and fabulously rich, having twice the calcium, thrice the magnesium, five times the nitrogen, seven times the phosphorus, and eleven times the potassium of the surrounding soil. There is also a fabulous lot of them. On a single hectare of ground, in the course of a single year, earthworms may deposit five hundred metric tons of castings. These are excellent media for the growth of bacteria and fungi, as are, to varying degrees, the excreta of other invertebrates.

When bacteria and fungi take over the work of decomposition, they do not all plunge in at once. First to get their licks in are specialists in easily decomposed stuff such as sugars and amino acids. These breed and flourish when a new flush of debris hits the ground, then die off as their food is used up. More patient, slower-growing creatures—such as *Arthrobacter*, the most common genus of soil bacteria, and various higher fungi—then go to work on their leftovers: lipids, alcohols, chitin, cellulose. Lignin, the toughest plant material, is left for the common mushroom and its relatives—the last decomposers to kick in.

Other soil residents, less directly involved in the work of decay, nevertheless speed it along. Mostly they do so by eating those who do the work. Experiments have shown that litter decomposes faster when protoctists are present. The reason seems to be that by preying on the bacterial population they keep it young and healthy. One can apply the same reasoning to other steps of the trophic pyramid that is buried in the soil, from the meat-eating fungus that catches nematodes in a nooselike trap, to the mole that eats the heads off worms and hangs the trunks on the walls of its burrow (hundreds of them, like carcasses in a meat locker) against the lean winter. Though we cannot be sure that each of them plays a major role in the economy of decay, we cannot safely assume that any of them does not.

Why is decay so important? The fact that it encourages a good soil structure is part of the story. So is the fact that, without decay, we would all drown in a sea of leaves, corpses, and dung. But the big story has to do with the fact that nearly all aboveground life depends on plants. Plants can get energy from the sun and carbon dioxide and oxygen from the air, but their needs for water, carbon, nitrogen, sulfur, potassium, phosphorus, and other nutrients must be met by the soil. And they are very particular about the forms these nutrients take. First off, the nutrients must be readily soluble in water, so that the roots can take them up and the plant's tissues can move them around. The complex organic compounds of which living tissue are made are not readily soluble in water. If they were, you would dissolve when you took a bath. In fact—since your body is mostly water—you would dissolve right now, leaving this book to soak in a puddle of broth.

If you took a newly dead human body, sterilized it with radiation, and planted a seedling in it, the seedling would not grow, not even if you watered it every day. Neither would it grow in a sterilized pile of leaves. Agents of decay—bacteria, fungi, and the rest—are needed to break complex organic molecules down into simpler molecules that are soluble in water and usable by plants.

In the case of minerals such as iron and manganese, some of the same agents perform just the opposite function. These minerals are readily soluble in water and readily taken up by plants—too readily, perhaps, since at higher doses they are toxic. Luckily, in well-drained soils they are oxidized by microbes so that they become less soluble, and are taken up only in the small doses that plants can handle.

It may sound as if the soil were an industrial facility, an interlocking set of assembly lines in which everything is regulated for the greater good of the daylight world. That is not so. As every gardener knows, some bacteria, fungi, nematodes, and insects can do great harm to plants. Some can harm humans directly. The soil is less a factory than a souk, a Casbah, a flea market, an economic free-for-all in which each buyer and seller pursues his or her own interest, and in which every scrap of merchandise—second-hand, seventh-hand, busted, salvaged, patched—is mined for its last ounce of value. Decay is good business because there are nutrients to be extracted and energy to be gained from the breaking of chemical bonds. If the net effect of the activity of the soil biota is overwhelmingly helpful—in fact, vital—to life on street level, it is not because nature has ordained it so, but because the various forms of life above and below ground have coevolved.

THE RISE AND FALL OF THE SOIL COMMUNITY

The city of dirt was not built in a day. Like a good many things we look down on, dirt is the product of millions of years of evolution. The dry ground that lurched from the hot oceans of the Hadean some four billion years ago was hard, sterile, and inhospitable. As physical and chemical weathering set in, and microbes tried their luck in this new setting, a faint foreshadowing of soil—what geologists call "paleosoil"—began to take shape. But it was not until vascular plants colonized the land, some 450 million years ago, that anything we would dignify with the name of dirt emerged. Themselves an alliance of algae and fungi, land plants were merely the most visible part of a far grander and largely subterranean alliance.

Since then the soil population has grown gradually, handily supporting the rest of terrestrial life, including, for the last two million years, man. Handily, because the organisms that (figuratively) fed upon it also fed it. Each generation of plants left a new carpet of humus, and a new balustrade

of roots to anchor the soil against wind and rain; each generation of
legumes bequeathed nitrogen captured from the soil air; each generation
of animals willed the riches of its fecal and mortal remains. The soil was as
if bottomless, a magic barrel.

But in the last ten thousand years—a quick nictation of the cosmic
eye—roughly half the earth's soil has been consumed by an alliance of
species that has failed to put back what it took out. Suddenly, intricate communities of plants were cleared away to make room for rows of a single
plant, repetitive as a housing project. Laid bare by plowing, its root systems
ripped to shreds, the soil remained bare much of the year—bare to wind
and rain, which carried off great gouts of it, and bare to sun, which baked
the nutrients out and ravaged the soil population. As if that were not
enough, the soil population was starved for lack of plant and animal detritus to ransack and recycle. Nor were crops the only problem. Grazing animals had been mildly destructive to the soil in their wild state, but their
numbers had been checked by the damage they did and by their own
predators. Now, under human management, they were finally able to
achieve their full potential as denuders of the earth.

Erosion is, of course, a natural process. Without the abrasion of rock,
soil would never have formed in the first place. As the hot urges of the
earth's interior push mountains up, the cool afterthought of wind and rain
soothes them down. Nature is always redistributing soil. Some of the best
soils on earth are alluvial, which means that water took their particles from
one place and dropped them someplace else. Nor has the buildup of the soil
population on the earth as a whole been so sure and steady as I have made
it sound. Twenty-odd times over the past two million years, glaciers have
scrubbed the soil from large parts of the Northern Hemisphere.

For all that, a retrospect of the past 450 million years does show a
gradual increase in the size and complexity of the soil community. In most
parts of the world, the rate of soil formation has exceeded the rate of soil
erosion by a comfortable margin. Soil that washed away from mountainsides tended to end up on floodplains. Even the glaciers may have done
more good than harm. By grinding up the bedrock into mineral pills and
powders, they ensured the new soil population (and its vegetable and animal guests) a healthier diet.

On the world's farmland today, the rate of erosion exceeds the rate of soil
formation by (on average) a factor of thirty. Rich alluvial soil is washed
from river valleys into the sea, where for all human purposes it is lost. Dams
and levees prevent its replacement by soil washed from the hills; instead
that soil, too, is flushed into the sea, or else becomes a nuisance called silt.

By rights, this sort of thing should have led to crashes of the human population, and often it has, briefly and locally. But often we have just moved on to virgin land, riding the wave of human-led expansion. Occasionally, we have learned to manage the soil in one place for a good stretch of time: to control erosion with terracing, contour plowing, windrows, and winter cover crops; to replenish the soil by leaving it fallow, rotating crops, intercropping, and using compost, animal manure, and "green manure" (legumes like clover and alfalfa, whose roots host rhizobia that draw nitrogen from the soil air).

Lately, grown cocky, we have used fossil fuel and chemicals to give a jolt of specious life to dying soil. The effect of this is to make it die faster. Manure is something the soil population can use; chemical fertilizer is an attempt to go over its head. The crops get plenty of nitrogen, phosphorus, and potassium, but next to nothing of the countless trace elements needed for plant and human metabolism. It's like raising a child on white bread. Indeed, by feeding our children crops grown this way, we *are* raising them on white bread. But this is a mealy-mouthed euphemism for what industrial agriculture is really doing.

THE FEAST OF ATREUS

Consider Iowa, the quintessential farm state, with the highest concentration of prime farmland in the world. For every bushel of corn that Iowa grows, it sheds two bushels of soil. After just a century of farming, the state's rich prairie soil, which took millennia to make, is half gone. What is left is half dead, the roiling, crawling life burned out of it by herbicides, pesticides, and relentless monocropping. Petrochemicals fuel its zombie productivity. Hospitable Iowans assure their guests that the coffee is made from "reverse-osmosis" water, since agricultural runoff has made the tap water undrinkable.

Let us step back and look at the United States as a whole. In the past two centuries a third of its farmland topsoil has blown or washed away. On nine-tenths of U.S. farmland, soil is being lost faster than it is being replaced—not slightly faster, but, on average, seventeen times faster. The cost of erosion to land, water, infrastructure, and health over the next twenty years has been projected at $44 billion.

Though government-sponsored conservation practices cut soil erosion in the United States by one-quarter between 1982 and 1992, this only brought the rate down to something like that of the Dust Bowl days of the 1930s. Moreover, conservation tillage, the government's favored method for checking erosion, has farmers tilling less and spraying more, swelling the poisonous runoff. It can't work in the long run, anyway, as chemicals tend to destroy the colloidal and living properties that hold soil together. Their effect on the living properties of people is not pleasant, either: Kansas farmers exposed to herbicides for twenty or more days a year are six times more likely to develop non-Hodgkin's lymphoma than are nonfarmers. The prevalence of that disease in farming regions has been linked to use of the herbicide 2, 4–D. Throwing chemicals in wholesale quantity at pests only makes them smarter. In 1948, at the dawn of the chemical age, American farmers used 15 million pounds of insecticide and lost 7 percent of their crop to insects; today they use 125 million pounds and lose 13 percent.

Salmonellae are getting smarter, too; the broad-spectrum antibiotics fed to cramped, factory-farmed livestock to keep them "healthy" and boost their growth are breeding hundreds of strains of resistant bacteria, many of which are finding their way into humans. Millions of Americans each year get salmonella poisoning, the Centers for Disease Control informs us, and 69 percent of drug-resistant strains come from food animals. Nor is food poisoning the only danger. The General Accounting Office has found over 140 drugs and pesticides likely to linger as residues in meat. Of these, 42 are believed to cause cancer in laboratory animals, and others are thought to cause birth defects or attack the nervous system.

Livestock and the feed grown for them (which account for 85 percent of U.S. farmland) produce ten times as much water pollution as people do, and three times as much as industry. Nitrogen fertilizer may be contributing both to global warming and to the thinning of the ozone layer. Farm runoff is causing algal bloom and eutrophication in the nation's lakes; nitrates in groundwater have been linked to blue-baby syndrome and cancer.

The problem is not just what we put into the water table; it is also what we take out. If we reckon by agricultural water use, each American man, woman, and child drinks 325 gallons of water a day. The Ogallala Aquifer of the Great Plains, which grows the feed for 40 percent of our beef cattle, among other produce, held enough water forty years ago to fill Lake Huron; in another decade it may be too low to pump. The great dams of the West are silting up. Overirrigation and poor drainage (worsened by the compaction that heavy machinery causes) have caused saline or sodic conditions on 10 percent of our crop and pasture land.

Worldwide, more irrigated farmland is lost to salt, groundwater depletion, and other causes than is won by new waterworks. A United Nations study published in 1990 found that, in the years since World War II, 552 million hectares of the world's farmland had been damaged by agricultural malfeasance—an expanse equal to 38 percent of the land now under cultivation. The projected cost of soil erosion worldwide, over the next two decades, has been pegged at $400 billion.

To say that we are snatching food from our children's mouths is to put it too gently; what we are snatching from them is the very possibility of feeding themselves. The nutrients that wash from our fields and flush down our toilets are stolen from their protoplasm. Instead of eating the recycled remains of our ancestors, which is the natural order of things, we are eating our children. Lavish as it may look, the banquet that industrial agriculture spreads before us is the same one Thyestes set before Atreus.

While different kinds of farming do different kinds of damage, the basic problem inheres in farming as we know it: the alliance between humans and annual grasses. Annual grasses, remember, are pioneer species which throw their energy into setting seed, rather than into their root systems. When a hillside that once hosted trees or perennial grasses is cleared and planted with annual grasses, the earth loses its moorings. For the soil community, it is like having one's city plucked away by a giant hand and replaced with cellophane tents. And since annual grasses do best in "disturbed" soil, we as their allies are bound to keep disturbing it: the more so in that we have to keep disturbing the other annuals—the ones we call weeds—that thrive in soil we have disturbed.

NOTHING SUCCEEDS LIKE FAILURE

How can an alliance that is so successful be so unsuccessful? How can an ecosystem spread so quickly when it is so unstable—when it is not really an ecosystem at all?

The alarming fact of the matter is that in ecology, as in other realms, nothing succeeds like failure. It is often the most unstable systems that expand most ferociously. A stable system, by definition, does not need to expand—though it may do so anyway, at its leisure. An unstable system is like a giant with tiny legs who must keep running just to keep from falling. An obvious example is fire, an aerobic creature that quickly wolfs down the food around it and lopes outward in search of more.

Agriculture is such a system. It depletes the soil it is on and moves out in search of more. The more destructive a form of agriculture is, the more quickly it expands. But even relatively benign forms can expand fairly quickly, because the number of humans can rise geometrically.

A similar logic guides parasites and pathogens. A pathogen is most virulent when it is able to move easily from one host to another, for then it can kill with impunity—milking one host dry, then moving on to the next. When it is confined to a small population, or finds its movements blocked in some way, it is forced to moderate its demands. It must exploit the host more gently, making sure that he stays able-bodied enough to keep himself—and his pathogens—alive. (A likely example of this can be seen in Japan, where the scrupulous use of condoms may have led to the evolution of milder forms of human T-cell leukemia/lymphoma virus type 1, or HTLV-1, a relative of HIV.)

In their expansion, most life forms follow the same rule as parasites. The greater their demand on their "host"—for which read "environment," both living and nonliving—the greater their need to expand. Conversely, the more room they have to expand, the more crippling the demands they can make on their surroundings. To reduce it to a nursery rhyme: The more you demand, the more you expand. The more you expand, the more you demand.

Sooner or later, you run out of room. At that point, you either learn to treat your host more politely, or you drag your host with you to extinction. Has that point been reached by the alliance of grass and man? In parts of the world where peoples were confined by geographical barriers or by the pressure of other peoples, it was reached some time ago. In some cases the result was a more careful husbandry of the soil, in others famine and the extinction of cultures. The same point will soon be reached by humankind at large, and the same choice will have to be made.

Some scientists and economists tell us not to worry, as there is still plenty of virgin land that can be brought under cultivation. They are wrong. Much of the soil in the humid tropics is lateritic, and bakes as hard as brick as soon as it is stripped of its forest cover. In drier regions, land can be made arable only by means of irrigation projects that often cause grave ecological harm. Mountainous regions, when pricked by the plow, hemorrhage soil. Even when virgin land seems to invite our seed, we had better think twice. We and our allies have already junked so much of nature that the machinery is starting to sputter. For the soil community is not the only one that we have trampled, and whose work we do not know how to do.

4

THE NEW PANGAEA

Taking over the world is something like taking over a company. Either you leave most of the workers in place, or you replace them with others who can get the job done. Here the jobs in question are ecological services—the cycling of water and nutrients; the flow of energy; the dispersal of pollen, seeds, and other propagules; the regulation of climate; the balancing of gases in the air—without which most creatures on earth, including the triumphant axis powers, would perish.

All nonhuman alliances have followed this rule, more or less. If they had not, you and I would not be here to tell the tale. Eukaryotes displaced most cyanobacteria in the open ocean, but not along the shore, and together they managed to keep oxygen levels more or less what they had been. When herbivorous dinosaurs and fermenting bacteria teamed up, they did not greatly disturb the soil biota that made their picnic possible. When flowering plants, mammals, birds, and insects swarmed over the earth at the dawn of the Cenozoic, they only fine-tuned the ecological services that gymnosperms, reptiles, and earlier insects (as well as fungi and microbes) had provided.

We, too, observed this rule—though not scrupulously—the first time we conquered the world. Our alliance with perennial grasses and grazing animals did not, in most places, do great harm to the soil community or keep it from doing its job. The grasses we encouraged were almost as good as the trees we discouraged when it came to absorbing and releasing moisture, and cycling carbon and oxygen. The ecosystems we sponsored were variants of earlier ecosystems, which they closely matched in richness and diversity.

We were not scrupulous. Some species we hunted to extinction. Some ecosystems we impoverished. But the fact remains that our genus traipsed across Africa and Eurasia for half a million years without doing any serious damage to the workings of nature.

Our record in the last ten thousand years has not been so good. The alliance of man and annual grasses has achieved a hostile takeover of nature. We have gutted the buildings, fired the employees, sold off the assets. Here again, science casts a cold light on the Eden story. To paraphrase our paraphrase of Genesis, the earth is cursed by our taking over.

What about all the shiny new equipment we have brought in? What about the new ecosystems we have installed: the wheat fields vast as sunlight, the seas of clipped lawn? Can't these new ecosystems be trusted to take over the jobs of the old?

No, they can't. And here is where the language of coevolution, which I have used so freely, misleads. In evolutionary terms, our new ecosystems are still wet behind the ears. Most ecosystems have been shaped and tested by hundreds of thousands of years of trial and error. Our new ecosystems have had at best a few thousand years of testing. You might say they are being tested on an accelerated schedule on the bench of human history and with the instruments of the human mind. For the most part, though, these instruments have addressed problems of immediate human concern: pests, weeds, the loss of soil fertility. How the larger workings of the biosphere will be affected as cornfields displace prairies and pulp plantations displace forests, we have hardly begun to guess.

By the time we know for sure what the effects are, it will be too late to do much about them. But a rough index of our impact on nature can be found in the rate of extinction of other species. For extinction is both a cause and an effect of the loss of ecological services.

BURNING BOOKS

The loss of biological diversity in our time has been likened to the burning in 272 A.D. of the Great Library of Alexandria. In truth, that ancient conflagration pales to a candle flame beside the present one. Measured in bits, the genetic information in the chromosomes of just twenty randomly selected creatures would fill some 400,000 scrolls, which is a good estimate of the library's entire holdings. No one knows exactly how many species we are losing daily; but if we accept the fairly conservative figure of eighty, it follows that we are torching four Great Libraries every day.

In 1945, Yuri Knorosov, an artillery spotter for the Red Army, watched the National Library in Berlin go up in flames. Able to save only one book, he snatched up an edition of the three known Mayan codices. It was a good choice. Over the course of the next decade, as a linguist in Leningrad, Knorosov made a series of breakthroughs that led to the decipherment of Mayan hieroglyphics.

A biologist watching the rain forest burn feels a bit like Knorosov. What shall he save? What can he save? It takes no great heroism to tuck a specimen in his field bag. He can press or pin it and preserve its morphology. He can freeze it in liquid nitrogen and preserve its DNA for eventual sequencing. He can interview the natives and learn how the species is used for food, fiber, or medicine. What he cannot do is preserve the furious ballet by which this one species is linked to a thousand others. That he cannot do except by preserving the stage, which is in flames all around him.

How do we do it? How does man manage to achieve single-handedly a rate of extinction that in times past required the wallop of a giant asteroid or the shivering of continents? First of all, we do not do it single-handedly, but as part of an alliance of species. Let us consider first our own direct effect, then that of our allies and fellow travelers.

A VERSATILE KILLER

It often happens in nature that a predator overhunts its prey to the point of near extinction. What usually happens next is that the unfed predator population crashes, giving the prey population a chance to recover.

Human predators and their prey have often ridden this seesaw. But often the seesaw has gotten stuck, so that the side that was down never got up.

How have we managed this? Humans are ecological generalists. Our versatility lets us hunt a species to the point of near extinction and then (by eating other species) survive to hunt it to total extinction. We can destroy one ecosystem, migrate to another, adapt to it, and destroy it as well.

We don't have to be civilized to do this. Though the mass extinctions of large mammals at the end of the Pleistocene used to be chalked up to a change in climate, most paleontologists now blame them at least partly on the march of well-armed humans across Eurasia and North America.

Having extinguished their principal prey, these predators should then— by all the laws of ecology—have gone extinct themselves, or at least suffered a population crash. Instead, they invented agriculture.

Not only do we eat more different things than most animals, but we also prey upon things we don't eat. We prey upon trees for the energy in their flesh, and simply to get them out of the way. We prey upon animals for their skins, which help us retain energy in our bodies. We prey upon many things just to indulge passing fancies. The beaver was brought to the edge of extinction in North America to make elegant men's hats, and various bird species were pulled over the edge to garnish women's bonnets. Whales were slaughtered for corsets that made ladies look like insects. The rumble of the

African elephant was largely silenced so that Chopin's waltzes could be heard in every middle-class home.

Generalists are not the rule in nature, but they are a fairly common exception. Mutualists are common as dirt. What makes man uncommonly dangerous is the combination of the two traits. A generalist can take big bites out of the environment in a number of places, but still must accept the environment as given. As a rule, it is only if that generalist is able to recruit other species that it can really conquer the world.

INVADERS

The way in which major allies such as wheat, corn, and cattle displace native species is well known. Since ecosystems built around these allies tend to be monosyllabic, the loss of diversity is plain: Rabelais replaced by Dick and Jane. But these major allies are merely the tip of the eraser we thoughtlessly wield.

Sometime in the 1870s, a New York drug manufacturer and drama buff named Eugene Schieffelin resolved to let loose in North America every one of the forty or so birds mentioned in Shakespeare's plays that was not there already. The starling is mentioned only once ("I'll have a starling shall be taught to speak nothing but Mortimer"), but this was enough for Schieffelin to set free forty pairs in Central Park in 1877 and another fifty pairs in 1890.

The first successful breeding pair roosted under the eaves of the American Museum of Natural History. Within a few decades, the skies of America seemed at times to speak nothing but starling. Loud, pushy, and omnivorous, the greenhorns evicted bluebirds, martins, and woodpeckers from the nooks they liked to nest in. In farmyards they swarmed like winged mice. In cities they were airborne rats, or great speckled bees, roosting in their thousands and tens of thousands on a single building and leaving a white and malodorous honey behind.

Schieffelin was not a lone crank. He was a member of an organization of cranks known as the American Acclimatization Society, one of several Acclimatization Societies that brought to America hundreds of thousands of birds of dozens of species. Some species were imported for sport, some to soothe nostalgia, some to kill pests, some for rarefied reasons like Schieffelin's. Skylarks, wood larks, nightingales, linnets, chaffinches, great tits were uncaged, took a few turns in the unseasonable air, and died. (Not all of them died right away. Skylarks, for instance, made themselves at home in Brooklyn but were killed off by the blizzard of 1888.) The few species that made it, however, made up for all the rest. A handful of English sparrows were tossed into Central Park in 1850; by 1902, the naturalist Frank Chapman was fretting that they would soon "take over the country." Introduced to control an agricultural pest, they also controlled such nonpests as martins, robins, cardinals, bluebirds, and native sparrows. Ancient sidekicks of European grains and quadrupeds, they delighted in picking the undigested seeds of the former out of the stool of the latter. This gave them the edge in towns, where their squabbling roosts threatened to drown out not only the songs of all other birds but the sound of human intercourse as well. On farms they ate insects and weed seeds, but they ate crops and crop seeds, too. Along with starlings and a few other birds, they made up the tardy air force of the axis that had begun its invasion of America centuries before. Only a change in the cavalry—the displacement of horses by cars, whose droppings no bird eats—saved America's cities from drowning in a streaked sea.

Nevertheless, the distribution maps for both the English (now called "house") sparrow and the starling completely blanket the lower forty-eight, something that can be said of very few native birds. This, too, is typical of the way invasions work. A native species evolves in particular ecosystems and learns to deal with particular rivals. Only rarely can it adapt to as motley a set of environments as are found in the continental United States. But to an alien species that has joined or tags along behind a conquering alliance—at least, one led by a species as adaptable as man—all doors are open. Having learned the ways of the conqueror, his humors, and where the scraps tend to fall from his table, it can follow him anywhere.

In more ways than one, invasives depend on man for a free ride. While some invasives are tough enough to invade healthy ecosystems, others can get a toehold only in ecosystems that are altered, fragmented, or otherwise weakened in some way. Here again, humankind is happy to oblige. Sometimes we do so knowingly, as when we plow up prairie sod to plant wheat or corn—annual grasses which otherwise would get no purchase in the tightly woven mat of perennials. But when we bare the soil or rend the fabric of the forest, we make holes which unlooked-for species rush to fill.

Many invaders are confirmed edge-dwellers and get a leg up when an ecosystem shrinks or divides. Some native species, principally those at the top of the food chain, need vast stretches of unbroken habitat just to feed themselves. And nearly all populations lose their vitality when their gene pools are divided into puddles that cannot flow one into another. Inbred populations lose the flexibility they need to respond to new threats— for example, to the threat posed by an invading predator, parasite, or competitor.

When we speak of invaders, we are not speaking merely of a few noisy or noisome birds. A 1993 study by the Office of Technology Assessment found forty-five hundred alien species that had made themselves at home in the United States. Of these, some 15 percent were known to "cause severe harm." A mere seventy-nine of the more notorious had, in the years from 1906 to 1991, caused documented losses of $97 billion. If full information on all the invaders were available—and if a real dollar value could be placed on ecological mayhem—the total reckoning would be much higher.

Invaders now make up one-fifth to one-third of all North American plant species, and their share of territory is even greater. In the middle of the continent, the allies and hangers-on of pale-faced humans have all but erased the native prairie flora. In Iowa, Illinois, and Indiana, four-fifths of the land is so far gone that it can no longer support native plants.

But the flow has by no means gone only one way. The European wave that has broken over the Americas is part of a general tempest in which creatures of all continents and almost all islands have been jostled and mingled together. Even without habitat loss or other disturbance, this moving of species from place to place would be enough to do great harm. To understand why, we must have a look at the way evolution has worked for the past 200 million years.

A PILEUP OF CONTINENTS

At the beginning of that period the entity known as Pangaea, which encompassed nearly all the planet's dry land, began its slow breakup into continents. With oceans as well as mountains and deserts thrown between them, the flora and fauna of the various regions were free to go their own evolutionary ways. Allopatric speciation—what happens when breeding populations of a species are separated in space, and in time form new species—was greatly accelerated. The six realms of life described in 1876 by Alfred Russel Wallace—the Neotropical, Nearctic, Palaearctic, Oriental, Ethiopian, and Australasian—were the result of this long isolation, as Wallace himself (the man whose independent discovery of the idea of natural selection forced Darwin to rush his own *Origin of Species* into print after a mere twenty-two years of work) well understood.

You might say that, since the breakup of Pangaea, nature has been somewhat less red in tooth and claw than it might have been. With the evolutionary game split up into local leagues, there were batters who could not hit a curveball, pitchers who knew nothing of stolen bases, and shortstops who could only field grounders, all of whom made perfectly respectable careers—rightly so, for by the rules and usages of their respective leagues they were all highly skilled. On far-flung islands there were flightless birds who got along very nicely, never feeling the teeth of a cat or the club of a human. There were vast temperate forests whose trees had never felt the lethal embrace of a bittersweet vine.

For roughly the last twenty thousand years, beginning with the invention of seafaring boats, humans have been repairing this unprofessional state of affairs. The wakes of our canoes, galleons, steamships, and airplanes have been so many silken threads sewing Pangaea back together.

Not since Pangaea was born of a pileup of ancient continents some 350 million years ago has there been such a ramming together of life forms. That crash, occurring at the stupefyingly slow speeds of plate tectonics, took some hundred million years from first impact to full squash. Our replay of it has been less violent but unimaginably faster. As a result, nearly every ecosystem is under attack from all sides at once. Even within Pangaea there were barriers of topography, climate, and sheer distance. In the man-made Pangaea there are no such barriers. A plant or animal of the southern temperate zone can take an airplane to the northern temperate zone, hopping over the inhospitable tropics.

When ecologists use their standard species-area curves—plotting the regression of species numbers against the logarithm of land area—to predict the number of mammalian species on the land area of the planet, they come up with something in the neighborhood of two thousand. But in fact the planet hosts at least forty-two hundred species of land mammals. For land species of all kinds and kingdoms the gap is even wider: there are four times as many species on earth as would be predicted on the basis of land area. The gap arises from the simplifying assumption that all the land is bunched up in a single mass. In effect, we are making that assumption a reality: a service to the model-makers, perhaps, but a fatal disservice to most species on earth.

Shortly after the first Pangaea formed, there followed the greatest slaughter in earth's history: the Permian extinction, in which from 77 to 96 percent of all marine animal species are thought to have disappeared. No one knows exactly why this happened. Some clues point to vulcanism, others to the loss of shoreline habitat, especially in warm inland seas. It is not unthinkable, though, that the mingling of previously unmingled marine species might have put the ecological web—already so complex as to dangle at the edge of chaos—in a hopeless tangle, so that any major shock would push it over the edge. Or, more simply, that a sudden free-for-all between species that had been largely separated for millions of years brought about a global shakeout in which weaker competitors went under. The mere possibility should give us food for thought about the New Pangaea we are now assembling.

Some lesser and more local waves of extinction present a clearer case. From the middle Eocene to the late Pliocene, the Isthmus of Panama was broken by several straits far wider than the present canal. For forty million years, South America's flora and fauna simmered in their own peculiar juice, producing such prodigies as giant sloths, giant armadillos, giant chinchillas, and a marsupial saber-toothed tiger, as well as a phantasmagoria of monkeys, birds, and plants unlike any others on earth. About three million years ago, the straits closed up and the species of the Nearctic and Neotropic regions were partly intermingled. Though the drastic shakeout of species that followed cannot be chalked up with certainty to the mingling, this does seem the likeliest cause.

BEETLES IN BUTTONS

Just how neatly some species can stow away in human conveyances, and even on the human person, is illustrated by a story Charles Elton tells in his classic book *The Ecology of Invasions*. A friend of a friend, just returned from Egypt, found small beetles hatching out of the buttons of a shirt he had bought there. The buttons were made of the nut of a kind of palm, and the beetles had "apparently passed through the manufacturing process without harm—rather like Charlie Chaplin in *Modern Times*." In 1929, a tropical entomologist whiled away his passage on a Rangoon rice ship from Trinidad to the Philippines by making a list of the animals on board, from pets to pests. He counted forty-one species, most of them insects—among them, the grain beetles he found when he unpacked his clothes at the hotel in Manila.

On occasion, incidents of this sort have shaken the world. In the late 1860s, an aphid that lived an obscure life on the wild grapevines of eastern North America found its way to Bordeaux on a shipment of vine stocks. Thence it spread through the vineyards of Europe and Algeria, infecting the vines with root galls to which they had no resistance, and which proved fatal. In France alone, three million acres of vineyards were destroyed; and the ancient winemaking tradition of Europe might have come to an end had not a Frenchman conceived the notion of grafting European vines onto resistant American rootstocks. Around the turn of the century, a fungus that lived parasitically but more or less harmlessly on Asian chestnut trees found its way to the New York area on nursery stock. Spreading by airborne spores, the fungus took about a half-century to reduce the native chestnuts of the Eastern United States—once the most numerous and among the most useful of its trees—to a few hopelessly sprouting stumps. The woolly adelgid now devastating Eastern hemlocks arrived on Japanese bonsai plants; the Asian tiger mosquito, a carrier of dengue fever and Eastern equine encephalitis, came to this country in a shipment of used tires. The ballast tanks of seagoing vessels have long served as diplomatic pouches in which freshwater organisms could pass through borders unmolested (except by screens, which fail to catch many species in the tinier stages of their life cycles). This is probably how mitten crabs from the rivers of North China took hold in the rivers of Northern Europe, from the Vistula to the Seine, beginning about 1912; and it is almost certainly how European zebra mussels and Asian clams found their way into North American waterways, and so into water mains and the water intakes of power plants, where they have cost taxpayers and ratepayers hundreds of millions of dollars.

Yet most of the ferrying—at least of creatures visible to the naked eye—seems to be deliberate, if not terribly well deliberated. Schieffelin, the Birdman of Avon, is only one of many names on the honor roll. When Leopold Trouvelot, an astronomer at the Harvard Observatory, brought the eggs of the European gypsy moth to his house in Medford, Massachusetts, in 1868, his laudable plan was to breed a better silkworm. When some of the creatures escaped, there were no immediate repercussions; twenty years later, however, Medford had an artificial winter as a sudden plague of caterpillars stripped its trees bare. Since then copper-and-arsenic, DDT, carbaryl, and Bt have failed to keep the gypsy moth from spreading at a rate of twelve miles a year, reaching as far as west as Michigan and as far south as Virginia.

In the early 1980s, the golden-apple snail was brought from Argentina to Taiwan and other Asian countries in the hope that it would prove a profitable source of food. No one knew that the snail would show a prodigious appetite for rice seedlings; that it would spread easily through irrigation channels; that within fifteen years it would infest many of the rice-growing regions of Japan, the Philippines, China, Vietnam, Thailand, Malaysia, Indonesia, Taiwan, South Korea, Laos, and Papua New Guinea. Asia's staff of life, already lamely lagging behind its sprinting population, has been further hobbled. In the Philippines, the golden-apple snail is now the number-one pest, costing rice farmers about $55 million a year, or 14 percent of their gross income. As lagniappe, the snail is an intermediate host of the lungworm, which causes a fatal meningoencephalitis in humans, as well as of trematodes that cause skin irritations; and the pesticides used to control the snail are toxic to humans and persistently harmful to downstream ecosystems. The final insult is that the snails have found no market. Export to Europe is cramped by health regulations, and Asian consumers just don't like them.

Similarly, white colonists in Africa could hardly have known, when they imported the Amazonian water hyacinth to decorate their garden ponds, that seventy years later it would carpet perhaps fifteen thousand acres of Lake Victoria, clogging ports and power-plant intakes, snagging the skiffs of native fishermen, fouling the spawning grounds of fish, and sucking oxygen from the water. In fact, Lake Victoria offers a particularly graphic lesson of the way invasions work. As Darwin learned in the Galápagos, remote islands are the best nurseries of new and unique species. Once their remoteness is bridged, however, the nurseries become abattoirs. For freshwater aquatic species, the equivalent of an island is a lake.

In the mere thirteen thousand years that Lake Victoria has existed, it has spawned a family of fish—the cichlids—that has diversified to fill its every niche. In the merer twenty years just past, the five hundred species of these polychrome fish have been reduced by half.

An obvious cause was overfishing by two outsiders: human beings and the human-introduced Nile perch. Two anomalies, however, led researchers to suspect a deeper cause. First, cichlids were vanishing even in sheltered, rock-bound corners of the lake; second, the most gaily colored forms were vanishing fastest.

The answer has emerged, literally, from the murk. In turbid water, the males' scaly plumage is obscured from would-be mates; in the ensuing muddle, closely related species mix and the brighter species are lost. As for the causes of the lake's new murkiness, they are tolerably clear. One is deforestation of the surrounding uplands, which has filled the lake with exiled soil. Another is predation by Nile perch on grazing cichlids that formerly kept algae at bay. And a third is the water hyacinth.

"When we try to pick out any thing by itself, we find that it is hitched to everything else in the universe"—John Muir's warhorse has been hitched to countless environmentalist carts and has done good service. Strictly speaking, though, it is false. The death of a star in a distant galaxy can have no effect on my wife, unless she happens to be an astronomer. Even on earth, organisms in ecosystems that are far apart rarely have any direct effect on one another. The interaction takes place between higher-order systems: ecosystems, watersheds, weather systems.

One of the ways in which a half-digested Gaian worldview can be dangerous is by making us think that the whole planet is really a single ecosystem, and that it is natural for every species to rub shoulders with every other. Anyone who wants to take literally the Gaian image of the earth as a single organism (which no present evidence warrants) should at least keep in mind that the various cells of the body are segregated in their own organs and tissues. The tender cells of the brain must be shielded from the raw blood that courses through the rest of the body, the contents of the kidneys are toxic to other tissues, and so on. Above all, the functioning of the various tissues depends on their separateness. Put a frog in a blender and the result will contain all the ingredients of a frog, but will not act like one.

THE LOSS OF CULTURAL DIVERSITY

Some say we ought not to grieve overmuch for the loss of biodiversity, since our own cultural diversity is just as good. But cultural diversity will not keep the rain falling or the gases of the atmosphere in balance. Moreover, cultural diversity is itself dependent on biological diversity.

As I sit at my desk, snow is falling outside; Sibelius is playing on the stereo inside. As plainly as snow is born of clouds is this music born of snow. From the spruce forests of Finland to the red deserts of Australia, every landscape in which humans have lived has sent up its characteristic shoots of song. Conversely, every novel, poem, painting, and play has roots, at however many removes, in a landscape (or in several). One need not swallow whole the doctrine of environmental determinism to see that it has a kernel of truth.

As natural diversity has been plowed under, cultural diversity has gone down with it. This was true in Sumerian times and it is true today. But it has not always been evident. As peoples mingle and empires stretch their limbs, it looks for a while as if diversity—both natural and cultural—were increasing. Tigers and hyenas appear in the heart of Rome; legionnaires are inducted into the mysteries of Isis and Mithra. Even in the newly conquered provinces there is a flurry of novelty, as statues of Augustus take their place beside those of Ra. If you count the hothouse plants of the New York Botanical Garden and the vegetables dished out in Asian restaurants, you may conclude that the biodiversity of New York City is greater now than it was when Peter Stuyvesant debarked. Arguably, the city's cultural diversity is greater than that of any like-sized parcel of land in history. But how long can this last? When New York has tightened its grasp on Thailand and Nigeria—when American TV and American hybrid seeds have done their work—how much that is truly Thai or truly Hausa will be left to enrich New York?

In nature and culture alike, the first impression one gets is of an explosion of diversity. We eat bamboo shoots in Zurich, hear soukous in Kyoto, see ginkgos on the streets of Boston. It is the same impression one would get if all the animals in a zoo were put in a single cage—an impression that would lessen as the minutes and hours passed and at last only the lion, the brown bear, and the python were left to divide the cage between them. (In real-world ecology, things shake out a little differently: the great, noble beasts die out first, while resourceful vermin flourish.)

In the case of domesticated plants and animals, the diversity we are losing is both biological and cultural. When the world is a single market, agricultural diversity in each region is reduced. It is not unlike what happens when we move living things—our allies and fellow travelers—around with us: the native species or variety is driven out by competition from an invader. The difference in the case of agricultural competition is that the invader is dead: a side of beef, a head of lettuce, a bushel of wheat, a gallon of milk. Or one might say that the competitor is alive but distant, that it invades by proxy, by action-at-a-distance, as when humans use long-range missiles. A cow in Minnesota sends its milk to Massachusetts, and a cow in Massachusetts is put out of business. A wheat plant in Australia sends its seeds to France, and wheat in France goes unsown.

The wide dispersal of seed and other propagules is nothing new, but in the past its effect has been to establish new colonies of living organisms. The present dispersal of seed has just the opposite effect: it wipes out populations of the same species. But often the species is not the same; and often the species, variety, or genetic line that is displaced can be seen, in a global context, as a competitor—a competitor for human patronage and thus for reproductive success. A maize plant in Iowa outcompetes an oat plant in Scotland, or another variety of maize in France, or the same variety of maize in New England.

Thanks to free trade, man-made landscapes that have evolved for centuries—cultures in the most ancient sense, upon which cultures in the broader sense are founded—are vanishing almost as quickly as natural ecosystems. Small mixed farms go under, replaced by factory farms producing the single crop that can be grown more cheaply here than anywhere else. In wealthy regions, farming often vanishes altogether, replaced by industry or by residential development.

We are squeezing the whole world into a single ecosystem, a single market, and a single culture. American prairie grasses must compete with European grasses; farmers in India must compete with farmers in Australia; moviemakers in Turkey must compete with Hollywood. True, differences in climate and soil will keep every forest in the world from looking exactly like every other forest; differences of language and faith will, for a time at least, have a similar effect in the cultural realm. But that is not to say that the forests will be healthy forests or the cultures healthy cultures.

A planetary culture that is good for the planet is a hard thing to imagine. No matter how pious its intentions, it cannot possibly know what is best for each part of the planet. Yet one is just as hard put to imagine how the spread of a planetary culture could be stopped. Woman and man have tasted the fruit of cosmopolitanism, a fruit that tastes like Anjou pear on Monday, kiwi on Tuesday, lichee nut on Wednesday, and Cape gooseberry on Thursday. It is a taste that is easily acquired.

No one has the right to stop people from being Westernized or cosmopolitanized if they choose to be. But there are many people on earth who do not so choose, but are forced. Land is taken from traditional small farmers and amassed in plantations. Local elites are, in effect, bribed to extend the world market and the cash economy into places that want neither. Loggers, ranchers, miners, oil drillers, and settlers level and scorch and scar other people's habitat: and the loss of habitat is something which cultural diversity no more than natural diversity can long survive. Nor is free trade a law of nature which only flat-earthers would resist. It is deliberately imposed by global elites on people who, given the choice, would rather protect their traditional landscape and way of life.

Maybe we flatter ourselves when we assume that the first glimpse of a television set or snort from a can of Coke must turn any sentient being into an addict of Western pop culture and consumerism. Those photons and those bubbles are powerful things, to be sure—especially when the photons spend half their time insisting that life without the bubbles is not worth living. But so is the breath of the forest; so is the voice of the ancestors.

The analogies I have been drawing between nature, culture, and economics must not be pushed too far. An ecologist speaking of invasive exotics—saying that they are tough, that they outcompete the natives, that they don't play by the rules, that they can destroy in a few decades a system that has taken millennia to evolve—often sounds like a bigot talking about Jews or Asians. But the ecologist's statements are based mainly on fact, while the bigot's are based on rumor, fear, and frustration. All humans are members of a single species. Human cultures (and for that matter, human "races") are not, like species, distinct entities; they are fluid and have mingled fluidly from the earliest times of which we have any knowledge. Syncretism has always been a prime source of cultural energy. My point is simply that the energy it releases is energy stored over centuries of relative isolation: and my fear is that the great explosion we are now seeing may end in cultural heat-death.

EVERY GREEN HERB

In the New Pangaea, both natural and cultural diversity are drying up. Let us look more closely, for just a moment, at the place where these two dwindling kingdoms meet: namely, the world of domesticated plants and animals.

If I have made agriculture seem like a single wave sweeping across the face of the earth and drowning every natural thing in its path, let me correct that impression. Agriculture is a multitude of waves arising in various places, meeting and mingling. The upland Near East, with its alliance of wheat, barley, lentils, peas, olives, vines, goats, sheep, and cattle, was only one of perhaps fifteen major centers of plant and animal domestication. Southeast Asia gave us rice, bananas, taro, and poultry. China furnished citrus and soybeans. From Africa came yams, sorghum, and coffee; from the Andes, tomatoes and potatoes; from Central America, maize and beans.

In each place, humans allied themselves with local wild species whose natural genetic diversity gave plenty of scope for selection. As selection went on, diversity was often increased. (Two striking examples are *Brassica oleracea*, a single species whose varieties include cabbage, kale, cauliflower, broccoli, and Brussels sprouts; and *Canis familiaris*, whose varieties range from the chihuahua to the St. Bernard.) As long as these allies kept some of their wildness—their diversity, their hardiness, their fondness for ecological settings like the wild ones in which they evolved—they could not do too much damage to their surroundings.

Unhappily, those supple and diverse systems of indigenous farming have often been swamped by new waves of domesticates. New plants and animals take over because they are favored by new human rulers or colonists, or because they offer greater short-term yields, or because they lend themselves to large-scale market agriculture. In our own time, a wave of crops emanating from the West, bred by scientific methods and marketed by huge corporations, is sweeping away the last remnants of diversity.

The loss of cultivated varieties of plants in this century has been as grievous, if not so much grieved over, as the loss of wild varieties. A century ago, campesinos in the Andes, the farm women of Szechuan, the tribesmen of highland Burma grew thousands of varieties of potatoes, radishes, eggplants. It was a good policy. If, in a given year, one variety was tortured by drought, another, with deeper roots, flourished. If aphids attacked one variety, another sweated a chemical that kept them at arm's length. Refined and toughened by thousands of years of human and animal selection, these varieties were, in aggregate, ready for anything.

North America was not among these ancient centers of plant domestication (American cuisine in the strictest sense consists of sunflower seeds, Jerusalem artichokes, and some berries). But in the last millennium we have made up for lost time. Our immigrants, pre-Columbian and post-Columbian, have made us rich. From Central and South America came the intricate complex of corn, beans, and squash that greeted and succored the Pilgrims. French and English settlers brought sweet peas; Germans, a wealth of cabbages. Italians passed through Ellis Island with the seeds of treasured tomato and pepper varieties stowed in their socks. Swedes brought brown beans fit to brave Wisconsin frosts.

Some of these seeds, as they proved themselves in their new environments, were handed down from father to son, from mother to daughter, by hands as carefully cupped as if they held rubies. Some found their way into the compendious catalogues of nineteenth-century seed houses. Some crossed or were crossed with others to make new heirlooms and new items in the catalogues. Whatever the demands of your eye or tongue, your soil or your weather, there was a variety—or two, or two hundred—to suit you. For home gardeners and farmers alike, these were golden years.

And then things got even better, for a while. Inspired by the work of the great Russian geneticist N. I. Vavilov, scientists raided the mountain strongholds of Peru, Anatolia, China, Malaysia—the centers of crop diversity that Vavilov had discovered—and came home with fistfuls of genes for their breeding programs. Selecting one line for pest resistance, another for yield, and refining them until they bred true, they then crossed them to produce hybrids of startling vigor and versatility, likely to grow as well in New Mexico as in Maine. There was just one catch. If you tried to save seed from these F1 hybrids, as they were called, you found that they either didn't grow at all or grew into inferior plants. For all intents and purposes they were as sterile as mules.

The seed companies rather liked this. If you wanted their exclusive variety you had to buy it from them, and you had to buy it every year. The houses started dropping their open-pollinated varieties, the ones you could save seed from. There was more money in hybrids. There was more money, moreover, in hybrids for thousand-acre farmers than in hybrids for forty-acre farmers or quarter-acre gardeners. Little seed companies were swallowed by big seed companies. Big seed companies were swallowed by multinational petrochemical companies (in 1986, according to a report by the Rural Advancement Fund, Royal Dutch/Shell controlled seventy seed companies, DeKalb-Pfizer thirty-four, Sandoz thirty-six). Like oil flowing downhill, their research money went into the development of hybrids that responded well to machine handling and to megadoses of the parent company's petrochemical-based fertilizers, pesticides, and herbicides. (The seed might even come cocooned in chemicals, in a single, premium-priced pellet.) Money flowed into the development, in collaboration with publicly funded land-grant colleges, of cultivars such as VF 145, the square canning tomato (which can be thrown twelve times against a brick patio without breaking), and T-cytoplasm hybrid corn.

Nothing imperils like success. In the United States, the success of T-cytoplasm corn nearly wiped out the nation's 1970 corn crop. A newly mutated fungus blighted 15 percent of the harvest—50 percent in some states—and only a timely dry spell kept the devastation in check. The National Academy of Sciences blamed a technology that had "redesigned the corn plants of America until, in one sense, they had become as alike as identical twins. Whatever made one plant susceptible made them all susceptible."

Luckily, the success of T-cytoplasm had not been complete: 20 percent of the hybrid corn grown in 1970 came from other strains. And there were still some open-pollinated varieties around, such as Virginia Gourdseed corn, once the common corn of the Southern colonies, which had been kept alive by seed savers in Accokeek, Maryland, and now became the sire of new, blight-resistant hybrids. Mayorbella maize from Central America was another source of resistance. So there are still, in the world, scattered puddles of diversity into which we can dip for new genes. But the Green Revolution is rapidly drying them up. Small farms growing traditional crazy-quilts of native plants have been smothered by huge plantations growing a single miracle crop, whose miraculous yields (like the miraculous stone soup of the folktale) depend on pricey chemical inputs. When the miracle proves vulnerable to some unexpected pest, scientists look to the traditional varieties for a resistant strain, only to discover that the traditional varieties are no longer being planted—often, that they are extinct. In India, the ancient paradise of rice, where thirty thousand varieties flourished only a few years ago, just a dozen varieties will soon blanket three of every four acres. In the Near East, the world center of diversity for wheat, Mexipak and Sonalika now account for 70 percent of the harvest.

In America, the shakeout began earlier and has been more thorough. Nearly all the vegetable varieties available from seed houses at the turn of the century are now extinct; even in the National Seed Storage Laboratory in Fort Collins, Colorado, the putative Fort Knox of plant diversity, only 3 percent survive. Some good hybrids have taken their place, but the fate of those hybrids hangs year by year on the whim of the seed conglomerates, and they are likely to get less good as they lose their ties with wild or primitive ancestors. The paradisiacal marriage of wildness and domesticity—the wolf lying down with the lamb, the wild oat with its cultivated cousin—has ended in divorce.

THE ASSAULT on traditional farming can be seen as the work of a third wave of human-led expansion. This third alliance, which will be the subject of the next chapter, is distinguished from the first two by the fact that its most important nonhuman members are dead.

5

THE HUMAN MUSHROOM

HALF A MILLION years ago, our genus formed an alliance with perennial grasses which allowed us to conquer the world. Over the past ten thousand years, an alliance of humans and annual grasses has conquered much the same ground in a fraction of the time, displacing or subduing not only other species but other humans, their allies, and their cultures.

Only a few centuries ago, a third alliance arose which is now very close to total hegemony over the living world. It has displaced or subjugated much of the natural world that survived the first two waves, as well as what was left of the waves themselves, including humans, their allies, and their cultures. The odd thing about this third alliance is that our most important allies have been dead for millions of years. They are cycads, ferns, giant horsetails, mollusks, plankton, and other creatures that flourished in the Mesozoic and Paleozoic eras, and which the bear-hug of the earth's crust has crushed into energy-rich carbon compounds.

Modern humans are merely the latest in a long and distinguished line of saprophages: creatures such as fungi, maggots, and various microbes that feed off decaying or decayed organic matter. In our case, the dead organic matter in question is wood, peat, coal, and oil.

Our form of saprophagy is more flexible than most, since energy is processed outside our bodies. We don't actually eat fossil fuels, except in those rare instances (which may become more common) when food is synthesized from them. But when used for fertilizer, traction, and pest control, they contribute directly or indirectly to the energy content of the food we eat.

The direct contribution of fertilizer is obvious enough, though the extent of it may not be. But tractors and pesticides contribute just as much. Thanks to them, our plant allies can devote to the making of carbohydrate and protein the energy they would otherwise have to use to establish themselves on crowded turf, and to fend off predators and competitors.

Indirectly, every American man, woman, and child ingests 160 pounds of nitrogen, phosphate, and potash fertilizer a year. Because most agricultural chemicals come from petroleum, coal gas, or methane, we invest three calories of energy for each calorie of food we consume. (If processing and transportation are included, ten calories go in for each we take out.) By contrast, Tsembaga farmers in highland New Guinea, using Neolithic slash-and-burn methods, invest less than a tenth of a calorie for each calorie they eat.

Nor is it merely the agricultural use of fossil fuels that entitles us to the name of saprophage. The crucial question is not what a creature eats, but where it gets the energy it uses. The calories burned by our cars, tractors, water pumps, and textile machines are calories we (and our animal allies) don't have to burn walking, plowing, carrying water, working looms, or keeping our body temperature up.

NIGHT OF THE LIVING DEAD

Saprophagy is a fine thing, in its place. The fungi, annelids, microbes, and other organisms that live on decaying matter do a great service by returning carbon, hydrogen, nitrogen, oxygen, phosphorus, potassium, sulfur, and other nutrients to life's marketplace. But when humans go into a business they tend to go in whole-hog, and saprophagy is no exception. While other saprophages are content to scavenge at or near the earth's surface, humans dig deep into her bowels.

When a major oil company used a brontosaurus as its mascot, it nodded to the fact that much of our oil is the expressed life of creatures of the Mesozoic: ginkgos, ferns, and mollusks as well as the odd saurian. The dinosaurs of *Jurassic Park* are perhaps no more than a wide-screen image of what we already do—dredge up life forms from the distant past and use them for our profit, without regard for the dangers they may pose.

An oil spill is a kind of night of the living dead, in which dead organic matter that we have called from its grave rises and strangles the living. But oil spills are the least of the problems that fossil fuels cause. For species not allied with man, this third wave is a horror show in which their own ancestors come back to haunt and harm them. Whatever humans could do before—strip forests, rip up soil, move themselves and their allies to the outermost corners of the world—they can now do more easily. As travel becomes a casual thing, compact settlement patterns uncoil and sprawl outward, smothering the landscape in asphalt, concrete, and clipped grass. The earth is punctured, gouged, and scraped to get at more fuel, and at the minerals that are used to make the machines that use the fuel. At various points in the use of the fuel, and in the industrial processes that it powers, noxious waste products bubble up. While some of these may be reduced by improved technologies, one thing is always produced when organic matter is burned: carbon dioxide. The release of long-buried carbon into the atmosphere thickens the roof of the planet's greenhouse, trapping heat inside.

But fossil fuel does not have to be burned to be baneful. Exhausted soil, its biota depleted and in desperate need of rest, is instead jolted into manic life by artificial fertilizers. It is further tortured by tractors, which cause both erosion and compaction. Large-scale monoculture—that raised-letter invitation to pests—is made viable by the use of pesticides, at terrible cost to beneficial insects, birds, and other bystanders. Surfeits of these pesticides poison groundwater; washed-down helpings of fossil nutrients gag rivers and lakes with microbial bloom.

Not all nonhuman creatures suffer from this exhumation of buried energy. Some of our old allies have joined the new alliance. Its main living members, apart from man, are certain varieties of wheat, corn, and other crops that have learned to thrive under a regimen of artificial fertilizers and pesticides. For the most part, these are the same hybrid strains that have smothered genetic diversity and (since both the seeds and the chemicals cost money) stifled peasant farming throughout the world.

It is an odd coincidence that wheat, the fuel of the first great explosion in human numbers and abilities, should come from roughly the same part of the world as oil, the fuel of the second. Of course, wheat and oil "come from" the Near East in very different senses. Though in both cases a concentration of plants was the source of energy, oil stores the energy captured by plants millions of years ago; wheat, by contrast, carries instructions for capturing and storing new energy from the sun. Yet in both cases a sudden rush of energy has destroyed biological information. The amount of energy (and protein) that wheat and other grains offer is so prodigious that we are tempted to replace nearly all ecosystems with grainfields. The staff of life becomes a crutch. Petrochemical fertilizers let us rely, for a while, on depleted soils and only a few varieties of each grain, so that genetic information is lost both above and below the ground.

A RISKY BUSINESS

Saprophagy is not a human invention. We have, however, done one really new thing: we have made matter into energy. All other living things are content to get energy by horse-trading electrons from one atom to another and from one quantum state to another. We have contrived to get energy from the heart of matter, the nucleus. Although this is done routinely in the sun and the stars, in the form of fusion, and although sustained fission of uranium may have been achieved accidentally some 1.8 billion years ago by uranium-isolating bacteria in Gabon, it was not until Fermi set off his chain reaction in 1942 that any living thing did it on purpose.

Digging energy from the heart of the atom is an even bolder and riskier maneuver than digging it from the earth. The problem of waste products is stickier. Though the burning of fossil fuels adds carbon to the atmosphere, it does not add to the sum total of carbon on the planet. Fission, on the other hand, actually creates large quantities of radioactive material that never existed before. So far, no one has figured out what to do with them.

Why does the search for energy so often land us in the mulligatawny? Getting energy is inherently a risky business, and has been since life first opened its doors. Respiration is a slow burn: a controlled form of oxidation. It must be precisely controlled, for oxygen is a poison. It was toxic to the first bacteria who produced it by photosynthesis, and it remains toxic to us. Although we and other aerobes have learned to control it, oxygen still tends to break down our cells, and it kills most of us in the end. As anyone can see who looks at a rusting bolt or a burning log, the nature of oxygen is to destroy things.

When the cyanobacteria pumped oxygen into earth's atmosphere, the other bacteria had the choice of diving for cover in the mud, or adapting. Adapting was harder but had its advantages. If you could pull it off, an aerobic metabolism could get you far more energy than an anaerobic one. Precisely because it was so volatile, oxygen gave more bang for the buck.

As a general rule, the higher the energy content of a fuel, the more dangerous it is. The dangers of burning fossil fuels (which is a form of oxidation) and of fissioning uranium (which is not) should therefore come as no surprise. No fuel is unequivocally "clean" or "safe" or "natural." No fuel is unequivocally the opposite. The point is to take the broadest possible view of the likely effects of each. Mushrooms, cyanobacteria, and our other comrades in energy hunger are not equipped to do that; they must sit back and let natural selection sort out their good and bad moves. We could do the same if we chose, but it would not be pleasant.

MACHINES AS ALLIES OF HUMANKIND

Since their coming of age, machines have been creatures of Prometheus. Even before the general use of steam power, fire was used to shape the metals on which most machines depended; now, of course, most machines are powered either by internal combustion or by electricity made from the burning of coal and oil. Since machines do work that would otherwise be done (or not done) by humans burning food, they are major accomplices in our saprophagy.

If the third wave of human advance includes dead allies, it might also be said to include allies that never lived. As the domestic animals of the second wave feed on grass, so machines feed on fossil fuels. They are most analogous to oxen, horses, and other animals that provide traction and transportation. They also resemble food animals in processing energy that we cannot use directly. But their relation with us is more intimate and more mysterious than these analogies suggest.

A biologist from a pedestrian planet, peering at some stretch of North America from a height of five hundred feet, will conclude that its dominant species is a shiny lozenge-shaped reptilian creature that alternately basks in the sun and sprints at great speed. It is host, he will note, to small endosymbiotic organisms which at intervals emerge, move about slowly, then reenter the host. Further observation reveals why the host puts up with these seeming parasites. They are devoted to the care and feeding of the host. They suck energy-rich organic compounds from the bowels of the planet and feed them to the host, something it is unable to do for itself. At times they even fight other colonies of their own species for access to the host-food. They make over ecosystems to meet the host's needs, replacing vast forests and grasslands with flat surfaces on which the host can bask or sprint more easily, and building hives or dens in which the host can take shelter from the elements. What they get in return is as yet unclear. Indeed, it seems possible the two organisms are forms of the same species, the lozenge being a sort of queen and the smaller creature a worker.

We know better, or think we do. Machines are not living things. They do not (yet) reproduce themselves. They do not follow Darwin's rules. Yet they follow a set of rules so similar to his that someone must have been looking over his shoulder. Each population of machines shows great variation, produced by mutations which, while not actually random, often might as well be. As in the case of domesticates, this variation is acted upon by natural and human selection. Those few mutations which are useful or otherwise prized are reproduced and magnified. Distinct species arise, have their heyday, become extinct. Their careers can be read in the facies and strata of dumps.

Like our living allies, machines coevolve with us. They shape our evolution, both cultural and biological, almost as thoroughly as we shape theirs. We think we control them; in truth, they have a life and a logic of their own. As in the case of our living allies, there is a great deal going on that we do not fully control or comprehend.

Our incomprehension is, as usual, most striking in the realm of ecology. To speak of "the machine in the garden" is misleading, for the machine is never just *in* the garden. Machines are not just superimposed on ecosystems; they displace old ecosystems and create new ones, in which they are often the dominant species. They set up new energy flows. They reweave the food web. Cars are the most glaring example, but there are many others. The lawn mower is a noisy grazing animal that has coevolved with a very small group of perennial grasses (a simpler pasture than any other grazing animal encourages); together they suppress the complex plant and animal communities that would otherwise claim that turf. The tractor is a huge animal that spurs the growth of opportunistic annual grasses while alternately flattening and upending the home of the soil population.

As with our living allies, some machines are not really allies at all, merely hangers-on. Video games, motor yachts, leaf-blowers, and about half the items in the Hammacher Schlemmer catalogue are the rats and pigeons of the machine world, nibbling at our energy supplies without offering much in return. Television sets are perhaps the cleverest parasites, since they encourage humans to multiply all kinds of machines, from assault rifles to television sets.

To describe machines as if they were alive is a conceit, but a useful one. It reminds us to keep an eye on them. For they are not simply tools that lie inert in our hands, but active members of ecological associations whose effects we have not yet learned to gauge.

6

LIFE ON THE EDGE

THE MASTERY OF fire, of plants and animals, of mechanical power: these are hailed by most people as triumphs over nature, damned by a few as violations of nature. I have tried to show that they are neither. Instead, they are examples of the kinds of strategies living things have always used in the struggle for energy, nutrients, living space, and offspring. They conform to the basic patterns of human nature and of nature in general. In a word, they are "natural."

But that word is not a seal of approval. As we have seen, deforestation, farming, and the use of fossil fuels have thrown a set of monkey wrenches, of various sizes, into the works of the planet. No doubt the planet will sort itself out eventually, as it always has done after major jolts. In the meantime, though, the interests of humankind may suffer in ways that we can ill afford. Our decisions must be based not on a judgment of how natural our actions are, but on the broadest possible view of their likely effects.

WHEN YOU HEAR a social or political or aesthetic issue framed as a contest between nature and culture, reach for your revolver. Whatever the issue, you can rest assured that nature is on both sides of it and that culture is, too. Hunting and farming, town and country, manual labor and automation, classicism and romanticism, tyranny and democracy, sexual license and sexual repression: each has roots in human nature, and each has been tended and fed by human culture. Man is born free; man is born in chains. Of the chains that bind him today, some are made of iron, some of paper, and some of nucleic acids.

If something makes people happy, there is a good chance that it has struck a chord in human nature. No further philosophical or scientific test is called for. It may, of course, strike a tinny and short-lived chord, or a cloying chord . . . but that is only to say that it will not make people happy in the long run, or as happy as something else might. If such a thing is suspected, then it may be useful to appeal to some notion of nature. If that notion is grounded in science, it may have some slight predictive value. In any case, it will have rhetorical value, since we feel better doing things when we think it is natural to do them. (Indeed, the rhetorical value may be all too great. If you can convince people that something is not natural, your prediction that it will make them unhappy may be self-fulfilling. It worked with homosexuality, though it may not work much longer.)

Except as shorthand for "likely to make people happier in the long run," does the word "natural" have any real meaning? To speak of things that are hard-wired in humans is unhelpful, for culture is one of those things. A human without culture is not human. Such a creature (if such a creature there be) is like a computer without software. As computers are designed to use software, so humans are designed (by natural selection) to use culture. The particular kind of culture, like the particular software, is left to the discretion of the user. There are constraints on the kinds of programs that will run, but they are broad constraints.

The ancestor of our word "nature" is the Latin *natura* (from *nascor*, to be born), which glossed the Greek *physis* (from *phyo*, to bring forth or to grow). For the ancients, nature was the nature of something: of a tree, of the gods, of man. Nature could be opposed to convention, or to technology, but not to the human as such. It was not a realm unto itself; all life was shot through with it. If nature came also to mean physical law, that did not exclude man, either, as anyone could prove by jumping out a window.

In English, the opposition between man and nature first appears in the eighteenth century. It seems to have been at once a reaction against the coal mines and steel mills and a cause of them. By this time an intimate and (in some ways) relatively stable marriage of man and the land had been achieved in Northern Europe, much of which had been farmed productively for centuries. Now, in this continent-sized garden, sibilant promises of godlike power were heard. As industrialization paved over the garden, sensitive persons found refuge in remnants of wilderness or in fantasies of a great wilderness across the sea. But the Romantic idealization of pure nature may have served, at times, to ratify the trashing of what was already impure.

The coolest exploiters of nature and its warmest defenders have had one assumption in common: that nature is one thing, humankind another. To the first, nature is inert matter that can be manipulated at will. To the second, it is a kind of antimatter that vanishes at a touch, and so must be kept in special receptacles called parks or preserves.

Oddly, the rhetoric of "being part of nature" is common among extremists of both stripes. In one lexicon, being part of nature means abandoning our own nature—giving up the habits of delving, building, and transforming that, more in degree than in kind, set us apart from other living things. In the other lexicon, being part of nature means doing what comes naturally to us, which as often as not is mindless exploitation. From this latter point of view, the fact that we are part of nature means that anything we do—no matter how ugly or willful or contorted—is by definition natural.

Neither side seems to appreciate the obverse of the fact that we are part of nature: namely, that nature is part of us. Another way of saying the same thing is that we have a nature—human nature—which is like and unlike that of any other thing.

The Hasidic master Simha Bunim of Pzhysha advised that every person carry in one pocket a slip of paper reading, "For me the world was made," and in the other a slip reading, "I am but dust and ashes." It is dangerous to forget that we are part of nature, but equally dangerous to forget how different we are.

THE HUMAN POSTURE

Life on the edge is supposed to be nerve-racking, but in some ways we prefer it. Humankind evolved in an edge habitat: the savanna, which can be thought of as a border or series of borders between grassland and forest. Though it may seem safer to be firmly in one place or the other, in fact being on the edge lets us take advantage of our human versatility—our ability to run like antelope, climb like monkeys, swim like otters—when hunting and when being hunted.

Edges are good habitat for many species. One reason for their great biological diversity is simply the overlap of species from two or more ecosystems. But that is not the whole story, for many species that live in wetlands, littoral zones, and the like live nowhere else. Some are drawn by the overlap of food species, as early humans were drawn to seacoasts and riverbanks. Others, such as frogs, make use of different microenvironments at different stages of the life cycle.

In another sense, the sense of the idiom, all species evolve "on the edge"—in the state of risk, peril, and uncertainty. They may not enjoy it, but all need some risk in order to flourish. Even domesticates need risk to keep them genetically supple and on their toes.

Living things may live on yet another edge: the edge of chaos. Oddly, that precipice of complexity may be the most adaptive place for a species to be. As a system approaches chaos, it becomes more and more sensitive to small stimuli: a very useful trait, up to a point. Beyond that point, the system becomes so sensitive that the slightest nudge can cause collapse. In the view of some complexity theorists, the edge of chaos is where "information gets its foot in the door in the physical world."

If all species may be said to live on the edge, humans have a special claim to that posture. We often find ourselves on the edge of some wave or other, ushering a new set of allies onto new ground; riding the wave of some human activity, or getting knocked around by it. Above all, we live on the wavering edge between nature and culture.

Yet culture should not be blamed for the edginess we too often feel. We like to think that man was once a happy animal, whole in body and mind. Then something happened: toolmaking, purposive thinking, language, gardening, farming, cities, machines—name your snake. The truth is that even animals are not happy animals in the sense imagined, least of all animals close to us on the phylogenetic tree. Chimpanzees appear to suffer self-doubt, depression, envy, and the fear of death, and it is a good bet that early hominids did, too. Undoubtedly the first true humans were at least as confused as we are about what humans are supposed to do.

If culture is (in some sense) a cause of this confusion, it is also a response to it. The confusion itself is perfectly natural in an ecological generalist that lives by its wits. A brain so big has ample room for demons.

As within, so without. An ecological generalist is never sure of its place in nature. It has many places. Its versatility lets it duck some of the usual controls on its numbers, which means that it can do more damage than most species. If, on top of this, it is quick to form alliances with other species, there is no limit to the places it can go and the damage it can do.

FARMING AND HUMAN NATURE

Let us take a closer look at one of the "natural" phenomena discussed earlier. Farming—some have argued—is deeply unnatural, a distortion of human nature from which all our later ills have grown. Others think it just as unnatural, but call it a splendid triumph over our animal nature, and the seed of all our progress.

In chapter 1, farming was presented as a case of coevolution. The main objection to this view is that farming seems to be a set of learned behaviors.

Since some peoples do not farm, it is clear that farming is not encoded in the human genome as such. Might it lurk in the gene pools of particular peoples?

Though no more than a hoe stroke in the long season of human evolution, farming has already changed the physical makeup of the peoples that do it. To take the best-known example, adults of European background are usually lactose-tolerant, while adults whose stock comes from East Asia or the Americas, where cows were not kept, get bellyaches when they drink cow's milk. There is every reason to assume that small but significant genetic changes have taken place in farming peoples' mental makeup as well. None of these changes would be big enough to make the innate mental capacities or emotional needs of a Bushman baby noticeably different from those of a Bantu baby or a French baby. Nor would they be big enough to imperil the basic equality of all humans, that lucky thread by which democracy hangs. They might, however, be enough to explain why we are not as unhappy in a world of cows and cornfields as four million years of hominid evolution say we should be.

Even if farming is not something any human group evolved to do, it may still have a basis in the human genome. Lots of things humans never "evolved to do," such as mathematics or the visual arts, are nevertheless thought to be part of our nature and our destiny. Selection for a particular function can produce "unexpected," unrelated, even dysfunctional results. Whatever our brain evolved for, it has a mind of its own. It has needs and abilities that have very little to do with survival on the African savanna.

To say that the last ten thousand years have been a fluke is to ignore the fact that they were not imposed on us. They flowed, somehow or other, from our nature.

As the view of our evolutionary past becomes clearer, the domestication of plants and animals comes to look less like a strange and sudden denial of our animal nature—a lion eating straw—and more like an offshoot of that nature. Winston Churchill is supposed to have said that two things are natural to man: war and gardening. Clearly there are some things about farming and gardening that appeal to our deepest instincts:

The instinct to nurture, as in keeping pets.

The instinct to control and manipulate, which is seen in many higher mammals.

The need to be secure in a predictable environment. The hunter-gatherer makes his environment predictable mainly by learning it, the farmer mainly by changing it.

The instinct to husband resources. Even the Bushmen will gather and hoard melons when they are abundant and will note a particular root and keep it in mind for a lean season. The saving of roots and seeds may have been the beginning of agriculture. In any case, husbanding leads logically to husbandry.

To call the rise of domestication a case of coevolution is not to deny that it is also a triumph, though not an unmixed one, of the human mind. (It is not a triumph of the bovine or gramine mind; that is one of many asymmetries in the story.) Farming is natural, but it is cultural, too.

Tradition often seems to have more weight in farming villages than in hunter-gatherer bands. At first glance that may be surprising, as hunting is a far older way of life. But it is just the newness of farming that forces it to grab hold so grimly. Its roots in human nature being broad but shallow, it needs the guy wires of culture to stay erect.

Though tradition did keep some kinds of farming in place and almost unchanged for millennia, on a global scale the effect was just the reverse. Like most mutualisms in nature, farming was by nature destabilizing, since it allowed a few species to multiply quickly at the expense of many others. It kept its balance by expanding; and when it clutched at culture for support, it clutched a well-greased wagon. Far from being steadied, it was caught up in a new and more reckless momentum.

Clearly farming is a cultural process. That does not mean it stops being a biological process, or that the laws (such as they are) of evolution and ecology no longer hold sway. A field of genetically altered tomatoes attended by machines and regularly doused with chemicals is still, in its fashion, an ecosystem which has displaced another ecosystem, and which indirectly has jostled many others. It is pregnant with consequences for the survival and reproductive success of humans, of members of the nightshade family, and of uncounted other species as well: the lupines that once carpeted the valley, the caribou flummoxed by Arctic oil drilling, the salmon muddled by the sucking of water from great rivers, and so on and on.

To claim that all our ills spring from the outraging of our nature by farming is silly, but suggestive. The tension between two layers of our being is always painful and almost always fruitful. To paraphrase Nietzsche: agriculture is a disease, yes—as pregnancy is a disease.

HOW MEN STOLE AGRICULTURE

As the first rays of history break over our fields and pastures, the interplay of nature and culture gets a little clearer. And one is struck by a sense that some pivotal changes in culture are determined by the contest of two halves of human nature, the male and the female.

Women, whose vital role in the Paleolithic economy was always somewhat scanted in the frenzy for meat, came into their own in the late Mesolithic and the Neolithic. As gatherers, they had always provided much of the sheer bulk of the human diet; but as gardeners and farmers they provided more sheer bulk and a good deal more protein. For a while they owned the secret of an ample and stable food supply. The female principle was thrice blessed, for the womb of earth bore food; the hands of women midwifed her; the wombs of women bore more hands to make more food, more feet to win more earth.

But as farming developed, men began to muscle in. More use was made of animals for food, fiber, and traction, and men as hunters had more experience with animals. More use was made of heavy tools, especially metal ones, and here again men had the edge. We can see the smirk of inevitable triumph in the face of the God with the Sickle, a figure found in Szegvar-Tuzkoves, Hungary, and dated to 5000 B.C. The copper sickle came first, then the bronze of plowshares and battle axes—both of which enhanced male power. An arms race driven by metallurgy might have given a peaceful village or tribe the choice of choosing a warlike chief or facing extinction. The arms race would be aggravated by crowding, for the march of farmers across Europe (for example) had finally to reach the Atlantic. Europe probably filled up with farmers the way a train fills up, with all cars filling to one-third or one-half capacity (each passenger two or three seats from the next) before the first car fills up completely. At some point, the easily arable land would have run out. It might be easier to take bottomland from a neighboring tribe than to clear dense forest or plow a mountainside.

When women invented agriculture, they displaced men as breadwinners. The delicate balance between the sexes achieved in the course of four million years of hominid evolution was knocked off kilter. Woman's new magic was more than men could bear. There had to be a reaction.

Both swords and plowshares, the two busiest and most emblematic implements of agriculture-based societies, were made of metal. A key had been extracted from the womb of earth which would unlock all her bounty—and the bounty of other peoples. The God with the Sickle shows man smug in his knowledge that metallurgy will in time make him master of the vegetable world, which formerly had been woman's domain. The iron plowshare will be a ten-inch nail in the coffin of the Great Mother. Hereafter she will re-emerge mainly as a vampire, witch, or whore.

Holding aloft a clump of turf to shake out the dirt as I break ground for a garden, I resemble Perseus with the head of Medusa. It may not be an idle allusion. Medusa was one of the numberless earth monsters slain by heroes erecting the new, rational, man-centered order. Could her hissing snaggle represent the tangled roots of the turf, decapitated by man and his plow? Among the roots would have been worms and maybe snakes.

Why would the sight of her be fatal? Perhaps because she represents the private parts of the Great Mother, which her sons must not look upon.

The Sphinx is a related figure. By destroying her, Oedipus denied the kinship of humans and other animals. He gave the right answer to the riddle, but for the wrong reasons. The Sphinx was talking phylogeny, Oedipus ontogeny. What the Sphinx meant to say was that man, in the grunting morning of his race, had walked on all fours; that he had then risen to proud erectness; but that now he was beginning to lose his animal vigor and to walk with a crutch. The crutch was what the Greeks called *technē*, what we call technology.

By applying the parable to one man's lifetime, Oedipus was able to dodge its cautionary point. He thereby proved to the Sphinx that man was too far gone, that his animal body would soon wither away. She, being an amalgam of human and animal, knew that her time had passed and in despair leaped to her death. Her worst fears were soon realized, for Oedipus, having denied his origin in Mother Earth, was now free to violate her.

To stifle our guilt about penetrating our mother, we deny that she is our mother. If peoples that take up plow agriculture give up the worship of the Great Mother, it may be from shame over their rape of her.

Other myths, not all of them so guilty, confirm the flow of farming from female to male. The golden apples of the Hesperides, which may be ensigns of grain, are stolen by Herakles fair and square (hasn't he slain the dragon—another servant of the Great Mother—that guarded them?). A young man named Triptolemus receives the art of agriculture from Demeter, no strings attached, to teach to the rest of mankind. Eve hands Adam the forbidden fruit, which certain of the rabbis claim is actually wheat.

Many cultures see culture as male, nature as female. When farming passes from female to male it may seem to pass, as well, from the realm of nature to the realm of culture. In truth, we have seen that there is no such boundary. Yet it is also true that, as farming progresses in male hands, its sites look less and less like natural ecosystems. Patches become fields, haphazard clusters become rows. Trees are more ruthlessly felled, soil is more deeply gouged, watercourses are commandeered. As farming becomes more productive, a host of cultural facts arise—cities, states, classes, bureaucracies—which swell and speed the waves of human expansion and make waves of their own. People find themselves perched on an ever-narrower edge, and feel an ever more urgent need for myths that will make sense of their experience.

In the next part of the book we will listen to some of those myths, and see if they make sense of *our* experience.

PART TWO

THE

MOUNTAIN

AND

THE TOWER

7

THE MOUNTAIN
OF THE GODS

TWO WAYS OF looking at the world arose in the ancient Near East
and are with us yet. For one, the heart of the world is wilderness.
For the other, the world revolves around the city, the work of
human hands.

A cartographer's quibble? Hardly. It is a fundamental dispute about the
way the world works and what our role in it should be. From the point of
view of ecology, there is no more important question one can ask about a
civilization than which of these views it adopts and acts on. Indeed, the
prospects of our own civilization may hinge on whether we can, at this late
date, change our minds.

The myths of the ancient Near East are the oldest myths we know of.
They come from the part of the world from which at least two great
waves—farming and civilization—moved outward to fill the world, and
they are the record of humankind's first attempts to make sense of those
waves. In a sense, they are the sandy shore in whose ripples one can read the
waves' passage. (Several earlier human and hominid waves passed through
the Near East on their way out of Africa, but it would be rash to claim that
they left their trace in any extant myth.) If in Part One we looked at the
waves of human-led change from the distant perspective of (let us say) the
extraterrestrial biologist, now we move in for a closer view, peering
through the jumpy eyes of humans buoyed or buffeted by the waves of
change, and watching as they try to still the choppy waters with the oil of
myth.

For anyone wanting to make sense of the modern West, the myths of the
ancient Near East are indispensable. Some have made a powerful impact by
way of the Hebrew Bible; for though the myth-scorning redactors ran a
tight ship, a number of myths managed to stow away almost intact, and
fragments of countless others were mingled in the cargo. Many found their
way to Europe by other routes, mainly by way of Greece. But even those

myths that never entered the Western mind can serve as keys to its inner chambers. They are clues to ways of thinking and acting that arose with the first civilizations and have not left us yet.

The last and best reason for listening to the myths of the ancient Near East is related to the other reasons, but stands on its own. It is simply that they are good stories, and they speak to us.

Of them all, the story of Eden speaks most compellingly. It speaks in many voices: childish, gentle, mournful, stern, sexy, venomous. To hear it clearly, though, we must hear it in the dry acoustic of its native landscape, amid its native sounds: the trickle of water, the scratch of the plow, the clink and babble of cities. We must set it in the context of the other myths of its world, and of the civilizations that thought them up.

The two great worldviews I mentioned at the outset belonged to two kinds of civilization: those of the hilly lands and those of the great river valleys. The first kind is typified by the Canaanites, the second by the Mesopotamians. The peoples of the hills, narrow valleys, and narrow coastal plains made their living from small-scale mixed husbandry. This was a much-refined but still-modest descendant of the earliest farming known, which had arisen in those same hills. The peoples of the great river valleys were more ambitious. They practiced large-scale, irrigated agriculture that was not so different, at heart, from what large corporations do in California today.

Tied to these different ways of living on the land were different economies, different social structures, different political forms, and different ways of looking at the world. Above all, the hill peoples and the valley peoples had different world-poles.

The world-pole is the axis on which the world turns. It is the heart of the world, the source of all life. Nearly every people has a world-pole, but they do not all agree on its shape. For the Canaanites, the world-pole was the Mountain: the wild place sacred to the gods, the font of life-giving water. For the Mesopotamians, it was the Tower: the ziggurat that rose in the midst of the city.

THE WORLD-POLE

If there is one thing all cosmogonies agree on, it is the need for division. Pine as they may for a time of perfect oneness, all peoples know that a world undivided cannot stand. For life to feed and reproduce itself, there must be division: between heaven and earth, male and female, man and beast and god. But for life to flourish—and on this point, too, all cosmogonies agree—there must be someplace where all these things are reconnected. That is the world-pole.

On the exact shape the world-pole must take there is less agreement, yet more than one might expect. Hindu texts tell of a mountain at the center at the earth called Meru, on which stands a tree called Jambu. Its fruits are as big as elephants. When they fall and splatter, their juice becomes the stream Jambunadi, drinking from which makes one proof against old age, vice, and body odor. Among the Buddhist Kalmucks of Siberia, Meru becomes Sumeru, a vast pyramidal mountain rising from the cosmic ocean. It sits on a sunken cushion of gold, which in turn sits on a tortoise. The cannibals of West Ceram (an island near New Guinea) say that the nine families of humankind emerged from the banana trees of Mount Nunusaku. The Norse Eddas sing of the great ash tree Yggdrasil, on whose trunk the heavens spin, and whose roots clutch the netherworld. The tree is a hive of activity: a great eagle perches on its crown, four stags romp and browse in its branches, a great serpent gnaws on its roots, and honeydew trickles from its bark. In the Zoroastrian Avesta, a mountain called Hara stands at the center of the world. From its peak flow all the world's waters, which course through the sea Vourukasha and water the seven regions of the world. Purified by the earth, they rise to the peak of Hara and start the cycle again.

What do all these places have in common?

As a rule, the world-pole is the source of life. Although I have used "world-pole" because it is plainer than Mircea Eliade's *axis mundi*, I would almost rather say "world-pipe," for the act of connecting heaven and earth would be meaningless if stuff could not move from one to the other. Like the trunk of a tree, the world-pole is something through which life flows. It is at once phallus and vulva, ram's horn and cornucopia. It is the uterus from which all creatures crawled and the teat from which they continue to suck. If a man or woman—a shaman, a hero, a prophet—would ascend to the heavens or descend to the underworld, here is the stairwell. Here the adept can powwow with gods and animals, even merge with them, as all of us used to do at the beginning of time.

For life to flow, for life to abound, things must be lined up. It is not a matter of lining up like things, as in a wheat field, but of lining up and reconciling unlike things. Heaven, earth, and underworld, of course, but also male and female; god, man, animal, and plant; organic and inorganic.

When the shaman finds or creates the world-pole and climbs up, his goal is not ecstasy alone. He seeks to draw down bounty to his people. Harmony yields plenty, as can be seen in hunting rituals that are meant to placate the spirit of the game. The primal harmony has not been wholly lost. The shaman can reclaim it; the rituals and technics of food-getting strive to approach it, within the limits set by man's place in the food chain.

Most of the world-poles we have mentioned so far have something else in common: they are natural rather than man-made things. This is true of the world-poles of most ancient peoples, among them the Canaanites.

THE CANAANITES

The fact that Canaan—the region now occupied by Israel, Lebanon, and Syria—contains the lowest dry land on the planet (the shore of the Dead Sea) as well as deserts, coastal plains, and steppe only makes its great mountains the more imposing. The ranges of Persia and Anatolia sprawl for so many hundreds of miles that it is hard for any one peak to seem a World Mountain; but the Lebanon, Anti-Lebanon, and Amanus ranges are so compact that they were spoken of in ancient times almost as if they were monadnocks. The Lebanon and Anti-Lebanon ranges reach heights of 10,131 and 9,232 feet respectively, the Amanus of 7,418 feet. Their splendor was a byword throughout the Near East, as was the price their timber could bring: in particular, the huge ancient cedars that grew on the upper western slope of the Lebanon.

It is no surprise, then, that in Canaanite poetry both the elder god El and the younger god Baal have their houses or tents on mountaintops. For that matter, so does Baal's sister and bride, the goddess Anat. The mountain of El is pre-eminent, being the place where the gods meet in council, to dine and haggle, and the fate of the universe is decreed. El lives at "the source of the Two Rivers, in the midst of the Pools of the Double-Deep." (The last phrase can also be rendered "headwaters of the Two Oceans" and may refer to the male ocean of the heavens and the female ocean that undergirds the earth.) From El's mountain flow the waters that bring life to the world. It is proof of his power that he has broken and yoked the primal waters, and that instead of breaking out and swamping the world they run dutifully in two rivers that give the world life. El does not merely sit by the waters like a poet; he sits on them, keeps a lid on them, and ladles them out.

Nevertheless, by the time of the great Ugaritic epics the waterworks are being handed over in part to a younger and abler god, El's son Baal. Baal's mountain is no slouch either. It is called Mount Zaphon (or Sapanu), from a root meaning "to look out" or "spy out." The Canaanite epics make it the site of Baal's struggles with the forces of chaos—Yamm ("Sea") and Mot ("Death"), as well as the Dragon and the Serpent—and his arsenal of lightning and thunder. But if Zaphon is a place of war and storm, it is also a place of wine, women, and song.

> He takes a thousand pots of wine,
>> Mixes ten thousand in his mixture.
> He rises, plays, and sings,
>> The musician plays the cymbals;
> The sweet-voiced youth doth sing
>> Of Baal in the Fastness of Zaphon.—
> Baal regards his lasses,
>> Looks at Padriya, daughter of Ar,
>> Also at Talliya [daughter of Ra]bb.
> . . .
> Henna of seven maids,
>> Smell of coriander and ambergris.

His "lasses" are in fact two of his wives, whose names might be Anglicized (or perhaps Native Americanized, in the Longfellow manner) Misty daughter of Bright Cloud and Dewy daughter of Showers. By and large, though, the poet conjured up by this passage is not Longfellow, but Coleridge. Like the Abyssinian maid singing of Mount Abora, our Ugaritic bard is drunk on the milk of paradise.

That Baal's mountain is a place where heaven and earth fruitfully meet is hinted at in a message he sends to Anat:

> Pour out peace in the depths of the earth,
> Make love increase in the depths of the fields. . . .
> The speech of wood and the whisper of stone,
> The converse of heaven with the earth,
> Of the deeps with the stars . . .
> Come, and I will seek it,
> In the midst of my mountain, divine Zaphon. . . .

Whether the "speech of wood and the whisper of stone" refers, as some think, to the palace or temple which Baal plans to build on his mountain or, rather, to the wooded mass of the mountain itself, the role of Zaphon as world-pole is clear. So is the role of Baal's mountain—and Anat's—as a place from which love and fertility flow. In his message, Baal calls Anat to his side. Love and fertility will be clinched, it seems, by a sacred marriage between Baal and Anat—icing on the cake of his victory and enthronement on Zaphon. There is reason to think that these events were replayed each year by the priests of Ugarit, perhaps with the help of statues and other visual aids.

Far from an empty rite, this was a fervent rooting for the home team in a match between life and death. For Baal himself has to fight one of his battles, at least, all over again each year. Yamm he dispatches once and for all (as Marduk does Tiamat in the Babylonian creation story, which is thought to be modeled on the Canaanite). His more Hydra-like enemy is Mot, the god of the dry season and of the underworld: the one being in time what the other is in space, a zone of sterility and death.

Baal descends into the underworld and is killed by Mot. As a result, Mot boasts, "lifebreath is wanting among men." El's mourning for Baal is a dumb show of futile husbandry:

> He plows his chest like a garden,
>   Harrows his back like a plain.
> He lifts up his voice and cries:
> "Baal's dead!—What becomes of the people?"

But Baal is reclaimed from Mot's clutches—perhaps from his very bowels, since Mot claims to have chewed him up and crushed him like a kid in his gullet—by his faithful sister and spouse, Anat.

> She seizes Mot the son of El,
> With a blade she cleaves him;
> With a shovel she winnows him;
> With fire she parches him;
> With a millstone she grinds him;
> In the field she scatters him.

If Baal's descent and chewing up is the sowing and plowing or trampling under of the seed, then his rescue is the harvest of the new grain. The havoc of harvest and processing is wreaked not on the grain god himself, but on the evil god who contains him. The freeing of the grain from dry husk and chaff is a rescue from the powers of sterility. Now El has a dream:

> The skies rained oil,
> The wadis flowed with honey.
> So I knew
> That alive is Mighty Baal!

In waking life, back on earth, the furrows remain parched until Baal lures Mot to his home stadium, Mount Zaphon, and there, in a nip-and-tuck fight, beats him. It seems likely that Baal dies each summer, bringing on the dry season, and that his rebirth and grudge match are in the fall, when—if he wins—the rains begin again.

Most of the poems we have mentioned are found on clay tablets from Ugarit, in northern Syria—the modern Ras Shamra—that date from about 1400 B.C. But the myths they relate may be far older. It now seems likely that the Canaanites were autochthonous—that they grew from the soil of Canaan. They were offspring of the first farmers in the region (the first in the world, as far as we know), probably of the first proto-farmers (the Natufians, whose villages were supported by the harvesting of wild grains), and possibly of the first modern humans. The seeds of the myths that bloomed so profusely in late–Bronze Age Ugarit may have been planted by farmers of the late Stone Age. Though many of the myths of Mesopotamia are far older in their written forms than those of Canaan, it seems likely that those of Canaan represent an older way of life. In any case, it is clear that, over the course of their long career, Canaanite farmers learned certain deep truths about the way the world works—truths that have been confirmed by the young science of ecology.

THE WORLD MOUNTAIN AS ECOLOGICAL FACT

The World Mountain is mythic shorthand for an ecological fact. There are certain places on earth that play a central role in the flow of energy and the cycling of water and nutrients, as well as in the maintenance of genetic diversity and its spread by means of gene flow. Such places provide many of the services that keep the ecosystems around them (and the biosphere as a whole) more or less healthy for humans and other life forms. They help control flooding and soil erosion. They provide fresh infusions of pollinating birds and insects, which grant continuing life to many of the plants (both wild and cultivated) we take for granted. They regulate the mix of oxygen, carbon dioxide, water vapor, and other ingredients in the air and keep its temperature within bounds. They are spigots for the circulation of wildness through places made hard and almost impermeable by long human use. All such places are more or less wild; many are forested; many are mountainous, and from them great rivers flow.

The mountains of Lebanon, Syria, and Armenia are the source of water for much of the Near East. From their slopes flow the headwaters of the Jordan, the Orontes, the Tigris, and the Euphrates. The pattern is copied on smaller scales as well, in the brooks, wadis, and underground aquifers that slide from the Judean hills to the coast.

Canaan as a whole, situated at the junction of three continents, has always been a maelstrom of gene flow. Even today its genetic diversity is dazzling, with flora and fauna of Europe, Africa, and Asia mingling in sometimes unsettling ways. A few thousand years ago, when the region was less bruised by human use, the mix was more dazzling still.

In this matter, too, the uplands have played a special role. During the Pleistocene Ice Age, when the locking up of water in glaciers made the earth as a whole drier and much of the Near East was arid steppe, the mountains gave refuge to species in flight from drought. Among these species were humans as well as some of the trees, grasses, and quadrupeds they would later tame. It was the expansion of these species, at the end of the last glaciation, from their mountain hideouts to the lower foothills of the Levant that set the stage for domestication.

It was also the slopes themselves. For the play of farming to get started, it was helpful that the stage be slanted. On hillsides a wide range of climates can be collapsed accordionlike within the space of a few acres. This produces a menu of variation in wild plants that fairly begs humans to pick and choose: that is, to select. It also encourages transhumance.

Transhumance, the practice of herding livestock to summer pasture in the hills, then back to the valley for the winter—or the other way around in some dry regions—can be seen as a telescoped form of nomadism. (That is not to say it arose later than nomadism; more likely it came first.) It is made possible by Humboldt's law, which states that climbing a hundred-meter hill yields roughly the same drop in temperature as trekking 110 kilometers (one degree of latitude) away from the equator.

Giving hard-earned grain to animals is a late and luxurious practice. The first domestic animals had to fend for themselves. Their wild forebears had followed the grass, the brush, and the seasons. If more or less settled farmers were able to keep them, it was only because those farmers lived in the hills, where the seasons crept up and down the slopes instead of (or as well as) gliding hundreds of miles north and south.

The same piece of legislation made the hilly flanks of the Fertile Crescent the ideal place for the domestication of grains. Variations in elevation produce variations in climate, which produce variations in plants; these in turn provide the raw material for breeding. Emmer, einkorn, and six-row barley were lining up along isotherms (lines of equal temperature) long before they were lined up by farmers. In effect, natural selection had set out upon the tablelands and in the bowl-like valleys of the Near East a smorgasbord from which human selection could take its pick. So the uplands of the Levant were the ideal setting for the domestication of plants and animals alike.

Myths of the world-pole say that the source of the first human life will be the source that sustains human life. I spoke (in the prologue) of the paradox that a wild or sparsely settled place should seem to be the point of origin of humankind, that is, Eden. But in fact Holocene humans do seem to have come out of the uplands of the Near East, descended into the valleys to build civilization, then edged up the hills again as their numbers swelled. Outside Africa, anatomically modern humans make their first clear appearance in the archeological record in the uplands of the Levant, in roughly the same place as the first proto-farmers. Some ninety thousand years ago, while Neanderthals had the run of Europe, *Homo sapiens* dwelt in the caves of Mount Carmel.

The Mountain was thus the source of one great wave of human advance—farming—and at least the proximate source of an earlier wave, made up of the first creatures whom, if they sat down next to us at a luncheon counter in modern dress, we would not hesitate to ask to pass the ketchup.

Going upriver, we go backward in time. That is what myth suggests and what ethnology confirms, whether the river in question is the Amazon, the Nile, the Hudson, or the Euphrates. The less accessible a place, the less it changes, and there is no place less accessible than Eden. Going upriver, we approach the source of all life.

FACED WITH THE fact that for five thousand years Canaan was a prize of contending empires, we might stop to ask ourselves why. Its soil was fertile in places, but less so than those of Egypt or Mesopotamia. Apart from some modest copper mines, it was not rich in minerals. Why, then?

There were two main reasons for the land's unenviable cachet. One was its site at the middle of the world, the crux of trade routes between Europe, Asia, and Africa; the other was its bristling mountains. Added up, these two reasons merge in one. The bone the big dogs tugged at was the world-pole. If this seems too spiritualized a reading, consider the fate of Canaan in this century. In four thousand years of being chawed and slavered over, it had lost whatever meat and gristle once clung to it. It was worn clean and white, a pure idea. Yet the dogs, big and small, are still tugging. (If agricultural improvements made by the Israelis have added value to the land, petroleum has added far more to its neighbors. In strictly economic terms,

Canaan remains some of the worst real estate in the Near East.) Two of the world's great religions make Canaan the center of the world. A third, while assigning that role officially to Mecca, nevertheless sees Jerusalem as a place of passage between worlds: from the rock over which the Al Aqsa Mosque now hovers, the Prophet ascended to heaven.

Now, however, we have strayed into the realm of historical accident, and away from the ecological point. In fact, all particular mountains, and even mountains in general, are beside the point. The real lesson of the Mountain has nothing to do with mountains as such. It has to do with wilderness.

The point is that man-made landscapes, from the wheat fields and vine-yards of ancient Canaan to the strip malls of New Jersey, survive only by courtesy of the wilderness around them and the wildness that remains in them. Energy flows, water and nutrients circulate, climate is kept within bounds, the ingredients of the air are kept in balance, the soil is made fer-tile. All these things are matters of life and death for us. All are done for us free of charge, in ways we do not fully grasp. Even if we knew how these jobs are done, we would be unwise to try and take them over. For we would then spend most of our time trying desperately to manage what used to be man-aged for us.

To use the language of Part One: If humans and their allies conquer the world too thoroughly, they may find themselves with more work on their hands than they bargained for. If the waves of human expansion go too far or run too deep, they may bring about their own collapse. The Mountain rises above these waves. They may tickle its flanks, but beyond this they must not go.

THE GOD WHO LOOKS AROUND

It is hardly surprising that people whose farming depended on infrequent rain or on rain-fed rivers should worship a sky god or mountain god. Nor is it surprising that that god should be a god rather than a goddess, for the analogy of rain to semen has escaped few farming peoples. (One that it did escape was the Egyptians, for a very good reason: not having seen the sources of the Nile in the rain-flushed mountains of Abyssinia and Central Africa, they thought it sprang full-grown from the depths of the desert. So they had a sky *goddess*, Nut, and an earth *god*, Geb. A marvelous painting of Nut arched above the prone Geb's towering penis arouses thoughts of, among other things, the causal nexus of ecology, ideology, and sexual posi-tions.) By contrast, in humid regions water is taken for granted, but the soil can be lateritic (tending to harden when exposed: from the Latin for "brick"), or clayey, or sour, and so unreliable. In such places the Great Mother has held on much longer. As the squeaky wheel gets the grease, so the undependable deity gets the fat.

The ascendancy of males over females (both human and divine) as farming develops has been noted. It is especially strong in the drier parts of the world, partly for the reasons given above. But the main reasons for this shift are more concrete. Among them are the usefulness of a hunting patrimony in herding and in controlling draft animals; the usefulness of upper-body strength in those jobs, and in such work as digging canals; and the rise in the birthrate, which keeps women's hands and wombs full. Above all, there is the bigger scope for bullies opened up by bigger surpluses and bigger social groupings.

One gets a sense of geological unrest beneath the surface of Canaanite myth. Anat's mountain has subsided, one feels, as El's and Baal's have risen. Behind the simple image of the World Mountain as world-pole, male and female imagery wrestle for control. This is true in other cultures as well. In the West, the Mountain gradually becomes less a swelling of earth— a breast, a naveled belly—and more a platform from which to look down on earth.

The male sky god or mountain god "looks around" (as in the Iranian Yashts), is "all-seeing." The goddess-struck Minoans, too, had sacred mountains, but these were for looking at, not looking from. They were notable less for their height than for the way they looked from afar (like breasts or horns). Cleft or twin peaks do not suggest a single vantage point. Warriors, hunters, and often herdsmen seek the high ground, the commanding panorama. Among the cattle-mad Dodoth of East Africa, the head of the household will have a seat or shelter on the hill above his compound from which he can watch and (when necessary) yell. Men in hunting cultures often have a casual "men's camp" on a summit from which they can watch for game. By contrast, when we think of mothers watching their children, it is usually from a few feet away. In "primitive" societies, children swarm over and around the women as they work.

Once the world-pole has become an observation and command post, it is free to lose its earthiness altogether. It can be man-made. Once the bounty that flows from the world-pole ceases to be a spontaneous flow, and instead becomes the product of surveillance and control, there is no obvious reason why humans cannot take over the job. That is what happened in the flatlands where the first cities arose.

8

THE TOWER
OF BABEL

WHEN FARMERS FIRST wandered down from the hills of Iran or the Syrian steppe, or wandered up the shore of the Persian Gulf, and gazed on the vast floodplain of the two rivers, they must have been intrigued but not altogether pleased. They knew that rivers were arteries of life, and here were the two biggest rivers they had ever seen. Yet the land before them was no well-watered paradise, but a patchwork of swamp and desert. Swept in summer by a wind like the blowback of a kiln, the dunes shifted irritably under a sparse cover of artemisia (a cousin of sagebrush) and other shrubs, while the remnants of the spring annuals dried up and blew away. What was a farmer to make of this? His wheat and barley would not tolerate either the wet or the dry. His livestock would founder in the marshes or go thirsty among the dunes.

The first signs of permanent settlement on the floodplain date to the sixth millennium B.C. Two thousand years later—the length of an afternoon nap in prehistory—Mesopotamia was a paradise. It was a man-made paradise, a thing without precedent on earth. Although there were still marshes in the south, and plenty of semidesert in which seminomads as well as villagers and cityfolk grazed their herds, a wide tract of land on either side of the Euphrates was generously spangled with grainfields, date plantations, fishponds, and gardens of lettuce, onions, lentils, garlic, and cress. Cities of sunbaked or kiln-baked mud brick sprawled like lions amid these spoils, outwardly reposeful but inwardly (like the lion of Samson's riddle) buzzing.

The magnitude of this achievement can stun us even now if we stop to think about it. These people—the Sumerians and their predecessors in the region, the Ubaidians—gave us wheeled vehicles, yokes and harnesses so that animals could pull them, animal-drawn plows, sailboats, metalworking (casting, riveting, brazing, soldering, inlay, and engraving in copper and bronze), the potter's wheel, the arch, the vault, the dome, surveying, mapping, and a rough-and-ready mathematics. Above all they gave us the process in which you and I are now engaged, even if we no longer use wedge-shaped marks on soft clay. On the debit side of the ledger (another Sumerian invention) we might place large professional armies, siege engines, war chariots, a rigid division of labor and status, imperialism, and bureaucracy.

All this is the more remarkable in that hardly any of the raw materials of civilization, apart from the clay to make bricks and tablets, were to be found in the place where civilization began. Metal, wood, stone, and other necessary things the Mesopotamians got in exchange for their agricultural surplus and the finished products of their craftsmanship. (By "Mesopotamians" I mean all the civilized peoples who lived in the valley of the Tigris and the Euphrates in ancient times: in chronological order, the Sumerians, the Akkadians, and the Babylonians. For the most part they were the same people under different rulers.)

What made it all possible was a series of trenches running alongside the rivers in a pattern like a chain, or like the braids of a young girl. (Later the pattern would be dendritic, sharing the efficient layout of a tree's branching or a leaf's veins.) The sandy soil, with its patchy drainage, had made a mess of the job of distributing the groundwater that seeped from the riverbeds. The canals took over that job. What had looked like desert now proved to be soil far richer than anything wheat and barley had known in their native hills.

In a sense, though, this *was* the soil they had known in their native hills. Each winter for millennia the rains had gouged the hillsides of Syria and Anatolia. In recent millennia the gouging had been especially cruel, egged on by axes and hoes. Periodic floods had spread the deducted soil over the Mesopotamian plain. Wheat, barley, and humans followed the soil downstream.

But the floods had not spread the soil evenly. When the rivers overflowed their banks, they dropped coarse particles first, fine particles last. As the rivers wound, unwound, and changed their courses, they laid down a patchwork of coarse and fine soils: the former draining too quickly, the latter too slowly. So the humans who now showed up to claim the humus they had lost found it less immediately usable than they might have hoped. The answer was irrigation.

In Sumerian, a single word denotes both rivers and canals. Both were supposed to be the work of the gods, which humans merely maintained. Many ostensible canals were in fact natural channels in which the Euphrates had sometimes run. Unlike the Tigris, whose short course and swift current let it cut a deeper and straighter path, the Euphrates—sluggish as a pasha, and luxuriously indecisive—would flow now here, now there, now both ways at once. As it dropped sediment, it made its bed ever higher, like a princess piling up mattresses. In fact, it flowed above the surface of the plain and was kept from overflowing only by the natural levee it built up on either side. But then, at some point, it would break through the levee and find a new channel. So the river took on a braided look. Although the first settlers on the plain had some experience of small-scale irrigation, by and large they took their cue from the river itself. Some canals were made by adapting the existing side channels, others by mimicking the process by which they were formed. Since the surface of the river was often above that of the plain, a simple breach in the natural levee would send water gushing into a new man-made canal. A breach in the canal's levee would suffice to drench a field.

THE CITY AS WORLD-POLE

At the heart of every Mesopotamian city was a sacred precinct, and at the heart of every sacred precinct was a ziggurat, a stepped pyramid of mud brick. Unimposing by our standards—the great ziggurat of Ur was about seventy feet high—they were by far the tallest objects, natural or man-made, to be found on the Mesopotamian plain. Oddly, some ziggurats seem to have imitated mountains in a fairly literal way. One of the first, in Uruk, stood on an artificial hill about forty feet high. Although in most later temples the hill was replaced by a platform, the tower itself was often called a "cedar-scented mountain." It is not clear whether this was a bare-faced metaphor or whether some planting was done to buttress the claim. Most likely cedar and cypress oils were used as air fresheners within the temple. In the reborn Babylon of the sixth century B.C., the fabled Hanging Gardens were planted in the steps or terraces of a ziggurat. Legend has it that Nebuchadnezzar planted them for his Persian bride, who pined for the hills of Ecbatana.

Every great temple claimed to stand on the *axis mundi* or elevator shaft of the cosmos, offering the gods a way station between the upper and lower worlds. The creation epic *Enuma Elish* assigns to Babylon the role of divine motel and convention center—a role that seems to have been competed for and claimed by other cities at other times. For the Mesopotamian gods are city slickers. If one or another has his or her "throne-seat" or "abode" on a real Cedar Mountain, it is evidently used for ceremonial purposes, or as a summer place. Compared with the city, the countryside is godforsaken.

It is a sign of the Mesopotamians' pride that they drew the gods—and paradise itself—down from the mountains and into their own cities. If the source of life is upstream, downstream is where the fat collects: the rich bottomlands, the canals, the cities, the good life. The mountains may give life, but in these matters it is better to receive.

The Sumerians were incapable of imagining a world without cities and agriculture—a world different from their own. They knew of Bedouin, whom they called Martu, but did not think them fully human. As for hunters and gatherers, the Sumerians did not seem to know of them at all. The first man was created a "savage" but remained one only as long as it took the gods to shove a hoe into his hands. Cities, agriculture, and irrigation were devised by the gods to serve their own needs. Humans were an afterthought—a source of cheap labor. So far were the Sumerians from imagining a preagricultural humankind that they saw humankind itself as a crop.

The Tigris and Euphrates were the gods' irrigation canals. What is more peculiar is that artificial canals were also supposed to be the work of the gods.

How deftly and quietly the "black-headed people" patted themselves on the back! So goodly were Sumer's fields and cities that the gods themselves abandoned their mountain aeries and moved in. Lest the gods envy or take offense, the Mesopotamians were quick to give them all the credit. You were here all along, they told the gods. You built these cities and dug these canals; we humans were brought in later to do the maintenance.

In preferring the plain to the hills, the river valley to the headwaters as a place to lay out fields and build cities—in noting that the fat collects downstream—the Mesopotamians had a point. They were in fact recipients of the hills' largesse. Their mistake was to forget that fact.

Giddy with prosperity and progress, they came to think they had done it all themselves. Instead of recipients, they came to think of themselves as the source of life and plenty. They controlled the waters, tapped the great rivers like kegs of beer. It was easy to forget that the water came from somewhere. They had agriculture down to a science. It was easy to forget that it had arisen among the savages of the hills. The storehouses spat out grain; the markets were littered with dates and slippery with oil. Surely the city was the source of all life.

One can hardly blame the Mesopotamians for wanting a world-pole closer to home than the distant and (to most of them) invisible mountains. They might have chosen some flatter but still-natural world-pole to match the world they knew. They might have embedded the gods in the rivers, or in the salt marshes at the rivers' mouths, which were great dispensaries of wildness and of natural wealth. Maybe they did, at first: Enki, for instance, seems to have spent a lot of time loitering among the canebrakes. But by historical times Enki, like the other great gods, is safely installed in a city.

For some peoples, religious feeling is the feeling that some things are beyond society's control, that the sources of good and bad are unplumbable and can easily drown the flimsy channels we make to contain them. For others, the man-made order is so firmly established that it seems god-made. Awe is stripped from nature and affixed to the social and technical order. On the whole the Canaanites were a culture of the first type, the Mesopotamians of the second.

CITY AND DESERT

In the story of the Tower of Babel can be seen the myriad lights and fractures of city life. Civilization brings people together—in the most literal sense—but also divides them. Indeed (and here the Bible gets it wrong), the city thrives on division: of labor, of classes, of ethnic groups. The cities of the ancient Near East were nearly as cosmopolitan as Paris or New York. They were melting pots with the heat turned low. Merchants, mercenaries, nomads bringing their goats to market spoke a stew of tongues. The towers rose anyway—if not to heaven, then high enough for most purposes.

Really there was no question of their reaching heaven. The confounding of tongues was redundant. Had God stopped to think about it, he would have seen that the engineers of Babel would fall short, having failed to consider the relationship of a ziggurat's height to its base. For its top to reach heaven, the tower's base would have to cover the earth.

Though modern engineers seem to have overcome this difficulty, they have done so only on the physical plane. The economic height of a city depends on the subjugation or smothering of vast stretches of countryside. A typical New York skyscraper seems trimly anchored in a quarter-acre of schist. In fact it rests on thousands of acres of land: not only the suburbs and oil refineries of the tristate area, but the wheat fields of North Dakota and the sooty cattle pastures of the Amazon.

If the effects of Babel's towers did not reach quite so far, they were just as profound. Today southern Mesopotamia is once again the patchwork of swamp and desert it was when the first settlers arrived. The main difference (apart from the fact that the wildlife is gone) is that much of the soil is no longer even potentially fertile, for reasons that will be taken up in a later chapter.

One makes this obligatory remark without any sense of satisfaction. No one can say "I told you so" to the Mesopotamians. No one told them so. What they did was done for the first time on earth. It seemed a good idea at the time, and in many ways it was. Then the edge of civilization moved on, leaving a desert behind it.

So the moral of the biblical story, if somewhat overdrawn, was not without truth. Shot like ballast from God's mountain, men conspired to build their own. In the long run, they were confounded.

Babel is the Tower seen from the point of view of the Mountain, or of people who believed in the Mountain. In other words, it is the Mesopotamian world-pole as seen by the Canaanites.

Implicit in what I have just said are two ideas that may seem odd. The first is that the Hebrew authors of the Bible were essentially Canaanites. The second is that Eden, from which humans were ejected, is essentially the World Mountain. In the next chapter we will take up both of these ideas, together with the vexing question of why we were ejected in the first place.

9

THE FIERY SWORD

T HE TWO BROAD types of culture in the ancient Near East were broadly matched to two types of agriculture: that of the irrigated river valleys and that of the hills. The first is typified by the Mesopotamians, the second by the Canaanites, *including the Israelites.*

Let us not be deceived by the Bible's polemics against the Canaanites. As anyone who follows radical politics knows, the sharpest barbs are always reserved for those closest to one's own position: the group one schism away. In ecological terms, it's the species one has just split off from that one must compete with for a niche.

The more we know about the Israelites, the clearer it is that they were Canaanite hill farmers who practiced a sophisticated and fairly sustainable mixed husbandry of grains, vines, livestock, and trees yielding fruit, nuts, and oil. They were neither desert nomads mistrustful of nature, nor proud hydraulic despots lording it over nature. They were good farmers living frugally on the margins and using the best stewardship they knew. They were dependent on rain and groundwater, neither of which was over-abundant, and on thin and rock-strewn soil, and had to use their wits to conserve both. They were not so different from present-day farmers of the Andes or of Szechuan. They were not so different, perhaps, from other peasants of the Mediterranean basin, past and present.

Why these particular Canaanites came to think of themselves as different from other Canaanites, how they in fact became different, and which happened first, is anybody's guess. Some scholars suspect the influence of an Egyptian elite, since many of the Levites had Egyptian names; this would salvage some flotsam of truth from the legends of the plagues and the Red Sea. Others speak of a league of Canaanite tribes that joined forces to fight the Egyptians. By far the most attractive view, at least to those with a weakness for "liberation theology," is that the Israelites were land-starved or tax-gouged peasants who threw off the rule of the lowland warlords and staked out their own territory in the hills, acting out the passion for justice and equality that would later find its voice in scripture.

When archeology first finds the Israelites—about 1100 B.C.—they are pioneers in the hills of Judea and Samaria, part of the central range that runs like a spine down the length of Canaan (its Apennines, one might say). There is scant evidence that this pioneering was prelude to a "conquest" of the valleys and the coastal plain ("destruction layers" of ash and debris are mostly absent from the relevant strata of the Canaanite sites mentioned in the Bible). Nor is there much evidence of a flight from Egypt. Yet is it true that these settlers had just escaped the pharaoh's yoke. They were not so much settling as resettling the uplands, which had been depopulated during the four centuries when Egyptian rule ravaged Canaan.

Since we have spoken only briefly of Canaanite farming, let us fill that gap by dealing at some length with a subclass of it—namely, Israelite farming. Since this is the way of life glimpsed fleetingly in the spare narratives of the Bible, it has great interest in itself. For our present purposes, though, what is important is the contrast between this way of life and that of Mesopotamia. It should be clear why this way of life nurtured, and was nurtured by, the vision of the Mountain.

THE HILL FARMERS

Having known both the ax and the torch in earlier times, and getting at the best of times only modest rainfall, the hills of Judea and Samaria were not clothed in what we would call "forest primeval." Where least disturbed, the landscape was the sort of open Mediterranean woodland known as high maquis, with evergreen oak, Aleppo pine, and pistachio (known in the Bible as terebinth) the most common trees. Elsewhere this would dwindle to low maquis, a mix of shrubs and herbs such as rosemary, sage, summer savory, rock rose, and thorny burnet. The settlers cleared a good deal of this forest for pasture and cropland, knowing that beneath lay the red soil now called *terra rossa*, the richest of all mountain soils. (In Hebrew the words for "earth," "human," "red," and "blood" are similar and may derive from the same root.) Of this process the Book of Joshua preserves a hint:

> And the children of Joseph spoke unto Joshua, saying: "Why has thou given me but one lot and one part for an inheritance, seeing I am a great people, forasmuch as the Lord hath blessed me thus?" And Joshua said unto them, "If thou be a great people, get thee up to the forest, and cut down for thyself there in the land of the Perizzites

and of the Rephaim; since the hill-country of Ephraim is too narrow for thee."

This new round of deforestation promised to be the worst yet. Besides the usual sheep and shoot-nibbling goats, these settlers had big animal-drawn plows with bronze shares, a loan from Mesopotamia. Although still only scratch plows, which did not turn the soil upside down as mold-board plows do, these could still be lethal to the soil of any slope steeper than a wheelchair-accessible entrance ramp. The red soil was rich but rarely more than a foot deep.

As the slopes began to lose their thin layer of rich red soil—and as popu-lation growth made land dearer and labor somewhat more plentiful—the Israelites began to build terraces. Whether this was their own invention is unclear; a word appearing in the Ugaritic epics may refer to the same device. Either way, there is a good chance that some group or other of Canaanites came up with the idea. As the first farmers of the Mediter-ranean basin, they devised many of the methods that its peasants use to this day.

To make a terrace was no small matter. Remaking a ramp as a series of steps meant moving a lot of earth and rock, though the naturally blocky karst limestone of some regions helped somewhat. (Today in the vicinity of Jerusalem it is hard to tell offhand whether a particular hill got its stair-case shape from ancient man or ancienter nature.) Pillars made of large boulders would be erected and the gaps between filled in with smaller stones. Behind these walls, it would not do just to pile up dirt any which way. Above the bedrock would be a layer of soil, on top of that a layer of gravel, and on top of the gravel another layer of soil. In this way water would percolate from one terrace to the next one down. Some of the soil in the terraces seems to have been hauled up from the valley floor, perhaps because erosion was already well along.

For all this effort the farmer got a number of benefits. He got a nice, nearly level surface to plow with his ox and his ard. He got the soil to stay put, at least for a while. And he got rainwater to tarry far longer than it would usually care to do on a denuded slope. (The same could be done with springwater: in Roman times, if not earlier, a system of channels might distribute the water of a spring near the top of a hill among the ter-races below.)

This last was the main reason he took the trouble, for water is the limit-ing factor in almost all farming in the Near East. Another way he tried to control this variable was by catching runoff from the rains in cisterns dug in the bedrock. To make the limestone watertight, he cemented the pits with lime: another technique that the Israelites or Canaanites may have pioneered. Evidently it worked, for the Talmud compares the prodigious memory of Rabbi Eliezer ben Hyrcanus to "a cemented cistern, which loses not a drop."

In general, the hill farmers dealt with the whims and capers of nature by hedging their bets. While they did not exactly have hedges, they did have plenty of borders between plots of different crops. The hills were full of nooks and crannies, of microclimates and microenvironments. As their natural flora and fauna were diverse, so (up to a point) were the flora and fauna the settlers put in their place. A single household might have fields of wheat and barley as well as lentils, peas, and other legumes; a vineyard enclosed by a wall of thorns or of stone, with the vines trailing on the ground or trained to form arbors; and orchards of fig, apricot, almond, and pomegranate. Any patch of hillside that was left would be planted with olive trees, whose fruit was not eaten (the art of curing being unknown) but pressed for the oil used in cooking, lighting, and grooming. A household would also have herds of sheep, goats, and cattle that would winter in dry areas, grazing on rain-primed seasonal growth, or in the upper hills, where they would help to degrade the maquis to the lower and sparser mix of shrubs and herbs known as garigue. In summer the livestock would stay closer to the village, nibbling on stubble and fallow weeds and paying the check with manure.

Nor was husbandry the only source of food. The women might forage for pistachios, acorns, herbs, and other wild foods, the men hunt the gazelles and wild goats that still roamed the hills. To the hearers of the legends from this period collected in the Book of Judges, it apparently did not seem odd that Samson should run into a lion among the vineyards. Even the poet of Canticles, who lived some centuries later, knew a world in which wild and man-made mingled, not least in the imagery of the human body itself.

When I say that a single household might do all these things, I mean a *bet av*, an extended family of perhaps ten to thirty persons led by a patriarch and living in a cluster of stone or mud-brick dwellings. Such a household shielded itself from nature's whims by sharing labor (at harvest, for instance) and food (at times of scarcity) with an even more extended family called a *mishpahah*, a kind of clan or tribe that might take the form of a village. (What is called a tribe in the Bible was a still more fluid grouping in which political allegiance was thicker than blood.) The command to "be fruitful and multiply," though common to most farming peoples, must have had double force in this frontier setting, where two new hands added far more food than one new mouth subtracted.

Land seems to have been held communally within the *mishpahah*, at least in the sense that a piece of land, even if "sold" to an outsider, would eventually revert to the clan. In contrast to the plantations that sprawled across the great river valleys of the Near East, the hills nurtured a world of small holdings, painstakingly husbanded. When, in a larger Israel, the royal houses and others began to amass great estates, the prophets were outraged: "Woe unto them that join house to house, that lay field to field. . . ."

In this respect the Israelites out-Canaanized the Canaanites, who were somewhat more prone to plantation farming and cash-cropping. The Phoenicians (as the Canaanites came to be called when, squeezed by Israelites on one flank and Philistines on the other, they bunched up in the cities of the Syrian coast and increasingly took to the sea) were both crackerjack farmers and peerless merchants. Though it is hard to know which role came first, farm produce—in particular, olive oil and wine—was among their primary wares. To the Canaanites the ancient world owed not only the alphabet and the art of seafaring, but much of its agricultural science as well. The bible of ancient farming was the work of a descendant of Canaanites, Mago of Carthage (Carthage started out as a colony of Tyre). Even the Romans, who were not likely to overpraise a Carthaginian, acknowledged Mago as the father of agricultural knowledge. The twenty-eight books of his treatise were translated from Punic into Latin by order of the Roman senate; though lost, they left their traces in passages cribbed by Pliny, Varro, and Columella. The farm belt of North Africa remained fertile as long as the Carthaginians were in charge. It was the Romans who ran it into the ground.

MOUNT EDEN

Being Canaanites—if Canaanites of a rather peculiar sort—the children of Israel might be expected to have some notion of a World Mountain. So they did. It took several forms, some of which we will deal with later; but foremost among them was Eden.

Today it is not common to think of Eden as a mountain. But in earlier times—from at least the sixth century B.C., when Ezekiel prophesied, to the seventeenth century A.D., when Milton wrote *Paradise Lost*—it was very common.

Although the Bible never specifies Eden's elevation, the fact that it is the source of four great rivers speaks for itself. Armed with the knowledge that water does not flow uphill, scholars from Philo's time to the present have placed Eden in the mountains of Armenia, or other mountains vaguely north of Mesopotamia.

These same mountains are vaguely north of Canaan, too. The Tigris and the Euphrates arise in the mountains of Turkish Armenia. While the Tigris runs straight into Mesopotamia, the lordly Euphrates adopts a more leisured route, taking in the sights and waters of eastern Syria. Some affluents of the Upper Euphrates start within seventy kilometers—a god's spitting range—of the Amanus Mountains. The closest thing in Canaan to what we would call a river, the Jordan, has its ultimate source in the mountains of Syria.

No wonder, then, that the peoples of the Fertile Crescent shared a firm if somewhat cloudy feeling that life flowed from the north. (As great dams go up, that feeling gets less cloudy and more anxious. An open secret of Near Eastern politics—closed until recently to the publics of the West—is that many struggles in the region, from the fight over the Golan Heights and the West Bank to the tussles between Syria, Iraq, and Turkey, have less to do with oil or blood than with water.)

To be the Mountain, it is not enough to be a mountain. But Eden has other qualifications. It is the source of life, in several ways. First, it is the source of water not only for the Near East, but for the known world. That is the import of the four rivers, two real and two mythic (or semimythic), whose hydrologically improbable courses extend to the ends of the earth. Two of these are labeled clearly as the Tigris and the Euphrates; the other two are identified by the rabbis with the Nile and the Ganges. The two rivers that run from El's mountain have been doubled—perhaps under Mesopotamian influence, perhaps in sheer one-upmanship—so that they may reach and refresh the very corners of the world, dividing it neatly in four quarters.

Like many world-poles, Eden crowns its mountain with a Tree of Life. According to the Midrash, this means "a tree that spread its canopy over all living things. . . . All the primeval waters branched out beneath it." To walk around its trunk would take a man five hundred years.

Eden is the source of life in another sense, too. It is the navel of the world—the first home of all creatures, both human and nonhuman. It is even a home of sorts for God, who walks in the Garden in the cool of the day. But whereas God and plants and animals get to stay in Eden, humans get the boot. And this, too, is a hint that what we are dealing with is nothing less than the Mountain of God.

BUT NOT FOR US

The failure of the author of the Eden story to mention that the garden was on a mountain makes one wonder what else got left on the cutting-room floor. Two passages from Ezekiel give some bright if jagged clues. One is a parable against the king of Tyre.

> Thus saith the Lord God: Thou seal most accurate, full of wisdom, and perfect in beauty, thou wast in Eden, the garden of God; every precious stone was thy covering. . . . Thou wast the far-covering cherub; and I set thee, so that thou wast upon the holy mountain of God; thou hast walked up and down in the midst of fiery jewels. Thou wast perfect in thy ways from the day that thou wast created, till unrighteousness was found in thee. By the multitude of thy traffic they filled the midst of thee with violence, and thou hast sinned; therefore have I cast thee as profane out of the mountain of God; and I have destroyed thee, O covering cherub, from the midst of the stones of fire.

In the second passage, the king of Egypt is compared to Assyria, which is compared to a cedar in Lebanon, which is compared to the trees of Eden: "All the trees of Eden that were in the Garden of God, envied it." Yet this great tree is cast down into the netherworld.

Apart from the clear identification of Eden with the Mountain of God, some scholars have glimpsed here the glittering sherds of a lost Hebrew epic of Eden. The Eden story in Genesis gives reason to think such an epic may have existed. Several terms in the narrative carry the definite article at first use—"the tree of life," "the tree of the knowledge of good and bad," "the cherubim and the sword of flame which turned every way"—which in Hebrew is done only if the reader is assumed to be familiar with the thing mentioned.

What that epic might have looked like is anybody's guess. Combining the passages above with the lines from Isaiah about Halal ben Shahar, "the son of morning," we can assume the action had to do with an angel or god who rebelled against Yhwh and was cast down to earth, or into the pit (an easy thing to do from Eden, since the world-pole is the shortest distance between the top and bottom of the world). It is a familiar plot, alluded to in the story (Genesis 6) of the sons of God who descended to earth and married with the daughters of men, and elaborated in the legend of Lucifer.

The drama would be played out in a garden of God or of the gods, a divine country club to which earthlings (if they exist yet) need not apply. Some of its trees would be gigantic mountain trees, such as cedars, while others would bear jewels for fruit—both being found in the garden of Siduri, the "barmaid of the gods" who entertained Gilgamesh.

In any case, the moral I want to draw from the lost epic is this: if even gods and angels were cast out of Eden, what chance did humans have of lasting out their term?

Modern scholars tend to picture Eden as a formal garden in the Mesopotamian style, irrigated to a fare-thee-well. Some of the sources of the Eden story may have had that squared-off and straitlaced shape, but others were a good deal wilder and woollier. In the Mountain of God, even the Garden of God, we have a vision of paradise as a forested peak—the summa and last resort of wildness in a region chockablock with cities, fields, canals, herds, and armies. Though the Hebrew word *gan* usually means an enclosed vegetable garden or fruit orchard, the phrase *gan Elohim*, "garden of God," seems to be meant as a kind of analogy: just as we might call the prairie "God's lawn," so the ancients saw the wooded mountain as God's private garden.

Such wild places were paradises not for humans but for gods. They were not meant for humans at all.

The cosmic center is not always thought of as a nice place for humans to live or even to visit. Nevertheless it is the source of life. "All the world is watered with the dregs of Eden," the Talmud says; and the dregs are as much as it can take. Humans cannot see God's face and live.

If God is the heart of nature, then to say that we cannot stand pure godhead is to say that we cannot stand pure wildness, except in small doses. We can stand (if sometimes just barely) the electric blue of the sky, the buzz of bees, the jolt of sex. Uncut, nature is too much for us. The main lines of wildness make us jumpy—and rightly, for an instantaneous surge can kill.

To think of living in Eden is to deny the primal sundering of heaven and earth, of god and animal and human. The world-pole is the one place where the sundering has not happened, or has been repaired. We must revere it, draw sustenance from it, keep it alive, keep the channels of wildness open. But to think of living in it—why, that's like wanting to live in the sun.

In a Bushman story of beginnings, the sun was on earth and hid its light in its armpit. It was so close, it was useless—darkening with excessive light. Hunter-gatherers agree that division is necessary, that you can't have heaven smothering earth. It follows that you can't live on the world-pole.

In fact, even shamans can only climb it once in a while, and soon slip off. It is slippery—wet with the water of life, greasy with the fat of offerings, alive with energy of all kinds.

The peak of the World Mountain is like the head of a pin on which only angels and animals can dance. It is the vanishing point of the trophic pyramid. There is room at the top, but not for us.

Nowadays most people (as opposed to scholars) like to imagine Eden as a wild place: a rain forest rife with orchids and lianas, a savanna rumbling with game. Conversely, they like to stick the words "Eden" and "garden," like Sierra Club stamps, on any wilderness that is not unlivably frigid or arid, especially if they have never been in it themselves. And while they are right to imagine Eden as a wild place, they are wrong to think that such places are still paradises *for us*. A brief backpacking trip is about as much of real wilderness as most of us can stand, and even that will seem like paradise only if nothing goes wrong: no rain, no grizzlies, no marmots eating our boots. After a week or two, we are glad to be expelled. And were we to stay— to become settlers, pioneers—we would soon transform the place, or at least our immediate patch of it, into something wildly different.

The World Mountain is a paradise only when seen from a distance, or with the moist eye of memory. Once, wilderness was our home. Looking back, we endow it with all the longed-for comforts of home. We see a garden: a place wholly benign, a place of harmony and plenty. We forget that the harmony, such as it was, was possible only because we were still beasts, and the plenty only because we were scarce. As soon as we become fully human, we begin to "fill the earth and subdue it." We begin to destroy Eden, and thereby expel ourselves.

We have spoken of the Mountain in two ways: as the place of origin of human-led waves, and as the place they have not yet reached: that is, the area outside the circle. When we look back upon Eden, this distinction evaporates. At the moment of origin, there was no circle. At present, wilderness is beyond the circle, and is not a place where humans are at home. But then it was all there was. And we were at home, being not yet far removed from our animalhood. Or so our burnished memory says.

Wilderness only becomes the earthly paradise when it is looked back upon as the point of origin. Perhaps the word "Eden" should be reserved for this wistful, over-the-shoulder view, and not squandered on mere circumambient wilderness. But almost any wilderness can be viewed in this way, at least for a few minutes. Even a patch of forest a few miles from a great city can seem, in the sudden silence as deep as organ chords, a holy place: the navel of the world, the birthplace of humankind. Though the oddly mingled sense of awe, comfort, and surrender that we feel at such a moment may not survive the third mosquito bite, it is real enough while it lasts.

WHEAT AND THE FALL OF MAN

So far I have spoken in the language of myth. But the myth of the Fall, like that of the World Mountain itself, is based on ecological fact. In fact, of course, the Fall was not a single event. It was a gradual slipping that, by degrees, snowballed into a full-speed charge downhill. Humans began to change their surroundings in a drastic way as soon as they mastered fire, but it was the second wave—the alliance with annual grasses—that sealed our self-expulsion. We and our allies moved outward, driving Eden before us. We stripped forests, troubled the soil, uprooted whole ecosystems.

On this point the Bible is clear: "Cursed is the ground for thy sake; in sorrow shalt thou eat of it all the days of thy life. Thorns also and thistles shall it bring forth to thee; and thou shalt eat the herb [grass] of the field. In the sweat of thy brow shalt thou eat bread, till thou return unto the ground; for out of it wast thou taken: for dust thou art, and unto dust shalt thou return." Agriculture as we know it: the earth is tilled, grain (a kind of grass) is planted, weeds interfere. The Hebrew word *lehem* means food in general, bread in particular. In place of the herbs and fruits of paradise, man will eat bread. As we have said, the culture of barley and wheat—first for beer and toasted seedheads, then for bread—did apparently begin in the uplands of the Near East, some ten thousand years ago. That it was woman, not man, who surely began it—being the foremost gatherer, she must have been the first farmer—may be dimly recalled in the story that it was Eve who first tasted the forbidden fruit, then handed it on to Adam.

What was the forbidden fruit? In the Midrash, a rabbi of the second century A.D. gives a remarkable answer:

> Rabbi Meir said: It was wheat, for when a man lacks knowledge people say, "That creature has never eaten bread of wheat." Rabbi Samuel ben Isaac came to Rabbi Ze'ira and asked, "Is it possible that it was wheat?" "Yes," he said. "But is not 'tree' written?" he asked. "It rose high as the cedars of Lebanon," he replied.

If Rabbi Meir is half joking—for this is a typical midrashic game of competitive whimsy, with other rabbis one-upping him by proving that the forbidden fruit was the grape or the fig—his half-joke has deep roots in the

Near Eastern mind. Wheat is the premise of civilized life. Whoever has not eaten bread made of wheat is a savage, at best a Bedouin. The Sumerians had a similar gibe for the nomads at the fringe of their world: "The Martu eat bread, but they don't know what it's made of."

Rabbi Meir's notion sounds less odd in Hebrew than in English, for the word translated "fruit" can mean any kind of produce. And if the forbidden fruit is indeed wheat, the role of the snake becomes clear: in ancient times, snakes were used to protect granaries from rodents.

If grains meant knowledge, they also meant hard work, and it was work man had not previously evolved to do. For five million years we had foraged, scavenged, hunted, gardened a little, which was hard work at times, but desultory, even leisurely. Adam's punishment, like Eve's, was a wrenching departure from the path of primate evolution. Eve was sentenced to pain in childbirth—outrageous pain by any reasonable mammalian standard—because infants were being selected for bigger brains than the pelvis of a bipedal female could accommodate. To be sure, Eve's penance began long before Adam's, millions of years before the start of farming. But farming, by raising the birthrate, made it necessary for her to bear that penance more often. Farming meant more children; children meant more farming. Adam's and Eve's punishments fed each other.

They fed other things, too. With time, with irrigation, mountains of grain became the foundations of cities. To protect the grain from marauding nomads, armies arose and enslaved those whose grain they protected. To dig the great irrigation canals and keep them clear, slaves were called for, and bureaucrats, and despots. (The first great emperor in history, Sargon the Great of Akkad, began his career as a gardener; fittingly, the last great emperor, Pu Yi of China, ended his career that way.) Humans were winnowed like grain, separated by function, wealth, power. Civilization arose, and writing, and real estate. All this happened just a few miles downriver from Eden.

Adam's fate was to outrun his own nature: to be dragged by his big brain, by his snowballing technology, into regions his body had never known. Like Eve, he was the victim of an overgrown head.

Of course, our big brain is our nature, too. But in eating the forbidden fruit, though we failed to become gods, we ceased to be pure animals. At that point, Eden could no longer be our home—for whatever place we made our home ceased to be Eden.

We noted earlier that the Hebrew phrase usually translated "Cursed is the ground for thy sake" can, with a bit of license, be read to mean "Cursed is the ground by thy passing over." As the waves of human expansion move across the earth, Eden is trampled underfoot.

Adam was put in the garden "to work it and protect it." The two jobs are complementary, but they are also contradictory. From what are we to protect Eden, if not from our own work? The more we work the earth—by which I mean not only tilling but the whole spectrum of human meddling, from setting grass fires to splitting the atom—the more we are obliged to protect it. If we fail to do either, we fail to be fully human.

These tasks were set us not just for our brief tenure in Eden, but for the whole span of our stay on earth. Indeed, by setting us the first task, God set us up for expulsion. For when we work the earth we work her hard, and the place we work ceases to be Eden. We move outward in waves of work—waves of improvement and devastation, of fruitfulness and waste. By setting us the second task, God set (or tried to set) a limit to the height and reach of those waves.

HUMANKIND'S ZIGZAG

"So He drove out the man; and he placed at the east of the garden of Eden cherubim, and a flaming sword which turned every way, to keep the way of the tree of life." The Tree of Life is the inner core of the world-pole: the heart of the heart of the world. Man must be prevented from reaching—and ruining—the source of life.

What exactly is the fiery sword? Is it our awe of wilderness? Our fear of its rigors and dangers? Our discomfiture in the face of its unearthly beauty? Whatever it is, it is the best friend we have. For only by keeping our distance from wilderness—some wilderness, at least— can we keep from fouling the wellspring of our own life.

A moment ago I spoke of a paradox: wilderness seems to be at once the center of the circle of human expansion, and the region outside the circle. We saw that the paradox vanishes when we look back to the instant of origin. But there is another, more matter-of-fact way of looking at this problem.

Is the Mountain our point of origin, or is it our destination? Is it where we come from, or where we are inexorably and ill-advisedly going? It is both. Our past has been a zigzag. Humans and their allied species come from the World Mountain, descend into the valley and fill it—becoming more or less domesticated in the process—then turn and edge their way up the slopes of the World Mountain, crushing its wildness as they go. Grasses edge out trees, domesticated grasses edge out wild ones, hybrid crops edge out hoary and varied domesticates. The beginnings of this reflux can be seen clearly in the ancient Near East—for instance, in the various resettlements of the Judean and Samarian hills. It is still going on. But as the hour gets later and space gets dearer, we find ourselves assaulting earlier and earlier world-poles: wild places in which were born not our species but the families from which we descend: even the family of life itself.

In recent years, we have set about dismantling our primate home, the rain forest, which is perhaps the greatest of terrestrial world-poles. As the uplands of the Near East are vital to the ecology of the Near East, so the rain forest of South and Central America, Africa, and Asia is vital to the ecology of the entire planet. Its role in regulating climate and maintaining bio-diversity is well known, and yet we do not know the half of it. It is also the only home of most of our primate relatives. That home was already shrunken by the drying trend of the Pliocene, at which point we moved out. Now we have returned and are hacking away at what is left, pushing our own cousins into destitution and death. Their only consolation is that, if we keep it up, we may be next.

What will be left after the rain forest is gone? Our oldest home, the sea? The continental shelf, with its host of atmosphere-regulating algae and microbes, may prove to be the most vital wild place of all—the ultimate world-pole. Yet densely settled seaboards ooze sewage into coastal waters, and rivers spew out the runoff of farms. Offshore drilling and aquaculture threaten to change the continental shelf as thoroughly as we have changed the continents. As usual, we do not know what we are doing.

Humans fail to grasp the fact that the world-pole cannot be grasped without choking off its flow of blessing. "I will not let thee go, except thou bless me," they say, as Jacob said to the angel; but they grip the angel's throat and throttle him speechless.

To destroy wilderness is to cut off the source of human life. Yet destroy-ing wilderness—at least, making it something other than wilderness—is something humans must do. These elementary truths stand behind many myths of paradise and the Fall. As soon as man becomes man, he must leave the place where he was born—namely, wilderness. But the expulsion takes a subtle form. For he leaves wilderness by entering wilderness. Each place he enters ceases to be itself, and so he is expelled from its true self: he is left with the husk of a place. The only way to avoid expulsion is not to enter at all. (Or to enter only briefly, treading lightly, breath held.) The fiery sword is a necessity which our own wiser selves acknowledge.

The lease on the World Mountain forbids major alterations of the premises. It forbids not only cats and dogs but all domesticated species, including humans. If we return to Eden, we destroy either Eden or our own humanity.

10

THE RIVERS
OF EDEN

THE FOUR RIVERS of Eden are symbols of the flow of wildness from the heart of wilderness to the rest of the planet. On that flow depends the health of the whole planet, even those parts of it most shaped by human hands. Yet human hands are blocking that flow. Not content with reducing the size of wilderness, we are keeping what is left of it in boxes.

But wilderness cannot be kept in boxes; even in big boxes, it withers and dies. In recent years, ecologists have learned that even our biggest national parks and wilderness areas are not big enough to keep their health when they are cut off from kindred ecosystems by barriers of urban sprawl or unbroken farmland. One breeding pair of spotted owls may need two thousand acres of old-growth forest to hunt in. Grizzlies need nearly as much elbow-room; even Greater Yellowstone's 13.5 million acres can't support a big enough gene pool to keep the species vigorous.

Nor is it only top predators who suffer when their habitat is broken into fragments. Warblers, thrushes, and other migratory songbirds need deep forest to nest in. The forest edge is infested with predators—raccoons, feral cats, egg-eating crows and blue jays—and with an even more dangerous parasite, the cowbird, which lays its eggs in the open nests of songbirds. Hatched, these unwanted foster children toss out the songbirds' own young or gobble up their food. When the forest is fragmented by human activity, edges are everywhere (indeed, even an unpaved road or a power-line cut can attract cowbirds and crows). And songbirds are driven to the edge of extinction.

For that matter, preserving the summer habitat of our migratory birds means nothing if the fields and waters of their flyways are poisoned or drained, or if their winter homes in the tropics are burned to ashes. Migration, the cycling of nutrients, the flow of energy and genetic information: these are the lifeblood of the planet. But the planet's arteries are hardening with concrete, with chemical salts, with embolisms of asphalt. To speak of the planet as an organism is to speak loosely, but here the metaphor fits. Much as we might like to preserve a particularly lovely limb or patch of skin, we find that, isolated, its blood cut off or poisoned, it gets gangrene and goes cold.

All of this becomes plainer when we shift our gaze from land to sea. The sea's wildness is diffuse and—not to put too fine a point on it—fluid. To carve out a wilderness preserve or park in the ocean seems a fatuous enterprise, like writing a treaty on water. So it is more common to protect a species—by banning or limiting fishing, for example—than to try and protect a region.

What is clear at sea may be hazy or hidden on land: that all wildness is fluid. Frozen, it dies. There are gulf streams on land, too, though we do our best to dam them. I am not saying that species should be protected instead of habitats: that is a sucker's game. Only that, in the long run, boxed-in wildlife refuges and wilderness preserves are death traps.

Take a lobster trap miles long and miles wide, with walls as fine as seine nets. Leave it open on the ocean floor and let it fill with life. Shut it. Wait a decade, and draw it up. The stink of death will knock you flat. A dumb experiment, you say, and so it is. But we are conducting its landlubber counterpart in hundreds of parks and forests worldwide.

We need pure wilderness. Our pure wildernesses should function, however, not as boxes, but as organs: hearts and lungs from which wildness circulates. So there must be arteries as well: corridors of wildness, underpassed or overpassed by the highways, allowing flora and fauna to migrate and expand. The network might embrace the wooded windbreaks of farms, city greenbelts, and disused railway right-of-ways as well as corridors created for the purpose. Like the hedgerows that once crisscrossed much of England—intricate living walls dense with trees, shrubs, small mammals, insects, and songbirds—such corridors would turn the walled garden inside out so that, like a Klein bottle, it would contain the universe. The wall between man and nature would fall, for the wall would be nature.

But even this would be an empty gesture if there were not also capillaries, branching and subbranching, touching nearly every cell of civilization with wildness. Just how this would work is harder to imagine the more encased we are in industrial life. Later in the book I will make that effort; in the meantime, here are some homely examples. A lawn in Tucson sporting mesquite and bloodroot amaranth rather than water-guzzling grass; a Massachusetts home encircled with nut trees and berry bushes instead of privet and a tomato patch; a toilet that makes compost instead of sewage; city sewage treatment by willows, duckweed, snails, and microbes in narrow greenhouses running along the sidewalk. At this point, of course, the network would no longer be visible as a network. At this point wilderness would not be a thing or a place, but a process.

MOUNTAINS ON EVERY SCALE

As the city on a hill is a light to the nations, so the wilderness on a hill is a light to nature: nature dispersed, nature exiled, nature clip-tailed and in harness. This is not so fanciful as it sounds. All of nature does in fact receive information from wild places, and send information back; that is what gene flow is all about. What is just as crucial is that wilderness should be a beacon and model to humans in their dealings with nature.

One of the reasons for having wilderness around is to see the way nature works when left alone. It gives us a model to work from. But a wilderness does not just show the way nature works in general; it shows the way she works in a particular place. For this reason (among others), we cannot rely on a single World Mountain, or a few big parks and preserves. We must have many small patches and strips of wildness as well. So the inability of most ancient peoples to fix on a single World Mountain, their eagerness to plant gods like flags on every bump in the ground, was a good thing and worth emulating (even if it did not always stop them from logging every bump in the ground).

Yet we mustn't put mountains, as such, on a pedestal all by themselves. Wetlands, prairies, seashores, rain forests, even deserts may be just as vital in nature's economy. Every ecosystem is a model of the biosphere, but none is a complete model; they complement each other in providing information to nature and to us. But local information is the most vital of all. The world-pole for a region of rain forest should be a rain forest. Preserving a swatch of cloud forest fifteen thousand feet up will not do. (On the other hand, the upper elevations of the rain forest often have the richest ecologies, having served as refugia for lowland species during the Ice Age.)

Wildness must be scattered fractally, with the same pattern of center and branches copied on every scale. (This can be done in time as well as in space—as in the case of Sabbath, sabbatical, and jubilee, which I will touch on later.) If one take this model seriously, then the idea that any place can be the world-pole no longer seems a literary conceit. In the human body, every living cell has access to the bloodstream (though for some, such as brain cells, that access is carefully limited). As far as the average cell is concerned, the heart is only hearsay, as real as the emperor of China. For all intents and purposes the source of life is the nearest capillary.

The analogy must not be pushed too far. We don't have the cell's luxury of ignoring the ultimate sources of life. Only in the desperately ill are the organs of the body as beset as are the organs of the biosphere today. If there is a central nervous system in charge, it has not yet figured out how to counter humankind's attacks. Nosy and even colonial as it may seem, the task of protecting the tropical rain forest falls to all of us.

Still, it may not be amiss to take a hint from the cell and pay attention to the world-pole in one's own backyard. It may in fact be in one's own backyard—in the garden, the trees that harbor birds, the compost pile. It may be in a vest-pocket park, or a marsh a mile down the road, or the young linden lately jammed into the sidewalk outside one's apartment building.

THE NEED FOR WALLS

Of course, when you look for wildness in your own backyard, what you find it not always what you had in mind. You find rats and mice, earwigs and cockroaches, crabgrass and knotweed, ailanthus and sumac, head lice and fleas and hookworm and salmonella. You find unspecified rodents tap-dancing in the ceiling. You find vermin, and weeds, and parasites. For the most part these are not wild species, they are accidental domesticates. They are the camp followers and cutpurses and ragtag irregulars in our army of allies. Some are common among hunter-gatherers, others wait for farming and cities to make their debut. But it is only at the height of civilization that almost all the wild things most people see are pests.

A rule of thumb: the more tightly strung a man-made ecosystem is, the more likely it is that a slight relaxation will allow nasty things to creep in. In other words, the more thickly we wall ourselves off from nature, the less we will like what sneaks in through the cracks. In part this is a psychological fact, since people suspicious of nature may see any wild thing as a pest. But it is also a biological fact. Of the many examples that might be given, one that seems especially apt is the comedy—comical for those who did not live through it—played out in Kern County, California, in 1926 and 1927. Having made their world safe for themselves and their livestock by diligently slaughtering every skunk, fox, badger, weasel, snake, owl, hawk, and coyote in sight, the farmers of Kern County found themselves confronted with an army of 100 million mice—the greatest rodent infestation in American history.

Walls are not always a bad thing, though. If the flow of energy, matter, and information were perfectly free, there would be no life at all. There would be entropy.

Any organism is really a wall or system of walls. The wall defines a living thing; in a sense it defines life. The first job of every living thing is to define and regulate its internal environment. Arguably, before organic molecules could master the trick of self-replication by means of nucleic acids, they had to learn to hide inside a membrane.

To the ancient mind, a human being was not a solid clay figurine but rather a vessel, mostly hollow. Break the vessel and the life spills out. Blood is not the only liquid in question. Modern biologists tell us that we are 70 percent water; on the cellular level, life is a soup of water and protein sloshing in a bag of fat.

Culture in its broadest sense can be seen as the making of walls. We do not feel secure with only our skins standing between us and the outside world (call it what you will: nature, wilderness, the unknown). We crave a second skin. The house, the city, the village, the garden are all attempts to grow it. Even Pygmies have their huts in the forest; even Bushmen make their rounded indentations in the sand.

There is nothing unnatural in this. Besides regulating the inside of the body, organisms try to regulate what is outside. They may manage to regulate or at least affect the ecosystem or even the biosphere as a whole; but walling off the immediate surroundings is a good shortcut. Hives, nests, and other homes are extended bodies. What makes us unusual is that our bodies are extended mostly by cultural means.

In view of all this, my ideal of the flow of wildness must be tempered. All living systems need walls or boundaries of some kind. All require some degree of closure, as well as some degree of openness to the outside world. In general, the degree of closure increases as one rises in the hierarchy of living systems. At the bottom end, the cell is constantly exchanging matter, energy, and information with the outside world. At the top, the biosphere as a whole, though dependent on outside energy, is (apart from the pelting of occasional meteors) almost a closed system as far as matter and information are concerned.

A step down from the biosphere are large bioregions defined by oceans, mountain ranges, and other barriers. While these systems are less closed than the biosphere, they still need a good deal of closure, especially when it comes to genetic information. They need barriers of rock, water, or sheer distance in order to keep their integrity. But move down another step, to the ecosystem level, and the need for closure is much less. For genetic diversity to be maintained, there must be a free flow of genetic information between organisms in local ecosystems.

We have got things exactly backward. We are blocking local gene flow with barbed wire, asphalt, and concrete, while making it all too easy for species from one continent to invade another and drive out the natives. In other words, we are blocking the flow between organisms in local ecosystems, but increasing the flow between large bioregions. As we carve up ecosystems that evolved to be seamless wholes, we stitch together continents that evolved to be separate.

As we noted earlier, the breakup of Pangaea is one reason for the diversity of species we have enjoyed until now. In the global hash that is the New Pangaea, there is a brief illusion of increased diversity. Soon, though, only the toughest, most weedlike species will be left—the same ones everywhere, to the degree climate allows.

Our man-made or man-allied ecosystems are blanketing the planet, leaving only pockets of earlier forms. So the species that know how to live with us, live for us, live off us, or just hitchhike with us, proliferate and have the run of the planet, while the holdouts of the old order are locked up and left to rot. But the revolution also eats its children, and many species that once shared our cockade have since gone to the dungeon or the chopping block.

Swiftly and surely, the planet is being gentrified. Soon only humans and their entourage will be able to touch the real estate. The only businesses left will be fern bars and Tex-Mex cafés. All the organisms that provide basic services, the ecological counterparts of laundries and hardware stores, will be priced out.

Much of the flow that humans encourage is not really a flow of wildness, but a flow of domesticity: that is, of our allies and hangers-on. Here is one hint that the problem we face is not merely one of global flow versus local flow. We must also consider whether we are talking about the flow of information, of matter, or of energy, and then exactly what forms of each. The flow of hydrogen, carbon, and oxygen through the atmosphere is vital to the health of the planet, but the flow of carbon from long-buried organic matter into the atmosphere may cause global warming. The trout and white pine of the Adirondacks reside in the largest wilderness in the lower forty-eight, yet they are dying, since New York State rangers cannot arrest sulfates that are released by smokestacks in Michigan.

These questions have no easy answers. The young science of ecology is scrambling to keep up with an object of study that is being scrambled and eaten up right before its eyes. The science (such as it is) of politics is just beginning to grapple with problems of global ecology. We don't know exactly what kinds of flow are good and what kinds are bad, and would have a great deal of trouble trying to police them if we did. Nevertheless, the effort must be made. If we can manage the problem of flow, we may be able to dispense with other, more intrusive means of management—means that leave wilderness deeply pitted with our handprints.

Unfortunately, we find it easier to manage wilderness than to keep our own house in order. We would sooner radio-tag and relocate elk or wolves than limit the actions of ranchers, developers, and civil engineers; sooner interfere with the laws of ecology than with those of the free market.

TWO NETWORKS

As the problem of flow comes into sharper focus, it becomes clear that we are dealing with two networks, not one. Each has its own center and sub-centers, its own vessels and subvessels. We may call them wildness and domesticity, putting the stress on the contest between humankind's allies and hangers-on and the rest of the living world. We may call them wildness and civilization, thereby giving more weight to the role of technology on the human side of the contest—though in that case we must bear in mind that the contest predates civilization in the strict sense. We may even, for convenience, call them nature and culture—but only if we take care not to fall into the sloppy rhetorical habits those terms encourage.

In a sense, the prophets of the Mountain and the priests of the Tower are both right. The Mountain is the center of the world, and so is the Tower.

How can a single entity have two centers? To a degree, the question may be an artifact of flat maps: after all, the surface of a globe has no center. But the real question is not spatial, it is functional. We have already used the image of the heart and bloodstream; now let us overlay on it, as in a deluxe encyclopedia, the image of the brain and nerves. If the Mountain is the heart, the Tower is the brain. Each is a kind of center, and each has its own network to do its bidding.

It will be objected, justly, that wild nature is not only a source of energy and a conveyor of fuels and nutrients, but a source of information. But so is the bloodstream: it courses with hormones, antibodies, enzymes of all colors and creeds. It is the metabolic Ma Bell that joins all cells, allowing the segments of DNA expressed in one cell to powwow with the different segments expressed in another. From the biologist's point of view, the circulatory system is a much more crucial conduit of information than is the nervous system, which regulates mainly our conscious and grosser unconscious behavior—coarse-grained, rudimentary stuff. Seen in this way, the brain is no more than the thermostat of a building in which tiers of the subtlest computers are housed.

The same may be said of human civilization. Even from the point of view of human survival, the informational and logistical network thrown out by our great cities—the roads, canals, cables, power transformers, gas stations, factories, grain elevators, even farms—is finally less vital than the network of wildness. (Arguably, this is true even if we include the genetic information contained in our allies.) From the point of view of the planet as a whole, the network of civilization is simply unnecessary, or even harmful, like a human brain and nervous system imposed on an oyster. When we think of civilization as the world's nervous system, we are thinking in terms of our own wants and needs.

To the idea of waves of human advance, the image of a network adds a degree of nuance by making clear that there is no border—not even a watery one—between wildness and domesticity. Instead they interpenetrate, each being shot through with veins of the other.

The arteries of nature and of culture are related in much the same way as the duck and the rabbit, or the wineglass and the kissing couple, in a field-and-ground picture. Blink, and a road becomes a wall. Our fences are highways for squirrels; our highways are walls that other animals scale at their peril.

Roadkills are a tax or tribute that nature pays to civilization. To us behind the wheel, the tax does not seem too steep; after all, we pay a similar tax ourselves in the form of traffic fatalities. But a squashed fox or a battered panther is not just a dead animal. It is, or may be, a breeding pair neutered and neutralized, a migration foiled, a population or even a species genetically lamed. For each animal that is hit, an unknown number may see and hear the rush of traffic and turn back, and the consequences of this may in the long run be just as bad.

In any case, roadkills are only the most visible tax that nature pays, and by no means the stiffest. As the flow of civilization waxes, the flow of wildness wanes. Roads bring loggers, miners, oilmen, farmers, and ranchers, together with their living and mechanical allies, into the heart of wilderness. Once a road has been built, the effects are swift and irreversible. (Sometimes they are legal: when a section of federal land in the United States has a road built into it, it ceases to be a "roadless area," and so loses its best chance to be protected as wilderness.)

Despite the mirror-image symmetry of the two networks, there is, or ought to be, a crucial asymmetry. Wildness must penetrate every cell of civilization; but there must be vast spaces of wildness that are nearly untouched by the hand of man. The reason for this is as plain as the nose on your face, and as often overlooked. Without wildness, civilization could not survive. The converse does not hold.

In part this is a question of evolutionary seniority. Human culture and biology are novelties; they have evolved in the ancient lap of planetary biology, on which they remain dependent. Without that larger shelter, all our brave devices would soon collapse.

Again, an analogy to the two major networks of the human body may be useful. If one considers the forms of metabolism and chemical signaling it makes use of, one can see that the circulatory system has perhaps a billion more years of evolution under its belt than the nervous system does. It has a corresponding priority of function.

There are tissues of the body that the nerves do not penetrate. The nervous system would upset, and be upset by, direct contact with these tissues, whose action is regulated by chemical signals in the bloodstream. But there are no tissues untouched by the blood; the closest to that status is the brain, which blood reaches only after crossing a complex and selective barrier. In the same way, wildness courses through all things, even in the fastnesses of the city—though here it is rightly filtered. But there are wild places from which civilization must be barred. Any effort to regulate these places by human design can lead only to disaster.

An occasional foray of culture into the heart of nature can be a wonderful thing, of course. Take the Sierra Club outings of the 1920s or 1930s in which a violinist played by firelight (a phenomenon captured more fully than one might imagine by the camera of Ansel Adams). Perhaps Thoreau played his flute atop Monadnock or Katahdin. But such things should be done in awe and trembling, and not often.

Or consider the Gore-Tex anoraks, Thinsulate-lined boots, and freeze-dried vacuum-packed rations that backpackers use. Here civilization invades wilderness in order to protect it. For without visitors (or the possibility of visiting) there would be little support for wilderness protection; and without backpackable provisions, visitors would have to live off the land and would soon bankrupt it. Roughly the same principle applies when ecologists insert their instruments into wilderness in order to understand how it works, and how its workings have been disturbed by human action. In either case, the invasions can be justified only if they are not too frequent.

WORLD-POLES AT HOME

We said that there must be world-poles on every scale, from the planetary on down. A property of several acres can run the gamut from unmanaged forest through woodlot, meadow, wild garden, lawn, and formal garden, right up to the window box. Even a quarter-acre lot can have its Mountain: its sacred grove, its tangle of shrubs and tall grass consecrated to the birds and rabbits.

Of course, all this is easier if your quarter-acre is joined by like-minded quarter-acres that share some wild land between them, or own it in common; or if there is public wild land nearby. When the public landscape has reservoirs of wildness, the private landscape can draw on them.

Few things are trickier, if you are a landowner, than managing the gradations and interpenetrations of wildness on your land. Few things are more rewarding when you get them right. The interval from each degree to the next can be sharp or gentle, augmented or diminished. Nor do the degrees of wildness need to arrange themselves in a straight line, like a scale, from the edge of the property to the threshold. They can twist and skip, making a kind of melody in space. Wildness and domesticity can dovetail and interpenetrate, just as they do on a planetary scale. A splash of wildness near the house can be invigorating, not least for birdwatchers; and there are few more pleasant surprises, even the hundredth time, than following a path through the woods and coming upon a secret garden.

The tricky part is not these *a priori* assignments, but the day-to-day pickles. Whether to put off the mowing until after the violets have bloomed, or the Indian paintbrush, or the Queen Anne's lace. How much weeding to do in the vegetable garden, and whether the purslane and clover are to be reckoned as weeds. Whether to uproot the thistles as soon as they rear their spiky heads, or to play the dangerous game of waiting for the moment between the flowering and the dispersal of seed. Whether to scavenge dead and fallen trees from the forest for cordwood, or let them rot and go hollow and play host to mushrooms and birds.

No wonder some people bridle at the idea of wild gardening. It seems to them that they get quite enough wildness in their gardens without trying. They want their immediate surroundings, both indoors and out, to conform to a human order. And they are not wrong. Theirs is a normal human want, though one that takes very different visible forms in different places. Wildness will not win many converts if it threatens to overrun our very homes. Wildness has its place in the home, of course—if you wanted to exclude it wholly, you would have to lock yourself out—but its place is limited. Its scope expands as we move outward from the hearth, the domestic world-pole.

The hearth is the most primal of man-made world-poles, and maybe the most begrimed with layers of irony. It is a place where trees—the most plentiful of natural world-poles—are consumed. While the Tower copies natural world-poles, the hearth eats them. The eating is done by fire: the wildest of all "species," yet the first to be tamed. In the heart of domesticity a wild thing ravens.

One might even say that the hearth is an inverted or inside-out world-pole, an antipole. Instead of a font of energy, it is a vortex or sinkhole into which stored energy disappears. From a strictly ecological point of view, the same might be said of the city. Yet from a cultural point of view the city, like the hearth, is the center from which good things radiate—a source of energy, of bounty, of life.

Nature and culture can be viewed as parallel worlds inhabiting either side of the same plane. On one side, we see energy being drawn into a vortex. On the other side, we see it bursting out, fountainlike.

Consider meat. When parts of dead animals are placed on the hearth, they shrink. Large amounts of energy and matter go up in greasy smoke. One can explain the practice in biological terms by saying that parasites are killed; that parts indigestible to humans are made digestible; that some of the work of digestion is done by fire, which like an Eskimo mother chews the meat for us. All true enough, but murky. Flip the plane over and look at the cultural world, and the picture clears up. What is brought to the hearth is raw matter; what is taken from the hearth is human food.

We love to stare into the hearth fire, and it bears staring at, for it holds in its wraithy depths the central mystery of our life. No wonder religion has its home here; no wonder humans have so often mediated their strange relation with nature by means of the hearth, which then is called an altar.

THE NEED FOR A FOCUS

If the world-pole can shatter into so many splinters, and send out so many lines of fracture—if it can be as big as the Amazon basin or as small as a thicket in my backyard—why do we need it at all? If wildness is (or ought to be) everywhere, why hypostatize it in a single place?

What is represented by the World Mountain is, in a sense, Gaia herself: the life-giving wholeness of the biosphere. It may seem odd that an enveloping wholeness should be reduced to a single point or axis. It may seem dangerous, even. As we have seen, the Mountain is easily hijacked for political ends. Shorn of wilderness or torn from it, the conical icon becomes a temple-crowned hillock, or a ziggurat, or a city. Even where, as often in America, the Mountain remains a wilderness, it is all too often kept like a trophy on the wall. Unable to circulate its wildness, it is soon as useless as a mounted moosehead, and as dead.

Despite its pitfalls, the idea of the world-pole is an idea we seem to need. Nearly every culture has it, in one form or another. What it expresses, perhaps, is the need for a focus.

You can never feel quite so cozy in a house with central heating as you can in front of a fireplace or a wood stove. You want a focus of warmth (*focus* being the Latin for "hearth"). Sunlight is the best kind of warmth there is, because the sun is the ultimate focus—the hearth of the solar system, the pole of nine worlds. What goes for warmth goes as well for other good things, such as water and food. The feeling that good things come from a single source is imbibed with our mother's milk.

To look at, or point at, or bow to the biosphere is not easy. What makes the World Mountain a good focus of our Gaian attentions is that it is the pipe through which the stuffs of the atmosphere, of the terrestrial surface, and of the soil (as well as the oceans) are pumped and recirculated. When it ceases to be a pipe, a pump, a spigot, the World Mountain might as well be made of dry wobble. It is then no worthier of our reverence than a wooden idol.

AS THE POSTMODERNISTS never tire of telling us, wilderness is a myth. What they fail to tell us, because they do not comprehend it, is that it is a necessary myth—necessary because, on a biological level that mutely resists deconstruction, it is deeply and urgently real.

True: few wildernesses are certifiably pure. True: all wilderness has a history, in which humans have generally played a part. True: the idea of wilderness has been used as an excuse for elitism, imperialism, and sheer complacency. But the trendy debunking of wilderness may breed even greater mischief. Advanced in the name of the people, it had been seized upon by corporate and political elements whose only interest in people or nature is to squeeze them dry. Even in the best hands, it leads us toward a slippery slope whose final declension we cannot measure.

Wilderness is a social construction. So is the guardrail at the edge of a precipice: and I would not gladly see either dismantled.

11

STORMING

THE MOUNTAIN

THE LAST TWO chapters adopted the point of view of those peoples who look to wilderness as the heart of the world. Tinged with reverence, gratitude, and regret, it is a view from which a number of useful ecological lessons can be drawn, and I have tried to draw some of them. Now, having looked up at the Mountain, let us try looking down from the Tower. How does it feel to stand at the center of the universe? How does it feel to know that the work of your own hands, and other hands like them, is the source of all goodness?

Not much imagination will be required to find out. By and large, the Tower remains our mental home. Yet there is much to be learned by looking through the eyes of the people who first built and inhabited it, for they felt keenly things that we take for granted. The thrill of civilization's successes was fresh to them; so was the bewilderment of its failures. Most bewildering, perhaps, was that sort of failure which seems as closely wedded to success as a bronze casting to its mold, as is often the case where nature is concerned.

IF WE THINK about our own lives, we may not be surprised that the view from the Tower is not entirely triumphal. Many of the doubts that trouble us now were first worried like hen's tracks in the clay tablets of Mesopotamia. Even the legend of Gilgamesh, the greatest of Mesopotamian heroes and the sire of half the heroes in Western literature, is pitted with the ironies of human designs gone askew. Gilgamesh is the first of many heroes who chafe at the bounds that nature imposes, and defy them with temporary success. In his person, the ancient scribes shrewdly anatomized the needs that drive men to conquer nature.

One way of going back to Eden is to take the place by storm: to fight fiery sword with fire and ax. In Gilgamesh's conquest of the Cedar Forest—"the Land of the Living"—we find the paradigm of all such attempts.

If the Bible is the best-selling book in our world, the best-seller in the Bible's world was the Epic of Gilgamesh. Although the epic was probably composed in Akkadian by a poet of the Old Babylonian period (circa 2000–1600 B.C.), much of its raw material can be found in Sumerian poems that are centuries older.

Gilgamesh, the Sumerian King List informs us, was fifth ruler in the First Dynasty of Uruk and reigned from 7981 to 7855 B.C.—that is, for 126 years. (He was a toddler next to the antediluvian kings, one of whom reigned for thirty-six thousand years, some quarter-million years ago. As in the Bible, the first ancestors tended to get a bit long in the tooth.) Modern historians do not doubt his historicity but place his reign about 2600 B.C. Although we are told that his mother was a goddess and that "two thirds of him is god, one third of him is human," this famous god-king is most famous for his mortality.

The name of the poem in Akkadian is *Sa nagba imuru* (He Who Saw Everything). It is taken from the opening lines—

> He who saw everything [to the end]s of the land,
> [Who all thing]s experienced, [conside]red all!

This passage is a retrospect of his career, yet it may describe, as well, an almost bodily sense of being coextensive with the world. At the outset of the story, Gilgamesh is like *Adam Kadmon*, the macrocosmic primal man of kabbalistic lore, whose body comprises all being. He is like Adam himself, who before his Fall possessed a holy light by which—the Talmud says—he could see from one end of the universe to the other (perhaps a way of saying that for Adam, since time did not exist, neither did space). Less gloriously, we might say that he is like an infant—or an animal.

A mystic can encompass the world without budging from his cushion. A shaman does not need to move around a lot horizontally; wherever he is, is the world-pole, by which he moves vertically from one world to another. (He may engage in spirit travel, often in animal form, but does so usually on specific errands, not as a tourist.) Gilgamesh may be such a shaman, may have that primal sense of oneness with the universe whose loss humans in all times and places bewail.

The poem goes on to marvel at the walls of Uruk, which Gilgamesh built. (Their massive stumps can be seen to this day.) At once a contradiction arises. The hero encompasses the world, yet encloses a small part of it in a defensive membrane. The wall is a parable of the self. If we felt no wall, and no need for a wall, between the self and the outside world, we would not fear death. Gilgamesh is invulnerable, a wild ox that romps the world at will—or does he?

The third millennium, the age when heroes such as Gilgamesh roamed Mesopotamia, was also the age when walls went up around the cities to protect the people from the heroes and the heroes from each other. The price of the hero's freedom is confinement. Gilgamesh's very wildness makes him an architect of the Tower.

THE KING AND THE WILD MAN

To have a king who is two-thirds god has its drawbacks. The town is too small for Gilgamesh. Once he has built Uruk's walls, he bounces around inside them like a billiard ball. Brawling and wenching with divine appetite, he does not "leave the young man to his father" or "the maid to her mother." The first clause may refer to some kind of corvée or forced labor, perhaps on the walls of Uruk, the second to a *jus primae noctis* or "right of the first night." The Old Babylonian version says, "He cohabits with the betrothed bride—he first, the husband afterwards."

The gods are moved by the people's outcry and decide to make a chum for Gilgamesh, a "double" who will absorb his energy and keep his hands full. Since no ordinary mortal can stand up to Gilgamesh, a special creation is called for. Enkidu is shaped from clay and blood by the mother goddess, just as the first humans were in Mesopotamian myth. He is called a *lullu*, or "savage," just as the first humans were. But whereas the first humans were instantly set to work on the canals and in the fields, Enkidu is set down among the wild beasts of the steppe. Shaggy like one of them, he runs with them, eats grass with them, drinks at their watering holes. The double of the man who is two-thirds god seems to be two-thirds beast.

A trapper, incensed that this strange wild man fills in his pits and destroys his snares, complains to the king. Unaware that the gods have sent Enkidu to snare him, Gilgamesh sends a hooker to snare Enkidu—to "make a man of him," as it were. She succeeds.

> After he had had (his) fill of her charms,
> He set his face toward his wild beasts.
> On seeing him, Enkidu, the gazelles ran off,
> The wild beasts of the steppe drew away from his body.
> Startled was Enkidu, as his body became taut,
> His knees were motionless—for his wild beasts had gone.

Enkidu had to slacken his pace—it was not as before;
But now he had [wi]sdom, [br]oader understanding.

Even today it would be hard to find a more wrenching image of man's estrangement from nature.

Is this a fall from Eden? From the Mesopotamian standpoint it is an ascent. Enkidu's state before his "fall" is the same one the Sumerians ascribed to the gods and (briefly) mankind before the advent of the goddess of grain, Ashnan, and the goddess of cattle, Lahar. They did not consider it a happy state. Officially, the Akkadian poet of the epic does not either. Yet his sketch of the wild man is (despite some ribbing) sympathetic and even a touch envious. Maybe he has been exposed to the West Semitic idea that will come to be called Eden. Or maybe he has found life as a court poet cramped and strained and has felt the tug toward "the simple life" that city-dwellers will feel in later times.

To us moderns, for whom (as Weber said) sex is nature's last gasp, it seems bizarre that sex should tear a man from fellowship with the animals. Yet by some lights that is what happens in Genesis. Nor is it utter nonsense from a biological point of view. What really separates members of one species from members of all others is the need to identify mates with whom viable and fertile offspring can be produced. Without a "species-specific mate recognition signal," no species can long survive. Sex not only splits nature down the middle, but slices and dices it into a million bits. As well, there is the male feeling (also adumbrated in Genesis) that women and children pull us away from our carefree, boyish, roving life in the woods and plains and force us to settle down, to be civilized, sissified, citified.

The harlot leads Enkidu to the city, where he confronts Gilgamesh in mid-carouse. In a titanic scuffle Gilgamesh gets the better of his double, but just barely, and the two become fast friends. It is at this point that Gilgamesh hatches his plan to conquer the Cedar Forest, a remote mountain realm guarded by a fearsome monster known as Humbaba or Huwawa ("Whoof-Whoof," to borrow Gaster's rendering). That was part of the gods' plan: Enkidu's presence would distract Gilgamesh from his wenching and launch him on ventures outside the city walls. The particular adventure he fixes on may not be part of the gods' plan, but neither is it chosen at random.

One explanation might go like this: Having met and (with some effort) vanquished the wild man, Gilgamesh is emboldened to pit himself against wild nature in a bigger shape. And the wild man will be his ally: Sherpa to his Hillary, Chingachgook to his Natty Bumppo. In fact Enkidu can serve as a guide, for in his wanderings with the beasts he has visited the Cedar Forest. More to the point, the wildness of Enkidu is like an inoculation against the wildness of Huwawa.

But this reading only skims the surface. The meeting of Gilgamesh and Enkidu is fateful not because they are opposites but because they are the same. Enkidu is said to be the king's "image" or double. Enkidu is a natural man until he meets Gilgamesh's agent, the whore. What is less obvious is that Gilgamesh is a natural man until he meets Enkidu. He is no Hamlet pinned like a butterfly by doubt (not yet, anyway), no fat epicene Ptolemy in a see-through gown. A city boy can be a savage, too. Even a city-boy king. Especially a king. It's good to be the king, as Mel Brooks has said, for you can eat, drink, fuck, and piss behind hedges to your heart's content. Bursting with primal energies and appetites, Gilgamesh is in a position to satisfy them.

Seeing Enkidu, Gilgamesh gazes in the mirror for the first time. His first instinct is an animal's: like a robin pecking at a car's wing mirror, he spars with his own reflection. When at last he finds that he can subdue Enkidu— in other words, that he can subdue himself—he becomes self-conscious. To put it the other way around, he prevails as soon as he sees that his glassy rival is himself. His primal instincts are reined in. You might say that he sees himself from the outside and becomes concerned with his "image," in the sociological or public-relations sense—or, as he would say, with his "name." But it goes much deeper than that. As he becomes self-conscious, he becomes human: aware of his mortality, his finitude in time and space. Before, he seemed to hold the universe in his broad chest. Now he sees that he has an outline, an edge, a limit—which only makes him more determined to swallow the universe whole.

Mastering Enkidu, the king masters his own nature, or nature in himself. And is intoxicated. Mastering nature when you are part of nature is a thrill, but a beast's thrill——an wild ox's, a lion's. Mastering nature when you stand outside nature is a thrill fit for a god.

RANCOR AGAINST NATURE

In the Sumerian story from which the main ingredients of this part of the epic are drawn, Gilgamesh explains to the sun god Utu his desire to enter the "Land of the Living" (as the Cedar Forest is called) in these words:

> I peered over the wall,
> Saw the dead bodies . . . floating on the river;

As for me, I too will be served thus; verily 'tis so.
Man, the tallest, cannot stretch to heaven,
Man, the widest, cannot cover the earth.

Neither Gilgamesh nor his extension, the walls of Uruk, can reach heaven. No man-made tower can. Peering in the mirror, we said, Gilgamesh sees that he has an outline, a limit. Seeing Enkidu, he sees that he is not unique. Although his mother is a goddess, he is finally a man like other men, bigger and stronger but not different in kind. And so he will die like other men. We are reminded of a passage from Psalm 82: "I said that you are gods, and all of you are sons of the Most High. Nevertheless you shall die like men, and fall like one of the princes." If the dying Enkidu will finally drive the point home, the living Enkidu has already given an inkling of it.

It is the knowledge of death that makes Gilgamesh want to make a name for himself by cutting down the cedars. Yet he might have chosen any number of other exploits. Why turn against nature?

Knowing that it is our fate to die—and, what is worse, that it is our fate to know that we will die—makes us furious at nature. Partly because nature seems to be the cause of our death (that it is also the cause of our life hardly mollifies us). But mostly because the rest of nature doesn't seem to have this knowledge, or does not seem to suffer from it. So we attack nature—partly in hopes of loosening its death grip on us, partly out of envy and spite.

The cedar, like other evergreens, is an emblem of eternal life. In the Psalms the cedar of Lebanon (probably the tree referred to in the epic, if not in the Sumerian tales) is a token of pride, strength, beauty, and permanence. What better way to vent one's anger at nature—as its endless self-renewal, its placid indifference to death—than by felling cedars?

There are, of course, less poetic reasons for felling cedars. Throughout the ancient Near East, cedar was the wood of choice for roof beams, lintels, and other structural parts of any sizable building, from Pharaoh's palace to Solomon's temple. Strong, long, ramrod-straight, and resistant to rot, it was hard to do without in an age when the arch was in its infancy. The cedar of Lebanon (*Cedrus libani*) was the genuine article, but the Sumerians, who had not yet come across it, prized such kindred species as the *Juniperus excelsa*, to which the Sumerian term *gish erin* probably refers. In the early years of Mesopotamian settlement these and other useful trees covered the nearby foothills of the Zagros range, and there were scrub forests in the valley itself. But by the time of Gilgamesh the need for timber, fuel (especially as kiln-baked brick replaced sun-baked brick), and grazing land had pushed the forest edge upward into the mountains proper. As so often happens, a man-drawn tree line crept up the mountainsides toward the true tree line. No tree could survive below the one line or above the other: the forests were squeezed in between. Then the nearer peaks were denuded and became obstacles on the way to the next. A logging expedition became the stuff of legend. This was all the more true when the Zagros range was picked clean and timber had to be sought farther afield, in Lebanon and Syria. A monument erected by Yahdun-Lim of Mari boasts that he "entered the great mountains, the Cedar Mountain and the Boxwood Mountain, and felled such trees as boxwood, cedar, cypress, and *elammakku*-trees."

If a logging expedition was the stuff of the Gilgamesh legend, it has been fairly well disguised—though less so in the Sumerian version, where the king brings with him fifty men from Uruk who saw up and bundle the fallen *erin* trees. In Sumerian the place is called either "The Land of the Living" or "The Land of the Cut-Down *Erin* Trees," and the latter name, together with the fact that the hero crosses seven mountains before he gets to the one he wants, has been taken to mean that his project is to glean the last few big trees in a region already logged over. But that seems less heroic than buzzardly. More likely "Land of the Cut-Down *Erin* Trees" is the name awarded the place after Gilgamesh's feat, and used sloppily by the poet. Be that as it may, even a buzzardly logging expedition to the Zagros range might have met with resistance from the people of Elam, in what is now southern Iran. No doubt timber was one fuel of the warfare that smoldered for centuries between Elam and Sumer. Since Huwawa has been identified with the chief god of Elam—one people's god being another's monster—his subduing by Gilgamesh may refer to a military success. Or it may refer to conquest of the wildness over which this mountain god presided.

As they stripped the nearby hills of timber, so the Mesopotamians stripped wild nature of its godhead. Any World Mountain worth its salt must be invulnerable. The Cedar Mountain had that reputation, but Gilgamesh debunked it. He unseated Huwawa, who resembles in some ways the mountain-dwelling storm gods of other peoples: not only the god of Elam, where the Sumerians seem to have located him, but the gods of Syria and Palestine, where later poets seem to have moved him. In a fragment of the epic the names of Wer and Addad, both of them mountain gods with Syrian cults, are invoked in the same breath as Huwawa's. Huwawa resembles Baal, Teshub, even Yhwh himself.

> E[nkidu] killed [the watchman] of the forest,
> At whose word Saria and Lebanon [trembled].

It is no great leap from this to Psalm 29:

> The voice of the Lord breaketh the cedars;
> Yea, the Lord breaketh in pieces the cedars of Lebanon.
> He maketh them also to skip like a calf;
> Lebanon and Sirion like a wild-ox.

Ordinarily it is the prerogative of such a god to "strip the forests bare," but now Gilgamesh and Enkidu have done it themselves.

If in early versions of the story Huwawa represents the storm god of Elam, it is reasonable to think that in later versions he represents the storm gods of Canaan. (The windy kinship of "Huwawa" and "Yhwh" is surely pure chance, though, unless both began as imitations of a storm god's voice.) And his humiliation is given a nice fillip by the fact that it is done by a storm. When Huwawa proves too tough for Gilgamesh, the sun god Shamash sends eight mighty winds which blind and bind the monster for the hero's convenience. It is as if Tom Mix were caught and hogtied with his own lariat.

Beaten (though not fair and square) by Gilgamesh, Huwawa offers to provide a sustained yield of timber if the hero will spare his life. It may be the Elamites or Canaanites who are making the offer. One can see the poor monster's posture in a relief from the great wall of Karnak, one of a series advertising the campaigns of Seti I (reigned 1318–1301) in Palestine. In this image the chiefs of Lebanon are cutting down cedars for the "great barque upon the river ... as well as the great flagpoles of Amon," while four of them kneel in supplication before the pharaoh. The one act seems to flow into the other, as if composing a single gesture.

Perhaps the Canaanites at first regarded these forests as holy but then, finding that they were being taken by force, tried to manage and market the wood themselves, on the theory that careful prostitution was preferable to being raped. But even this was not enough for the haughty lowlanders. The Egyptians and Mesopotamians wanted all the wood and they wanted it now. Thus Gilgamesh and Enkidu reject Huwawa's offer out of hand by cutting off his head.

NATURE'S OFFER SPURNED

But Huwawa's offer and its rejection have a meaning deeper than petty geopolitics. This is a turning point for the Western mind. Nature offers its services: flood control, pest control, climate control, gene banking, resources renewing themselves without end. Of all these boons there is no better symbol or dispenser (at least in the visible realm) than the forest. In brief: energy, information, and the two combined in ten thousand kinds of work done for us absolutely free of charge—"on the house," we might say, with a nod to the root of "ecology."

An offer we couldn't refuse, you might think. And what is our answer?

Off with its head. Nature is decapitated, its energy cut off from the information that once directed it. We are left with a thrashing trunk, dumber than an ox and more sullen than Caliban. And off in a corner the head blinks its code, which no one reads.

Noteworthy is the fact that it is Enkidu who is eagerest to finish the monster off. No zealot like a convert—and no pillager of nature like the "natural man" who has had a taste of Western technology and lifestyle. From ancient times to the present, one of the basic tools of imperialism has been to make subject peoples its accomplices in the rape of their homelands. With any luck, they will be the ones who suffer the consequences. In the epic, indeed, it is Enkidu whose death is brought about by the train of events that the conquest of the Cedar Forest sets in motion.

Yet here, as elsewhere in the story, Gilgamesh and Enkidu may be seen as two sides of the same personality. When we reject nature's willing services in order to take her by force, it is not the civilized man in us that insists; it is the primitive. The truly civilized man would act rationally, and would accept nature's gifts. But the denatured savage, to salve the pain of his loss, savages what he has lost.

Now it might be objected that man, even the notorious subspecies known as Western man, has not slain nature outright but, rather, has bent it to his will. He has even made some use of its intelligence, as that is incorporated in plants and animals, both domesticated and wild. But he has made use mainly of the intelligence of organisms, not that of whole ecosystems. In general, the use he has made of nature's intelligence is so much less than it might be that the image of decapitation is very apt. Of course it also mimics the act of the lumberjack, except that what is left below is a trunk rather than a stump.

This is not to say that our ideal relation to nature is what Gilgamesh's might have been to Huwawa, had he suffered him to live: that of master to cringing, fawning servant; or of master to sullen, resentful servant, plotting our overthrow (like Caliban). That is more what we have now. We feel more and more these days as if nature were plotting against us, were indeed in open rebellion—"fighting back," as the newspaper headlines say.

We are emboldened to read the story in these seemingly modern terms, rather than as a chronicle of ancient wars, when we see that it is not only alien gods that are mocked on the Cedar Mountain. Apart from the fact that the authors of the epic were writing in a Semitic language and might have had some scrap of awe left for an ancestral pantheon, we note that Gilgamesh "opened up the secret dwelling of the Anunnaki," the rank and file of Akkadian gods. All of this is in keeping with the hint of impiety that can be heard—though the tone is admiring—in the fifth line of the epic: "The [hi]dden he saw, [laid bare] the undisclosed."

If we listen only to its brassier tones, the story of Gilgamesh and the Cedar Forest seems to trumpet the victory of the Tower over the Mountain. It seems a mirror image of the Towel of Babel. As that story gave the hill people's view of the valley people's world-pole, so this story gives the valley people's view of the hill people's world-pole. In each case, the alien world-pole is humbled.

Here Gilgamesh, builder of cities, seeks out the last and most awesome redoubt of wildness, and lays it low. And why not? If the city is the center of the world, what is to stop its expansion, or its plunder? The rest of the world is rightly subject to its rule.

If we listen closely, though, we hear whispers of misgiving. It is not only greed—the poet hints—or hunger, or the lust for power, or even our old grudge against the Holocene forest that makes us want to strip and plunder the wildest place in our ken. Even if we reject its claim to the title of world-pole, the Mountain grips us. Deep down, we know that this is the Land of the Living. We want to seize its secret, or crush it to death.

In brief: we want to go back to Eden, even if that means destroying Eden. Since Enkidu leads the way—and insists on the *coup de grâce*—it is as if Adam should return to the Garden, fell the Trees of Knowledge and Life, and lop off God's head.

Does the doer of such a deed get off scot-free? Not quite. Let us fast-forward through the rest of the epic. The conquest of the Cedar Forest seems to inspire the lust of the goddess Ishtar. When Gilgamesh spurns her advances, she sends the monstrous Bull of Heaven down to earth as her avenger—only to see it slain by the two heroes. Ishtar demands that one of them pay for this crime with his life, and the gods pick Enkidu. Seeing the death of his double, Gilgamesh knows that he, too, must die. The conquest of the Cedar Mountain has not made him immortal, and he casts about for some other remedy. His quest takes him to Dilmun, a holy and remote place where Utnapishtim, the Mesopotamian Noah, has lived since the Flood, enjoying eternal life. Here Gilgamesh acquires an herb that is advertised to grant eternal youth. But while he is taking a swim, the herb is stolen by—of all things—a snake. At this point, Gilgamesh gives up. He goes back to Uruk, resolved to eat, drink, be merry with his family, and build up the walls of his city—walls whose praise ends the poem as it began, enclosing it like a wall.

Gilgamesh is back where he started. What has he learned? What have we learned?

That immortality belongs to nature, and cannot be taken from her by force. That, by destroying nature, we express a part of our nature: but kill a part of our nature, too.

A GODDESS SCORNED

The Mesopotamians did not lose much sleep over the fate of the highlands they conquered and stripped. If the nearest mountain range was bare, they moved on to the next. At some point stiffer armed resistance, or sheer distance, would end the process; at some point there would be no cedars to speak of in the whole Near East, and palaces would be cramped because the roof beams were poplar. But that day was far off. As for the ecological effects of deforestation, those were easy to ignore.

The people of the great valley did feel some anxiety, though, about what they were doing to their own soil. On some darkling level they sensed that their greatest triumph, irrigation, was bringing about their greatest disaster, salinization. Their history ought to make us think twice about the triumphs of our own scientific agriculture.

Disguised as a slapstick sex farce, a Sumerian poem called "Enki and Ninhursag" can be seen, on closer reading, to grapple with these matters. Enki, the water god, is a beloved figure in the Sumerian pantheon, at once a trickster and a sugar daddy to humankind: a bringer of technology. Ninhursag is the mother goddess, hoary and awful. Ninhursag asks Enki to "bring sweet water from the earth" to water her city, Dilmun: a place that some have called the Sumerian paradise. (It is here that Gilgamesh found Utnapishtim, the ageless survivor of the Flood.) Enki complies by masturbating, flooding with semen the dikes and marshes. In the process he impregnates Ninhursag, who nine days later gives birth to the goddess Ninmu. Soon Enki finds his daughter Ninmu walking by the banks of a river or canal, and lies with her; she gives birth to the goddess Ninkurru. Ninkurru then takes a similar walk, with similar results, and gives birth to the goddess Uttu.

By now Ninhursag has caught on. She warns Uttu against her spry grandfather and locks her up indoors. Enki manages to win her favors anyway with a gift of grapes, cucumbers, and apples, procured from a gardener whose arid land he has watered. But Ninhursag takes his semen from the girl and places it inside herself—that is, in the earth. The upshot is the sprouting of eight plants, among them the "tree plant," the "honey plant," the roadweed, the caper, the cassia, and the thorn—all of them, it seems, creatures of dry places. Enki, wishing to know these plants and decree their fates (which is after all his job), immediately eats them all up. Apparently this was not what Ninhursag had in mind, for she curses him and he becomes deathly ill—perhaps because these plants are actually his own off-spring. More likely his illness has a natural cause: he is pregnant with his own offspring but, being a male, cannot give birth. But the air god Enlil, with the help of a fox, somehow gets Ninhursag to relent. She seats Enki in her vulva and gives birth, on his behalf, to the eight divinities with which he is so painfully swollen. One of these is the god Enshag, who becomes "the lord of Dilmun" and thereby gives the story a welcome closure.

TIAMAT'S REVENGE

What on earth are we to make of this story? Before we can make anything of it, we must have some sense of the role of water, both sweet and salt, in Mesopotamian myth and life.

Water is necessary in Dilmun not only to make the land fertile but also to make it (as the poem says) "clean," "most bright." Cleanliness is hard to reconcile, by our lights, with a flood of jism. Nor do we naturally think of semen as sweet water. But in Babylonian cosmogony, the two primal sub-stances from which all things arise are salt water, personified by the goddess Tiamat, and sweet water, personified by the god Apsu. It is by slaying the monstrous Tiamat that the young god Marduk establishes order on earth. Fresh water is often thought of as male, since it seems to impregnate the earth. In the poem known as "Enki and the World Order," in which that randy but kindly god hands out blessings to the peoples of the world, the author tells us that

> He stood up proudly like a rampant bull,
> He lifts the penis, ejaculates,
> Fills the Tigris with sparkling water.

Similarly, a hymn to Ningirsu, the god of the yearly flood, speaks of "semi-nal waters reddened in the deflowering"—reddened, that is, with mountain soil flushed out by the spring rains.

In the very beginning, the fresh and salt waters "commingled as a single body." Their separation (a trick that scientists are still hard put to perform on a large scale) is what made all further creation possible. And their remixing could make certain kinds of creation impossible, most notably farming.

Two great oceans, sweet and salt, ran beneath the earth. In the natural order of things, the salt ocean would emerge only offshore, while the sweet ocean would run through the heavens and through earthly rivers and springs. But at times this pretzeled grid of plumbing (the texts speak of the "water pipes" of the ocean) would malfunction.

It was not always an accident. When the gods, annoyed by the racket of humankind, decided to destroy them, one of their first expedients, before they thought of a flood, was salt. "During the nights the fields turned white. The broad plain brought forth salt crystals, so that no plant came forth, no grain sprouted."

But it did not take divine pique to do this—man's improvidence sufficed. In arid climates, the groundwater is often brackish. As long as it stays below the level to which the roots of crops penetrate, it is no problem; but when a field has been used for a while, irrigation without proper drainage can raise the water table. Crops can filter out some of the salt when they drink, but in so doing they make the remaining water that much saltier. Eventually it catches up with them. Falling yields are the first sign of trouble. Then, when the saline groundwater has nudged within a few feet of the surface, capillary action starts to lift it the rest of the way. At the surface the water evaporates and the salt "blossoms out in mockingly beautiful floral patterns." The Mesopotamian idea that a spiteful Tiamat was rising up was not far wrong.

Having a general sense of what the problem was, farmers would fallow a field or else try to flush out the salts with more water. Fallowing might work for a while, because salt-tolerant weeds would move in, suck up some water, and lower the water table. Flushing would work very briefly and then backfire, by raising the water table and with it the salt.

We can see here an ongoing struggle to keep the female principle submerged. At least, to keep its barren half submerged: for apparently the female principle is not wholly life-giving. There is a barren, even a death-dealing substrate that must be kept down. The common idea (common, at any rate, in patriarchal cultures) that the goddess can be monstrous, can swallow us as readily as she gave us birth, is here given an agronomic slant. The very fertility of the earth may be tainted. Salt lurks just beneath the surface.

Among the many favors that pots do for archeologists is to retain, under certain conditions, marks of the kernels of grain that were stored in them.

Pots dug up in southern Mesopotamia suggest that from 3500 B.C. onward the ratio of wheat to barley in the harvest steadily shrank. By 1700 B.C. no wheat was grown at all. In an age before advertising, tastes in food were very stable, so, barring a major rise in beer consumption, it seems likely that the shift to barley was a matter of necessity rather than choice. One factor may have been the growing importance of wool textiles for export, which meant that more barley had to be grown for fodder. But the evidence points to a necessity rather more dire.

As befits a poor cousin, barley is on the whole less finicky than wheat about where and how it grows. In particular, it is far more tolerant of salt. Given the crust of salt that covers so much of Mesopotamia today, and the half-comprehending references to the problem that can be found in ancient texts, it has been suggested that salt must have caused failures of the wheat crop, forcing a shift to barley. Estate records from the period show a steady decline in yields of wheat and a lesser but vexing drop in yields of barley. As the soil turned to salt, the economic base of Sumer fissured and slowly crumbled. This helped make it vulnerable to the growing power of its northern neighbor, Akkad.

Over the course of millennia, the center of power and population crept surely if unsteadily upriver: from Sumer to Akkad and Babylon (with a southward detour to Ur), from Babylon to Nineveh. There were many reasons for this, such as the growing importance of the Mediterranean trade routes as against those of the Persian Gulf; the periodic incursions of warlike seminomads from the steppe; changes in climate; and fluctuations in the flow of the Euphrates. The last two factors might in turn have depended on the deforestation of neighboring mountains, in which case they might be thought of as Huwawa's revenge.

Yet it may be that the main reason was salt. The lower-lying regions, with their higher water table, were more susceptible to begin with, and they bore the brunt of long settlement. What we have, then, is the familiar swelling circle of human-led change squeezed into an oblong by the shape of the great river valley. Only this time the advancing edge is not the red of fire or the reddish brown of eroding soil, but a glistening white: the white of snow, of bleached bones, of a second and maybe permanent virginity.

The edge is still advancing. As so often happens in natural history, a pattern in time can also be found in space. In modern times, farmers north of Baghdad grow nine times as much wheat as barley, while farmers near the Gulf grow nine times as much barley as wheat. For both crops, yields are far lower than they were in ancient times.

"Enki and Ninhursag" can be seen as a cautionary tale about the limits of unnatural fertility. The story begins with a great act of irrigation, which makes Dilmun (as it makes Sumer) great. There is no question of the rightness of this act: earth herself has cried out for water. What follows, though, becomes more and more problematical. The begetting of daughter upon daughter might be the branching of a big canal into smaller and smaller ones. (Enki's couch, after all, is the banks of a river or canal.) Or it might be the successive flooding of the same field. There may be a sense of sharply diminishing returns as the system of canals snakes into the remote desert, becoming harder to control and more prone to silting. There may be a sense of something unnatural and monstrous, like incest, as a trusted field bears salt-flowers instead of wheat or barley. Either way, something has gone wrong.

The story of Enki and Ninhursag seems to assert the power of the Great Mother. Enki is punished for his presumption in thinking that his semen—that is, water—can do anything it wants, regardless of whether the earth is willing. He has made his penis the world-pole. His incest-upon-incest becomes, in the end, a kind of onanism, and threatens to be just as sterile. But the goddess's triumph is short-lived. Her indisposing of Enki seems to be her last hurrah. In relenting, she shows a failure of will that foretells her decline. The creature that makes her relent is the fox, the symbol of craft—and hence, we might guess, of technology.

All sex has some tincture of onanism, for in one way or another the imagination must kick in. Similarly, the farmer must imagine an earth more pliant and nurturing than the one in which he thrusts his hoe. And imagining is just the first step. The farmer—like the gardener, the architect, the engineer—goes on to fashion the nature of his dreams, bring her to life, and enter her. The Greek myth of Pygmalion is not about man and woman only, but about the remaking of nature.

Nature is game for anything. She will let herself be incarnated in shapes of the most deadly and unnatural perfection—for a time.

12

THE HIGHWAYS
OF ROME

BETWEEN THE MOUNTAIN and the Tower our culture shuttles
like a cable car. If we slide more often toward the Tower, both ideas
retain a strong attraction. In the last few chapters we have tried to
understand the ecological settings in which the two myths arose, and to
unearth the ecological lessons that have been buried in them for the past
four thousand years. Those lessons would be useful to us even if our civilization had crawled out from under a rock sometime last week. But the
ideas of the Mountain and the Tower have even greater meaning for us
because we have a history, of which they are a part.

How have these ideas come down to us? How have they jockeyed for
position—first one and then the other ascendant, like Jacob and Esau scuf-
fling in the womb? What other ideas have affected the contest? Why has the
Tower come out on top?

The first place to look for answers to these questions is the Hebrew Bible,
the main route by which the ideas of the ancient Near East have found their
way to the modern West. To understand the ascendancy of the Tower,
though, we shall have to look also at the role of Greece and Rome, who were
equal partners with Israel in shaping the Christian West.

A HYBRID OF MOUNTAIN AND TOWER

In setting out the contrast between the Canaanites and the Mesopotamians, we have so far placed Israel squarely in the Canaanite camp.
Roughly, that is right. But the crucial role of the Hebrew Bible in shaping
the Western mind requires us to be more exact.

Israel was not sheltered from the metaphysical winds that blew from the
great river valleys. The way of thinking of the Canaanite hillbilly, goatherd,
and dirt farmer was altered by Egyptian and Mesopotamian influences.
The wild Mountain of God was partly tamed and was crowned with a

temple. The result was a vigorous hybrid of Mountain and Tower known as Zion.

The name Zion referred at first to a hill in the southeastern part of Jerusalem, between the Tyropoean and Kidron valleys, on which stood the Jebusite citadel that David seized and made his capital. In the Bible and Talmud the name was used of the Temple Mount, or else of Jerusalem as a whole; nowadays it is wrongly pinned on a hill in the southwestern part of the city. Whichever hill one picks, the fact remains that from the geographer's standpoint Mount Zion is no great shakes. At something less than twenty-six hundred feet above sea level, it is overshadowed even by some of its Judean neighbors. But for the Hebrews it is the navel of the earth.

> Great is the Lord, and highly to be praised,
> In the city of our God, His holy mountain.
> Beautiful in height, the joy of the whole earth;
> Even Mount Zion, the reaches of Zaphon,
> The city of the great King.
> God in her palaces
> Hath made Himself known for a stronghold.

The phrase *yarketei tsaphon*, usually rendered "on the sides of the north" or "the uttermost parts of the north," would be read by an ancient Israelite as a reference to Mount Zaphon, the mountain from which Baal held sway. It is a bald admission of the psalmist's project: to make Mount Zion the new cosmic mountain.

Later Ezekiel, an exile in Babylonia, has a vision of the rebuilt Temple. From its east gate runs a trickle of water which, some four thousand cubits downstream, has become an unswimmable torrent with a gigantic tree (or perhaps a large forest) on either side. His guide, a man "with the appearance of brass," explains:

> "These waters issue forth toward the eastern region, and shall go down into the Arabah; and when they shall enter into the sea, into the sea of the putrid waters, the waters shall be healed. And it shall come to pass, that every living creature wherewith it swarmeth, whithersoever the rivers shall come, shall live; and there shall be a very great multitude of fish. . . . And by the river upon the bank thereof, on this side and on that side, shall grow every tree for food, whose leaf shall not wither, neither shall the fruit thereof fail; it shall bring forth new fruit every month, because the waters thereof issue out of the

sanctuary; and the fruit thereof shall be for food, and the leaf thereof for healing."

There was, in fact, a spring on the eastern side of Mount Zion, known as the Gihon. While it was neither unswimmably deep nor swarming with fish, it did supply most of Jerusalem's water. And it was holy. It was here that Solomon, at David's behest, was taken to be anointed as king.

Gihon is also the name of one of the four rivers of Eden. Although scholarly opinion has only recently begun to admit it, it is clear that the two Gihons are one and the same. In the Book of Genesis the author known as J, who is thought to have lived around 800 B.C. under the reign of a descendant of David, links Eden not only to the World Mountain of the north from which the great rivers flow—the Tigris and the Euphrates—but to the new World Mountain, Zion. For the sake of symmetry, and in order to slice the world into four quarters (a plan favored in the Near East as elsewhere), a fourth river is added, the Pishon. "Gihon" means something like "Gusher," "Pishon" something like "Bubbler." After the temple has in fact been rebuilt, Zechariah is emboldened to add an extra stream to Mount Zion itself: "On that day living waters shall issue from Jerusalem, half toward the eastern sea and half toward the western sea, both in summer and in winter." Besides making sure that all of the land of Israel, not just the eastern half, gets a share of the life-giving waters, this mirror trick puts the Mountain of Yhwh on a par with the Mountain of El as a "source of the Two Rivers."

The rabbis make Yhwh's Temple, like El's tent, the springhouse of the world, in which he controls and doles out the primal waters. They call Zion the navel of the universe, the point from which the world was created.

As I said earlier, the peoples of the Near East had a strong sense that life flowed from the great mountains of the north. To make it seem to flow, instead, from a skimpy hill on one tattered leg of the Fertile Crescent required oceans of poetry; but this was something the Hebrews had plenty of.

THE MOUNTAIN AND THE DESERT

What has happened to the Canaanite world-pole? Though Baal's mountain, too, had a temple on it, that temple was not shaped by human hands or visible to human eyes. Mount Zion, on the other hand, seems so thoroughly civilized that it might as well be a ziggurat. Has the Mountain been hijacked to serve the needs of priests, kings, and Levites? Are the Hebrews in danger of forgetting the role of wildness?

As if worried that the integrity of the wild Mountain might be at risk, the Hebrews set aside another peak which will perform that role in its unmixed form. Whatever happens to Zion, Sinai remains untamed and untamable. That is partly because it is in the desert; partly because the Israelites were from the outset forbidden to come near it, on pain of death (after Moses, only one man is said to have climbed it, the prophet Elijah); and partly because no one knows where it is. To this day, Sinai upholds fiercely the West Semitic idea of a wild world-pole.

The desert is hardly the source of life, as one expects a world-pole to be. Yet it is the source of what the rabbis called the "tree of life," the Law. The desert was where revelation erupted, where a covenant was welded, where a nation was tempered—plunged hissing into sand, as if into water. If Zion is the eternal city on a hill—"Jerusalem of Gold," as the popular Israeli song has it—then Sinai is the place of rejuvenation.

In all of ancient Near Eastern literature, the Bible is one of the few texts that knows wilderness as a place of majesty, a place where God lets himself be known. And this is wilderness in the modern sense: not only desert but forest, mountain, even (as in the Book of Jonah) the bowels of the sea. Far from rejecting nature, the Hebrews embraced her as a whole, thorns and all. No place is godforsaken, unless it be the cities of men. However bleak, the sand and stones have sermons to preach. Nor do they preach self-mortification or the vanity of this world. What they say is: Go back to your fields. Go back to your vineyards and orchards. Only this time do it right. This time, let there be justice and peace. "They shall sit every man under his vine and under his fig tree; and none shall make them afraid; for the mouth of the Lord of hosts hath spoken it."

The prophets, it has been said, voice a nomadic nostalgia of sorts, an urge to abandon the stink and swank of the cities, the oily wealth and sweat of the plantations, and head out into the desert. If they seek God in the desert, it is not because they think he is absent from the fields or the vineyards. God is everywhere; what is useful about the desert is that there there is nothing but God.

When the writers of the Bible adopt the viewpoint of nomads it is not because they are nomads, but because they are not. It is an alienation device, like Brecht's in his plays or the guises of Swift and Montesquieu, and has a similar moral intent. The desert is a corrective. It is an astringent or desiccant applied to the overripe smugness of settled peoples, and above all to the decadence and greed of their rulers.

This is not to say that there were no seminomads among the Israelites. There were such clans in the federation, most of them in the arid south. But the distinction between peasants and seminomads was rarely hard and fast. In the Levant as elsewhere in the Near East, seminomadism was an escape hatch. When a peasant household found itself so beset by tax collectors or armed ruffians (often one and the same) that the game was no longer worth the candle, it could leave its fields, take its flocks, and head for the steppe. In a sense, that is the story of the Exodus.

A nomad bends the knee to no prince. Even a tribal chieftain or sheikh rules only at the sufferance of his tribesmen, any one of whom can vote with his feet, or his camel's hooves. A nomad depends only on God: which is to say, on nature.

The jubilee year prescribed in Leviticus 25 can be seen as a momentary return to the (legendary) nomadic condition of the tribes when they entered Canaan. Every fifty years—seven times seven, plus one—the Israelites shall "proclaim liberty throughout the land unto all the inhabitants thereof." Slaves are manumitted, debts forgiven, fields returned to their original owners. We might look at it this way: For a split second, ownership of the land reverts to God. Then the assignment of land to tribes and families is repeated, no jot or tittle misplaced. Every stone, tree, and square inch of turf lost to sloth or greed goes back to its hereditary owner. We do not know if the jubilee was ever observed, but the mere idea is revolutionary. The trumpet blast of jubilee is like a burst of music in a game of musical chairs. The portly and comfortably settled are forced to get up. For a while, everyone is on an equal footing. The world melts into primal chaos for an instant and then is re-created—and reapportioned.

The same basic principle recurs on several scales: every seventh day, every seventh year, every seventh seventh year. On the Sabbath, both humans and animals are freed from the grind of domestication; all technology, right down to the kindling of fire, is taboo. In the sabbatical year, the land itself is allowed to revert to a state of wildness. Sabbath, sabbatical, and jubilee are all eruptions of wildness into the humdrum of the technical and economic order. Earth, plants, animals—even humans—are free to do as they will. So the rivers flowing from Eden leave puddles of paradise in time as well as in space.

Far from forgetting the role of wildness, then, the Israelites expanded that role. They gave the wild Mountain, and wilderness in its most unfriendly form, a social and spiritual function that the Canaanites had barely hinted at. This reverence for wildness, though often submerged by other tendencies in Western culture, would bob up again and again. Some of our greatest ecological prophets, from St. Francis to John Muir, Bob Marshall, and Edward Abbey, have been heirs of Moses in the wilderness, even if the modern vision is of a bush that *is* being consumed.

THE STILL SMALL VOICE

One of the things that have kept the Bible fresh, while most of ancient Near Eastern literature withered or was embalmed, is the gust of country air that hits the reader the moment he opens the book—the smell of cedar, sheep dung, sun-baked wheat, and olives bruised beneath one's sandals. There are passages in Sumerian wedding songs for Inanna and Egyptian love poetry that have the heady, spicy odor of the Song of Songs, but most of them reek of decadence. They are city boys' vegetable wet dreams. The Song of Songs, despite their manifest influence upon it and its own royal trappings, has an innocence they lack. There are Mesopotamian laws and proverbs that deal sagely with agricultural problems but do so in a language dried out on the desks of scribes and administrators. In all the ancient Near East, only the Hebrew writers have dirt under their fingernails. Amos drove goats ("a herdsman and a piercer of figs," he called himself, referring to the gashing of sycamore figs to hasten their ripening). David was tending sheep when Samuel discovered him. One can almost believe that the trend toward universal literacy which would make the Jews peculiar among nations (the sort of thing that would later impress Crèvecoeur about the farmers of New England) began, if haltingly, in the time of the First Temple. Indeed, prophets such as Amos may present some of the few cases in the ancient world of an honest-to-God rural point of view finding its way into writing.

In looking at the Hebrew view of nature, it is a common mistake to pay more attention to the form—the bare idea of a single transcendent god— than to the content, both legal and poetic. The content of the Bible shows, as the great nineteenth-century naturalist Alexander von Humboldt noted, a greater and more sweeping sense of the grandeur of nature than is found among the Greeks, even at their most "pagan."

In the second volume of his great work *Cosmos*, Humboldt gives a conspectus of sentiments displayed toward nature by the great cultures of the world, past and present. Having chided some Greek and Roman writers for skimping on descriptions of nature, he goes on to praise the "profound sentiment of love for nature" to be found in the Hebrew Bible. "It is a characteristic of the poetry of the Hebrews, that as a reflex of monotheism it always embraces the universe in its unity, comprising both terrestrial life and the luminous realms of space." Hebrew lyrical poetry "develops a rich and animated conception of the life of nature. It might almost be said that one single psalm (the 104th) represents the image of the whole Cosmos:—'Who coverest thyself with light as with a garment . . .' "

Humboldt's testimony to the love of nature evinced by the Hebrew scriptures is not just anybody's testimony. Evidently the spirit that drove the first ecologists was not new or exotic, but had lived all along (if partly dormant) in their own culture's central book. Nor was it only the profusely detailed love of nature that inspired them, but the very monotheism of which that love was (as Humboldt says) a "reflex." Monotheism, with its faith in nature's harmony, sustained such Christian pioneers of ecology as Humboldt, Gilbert White, Louis Agassiz, and George Perkins Marsh—until they were sent reeling by Darwin's bloody polytheism. Although Darwin considered himself a Christian, his view of a world shaped by Sex and Strife, Eros and Eris, is deeply pagan. The odd thing is that these two opposing ways of looking at nature pioneered in the nineteenth century—evolutionism and ecology—both turned out to be right. One saw harmony, the other conflict, and between them they saw reality in depth. The same may prove to be true of such older and murkier pairs as monotheism and polytheism, or earth cult and sky cult. Some sense of this resolution can be seen already in Darwin. In fact, it can be seen in Homer. The role Darwin assigns to divine providence is the role Homer assigns it: at the end of the day, after the battle royal of gods, demigods, and humans, "the plan of Zeus is accomplished."

On balance there is little reason, either logical or historical, to think that monotheism is less friendly to nature than is polytheism. One underlines the unity of nature, the other its diversity. Both are right. As for transcendence, that is a concept of more interest to theologians than to tinkers or tailors. To the man in the city street, a nature god is more transcendent than a city god, for whatever is outside the city gates transcends his daily experience. For the peasant, just the reverse is true. The schoolbook definition of a transcendent God—one who is beyond nature rather than part of nature—is at once too abstract and too simple-minded to match up with the religious experience of real people. To get a sense of the kind of transcendence the Hebrews knew, the best place to look is Elijah's vision from the mouth of the cave at Horeb, "the mount of God"—Horeb being another name for Sinai.

And, behold, the Lord passed by, and a great and strong wind rent the mountains, and broke in pieces the rocks before the Lord; but the Lord was not in the wind; and after the wind an earthquake; but the Lord was not in the earthquake: and after the earthquake a fire; but the Lord was not in the fire; and after the fire a still small voice.

This sequence sums up the Jewish understanding of God, man, and nature. It is the keystone of an arch that reaches from the Yahwist to Buber and beyond. The still, small voice makes the best case that can be made for a God that is not in nature. Yet nature is not diminished. God is not absent from nature, any more than Shakespeare is absent from *The Tempest* (any more than Prospero is, for that matter). God is not outside nature so much as unfathomably deep within it: the essence of nature. Forced into an abstract form of transcendence by later rationalists (themselves influenced by Greek thought), this protean God would escape his cage and burst back into nature time after time in Jewish thought: in kabbalism, in Hasidism, in the pantheism of Spinoza. Even those among the rabbis and philosophers who insisted on a strict division between God and nature often regarded the earth itself as a living being.

THE PERILS OF UNIVERSALISM

Composed by persons living under various political and economic regimes, in various lands, over the better part of a millennium, the Hebrew Bible laid before the Western world a kind of rijsttafel from which it might pick whichever ideas smelled good or promised to be useful. There was no single "attitude toward nature" in it, but a shifting set of thoughts and feelings that sprang from the lifeways of hill farmers, herdsmen, and city people. Mixed with the rough Canaanite ingredients were Mesopotamian and Egyptian spices and others yet more recondite.

The Western world that did the picking had its own values. These had been shaped successively by two empires, the Greek and the Roman. The Greeks had much in common with the Canaanites and Israelites. They lived between the mountains and the sea. Their trades were the trades of the eastern Mediterranean basin: the raising of sheep and goats, the culture of the vine and the olive, small manufactures, and all kinds of merchancy. From earliest times there had been intercourse between Greeks and Levantines, fruitful if not always friendly (the Philistines were mostly Greek). The Greeks and Hebrews shared democratic leanings, a taste for logical rigor, and a passion for the word.

But the Greeks had much in common with the Mesopotamians, too. They were confirmed city-dwellers, though their cities were small and had a more democratic complexion. They shared a taste for domination, though (at least in early times) they preferred piracy, trade, and the establishment of colonies to the daily grind of empire. They were cool toward wild nature, and put great stock in the work of human hands and minds. Perhaps because they felt less of a need to mystify subject classes and peoples (or, rather, used the language of democracy to do that), they did not so often credit their own achievements to the gods. Thus they introduced the way of thinking we call humanism, making man (in the words of Protagoras) "the measure of all things." It was not quite the calm and sunny humanism that Victorian scholars so admired; amid the terrors, both natural and supernatural, of ancient life it was almost a kind of whistling in the dark. (Even the famous praise of humankind in the *Antigone*—the chorus that begins, "Many things are wonderful, but none more wonderful than man . . ."—is harshly lit, and leaves us wriggling helplessly before the final blackout.) Still, it did assert the right of humans to make whatever use of nature they chose, with or without a green light from the gods.

People who thought in this way would jump at the idea of filling the earth and subduing it. (One can almost see the cloud of small farmers, businessmen, and merchants swarming like bees across the landscape.) They might be less attuned to the idea of stewardship implied a few lines later ("to work it and protect it"), or the awe of wild nature coursing through the book. As for the idea of the unity of nature—of a cosmos ordered by law—this was something to which the Hellenized mind needed little prompting. Greek philosophy had already gone much further in this direction than the Hebrews had done. And by keeping gods out of the picture it had been able to make nature lie supine on the table of science.

But the world that came to the smorgasbord of the Bible was not just Hellenized, it was Romanized. Rome was the true heir of the Tower. Only now humankind's right to use nature just as it liked was no longer clouded in priestly mumbo-jumbo, or restrained by Greek good taste. Beneath a thin veil of piety and poetry, nature lay bare: a stack of what we would now call "resources." The inherent pragmatism of the Tower, its rationality of means and purely human ends, could now express itself openly. Sextus Julius Frontinus, appointed water commissioner of Rome in 97 A.D., boasted that, unlike the "indolent pyramids and the celebrated but slothful works of the Greeks," Roman aqueducts were useful.

Not content to imagine itself the center of the world, Rome *made* itself the center. This it did not merely by conquest, but by consolidation: by throwing out a network of civilization woven of uniform fibers. Rome made the world over in its own image.

We spoke earlier of two networks that gird the world, that of wildness and that of civilization. The network of wildness is woven of millions of varied threads: various species, ecosystems, metabolisms, trace elements. The image of the Mountain reminds us of the deep unity beneath this diversity, but the diversity is real, and it is necessary. So is it necessary in the network of civilization.

Both necessities Rome denied. Not only did it try to replicate itself and its institutions all over the known world—the make civilization a single thing, made of standard parts; it tried to do the same to nature. The highways of Rome paved over the rivers of Eden.

For lack of a better word, we may call this universalism. It was not entirely a new idea, being implicit in the idea of empire. In embryo, it could be found among the Assyrians, the Babylonians, and even (in a different form) the Hebrews. Perhaps it came to manhood in the person of Alexander of Macedon, who seems to have believed that Greek culture would stretch to accommodate all members of the human race. But it reached full maturity in Rome. For it is one thing to think, as the Hebrews did, that a deep order underlies the hubbub of nature, and quite another to impose a uniform order of one's own. No Hebrew would have dreamed of subduing the earth in this way.

The Romans were the first globe-spanning technocrats, the forebears of Bechtel Corp. and the U.S. Army Corps of Engineers. They were not especially good engineers; most of their technology was borrowed, for they were too bored with abstract science to make any great strides in what was concrete. (They did, however, make the first major use of concrete.) Centuriation—the dividing of land, without regard for the lay of it, into plots 710 meters square—was the model for the waffle iron that would be used on the American West. From Britain to the Levant, new Roman cities were built on the same rectangular plan. If all roads led to Rome, it was because so many roads came from Rome (that is, were built and maintained by Rome). The same could be said of aqueducts and sewers. The world-pole no longer had to be a real mountain that could snag the clouds of heaven and gouge them like gray oranges on a juicer. Instead it could be a city with excellent plumbing—a city whose pipes, or whose idea of pipes, would reach every cranny of the known world.

Mesopotamian irrigation may have begun this process, but it could only go so far. A canal is dug in the earth and must beg the earth's indulgence. A canal is a baby river still hugging its mother's side. A dike or canal is always dissolving into the muck of its makings. Unless tirelessly spooned out, mud chokes the canals, scales the dams and levees. An aqueduct, in contrast, hardly muddies its feet. It soars above the landscape like an elevated superhighway. Each arch is its own world, complete with its own gravity.

The clash between the rivers of Eden and the canals of Babel reappears when we look at later waterworks—the fountains of Tivoli, dams on the Colorado, levees on the Mississippi. If a river is to bring the water of life, and not just dihydrogen oxide, it must remain alive. It must be an artery of wildness.

Water loves chaos, uses chaos to find its way. Forced to be linear, it is blinded and crippled. Water wanders, loops, meanders; confined to the straight and narrow, it stultifies or else rebels. When a complex system is suppressed, it either shudders and dies, or—if it has great power, as a great river does—spirals out of control.

In mathematical terms, chaos can be seen as the expression of shapes in imaginary space, where one or more axes are marked off in imaginary numbers. Where one of these shapes intersects the real-number axis, the sequences it makes seem pointless. Blind to those hidden shapes, we try to drill the sequences we can see into parade-march lockstep. And the delicate unseen shapes are tugged and squeezed to death.

Here is a geometer's metaphor for much of what we do to nature. When we replace a living forest with a yardstick-spaced plantation of Douglas fir, we are doing the same thing—except that the shapes of ecological ties between trees, understory, and wildlife are not imaginary in any sense. They are, or were, as real as your bloodstream.

A river is an artery of wildness in a more straightforward biological sense as well. Even in America, where nearly every stream deeper than a swimming pool has been dammed and most have houses and farms lining their path like a Fourth of July crowd, rivers remain the best corridors of wildness. Fish still find the way to their spawning grounds, birds and mammals and reptiles still manage to make their way along the banks. In most of the country, these are the only places where an animal can move in one direction for a long time without running the risk of becoming roadkill. Estuaries, marshes, and wetlands of all kinds are motels along this highway, and not just motels but nurseries, taking in vast amounts of life and releasing it vastly augmented. As blockades of concrete and asphalt rise up both in the water and out, all of this becomes more and more problematic.

Although I have talked about water, I might have talked about any number of things. The lengthening of the Tower's reach by the logic of universalism has had profound effects on the modern West and hence (by the same logic) on the modern world. This logic takes various forms. In the realm of economics, it takes the form of market capitalism. In the realm of science and technology, it asserts that physical law applies identically in all places. A falling stone obeys the same equations in Capetown as in Cheapside; a steamboat moves as swiftly, given the same conditions, on the Niger as on the Thames. But this belief causes problems when we transfer it to the realm of biology. For biological law does not always apply, on a useful level, identically in all places. A tractor obeying the same physical laws in Illinois and Brazil produces in the one place a fertile field, in the other a desert.

A CHRISTIAN PANGAEA

When we talk about the West picking the bits it liked from the Hebrew Bible, we are, of course, talking about Christianity. By and large, the bits Christianity chose—unsurprisingly, given its early constituency—were the ones that sat best with city people; city people, moreover, who had been Hellenized and Romanized. And it added some, such as the notion that humans could use nature as they pleased. Above all, it translated Roman universalism into the realm of the spirit.

There was, of course, a kind of universalism implicit in the Hebrew Bible, particularly in the prophetic books. But though the Hebrew god may be a single and transcendent god, he is still a god tied to a certain place. If he floats above nature he is nonetheless tethered to it like a hot-air balloon. He is a community god. He demands that we account for what comes in and what goes out. This obsession with what comes in and what goes out, with purity in eating and sex; this insistence on neat walls between peoples, between species, between bodies—all this has its nasty and neurotic side, no question. Parodied by those who became the Jews' hosts, it did them great harm. Properly understood, though, it can be of great ecological use.

As humans cannot bring themselves to care about nature in the abstract, so they cannot quite believe that God does. A local god, a god who may rule the world but who keeps house on some local peak, has a stake in the local landscape. A universal god is like an absentee landowner.

The Jews never tried to impose their culture on others, but only to keep it alive themselves. In this effort they had the advantage of letters, and of a text they could clamber aboard—a double-scrolled outrigger, a lifeboat. Ironically, that same text, fitted out by Christianity with a steam engine and a steel hull, has become a dreadnought ramming and sinking the struggling hulks of other cultures. In fact, many of the world's peoples got letters and Christianity in the same package. Their first book was the Good Book. Some, such as the Irish, were able to write down and preserve their own stories. Some have preserved their languages. But in a deep sense their cultures were lost, as others continue to be lost—acres of culture every hour, bulldozed by missions that are bankrolled, in turn, by Americans sitting with beady eyes before their TVs, greedy for healing and grace.

No wonder Christianity and capitalism have marched shoulder to shoulder—Christianity is the cash economy of the spirit. As all things can be changed into money, and all moneys into all other moneys, so one can and must translate the Gospel into all tongues. The basic blurb-length version has been converted into more than eleven hundred languages, twenty-five of which can be found on the first pages of the Gideons' Bible: "Want so lief het God die wereld gehad, dat Hy sy eniggebore Seun gegee het. . . ."

This is not to question the depth and beauty of the Christian message; it is only to point out that no message can be made into a universal medium of exchange without being debased, and thus doing as much harm as good. Jesus, though a carpenter's son, lived in a world that for two thousand years had been shaped by emperors: Akkadian, Egyptian, Assyrian, Babylonian, Persian, Greek, Roman. His followers built an empire of their own. Words that had one meaning when spoken from a hilltop have a different meaning when spoken from a balcony of the Vatican or beamed to the world by satellite. Many Christian denominations have seen this and have given up the claim to universality, or at least the will to impose it. But the others are more than busy enough to doom every shred of indigenous culture on earth.

If the New Pangaea is partly an accidental result of the waves of human expansiveness, it is also something we have willed. In the cultural realm, the various universalisms now at work—Christianity, scientific humanism, Marxism, capitalism—may not agree on much, but each agrees that it has the answer for everyone else. Though it is hard to know what their net effect will be, one effect is clear enough: a wave of cultural extinctions.

We must take care that the new Gaian consciousness does not become a new universalism as crushing as the old. Once we crushed all the peoples of the world to the bosom of the Mother Church and the belly of the Market. Now we may crush them to the dugs of Gaia, too. If the world is a single organism (the logic seems to go), clearly we who are the brain, we scientific Westerners, have the right to tell all its other parts what to do. What works here, must work there.

True, all natural systems are linked; the smoke of the rain forests will sear our lungs. What is galling is our hypocrisy. Formerly we dispensed Bibles with the right hand and picked pockets of mineral wealth, coffee, and tobacco with the left. The left hand has not changed much; the right, however, is now dispensing ecological nostrums to people who don't have cars to stop driving. These people do not have to be told to recycle tin cans; they fight over them like stray dogs over scraps of gristle.

With one hand we tear down their huts and their cosmologies and, when they stand naked before us, entice or drag them into the market economy. With the other hand we slap the wrists that reach for the "natural resources" we have taught them to pull up and sell. We slap the fingers that scrabble for a bare living when the market economy has taken a man's land, annulled his skills, or otherwise betrayed him.

Even anthropologists and ecologists, in their haste to document the diversity of human and nonhuman nature before it is erased, hasten its erasure. With the best of intentions, they find themselves working hand in hand with missionaries, oil geologists, and the like on matters of logistics and linguistics. Trying always to stay a step ahead of development, they become its unconscripted shock troops. Scampering ahead of the road crew, they beat a path that others will follow. Their hands held up in protest are signposts for those whose actions they protest. Their cries of indignation draw the jackals. (Many of them know this, but figure it would happen without them, too.) The need to know all cultures can be nearly as deadly as the need to save all souls.

No doubt universalism in some form or other was a necessary step. Without it we could not have imagined science as we know it, or democracy, or the kind of community of nations (still only imagined) that a crowded world requires. But maybe the time has come for the next step.

The worldview of modern ecology may point the way. If this view is to work, it must be particular. Our planet is one of many. It is not the center of the universe. If we think of a dialectical movement in which particularism comes first and universalism second, this may be the third and unifying stage.

Oddly, the eclipse of geocentrism lets us prize the earth all the more. Few things have done more to arouse our neighborhood pride than the photographs taken from space of this blue ball, fragile as a robin's egg, on which we build our infinitesimal nests. Whether that pride will be matched by concern remains to be seen. And, admittedly, this vision of the planet as a particular place does not, of itself, solve any problems. It does help to keep some problems in perspective—for, once you can see that life on earth is an odd and somewhat provincial affair, you are less likely to look down your nose at a little healthy provincialism on smaller scales. But to promise a single answer to all conflicts between the universal and the particular would itself be a glaring instance of universalism. Once again we find ourselves on the edge, trying hard to keep our balance.

THE SAME GOES for the conflict between the Mountain and the Tower. Each is a necessary idea. Though the West has undoubtedly placed too much faith in the Tower, only a madman (or a Deep Ecologist) would demand that it be razed.

But accepting the rift between the two world-poles, and keeping them balanced in our mind and in our world, is not so easy. Many of us would rather dream of recapturing, in one way or another, the wholeness lost when we left Eden.

Suppose you are fed up with city life but know that wilderness is no place for people. The obvious next step would be to look for someplace in between. You would look for a place that was softened, but not yet spoiled, by human settlement. You would look for a way of life that used just enough technology—not too little, not too much. You would look for a place nestled neatly in the trough of the wave of human advance. In a word, you would look for Arcadia.

PART THREE

IDYLLS

13

ARCADIA

NOW IS THE TIME to recall the person from Porlock who—calling at Coleridge's house on business—shattered the vision of Xanadu "like the images on the surface of a stream into which a stone has been cast. . . ."

A stone has shattered our happy view. Some of us try to look beyond the ripples, some between them. Others squint at the dead center (the eye of the storm), while still others shut their eyes and try to recall the image that must have floated there before the pebble struck. Rarely does anyone stop to think that the ripples might be a permanent fixture (or should I say fluxture) and that, far from obscuring the picture, they might *be* the picture.

In the last part of this book I focused on two apparently fixed points, the Mountain and the Tower: points from which waves of change radiate or toward which they tend. Of course they are not really fixed, but it is easy to keep up the illusion that they are. As we move toward the middle range—the great stretches of the earth's surface that are neither wild nor metropolitan, but in the process of changing from one state to the other—the illusion of stasis becomes harder to maintain. But the human mind rises to such a challenge. Its dodges take many and artful forms: a pastoral poem, a suburban subdivision, a pleasure garden, even a scientific theory.

Each of these idylls—from the Greek for "little picture"—is a way of denying or declawing change. But each tries to deny something else, as well: the gaping rent in the world and in the soul that opens up when the Mountain and the Tower claim their own realms. Although this rent really begins to widen with the start of civilization—visibly in the story of Enkidu, the wild man tamed by a whore, who then finds that his old mates the animals run from him—it must have begun much earlier. In the soul, it must have begun as a run in the seamless fabric of hominid unselfconsciousness. In the world, it must have begun as a shifting border between the campfire's light and the peltlike darkness beyond.

Be that as it may, by the time the idylls I am talking about become important—mainly, from late antiquity to modern times—the fissure has become a canyon. Though Western thought is trendily condemned for its dualism, the truth is that no map of the world or of the self is really comfortable with a split right down the middle. In fact, the Western mind is even more allergic than most to certain kinds of division. We like our world, and our self, to have one center, not two.

But that is not always easy to arrange. If we make the Mountain the center, our human stature is cramped. If we make the Tower the center, we feel estranged from the rest of nature. So the idylls try to close the gap, one way or another: by finding a new center at the midpoint between Mountain and Tower; by juxtaposing the two, or including them in a tight symbolic whole; by claiming a new continent in which the gap is willed away.

Beside the search for stability, then, is a search for wholeness. Both can be seen as forms of the same thing: a refusal to accept the exile from Eden. If that refusal causes all kinds of trouble, it is also the yeast for all kinds of human creativity. It can never be wholly suppressed, nor should it be. But if we are going to deal like grown-ups with our present problems, we must try to put that refusal in perspective.

The natural place to start is with the original idyll, which was painted by the city-weary poets of the aging classical world and still tempts us today. If we cannot live in Eden, can we live in Arcadia? Will Arcadia stand still long enough to be lived in? What if everyone tries to live there? What ecological wounds chafe beneath Arcadia's green cloak?

From Arcadia, we move on to the pleasure garden. A stroll down the garden paths of three millennia, with frequent glances over the garden wall, is a pleasant and revealing way of traversing the ecological history of the West. Which facts about their relation with nature and with other people did Roman lawyers, Renaissance mercers, French tax farmers, and Whig industrialists want mirrored in their gardens, and which clouded over? What hidden strands joined the villas of Florence to the denuded plains of Spain, the avenues of Versailles to the famine-stricken provinces, the boundless and (apparently) fenceless turf of Blenheim to the private enclosure of common land?

America took up the quest for paradise where the gardens of Europe left off—or should we say, the garden of Europe: for in a sense all of Europe was a garden, thickly settled, intensively cultivated, and dense with the serried icons of past cultures. The dream of Eden tested in the New World—the dream of meeting nature face to face, all culture's masks cast off—was a daring one. Even more daring, perhaps, is the modern idyll with which this part of the book ends: the Gaia hypothesis, which claims that the biosphere as a whole is a self-regulating system. How much of this is science and how much myth? Is Gaia an indulgent mother who will care for us no matter how badly we behave, or a jealous goddess who will strike us down in our pride? These questions will bring our story up to the present, preparing us to deal with some very present dilemmas.

THE PERFECT MIX

We begin, though, in Arcadia. Even those of us who have never read a line of pastoral verse know some of its tropes: the babbling brooks, the dallying nymphs, the shepherds piping and pining. When teachers want to make their courses "relevant," this is not the stuff they reach for. Actually, though, pastoral has a great deal to tell us about the way we live now. Never before have so many people tried to live in Arcadia.

Although hints of it can be found in texts as old as the tablets of Sumer, pastoral is usually said to have begun with the *Idylls* of Theocritus, a Greek poet of the third century B.C., and to have found its standard form two centuries later in the *Eclogues* of Virgil. For over two thousand years, pastoral was taken very seriously by some of the West's greatest poets, among them Petrarch, Tasso, Spenser, Shakespeare, and Milton. In the Renaissance, pastoral stood alongside epic, tragedy, and satire as one of the major poetic forms. Nor was it confined to the three-walled prison of verse. Pastoral dramas, operas, ballets, prose romances, and paintings were turned out at a great pace. Arcadian societies sprang up in which bored courtiers gathered in gardens, clad in lambskins, to play panpipes and sing "rustic" songs.

Since the Latin *pastor* means "shepherd," it is no surprise that shepherds are the main actors in pastoral. They are joined by cowherds and goatherds, shepherdesses and nymphs, cyclopes and giants, the god Pan and his satyrs, and the occasional visitor from the big city.

What happens in a typical pastoral idyll? Not much. A shepherd rebukes the object of his affection—a nymph, a shepherdess, or a boy, as the case may be—for her or his hardheartedness. Two shepherds engage in a poetry slam, the prize being a heifer or a carved cup. A shepherdess uses herbs and magic charms to win back the swain who has jilted her. A visitor from the city falls in love with a shepherdess. A shepherd dies of love.

If today's readers find all this hard to take, they are not the first readers to feel that way. The bare-knuckle critics of the eighteenth century gave pastoral a beating from which it has never recovered. Dr. Johnson said the form was "easy, vulgar, and therefore disgusting." Although Milton was just following the rules when he claimed in "Lycidas" that he and his friend "drove a field" and "battened their flocks with the fresh dews of night," the good doctor would have none of it: "We know that they never drove a field, and that they had no flocks to batten." Pastoral was useful only as a playpen for young poets, since, "not professing to imitate life," it required "no real experience."

Despite this trouncing, pastoral has had a way of popping back up in various disguises. From *Alice in Wonderland* to *Walden*, from Westerns to the trifles of P. G. Wodehouse, a broad assortment of literary works—to say nothing of movies, Broadway musicals, ballets, symphonies, cartoon strips, and television shows—bear the telltale marks of the genre. As the old clichés are phased out and replaced with new ones, the basic logic endures. It has to do with the effort to find the perfect midpoint between wilderness and civilization. This is one way of healing the rent between the two: a rent in the world and a rent in the self.

People go to great lengths to find, or dream up, places and times in which nature and culture are mixed in just the right proportions, like a good martini, but without the necessity of being either shaken or stirred. Which is the gin and which the vermouth is a matter of taste; the principle stays the same.

For most tastes (in imagination, at least) the perfect mix comes after the rigors of pioneering, but before the constraints of the fully civilized life. Though the wild Mountain is the true and original paradise, it is not paradise *for people*. Despite the burblings of nature romantics, few of us—even "savages"—are happy in a landscape wholly untouched by man, and fewer in one that is wholly unfamiliar. We have to adapt to it, and adapt it to us. We want to make it over in our image, in the image of the savanna or edge habitat that was the lap (a smoothness here, a wooded mound there) in which we were dandled. And that is just the beginning.

Minor tampering with an ecosystem can often be kept up indefinitely. Even major remakings can be sustained if they are limited to strips and patches, so that the larger ecosystem can keep doing the thankless jobs that keep the little man-made ecosystems alive. Arcadia is a place, or a stage, in which nature has been remade enough but not too much. Or, if it has been remade too much, the effects have not yet come home to roost. In that case the Arcadian is living on ecological capital.

WHY THE LONG FACE?

In any roundup of pastoral plots, one question jumps out at you: Why are these "idylls" so often sad? The obvious answer has to do with unrequited love, something I will talk about a little later; but to take that answer at its face value only begs the question. Why should poets go to all the trouble of creating a perfect world, only to pack it with broken hearts?

What is kept locked up in the green cabinet of Arcadia, and is the most secret cause of its melancholy, is the knowledge that this stage, too, shall pass. Maybe it has passed already. Although the setting is "idyllic," there are vexations both petty and large, and sometimes one hears a shepherd complaining that the Golden Age is gone. In fact, a whole subclass of pastoral, the pastoral elegy, is all about the passing of the Golden Age, usually personified in a handsome young shepherd who is dying or has just died.

Even in Arcadia there is death—as, in Renaissance iconography, the death's head with its motto *Et in Arcadia ego* attests. It is not only the death of the person but the death (if only piecemeal) of nature. While nymph and swain tussle in the shade, his goats nibble the shoots that would have made the shade immortal. If his goats are curbed, someone else's will sooner or later show up. Beneath the shimmer of Arcadia is the surge of the waves of human-led change. Humans and their four-legged allies (as well as various herbaceous ones) are moving outward, changing ecosystems and displacing species.

All of this becomes clearer when we look at the actual ecology of the world that pastoral depicts. In much of the ancient Mediterranean world, towns sat in the coastal plains or in valleys between the mountain ranges. In the bottomland around each town was a band of grainfields. As the terrain grew steeper and leaner, this shaded into a ring of olives and vines. Next was the patchwork of woods and fields in which herdsmen grazed their flocks in summer, in the practice known as transhumance. Last of all was wilderness: the forested mountainsides and rocky crags ruled by gods and wild beasts.

The third band, the patchwork of trees and fields—to be exact, of trees and the various communities of shrubs and herbs, known as maquis and garigue, that spring up in the Mediterranean uplands when trees are cut down—is the world of pastoral. It was, and in many ways still is, a very nice world, if not quite as nice as the poets said. Open enough for good grazing, it was still wooded enough in places to afford shade to herdsman and herd, to keep streams clear and pure and not too erratic, and to let many of the native flora and fauna go about their business. Add leisure for singing, piping, and making love, and you might well think you had found the perfect midpoint between the Mountain and the Tower, the place where people and nature were in harmony.

Unfortunately, the midpoint did not stand still. As flocks multiplied, the ring of grazing land crept uphill, tightening like a noose around the last islands of forest. Loggers swarmed about, felling timber for ships or cord-wood for smelting. Before the woods could grow back, herds of goats moved in to eat the saplings. Short of good pasture, herdsmen set more fires to open the way for fresh grass. Some lines from a Greek comedy of the fourth century B.C. give some idea of what really went on in the pastoral landscape:

> Next morning, I was minding my flock again
> When *he* showed up—he's a charcoal burner—
> In the same spot, all set to saw out stumps.

The net effect in Attica is grimly summed up by Plato:

> By comparison with the original territory, what is left now is, so to say, the skeleton of a body wasted by disease; the rich, soft soil has been carried off and only the bare framework of the district left. At the time we are speaking of these ravages had not yet begun. Our present mountains were high crests, what we now call the plains of Phelleus were covered with rich soil, and there was abundant timber on the mountains, of which traces may still be seen. For some of our mountains at present only support bees, but not so very long ago trees fit for the roofs of vast buildings were felled there and the rafters are still in existence. There were also many other lofty cultivated trees which provided unlimited fodder for beasts. Besides, the soil got the benefit of the yearly "water from Zeus," which was not lost, as it is today, by running off a barren ground to the sea; a plentiful supply was received into the soil and stored up in the layers of nonporous potter's clay. Thus the moisture absorbed in the higher regions percolated to the hollows and so all the quarters were lavishly provided with springs and rivers. Even to this day the sanctuaries at their former sources survive to prove the truth of our present account of the country.

That Plato was right—and that not only Attica but most of Greece suffered grievous soil erosion over much of its history—has long been obvious. In recent years it has become clear that much of that erosion was man-made. Soil profiles in the valleys show layers of alluvium washed down from the hills; pollen sediments from lake bottoms show oak giving way to hornbeam, pine, scrub oak, and heather, which typically spring up on cleared land. Since the dates of strata in different regions do not match each other, but do match the archeological and literary evidence of dense human settlement, it is plain that the main culprit was not climate, but a two-legged creature and its various allies. And this has been going on for the past eight thousand years—since shortly after the start of farming in Greece.

Change did not always take such a gloomy form. Arcadia might be threatened by prosperity instead. As a city flourished and grew, each of the rings around it might move outward, pressing on the next. Herdsmen might find themselves jostled by farms and plantations, or by the summer homes of the rich. In practice, the gloomy and nongloomy forms of change might be hard to tell apart. Sometimes it was just a matter of time before one turned into the other.

Either way, the dream of pastoral—to find the perfect midpoint between the Mountain and the Tower—is foredefeated. There is no fixed scale of gradations between civilization and wilderness; there are only the waves of human-led change. If the setting of pastoral seems fixed, that is just because it is the placid trough behind the wave. The last big wave passed by some time ago, and people and nature have had time to adapt. For the moment, things seem to be just right. This state can be thought of as a place or as a time. In pastoral the place is called Arcadia, and the time is called the Golden Age.

The waves in question can be any of the human-led waves of change I talked about in Part One, or they can be other, subsidiary waves. In the classical setting of pastoral, the wave that came a while ago is primitive mixed farming (crops and livestock), with the stress placed on the four-legged part of the mix. The wave that has not yet arrived could be a number of things: intensive terraced farming, plantations, large-scale logging, the proliferation of villages or of country estates. The upheavals of the first wave—the clearing of forests, the displacement of indigenous peoples, the conflicts over grazing rights, the running war against wolves and bears—have settled down somewhat, and those of the next wave have not yet begun. Sooner or later, though, the Golden Age must pass (if it hasn't already), and Arcadia must cease to be Arcadia.

Hesiod, a Greek poet of the eighth century B.C., gave the Golden Age its name and its classic description.

> The gods, who live on Mount Olympus, first
> Fashioned a golden race of mortal men;
> These lived in the reign of Kronos, king of heaven,
> And like the gods they lived with happy hearts
> Untouched by work or sorrow. Vile old age

Never appeared, but always lively-limbed,
Far from all ills, they feasted happily.
Death came to them as sleep, and all good things
Were theirs; ungrudgingly, the fertile land
Gave up her fruits unasked.

As Hesiod notes, in the Golden Age the world was ruled not by Zeus but by his father, Cronus, known to the Romans as Saturn. In ancient lore, Saturn is the god and planet whose influence makes people melancholy: that is, "saturnine." At first this seems odd. But the happy age of Cronus is past, and has been past for as long as anyone can remember. Naturally the thought of it gets us down. Not only is the Golden Age past, but it was always destined to pass. As Hesiod says, at length this golden race "was buried in the ground." No reason given, no blame assigned. Change— usually for the worse—is just in the nature of things.

A GORGEOUS PARALYSIS

Yet change is what the pastoral poet spends much of his ingenuity covering up. Pastoral would say to the moment, "Stay, thou art so fair." Pastoral, the pleasure garden, the myths of Eden and the Golden Age—all show a craving for a kind of gorgeous paralysis. In the Isles of the Blessed, winter never comes. In pastoral, by the logic of transhumance, it has to stay spring or summer so that the shepherds can stay in the hills. The time is usually noon—when it is too hot to move about, the sun seems undecided whether to keep going or turn back, and all of nature shares a convivial yawn. It is time of *otium*, of a leisure as thick and palpable as honey. One can almost believe that the ripples in the stream are mesmerized in place, like furrows in a field. So perfect is the calm that one half fears it may be shattered by a storm—by the arrival of Pan, who is known to walk abroad at noon.

Beneath the unruffled surface of pastoral, conflict roils. Even in Arcadia there is ego. If survival is easy and power beside the point, the other great agon—the competition for mates—absorbs all energies. The shepherd may seem to sing for a cunningly worked cup, but in truth, like a bird, he sings for a bride. Singers contending, lovers scorned, lovers pining and dying for love: all this is as natural as natural can be—Darwin with flute obbligato.

In pastoral, nature and humans converge on the common turf of love (read: lust) and emulation. The maids are lovesick as heifers, the shepherds horny and rambunctious. Eros and Eris—Love and Strife—are the two great principles.

Goat chases moon-clover, wolf chases goat, the crane
Chases the plow—and I'm crazy for you.

In Theocritus there is frequent crosscutting between the sexual antics of
the flocks and those of their keepers. Humans and animals excite each
other. What equips shepherds and goatherds to be the subject of idylls is
not only their imagined leisure, or their straddling of borders, but also their
intimacy with animals (by which I don't mean bestiality, though that may
lurk in the background).

Pastoral is Darwin idealized. The process of natural selection is shown
but not the result, which is change. The poet has the slippery task of depict-
ing people in nature while squelching both history and natural history.
Inevitably, change lurks in the shadows. And this, as I said, is one source of
the melancholy one finds in pastoral. The others are death and unrequited
love, each of which is crucial to the Darwinian process.

In pastoral, music and sex are the human necessities. Food and drink are
taken for granted, clothing scanted, shelter ignored. This reduction of life
to music and love is at once sophisticated and primal.

By reducing human concerns to one or two, pastoral shows that a time
or place of perfect happiness is impossible: people will always suffer for
love. Arcadia is a place where nature's constraints are eased and the constraints of the fully settled life have not yet gelled. But even here we come
up against the fundamental fact that a female can bear the offspring of only
one male at a time. As long as there is competition for mates, someone is
going to get hurt.

SEX WITHOUT BIRTH

The stage whispers of wind and water, the tootling of birds and flutes, the
reassuring racket of livestock: all these we hear in pastoral. We do not hear
the bawling of infants. That is funny, for pastoral is all about courtship, and
takes place against a backdrop of animal breeding. There are plenty of kids
and yearlings about—but no children. (There are boys in Theocritus, but
since they are not sexually *hors concours* they function as adults.) Except in
introjected myths, no nymph or shepherdess ever gets knocked up. Greek
homosexuality may explain this abstraction of sex in part, but only in part.

A glance back at another myth may be helpful. If we think of the Garden
of Eden as God's game preserve, stocked with a breeding pair of each kind
of animal, including humans, and with the snake as warden, we must ask
ourselves why the humans don't breed until after they leave. It may be that,

from the male's point of view at least, the Fall is having children. That would explain why he sees woman as temptress and agent of his Fall. In pastoral, too, the hymning of nature's fertility rarely extends to humans. The poet must sense that their weight will finally sink his pleasure barge. Bountiful as his Arcadia may be, he can't kid himself that its bounty is infinite. Too many mouths will wear it out. At best, they will force a more intensive use of the land. Either way, the wave will move on and take Arcadia with it—a fact of which the poet can allow only furtive and far-between glimpses.

If you think I am projecting a modern concern with population growth onto ancient people, consider the common Greek practice of exposing newborn infants. Both Plato and Aristotle approved of infanticide not only as a way of culling out the weak and deformed, but as a means of population control. (Before resorting to exposure, a Greek or Roman woman would use giant fennel, Queen Anne's lace, pennyroyal, or pomegranate seeds as contraceptives or abortifacients.) Exposure makes a good plot device for tragedy, but the pastoral poet uses the gentler method of simply erasing infants from the picture. Sex without birth is one of the conventions by which pastoral tries to pretend that animal husbandry (and farming in general) is not unstable, that the wave is not moving, that change is only a nasty rumor.

The big reason why infants are not allowed in Arcadia, though, is that they would keep the grown-ups from being infants. Our imagining of paradise has been drawn to pastoral scenes partly because of the pull of milk. When you get right down to it, paradise is a teat.

Of course, the state of the nursling is not really idyllic. The infant spends a good deal of time at the breast but a good deal more squawling for it. As Gregory Bateson and Margaret Mead have shown in their photographic studies of Balinese mothers and children, the patterns of need, frustration, relief, and renewed need of which our adult lives are stitched have their origin at the breast. But then, every paradise longs for an earlier one. The civilized man longs for primitivity, the primitive man for childhood, the weaned child for infancy, the infant for the womb.

One need not go all the way back to the womb, though, to reach an age that is free of responsibility. And responsibility is what the Arcadian most wants to be free of.

With this we come to the real fly in the pastoral ointment. Let us review: If we cannot live in Eden, can we live in Arcadia? No, because Arcadia is not a place, it is a phase. But suppose nature and culture were kind enough to stand still for a while. *Then* could we live in Arcadia?

Not even then. The Arcadian's big mistake is that he tries to solve the puzzle of man's place in nature by finding the place where it is already solved. He wants to stand at the perfect midpoint between Mountain and Tower, basking in their perfectly blended radiance. Clearly his solution, such as it is, is only for the few. Give me a place to stand, he says—*the right place*—and I will balance the earth and humankind. But how many people can stand in that same place? For him to live rightly, it is necessary that most of humankind live wrongly. He puts almost as high a price on his human fulfillment as Nero did when he displaced thousands of Romans to build the Golden House, his Arcadia in the heart of Rome.

Most of us do not have Nero's power, of course. When we find our perfect place, we have no way of keeping others from finding it, too. So we become part of the wave that sweeps Arcadia away.

SUBURBIA

If they never strayed beyond the fold of poetry anthologies, Arcadian delusions would be fairly harmless. But in fact they spread their woolliness through vast tracts of daily life. Suburbs and summer places are the main would-be Arcadias of our day. I want to look at each in turn, asking how the vision of Arcadia leads them astray. Then I want to return to a much earlier Arcadia—the first one, in fact—and see if it might not offer some hints toward a better way of balancing nature and culture.

When I speak of suburbs here, I mean middle-class, more or less American-style suburbs, not the working-class industrial suburbs that have long ringed European cities. Middle-class suburbs can be said to have their roots in the upper-class suburban estates of ancient times, which had their first great flowering around the time Virgil was writing his *Eclogues*. Though more dominant in the United States than anywhere else, middle-class suburbs have long been common in England and are becoming common in many other countries as well.

The suburb is meant to be a mean between city and country. By virtue of its perfect perch, it is meant to combine the simple pleasures of rural life with the amenities of civilization. In other words, it is supposed to be a workable, modernized version, shifted somewhat cityward, of the *locus amoenus*—the "pleasant place"—of pastoral.

The suburb tries to solve the problem of our relation to nature by finding a place where it is already solved. So the Arcadian error finds its way into real life. And it is all the more dangerous because everyone is making it at once.

The source of the error might seem to be a too-literal understanding of the middle landscape—as if Arcadia could be defined by a certain ratio of asphalt to grass. But perhaps the real problem is an understanding that is not literal enough. On the most blitheringly literal level, you can't have a middle unless you have both ends. If everyone lives in the middle, it is no longer the middle in any useful sense. What makes the middle landscape alluring is not any precise set of qualities—for these, as we have seen, change with time and place—but the quality of middlehood, of being in between. Arcadia is a dappled place, with the shadow of the wooded mountain on one side and the glimmer of the town in the valley below. Without the whispers of possibility from each end, the middle loses much of its charm.

It loses much of its friendliness, too. Humans, we said, are fond of edges. Many of us feel safest—physically, psychically—on the edge between civilization and wilderness. Either one in its pure state can turn demonic. We want to be able to run and take shelter from one in the other. If the edge spreads endlessly, we can't.

Why is the babbling brook so key a player in pastoral? Partly because it is pleasant, and gives both people and beasts a place to drink and bathe. But also because the brook is a bush-league River of Eden, joining Arcadia to the Mountain on one side and the Tower on the other. Beside the brook one feels sheltered, yet connected: linked, not lost. In suburbia, streams and their wetlands are channeled and filled in. Where a stray brook survives, hardly anyone knows where it comes from or where it goes.

Presumably suburbia is meant to be one vast edge, one endless middle—a logical impossibility. But maybe I overestimate the extent to which suburbia is "meant" to be what it is. As suburbia spreads, it falls prey to the hazard of all Arcadias, the motion of the wave. The worst thing is not the motion itself, but the change of scale. As a ripple moves outward from a center, the wave form broadens and softens. Distances swell, not only from center to periphery but from any point to any other. The first suburbs are close to city and countryside alike, and sometimes even to wilderness. But as the city expands, and the countryside retreats, and suburb joins suburb, the scale of the whole thing is blown up beyond recognition. Like a razor blade under a microscope, edges blunt and blur into vast tracts of sameness.

Up to a point, faster transportation makes up for this change of scale. But only up to a point. For instead of a public-transportation system that would clarify and sharpen the geometry of a region, we too often have (in the U.S. above all) a varicosity of roads that adds to the slackness. And even a perfect system of public transportation would not make up for the loss of a varied landscape close at hand—a landscape to walk in, bicycle in, play in. It would not make up for the deadening sameness of the horizon, shorn of the wild mountain on one side and the city's spires on the other.

It would be one thing if city and country had simply receded over the horizon and could be still be found, safe and sound, if you went far enough afield. But as suburbia spreads, it eats up countryside and drinks up the sap of the city. If there were always plenty of healthy countryside just beyond the last chunk that was chewed off, that eating up might be bearable. At a certain point, though, you push beyond the long-settled farmland into sparsely settled, wilder regions. Farmers displaced by the suburban economy often find it easier to melt into the metropolitan labor pool than to start all over in the next county. If they do want to keep farming, they may decide to move to another part of the country altogether. In any case, healthy rural communities are not made overnight. Certainly they are not made at anything like the rate that suburbia is.

Meanwhile, the city loses people, businesses, and much of its tax base to the suburbs. It loses much of its vitality, too, which the suburbs do not acquire, but merely dissipate. The city may try to save itself by stretching its borders to keep the suburbs inside, something cities have been doing for thousands of years. It works, up to a point—a point that is quickly passed when you are dealing with modern suburbs shaped (or unshaped) by the automobile. Though the maneuver may stave off financial ruin, it only confirms the death of the city as a city.

Without the city, Arcadia loses half its charm. Its mental weather turns chill, its tunes are so much whistling in the dark.

LIVING ON CAPITAL

When a suburban area is young, it is often fairly rich in flora and fauna: richer, for instance, than farmland where large-field monocropping is practiced. In a young suburb the species of former ecosystems are joined by exotics, among them birds, ornamental plants, vegetables, and weeds. They are joined, too, by a host of native opportunists. These are lured by garbage cans, bird feeders, gardens, and other welfare programs; by eaves, chimneys, storm sewers, and other housing programs; or simply by the edge habitat of which suburbia is chock-full. In North America, raccoons, skunks, white-tailed deer, cottontail rabbits, gray and red squirrels, chipmunks, red fox, cowbirds, crows, and blue jays are some of the big winners in the animal kingdom. Take raccoons, for example: while in rural areas of

Ohio there is one raccoon for every 12 to 45 acres, in suburban Cincinnati the density can be as high as one for every 1.4 acres.

But there are losers, too. The short-term loss to humans is clear when we think of the vandalism some of these species indulge in, and clearer when we expand the list to include such fellow travelers as the deer tick, the Lyme spirochete, and the rabies virus. The real loss, though, is long-term, and the real losers are native ecosystems.

Because suburbia is, in a sense, one vast edge, it has almost no room at all for species that need large stretches of unbroken habitat. Migratory songbirds such as the wood thrush and ovenbird need something like deep forest to nest in. In suburbia, their nests are preyed on by crows, blue jays, raccoons, and chipmunks and parasitized by cowbirds. Woodpeckers and other birds that need dead trees to live in find themselves evicted by the chain saws of tidy homeowners. As smooth-shaven, chemical-splashed lawns replace grassland, brush, and woods, insect numbers decline, and so do the numbers of insect-eating birds such as swallows and flycatchers. Meanwhile, groundfeeders and seed eaters such as house sparrows, starlings, chickadees, and finches do land-office business, grabbing the nesting sites that rarer birds need to survive.

While the sheer number of warm bird bodies is often greater in suburbs than in wilder places, the number of species is generally less. In the suburbs of Tucson, native birds such as the black-tailed gnatcatcher, pyrrhuloxia, brown towhee, and black-throated sparrow, which lived and dined in the paloverde and saguaro cactus, were driven out by starlings, house sparrows, and Inca doves. Not only did these birds claim the flat, simple structure of suburban lawns as their own; their population boom propelled them into the remnant saguaro cactus, where they stole nest sites from native woodpeckers.

Thus, by competition and predation, the go-getters, both native and foreign, are driving many gentler native species to the brink of extinction and sometimes beyond. Though I have spoken so far mainly of animals, plants may fare even worse. That stands to reason, since much of the native vegetation is purposely ripped up by builders, landscapers, and homeowners. But plants also have to deal with the suddenly numerous browsers, above all the white-tailed deer, whose gentle but insistent nibbling can keep many plants from regenerating, erase rare species, and leave whole ecosystems poorer.

Compounding this impoverishment above ground is impoverishment below ground. One survey of a typical suburb found that on 80 percent of its land area the soils had been physically changed, almost always for the worse. Beneath the front lawns, much of the "topsoil" was in fact subsoil cast out from the digging of basements: hard, barren stuff devoid of organic matter. The investigators guessed that it might take a thousand years for nature to restore the soil. She would not get much help from the grass, with its shallow roots and neatly removed clippings. Of course, the soil buried under parking lots will take even longer to recover—if it ever gets the chance. Most of it is biologically dead, its soil communities having been exterminated by being buried alive.

Nor is it only on land that pauperization proceeds. As sewage overflow and lawn-fertilizer runoff find their way into the remaining streams and ponds, a bloom of well-fed microbes sucks oxygen out of the water. Trout and whitefish, short of breath, give way to bass, perch, and pike, which in turn give way to carp and sunfish.

Like Arcadia, suburbia lives on ecological capital. As the wave moves outward, new suburbs take on the freshness of the forests and fields they invade, while older suburbs sink into biological torpor. There are exceptions (of which more later), but this is the general pattern.

Not only does the suburb live on ecological capital; it lives on social and moral capital, too. The values that vintage TV programs ascribed to the suburb—values like neighborliness, family civic spirit, hard work, thrift—are not the values of the suburb at all. They are values that evolved over centuries of face-to-face contact in streets, marketplaces, grainfields, and threshing floors. Now they are blowing away like chaff. They cannot flourish in the aquariumlike, cathode-ray-lit homes, fluorescent supermarkets, and untrod streets of the suburb. As politicians decry the loss of these values, the swelling suburban vote keeps those same politicians in office. As long as they look for "family values" in the individually wrapped nuclear family, it is a safe bet that they are not going to find them.

Even in its heyday (which in America was the 1950s) suburbia was a fiction. It lived off the ecological capital of a wilderness only lately subdued; the economic capital of bustling cities; and the social capital of small towns and immigrant ghettoes, with their extended families and close-knit communities. By sapping the strength of all these, suburbia's success brought about its own impoverishment.

In suburbia, idyll quickly shades into elegy. Nostalgia sets in before the paint is dry on the picket fence. The nostalgia is for something that never was, or was for only a moment, here and there. An unstable compound: the warmth of the village and the cool privacy of the country estate; the freshness of nature and the comfort of gadgetry; the specialness of the edge and the lulling seep of conformity.

ARCADIA FOR HOW MANY?

As I said, there are exceptions. Some suburbs manage to hold together many, if not all, of these mutually antagonistic virtues. Usually, they are suburbs of the type called "exclusive."

But the most beautiful of suburbs are at best Potemkin heavens. The beauty and ease they offer depend on a hidden structure of cities, factories, and factory farms. Because their economies are make-believe, so are their ecologies. A real paradise must be more or less self-sufficient, producing the fruit its denizens eat. When we include within their borders the slums and waste dumps and chemical-singed fields that service them, our Marin Counties and Locust Valleys, our Versailles and Bedford Parks don't look so idyllic.

Not everyone can live in the middle landscape. If the main reason for this is Euclidean, there are other reasons as well. Exclusive suburbs have, since Roman times, wholeheartedly jumped at this conclusion. They stake out their turf and make the most of it, unfazed by the subsidy they are getting from the rest of the world.

The one question the resident of a deluxe suburb (or of any Arcadia) must never ask himself is: What if everyone lived this way? In fact, this is a question that no resident of any American-style suburb must ever ask himself. But though it has not been asked, the question is now being answered. Nearly half of all Americans live in suburbs. If most of those suburbs are far from deluxe, they still manage to use resources and generate waste at a rate never before achieved by large numbers of people anywhere on earth. A good measure of this is carbon emissions: the average American exhales (directly or indirectly) over five tons of carbon a year, or roughly five times the world average. While this figure is in part a mirror of wealth, it also reflects the peculiar facts of the suburban lifestyle. A single-family house that stands alone, naked to the elements, needs a great deal of energy for heating and cooling. (A row house uses 30 percent less, an apartment 40 percent less. In general, for each doubling of density in a metropolitan area, energy efficiency increases 30 percent.) No one can go anywhere except in a car. And since both the wildness of nature and the wildness of the city are far away, the main form of adventure is consumption.

The growth of the middle class has meant the growth of the middle landscape. Against all logic, and in a way that would shock Theocritus and Virgil, Arcadia is being universalized. So far, though, that universalizing has been limited to North America and certain parts of Europe, and to this half-century. It cannot go much further. Already it has imposed great costs on other places and other people, born and yet unborn. If by some malign miracle every family on earth were suddenly to find itself in a ranch house with two cars, a freezer full of meat, and withal an urgent need for semi-weekly trips to McDonald's—were suddenly, that is, to adopt American patterns of consumption—the effect on the planet would be far from idyllic.

THE WORD "idyll" comes from the Greek *eidyllion*, meaning a little pic-ture: an ordered whole, frozen in time. The word "eclogue," which Virgil chose for his pastoral poems, comes from the Greek *eklogē*, meaning "selec-tion." Like other paradises, the suburb defeats change and division—or tries to—by means of a narrow selection from the gamut of reality. As often happens, we get to wholeness by way of exclusion.

Does this mean the whole idea of Arcadia is pure tripe? To draw such a conclusion at this point would be premature. Let us continue our survey of modern would-be Arcadias, and then—turning at last to the actual, geo-graphical Arcadia—let us see if we can find a more hardheaded way of deal-ing with change, and of negotiating the shifty, pebbly terrain between wilderness and civilization.

14

LOST ILLUSIONS

THE MIDDLE LANDSCAPE is sore beset when we ask it to do the job of both ends. Required to keep a modern person happy year-round, the middle landscape either fails or else succeeds at great (though often hidden) cost. But what if we use a somewhat more rustic, more old-fashioned middle landscape as a place to summer or weekend in? What if we move back and forth between city and country, making full use of each?

Suburbia is just one shape Arcadia takes today. Another is the summer place. Summer people, too, strive to find an ideal place in the sun somewhere between the Mountain and the Tower. But they do not put so heavy a weight on that place, for they do not expect it to meet all their needs. They admit that the split between wild and civilized—in the world and in their souls—is real, and the best they can do to heal it is to go back and forth like a suturing needle.

If going somewhere for the summer seems an indulgence of the rich, one has to remember that its pedigree is both ancient and humble. A story from Plutarch makes the point. On a visit to Lucullus's villas near Tusculum, which featured vast open banquet halls, his friend Pompey observed that they were perfect for summer but would be uninhabitable in winter. Lucullus burst out laughing. "Do you think I have less sense than cranes and storks, and don't change my address with the seasons?"

Thousands of years before the Rockefellers began summering on the Maine coast, the Abenaki Indians were doing it. Their pattern of hunting in the inland woods in winter and fishing and foraging by the sea in summer is typical of many peoples indigenous to the colder temperate zones. Wherever there are seasons, hunting-and-gathering peoples engage in some kind of seasonal migration. So did our ancestors during the Ice Age. Our minds and bodies have been shaped by millennia of going to and fro, following the game that was following the grass.

This brings us back to the storks and cranes Lucullus took as models. Migration has been called the breathing of the planet: inhaling toward the equator, exhaling toward the poles. Humans migrate *as* other animals do, and also *because* other animals do.

But it is not only hunters who move with the seasons. Among pastoralists there are nomads and seminomads, and among the latter the transhumant herdsmen of pastoral. They, however, are not the real authors of pastoral, or its real subjects. For the last twenty-three hundred summers, the pastoral genre has been a kind of journal kept by "summer people" in which they have set down their pleasures, dreams, and disillusionments.

When Virgil wrote his *Eclogues*, Rome was a city of almost a million persons, many of them wedged into ten-story concrete tenements that tended to collapse without notice. (As Juvenal later said, much of the city was "held up by slender props.") Even in a palace one could not escape the summer heat, still less the heat of lawsuits and intrigue. Like all big cities, Rome was often a place to flee from. One fled to the suburban hills, or to Tivoli or Tusculum or the Bay of Naples, in hopes of rustic solitude (a fashion that might be said to have begun in 184 B.C., when the Elder Scipio retired to his estate on the Campanian coast to escape prosecution for bribery). But soon everyone else had fled there, too, and had to be entertained, just as in Rome. Political contacts had to be kept up, egos stroked, palms oiled. Deals were made poolside, as in Hollywood. By degrees the *villa rustica*, a roof over one's head while one oversaw the sowing of wheat or the harvest of grapes, became the *villa pseudourbana*, boasting all the comforts of the city and then some.

What happened near Rome has happened near every great city. Most people want to escape to a place that's somewhat civilized. They recoil not only from wilderness but from heavy-duty rusticity, the boonies, the sticks. When the urban poet actually encounters the "simple life"—as Ovid did while in exile among the Getae on the shores of the Black Sea, to take an extreme case—he is rarely a happy camper. Theocritus or Virgil, plunged into the depths of a virgin Mediterranean forest, would have been terrified. As we have seen, the Golden Age or Arcadia is not generally the most remote time or place, beyond or sheltered from the concentric waves of progress. More commonly it is a placid trough between the waves. It is a phase—not the first, but not the latest, either.

So it is with the urbanite who seeks a summer place or a villa or a homestead. I say I want "real country," but do I really? Of course not. I want a few friends from the city, a modicum of culture, a few conveniences. A country store with Swiss-process water-decaffeinated organic French-roast beans.

But it should still look like real country, and be priced like real country, at least when it comes to raw acreage.

This doesn't describe a place; it describes a phase. You can find the perfect place, but it won't stay perfect. (In fact, you want to find it just before it gets perfect—before land prices shoot up—just as you would lead the receiver when throwing a pass.) No doubt one person's perfect is another's spoiled, and a third's backwater. One person's middle landscape is another's middle-of-nowhere. Nevertheless, in some sense they all are seeking the same thing.

In a deeper sense, though, you can never find the perfect place, not even for a day. If for a day, a week, or a month it does seem perfect, that is only because the tradeoffs have not yet made themselves felt.

"The city-dweller," writes Harry Levin, "if he escapes to the countryside, must overhear the countryman bemoaning his lot, and thence conclude that there can be no escape from the liabilities that flesh is heir to." Elsewhere he says, "Any period of the past, so long as it was the present, had its own discomforts, and . . . its own yearning for still more distant periods."

Though the first of these complaints is lodged in space, the second in time, both are complaints about the same problem: a problem caused by the waves of change. The countryside to which the city-dweller escapes is by definition accessible and so is on its way to being spoiled. In some other place the Golden Age still reigns, but that place is *not* accessible. As the farmer tells the motorist, you can't get there from here.

The summer person or transported city person feels either that he has arrived too late and the place is already spoiled, or else that he doesn't really belong. Any Arcadia that would have him he wouldn't want to join. As Gallus cries in Virgil's Tenth Eclogue,

> If only I had really been one of you—a shepherd
> Of one of your flocks, or a dresser of your ripe grapes!

For the city person, any country place must seem either too raw or too refined. If it is too raw, he can go home. Or he can sit tight while time (and the influx of his friends) does the refining. The rub is that by the time he is fully comfortable the place will be spoiled. If, on the other hand, the place is too refined, his only recourse is to strike out—or imagine striking out—for wilder places.

From ancient times, the ideal of the country home has waffled, striving now for rugged simplicity, now for civilized privilege. In nineteenth-century England, great manses were called cottages. In the Adirondacks, vast estates were called camps. In modern New England, the idea is not to be a country squire surrounded by peasants, but to be a yeoman surrounded by neighbors. Some of the neighbors might be city people like yourself, but most would be authentic country folk who heartily welcomed you into their midst, freely sharing the wealth of their traditions, their canniness about land and weather, their oneness with nature. They would be folksy but enlightened, not rednecks. A place like southern Vermont is set up to feed that fantasy; like painted corpses, its farm communities become more quaintly lifelike (the pastures greener, the barns a brighter red) as the farming disintegrates.

The Arcadian, I said, tries to solve the problem of humankind's place in nature by finding the place where it is already solved. In the case of the summer person, it is a matter of avoiding responsibility by taking over someone else's Arcadia. You find a landscape that has been shaped by centuries of steady, often testy debate between humans and nature, and you try to stop it in mid-sentence. You find a working landscape, put the workers out to pasture, and expect it to stay just as it is. The landscape becomes a "landscape"—an aesthetic object, sprayed with fixative. Beneath the glossy surface, reality decays.

STRIKING A BALANCE

Even if one could have Arcadia to oneself and one's boon companions—even if one could keep the population of nymphs and swains at the optimal level—fulfillment would remain elusive. For standing in the right relation to nature and culture is not a matter of standing at a certain point in time or space. At almost any point one can stand in a right relation or a wrong one.

To understand this, it helps to shift our gaze from the dimension of space to that of time. We often think of the Golden Age as something that happened a long, long time ago, when the human race was young. But we also have a sense that if you go back too far you get too primitive: a view embodied in myths about culture heroes who gave humans the tools that raised them above the animals. These two ways of thinking can cohabit in the same culture (Prometheus and Cadmus on the one hand, Pandora and Cronus on the other). They can even cohabit in the same myth or mythic cycle. Mesopotamian, Egyptian, and African origin myths (among others) tend to follow the sequence: creation, gift of technology, fall.

From at least the fifth century B.C., the ancient world wavered between a belief in the degeneration of man and nature on the one hand, and a belief in technological progress on the other. There was plenty of evidence for both views. Arguments for both could be drawn from myth, philosophy, and the testimony of the senses. The atomist poet Lucretius in particular found himself over a barrel, there being good Epicurean grounds for both views. Lucretius opts for a commonsense compromise between primitivism and antiprimitivism, which could lend itself to fixing on a certain phase as a Golden Age, or the closest to golden that man can attain. A pelt is better than nakedness in cold weather, but a robe of purple embroidered with gold is unnecessary. People don't know *quoad crescat vera voluptas*: in other words, when to leave well enough alone.

Arcadia represents a balance between nature and culture. In Arcadia men know just enough of the arts of civilization to feed their bodies and souls. They have music, they have rhythm, they have their girl or hope to have her presently.

While pastoral seeks an in-between place where this balance is achieved, the Golden Age myth may seek an in-between time. The error in each case lies in thinking that any one time or place can fill the bill.

When people look at history and try to find the Golden Age, the time when things were just advanced enough but not too advanced, they tend to forget the tradeoffs—that something is always lost when something else is gained. We commonly think of a certain phase of a nation's history as a Golden Age, as in the case of Elizabethan England or Periclean Athens or Jeffersonian America. Culturally this may be correct, but there is usually a seamy side that is ignored. We conveniently forget about the drafty houses where a glass of beer could freeze in your hand while you stood with your back to the fire. We forget about infant mortality and the restriction of suffrage to men of substance. We forget about slavery in Virginia and Athens, poverty and disease in Merry Olde Englande, deforestation in all three. Would we give up our vaccines and compact-disc players for all the glories of those golden ages? For that matter, would we trade the music of Mozart, Bach, and Beethoven for all the reedy felicities of Arcadia?

People miss what they've lost (or think they have lost), not what they have never imagined. Lucretius's semiprimitives were happy not by virtue of some innate nobility of character, but because they didn't know what they were missing. Virgil himself admitted as much: the Arcadians may be happy, but we can never be happy among them. And so we can never be happy at all.

At least, not in the way pastoral hopes. For there is no single social or technological phase or condition in which humans are in harmony with nature or a perfect balance between nature and culture is struck. Various societies and persons achieve the equilibrium to a greater or lesser degree at various times. Hunters, pastoralists, farmers, even city-dwellers may find it, may lose it.

Does this mean we must agree with the poets and critics who say that pastoral is a state of mind? That Arcadia is within you?

Not necessarily. Rather, that an equilibrium is something that must be actively maintained. Since neither culture nor nature stands still, there is a constant need for tinkering and rethinking. Hunting, herding, farming, even manufacturing can be done ill or well, mindfully or heedlessly. What matters is the relationship with nature, and the degree and kind of interference with nature's workings.

THE REAL ARCADIA

Oddly, the fact that there is no one Arcadia may best be brought home by looking at the one Arcadia you can find on maps.

On maps, you will find Arcadia in the mountain-hemmed middle of the Peloponnese, the huge near-island that makes up much of Greece's land mass. (Sparta sits on it, which is why the Athenians called their conflict with that city the Peloponnesian War.) Apart from being off the beaten track, what qualified Arcadia to become the official setting of pastoral?

It was Virgil who made it so. The first pastoral poet, Theocritus, set most of his idylls (or most of those whose setting is explicit) in Sicily, where he was born, so his landscape is perhaps the longed-for landscape of childhood. In Virgil's day Sicily, though still a breadbasket, was largely deforested and well on its way toward its present magnificent bleakness. Nowadays Arcadia is nearly as bleak, but in the first century B.C., when Virgil wrote, it was still "distant and unknown and 'unspoilt.' "

It had other qualifications. The economy of Arcadia was in fact largely pastoral. So was the landscape, which was known for its streams and springs, its forests, and its fine sheep. The Arcadians were an ancient people, rumored to be "older than the moon." They were said to be the first humans to grow from the earth, just as the oak was the first plant. They happily ate bread made from acorns, as men did in the Golden Age. Some of the reputation for rugged virtue which clung to faraway peoples like the Hyperboreans and Ethiopians clung to them as well. Certainly they kept alive in their mountain fastnesses many rituals and cults of Bronze Age and even earlier origin.

Arcadia was the site of the oldest cult of Artemis, whose haunt was wild nature and who may have been a Minoan or pre-Greek "mistress of animals," perhaps an earth goddess. Hermes, too, had his oldest cult here and was said to have been born here. (Both goddess and god are more commonly known by their Roman names, Diana and Mercury.) Hermes got his start as a heap of stones, being the god of cairns used as landmarks and boundary markers, often for pasturage. From this he evolved into the god of travelers and of the common people and so of herdsmen, who are both. He was also, like Artemis, a master of wild beasts, meaning that if you asked him nicely he would keep them away from your sheep. He was called Nomios, protector of flocks—though the title may at first have referred to his role as overseer of grazing rights. The word *nomos*, meaning "law" or "convention" and used in opposition to *physis*, "nature," is believed by some to have meant originally the allotment of pasture. *Nemō* means both "to divide up" and "to graze."

Hermes was both a messenger and a guide, and so a mediator between realms: divine and human, living and dead, wild and civilized. In keeping with the speed and cleverness these roles required, he was also known as a trickster and the patron god of thieves (his best-known heist concerned the cattle of Apollo).

The chief god of Arcadia, though, was Pan. Half man, half goat in aspect, he was a perfect glyph of the pastoral realm, where nature and culture mingled. A protector of herdsmen and country folk, he appears on vases bearing a crook in his hand or an animal on his back. In Aeschylus he is the avenger of mistreated beasts. Although small-time by Olympian standards, he is the unquestioned ruler of all leafy and grassy places. Nymphs, satyrs, even the great Hermes are part of his entourage.

For Romans, Pan easily merged with Faunus, the goatlike wood spirit of Italy. Helping the merger along was the legend that the banks of the Tiber had been settled by Arcadians—a legend Virgil used in the *Aeneid*, where he makes the Arcadians allies of the Trojans and, in a way, cofounders of Rome. By moving the action of pastoral from Sicily to wooded Arcadia, Virgil is at once bringing it home to his own wooded Campagna, where he can keep tabs on it—and moving it so far away that nobody else can.

DISCORD AND HARMONY

The key fact about Pan, though, is that he is a musician. Like many mortal shepherds he is an adept of the syrinx, also known as the panpipe. Nor was Pan the only Arcadian who could play it. Arcadia was known to be a hive of musical skill. Polybius, a Greek historian of the second century B.C., says that music is important for all peoples but necessary for the Arcadians. The reason is the chill and gloom of their climate, which requires of them austerity and hard work. "Wishing to smooth and soften the stubbornness and hardness of nature"—their own nature, which had been formed by the nature around them—they instituted musical exercises, games, and festivals.

By the time of Virgil, an opposite rationale seems to be taken for granted. Blessed with a mild climate and abundant leisure, the Arcadians have nothing better to do than make music. That is, they can afford to spend their time making music; whereas Polybius would say they can't afford not to—if they did not make music, their hard life would make them savages. (The only Arcadians who failed to make music were the Cynaetheans, who made war instead. They were so famous for brutality that other cities would not receive their emissaries, and sometimes went so far as to purify the streets after they had passed through.)

However much at odds the poet and the historian seem on this score, they agree on a fundamental point. Both take an interest in the moral function of music, an interest that is deeply Platonic. Music teaches the Arcadians how to resolve the disharmony between people and nature, and between people and other people. Even the contests Polybius talks about make for social harmony—just as they do in Virgil's Arcadia, even if the implied prize is often a woman's favors.

What do we learn from these tidbits about the real, geographical Arcadia? In the lore of antiquity, Arcadia is an out-of-the-way place where people still cleave to the simple ways of the Golden Age. Though close to wildness, it is a place where culture has made its mark: grazing rights are apportioned, ravishing songs are sung. So nature and culture are mingled, as they are supposed to be in the middle landscape.

But when we look at these particular instances of culture, we find that they are not chosen at random. Grazing rights are not property lines that are set once and for all. They are *ad hoc* arrangements that may change in response to drought or plenty, or the retreat or advance of wilderness, or changes in the population of a village or a clan. That is why you need a real-live god on the job, not just a heap of stones. No permanent wall divides pasture from sown land on the one side, or from wilderness on the other. If a hillside has been grazed too hard and shows signs of strain, a good shepherd takes his flock elsewhere. In short, the balance between nature and culture must be actively maintained.

The same process is played out in music. Music is not just the culture that is mixed with nature; it is itself a parable of the mixing. In music, harmony is not established once and for all. Concord grows out of discord, and needs new discord to keep it fresh. A permanent tonic chord does not interest us, partly because it has nothing to do with life.

The ancients saw less conflict between conflict and harmony than we do.

Their belief is borne out by music—and by ecology. Here is one of the things that have kept pastoral reasonably fresh, in spite of shameless canning and freezing, through the ages. Its paradise is not a place without conflict (which would soon pall, as the Christian heaven must) but a place where conflict is sublimated—through not too much!—and resolved in a kind of harmony.

Trying to strike a balance between nature and culture is not in itself an error. In fact, it's a necessity. It is part of the search for wholeness, for a healing of the breach we feel in the world and in our selves. Of course, the wish to have the best of both worlds may involve a quantum of self-deception or hypocrisy. But it can also be honest and rigorous, as in the case of Thoreau and his cabin by the pond. Forget about the laundry done for him by Mrs. Emerson—that's nitpicking. But he had his books, his pen and paper, his flute, his visitors and their conversation, his scientific studies inspired by Agassiz. What was wrong with that? The Benedictine monasteries, the early kibbutzim also aspired to have the best of nature and culture. Many people have recognized the need to strike a balance. One can see this as a contradiction, as a compromise, or as an ideal in its own right.

Striking a balance is a trick—not only in the sense of being hard, but in the sense of requiring light fingers, quick thinking, a sense of humor, even a kind of guile. Here is another reason to invoke Hermes—the divine cat's paw, the trickster. Hermes is the guardian of flocks, but he steals them, too: most famously Apollo's, whose values—order and clarity—he shows to be less immutable than they look.

Where the error lies is in thinking the balance can be struck once and for all. Try to lock yourself into the perfect place or the perfect time, and you not only shirk the real challenge; you bait the demons of change. Jammed tight like a spring in the psyche, change waits to explode. Here the figure of Pan appears again, in a less amiable light: not as lusty goat-man, not as merry piper, but as the god who rends the stillness of noon with a bolt of frenzy that makes flocks stampede and herdsmen go mad. Pan, the god of panic.

In modern times, the same god puts his mark on the mad housewives and clenched commuters of suburbia. The attempt to freeze in place the world of the Cleaver family—the ideal world of the 1950s that never really was—has kept many pharmacists in business. A world that smothers both the wildness of the Mountain and the human wildness of the Tower is a world that invites revolt.

A LEAF FALLS

Pastoral is a dream troubled by shouting in the street. We can pull the pillow over our heads, but the noise of history will wake us in the end.

Read between the lines, and you can find in pastoral some useful hints about how to strike a balance between nature and culture. Read the lines themselves, and you can learn a lot about what not to do. Though pastoral is an art of evasion, it is also an art of disillusion: of the slow sinking in of reality.

If the dream of paradise is deeply lodged in the human brain, so is its denial. As myths of union are balanced by myths of division, so myths of Eden are checked by myths of a Fall. Poets flock to Arcadia and, one by one, are shorn of their illusions. A leaf falls and the endless summer is over.

Well, we grow up. We face reality. But there is more to it than that. In any earthly paradise worth its salt, humans and nature are in harmony. Something in us senses, though, that nature's harmony is not our harmony—that, if we try to tune her strings to our scale, they may snap. Once that is understood, the quest for harmony becomes a good deal trickier.

Whether we seek paradise in a return to the heart of wildness or, instead, in a dalliance at its edges, we end up wrecking what we seek. In the first case we stick in nature's craw. In the second, we steadily press upon her; and the more faithfully we seek a gorgeous stasis—the more we try to keep an exact and ideal distance—the more relentlessly we hem her in. For, by definition, she withdraws as we advance; and by inclination, we advance as she withdraws.

Change, it turns out, is a form of love's unrequiting. We say to the moment, "Stay, thou art so fair"—but it never does. Nature meets our advances by withdrawing. If this seems a betrayal—as by a mother who leaves us bawling in the crib—it is also the condition of her bounty. To her, it must seem that we requite her love with violence.

15

THE WALLED GARDEN

A PLEASURE GARDEN gives pleasure to the senses of sight, sound, smell, touch, and sometimes taste, but most of all it gives pleasure to the soul. The soul is troubled by change and its attendant demons; the soul is troubled by the rift in the world, and in itself, that is made when Mountain and Tower each exert their pull. A pleasure garden responds by offering a microcosm of the world, and a macrocosm of the self, that is whole and unshakable.

The study of gardens, then, is not purely a matter for sturdy old ladies in tennis shoes who want to know when the Souvenir de la Malmaison rose was first grown in Rhode Island. You can learn a lot about successive Western civilizations and their relations to nature by seeing how they created their ideal worlds, and what they chose to wall out.

WHEN THE WAVES of change come tumbling toward you, one very natural response is to build a wall. After all, a wall does not just protect a region of space; it also protects a region of time. A walled place can be an island in the current of time.

In most Western languages, the word for "garden" comes from a root meaning "to enclose." Maybe that helps explain why, for many of us, a beloved garden is as close as we ever get—in the present tense—to paradise. But though the garden wall is comforting, it is not a simple or wholly benign thing. Sometimes it is not a thing at all, but a mental device that frames or screens our relationship to nature and other people.

Our word "garden," which has the same root as "yard" and "gird" and "girdle," is not exceptional. The Hebrew word *gan*, used for the garden in Genesis, also means "enclosure." To Greek it, the Septuagint chose *paradeisos*, a word introduced by Xenophon to describe the pleasure parks of the Persian nobility. He took it from the Old Persian word *pairidaeza*, the roots of which suggest mounding or forming a wall around something.

Here the ambiguity begins, for *pairidaeza* could mean either a formal garden whose walls kept wild plants and animals out, or a hunting preserve whose walls kept wild animals in and unwanted people (poachers, wood-cutters) out. By one wall man was protected from nature, by the other nature from man.

The ambiguity is not just Persian. Every garden wall is two-sided. Maybe nature is outside and the garden is "culture," pliant to the hoe and safe from wild beasts. Or maybe the garden is the preserve of nature in a world over-run by human feet and plucked clean by human hands.

A garden can be part of a wave of human-led change, in which case its wall is meant to consolidate that change on a small piece of ground, shield-ing humans and their allies from any counterattack by the forces of nature. But a garden can also be a piece of ground shielded against a new wave of change. In the first case the garden is an island of futurity, in the second case an island of pastness.

On one level this is all very down-to-earth, since garden walls really do perform both functions. On another level, though, it betrays a kind of wishful thinking—a yearning to have the cake of nature and eat it, too. A good way to see this is by considering what many would call the first of all garden walls.

Until recently, poets and painters have felt the need to put a wall around Eden, even before the Expulsion. No doubt the main reason is that it was a garden, and gardens (as everyone knew) had walls. But these artists may also have been engaged in a kind of retrospective prophylaxis. Perhaps a wall would keep evil out and forestall the Fall.

They should have known better. In Milton, the ramparts are leaped by Satan "at one slight bound." How could they have kept him out, when the possibility of evil was already present in the human heart?

Let me rephrase this in terms I have used before. Humans, the Bible says, were put in the Garden "to work it and protect it." But from what do we protect Eden, if not from our own work? We move outward in waves of work, changing Eden into something else. This is the koan that we must solve, at peril of our lives: how to protect wilderness from our work. Now, to think that we could prevent the Fall by building a wall around Eden is to miss the point of the riddle entirely. It is as if we thought we could stay hap-pily in Eden by walling ourselves out.

What is the Fall but change in its most abrupt and brutal form? And change is brought about by something within the walls: not just the walls of the Garden, but the walls of our own skin.

THE PARADISE PARK

The enclosed park may have been invented by the Persians from whole cloth, or it may have had Mesopotamian sources. In any case, it is the sort of thing that is bound to arise in any place where hunting is a general practice and people come to outnumber the game. At some point the ruling classes must protect their sport from the people's hunger. In Mesopotamia hunting was a fringe activity. In Egypt it was a passion of the rulers but not of the fellahin, who were stuck in the mud of their fields. The Persian highlands, on the other hand, bred tribal, seminomadic herdsmen and farmers who were also hunters. Like the American colonists, they were redoubtable marksmen and they devastated the wildlife around them. Hence the need for walls.

Within the paradise park was a second enclosure, the formal paradise garden. The two parts of the Persian paradise had very different functions. In the park one rode, ran, shot, swam, and engaged in all kinds of strenuous activity. In the garden one did nothing. Or, rather, one sat, and thought. And maybe talked, wrote poetry, made love.

The outer wall around a Persian *pairidaeza* was meant to keep nature (in the form of game) in and humans (in the form of poachers) out. But the inner wall around a paradise garden was meant to keep all but a finicky selection of nature out. (Psychologically, the two walls may have had the same function. To make us feel safe and sheltered, a wall must not be too far away. If the walled-in area is too big, it can harbor dangers of its own. And then we need a second, inner wall.)

Inside the inner wall would be a wall of poplar or cypress to keep out the sun and wind of the steppe. As is common in hot, dry climates, the garden was asked to grow its own protective membrane, like a cell. Nature walled itself in (or out). Surrounded by this wall of trees and then by a plantation of trees, the garden would feel like a clearing in a forest, reversing the figure and ground of Persian reality.

As pollen records attest, the Iranian highlands were once well, if unevenly, wooded. Their denudation was the work partly of marauding foreigners like Gilgamesh, but mainly of the inhabitants themselves: not only the Iranians, who might perhaps be thought of as marauding foreigners, but the earlier inhabitants as well. Probably the first paradise-makers had only to put walls around patches of ancient forest. Their successors had to make the contained as well as the container: that is, they had to plant trees. They tended to plant them in staggered rows: a quincunx pattern, which is the most efficient way to pack round things, in this case the canopies of trees, into a given space, and thereby maximize shade. The effect was probably much like that of a *bois* attached to a French château, with intersecting avenues down which hunters would chase their prey as if following a getaway car.

The Persians made paradises as refuges from the unblinking sun. But this very sun was a monster of their own making. For the treelessness (and so, to a degree, the cloudlessness) of their land was partly their doing. By making paradises they merely restored a few tatters of the shade of which they had robbed themselves.

The wall of the paradise park, then, was a wall against time, by means of which the rulers of Persia created an island of pastness. It was also a cultural frame which placed them in the idealized world of their ancestors—a geometrical wilderness in which nomadic nostalgia could be freely and safely indulged.

THE PARADISE GARDEN

The basic form of the paradise garden was a square or rectangle divided in four parts by two intersecting channels of water. In Persia, in the rectangles of ground between the channels—in the spaces unshaded by almond, cherry, apricot, peach, plum, fig, orange, lime, cypress, or plane trees— flowers spangled the grass, or were massed and mingled as in modern English gardens like Sissinghurst (whose maker, Vita Sackville-West, lived for a time in Persia and was an expert on its gardens). The formal beds and parterres to be seen on later carpets seem to have been an idealization and a convenience for the weaver. In time, the gardens would copy the carpets and adopt, in much of the Islamic world though less so in Persia itself, a static and textilelike decorum.

If the form of the Persian garden impressed itself on the West, so did its content. The clearing of Persia's ancient forests had perhaps had this one saving grace—that it made room for the spread of wildflowers. In spring Iran was, and is, a pandemonium of roses, delphiniums, hyacinths, tulips, jonquils, daffodils, lilies-of-the-valley, marigolds, gillyflowers, and ranunculi. (The slopes of the Mediterranean world, similarly shorn of wood, had covered themselves in a humbler garment of garigue, as showy to the nose but—despite anemone, narcissus, and some other standouts—less so to the eye.) Given such a wealth of color, hyperformal arrangement was beside the point. So was any kind of fancy breeding. The flowers spoke for themselves. (As Jean Chardin remarked after two visits to Persia in the 1670s, "I have found it to be a general Rule, that where Nature is most Easy and Fruitful, they are very raw and unskilful in the Art of Gardening.")

The Persian garden was meant as a distillation of spring which, with some care, might keep for a few months more. In the countryside spring was cut brutally short by the heat and drought of summer. In a garden flowers could be sustained with a little shade and a lot of water, often sucked from the gravel aquifers of the mountain slopes by means of underground aqueducts called "qanats," whose digging and periodic unclogging drank up vast amounts of labor. Since no cellulose garden could truly make spring eternal, the job was transferred to wool: and so the Persian carpet was born, its mission being to keep the eye, as well as the feet and buttocks, warm in winter. To the best of its ability, though, the wall of the Persian paradise garden was a wall against cyclical as well as linear time.

WALL AND WATER

From the beginning, the two defining elements of the formal garden—apart from plants, which change from place to place—have been wall and water. And from the beginning these two have stood in delicate strife. Wall separates, water connects.

Water is the medium in which life surely began and in which, on a cellular level, it still goes on. Within the cell, water is the reservoir of chemicals and the conduit by which they get where they are needed. But water left to obey its own nature is promiscuous, oblivious, letting anything go anywhere, dissolving all things in a cold entropic soup. Water is an agent of entropy. Agents of separation are needed, too. One of them is the wall, known to the cell as a membrane.

It is often said that we go to the garden to escape the self—to merge with the mother of us all. That may be so. But sometimes we go to find the self. The wall helps us mark the boundaries of the self; the water helps us see its links with other things human and not human.

There are many reasons why the paradise garden—and the formal garden in general—arose in a fairly dry part of the world. First, in a dry place the garden is tacked onto nothing rather than hacked from something, so the mind's geometry is less constrained by what is given. Second, the starkness of light in a dry place favors clarity and balance, a fact often remarked in the realm of architecture. Third and most important are the demands of irrigation. If you have a plot of land that you want to irrigate, the easiest thing is to run a channel down the middle. Or, to look at the thing the other way round: if you have a stream or canal at your disposal, it makes sense to range your garden evenly on either side. Aesthetics are the last thing on your mind. Nonetheless, you have just committed an act of bilateral symmetry.

Like most things, the garden begins in division. As the first living cell earned that name by being walled off with a membrane from the chemical free-for-all that had formed but would as gladly unform it, so the first garden (in the Western sense) was a part of the earth's surface divided from all the rest. And as the first cell with real prospects divided in two, and those two into four, so the progenitor of all the formal gardens of the West (and many of those of the East) divided itself in half, then in half again.

The fourfold square was a template that would stamp the rolled-out dough of earth from the Ganges to the Rio Grande. And it was itself a figure of the world. Muslims and Christians were frank about this in their paradise gardens, whose four channels of water betokened the four rivers that flowed from Eden to water the ends of the earth. They seemed to think they had derived the form from scripture, by a logic that Sir Thomas Browne explains as follows: "Since even in Paradise itself, the tree of knowledge was placed in the middle of the Garden, whatever was the ambient figure, there wanted not a centre and rule of decussation." But the form of the paradise garden is older than Christianity or Islam, probably older than the Book of Genesis. It can be discerned in ruins and accounts of ancient Persian gardens, and its ultimate source may be Mesopotamian.

Clearly this form was linked early on to the notion of a world-pole with rivers flowing from it. But nothing in that notion insists there be *four* rivers. The number does match the points of our compass, but there is no compelling reason why we should have four main compass directions rather than three or five or six.

The Akkadian rulers styled themselves "kings of the four quarters"— meaning the four quarters of the world. It is hard to see how anything in the landscape they knew could have led them to carve up the world in just that way. The great rivers of their acquaintance ran parallel to each other, not perpendicular. Nor did the canals (which in the Mesopotamian mind were just as primeval as the rivers) run at right angles to the rivers or to each other. Water does not run that way, and the Mesopotamians knew better than to force it. Their canals followed the lead of the backwaters and ephemeral side-channels of the Euphrates, running in a braided, later a branching pattern alongside the river.

The motif of the square or circle sliced from top to bottom, from side to side, or both ways at once is found on some of the oldest pottery of the Near East and Europe. It is found in Upper Paleolithic art and in the products of present-day hunting-and-gathering peoples, and so is older than the first garden.

Where did this form come from? Maybe from the brain itself, whose two hemispheres nicely body forth its habit of dichotomizing. But those who want to look for models in the external world will not have to look far. Bilateral symmetry is so common on this planet that a visitor from a world lacking it might think himself in a hall of mirrors. Leaves have it, except for freaks like sassafras and mulberry. Bees have it; birds have it, as do all other chordates. Violations of symmetry by certain of the vital organs are done surreptitiously, under the skin. Radial symmetry—the symmetry of a spoked wheel or a sand dollar—is also common, and one could say that the quadrisection—the basic principle of the paradise garden—is the point at which bilateral symmetry crosses over into radial. It is also the point at which the human mind asserts itself. Simply to mimic bilateral symmetry is hack's work, but to do a half-turn and repeat the operation shows some initiative.

In nature as in culture, form often follows function. In bodies as in irrigated fields, symmetry is often a response to the need for an even distribution of water, energy, and nutrients. In bodies, it is also the product of genetic programs and physical laws that cause complex creatures to grow from relatively simple germs. But the same kind of thing happens in the realm of culture, and the development of the formal garden is a fine example.

The kind of division water brings about is just the converse of connection. In any living system—a cell, an organism, an ecosystem, or a man-made ecosystem such as a garden—the principle of flow can be realized only by means of structure.

A WHOLE WORLD

For all this talk of division, the main job of water in the garden is still to connect. If the paradise garden copied the World Mountain, what it copied was not wildness (at least, not wildness in our sense) but the principle of flow. Yet here again there was a touch of ambiguity.

The paradise garden could be seen as the world-pole, from which the four rivers flowed to the corners of the earth. The canals could be projected beyond the walls of the garden, either in fancy or in fact. In fact, it was often so. In Persia, the overflow from the nobleman's garden might be channeled to serve nearby farms and villages. The same thing would happen in Renaissance Italy. Water then overcame the wall, joining the garden to the whole world.

But the paradise garden could also *be* the whole world. If the garden was a microcosm, only the central font was the world-pole. The four channels were still the four rivers, but they ended (or were imagined as ending) at the garden wall, whose four corners were the four corners of the earth.

All gardens share this confusion, which is tied to the question of what is in and what is out. There is no simple answer. Maybe the world is outside and the garden is a refuge from the world. Maybe the garden is the true world, the distilled essence of the world—all the world that matters. Maybe the garden is the nerve center from which the world can be controlled. Or maybe it is a miniature world that we can alter to taste. This last option is especially tempting: the garden as a doll's-house world that can be dressed up and possessed. We can put in it everything in the world that we like—rivers, lakes, mountains, forests, prairies—all in miniature. For some people, that is what paradise is: a small piece of the earth's surface that can be made over to match our dreams.

But the urge to make an ordered, pint-sized world does not always come from the clutching, controlling ego. Often its source is deeper. The rift between wildness and civilization, in the world and in the self, leaves us with a craving for unity: for an ordered whole that fuses both extremes.

If we take the paradise garden as a model of the self as well as of the world, its fourfold form would seem to be a mandala: according to Jung, the universal archetype of the self, which he found in sources as varied as Rhodesian rock drawings, the yantras of Tantric yoga, and the dreams of Swiss neurotics, as well as in the myth of Eden. In that case the garden would not only help to define the limits of the self, and show its connection to the outside world; it would also help give the self structure, and hence stability. Above all, it would give the self a unity that it might otherwise seem to lack.

The mandala, Jung said, represents "a psychic center of the personality that is not identical with the 'I' " (or ego). When it surfaces in dreams in one of its many forms, it shows that the sleeper is hard at work trying to "integrate the personality"—to cobble together conscious and unconscious, male and female, light and shadow in an ordered, centered whole. According to certain mystical traditions, the same process is going on in the universe at large. Pythagoras saw in the tetrad the structure of the cosmos, which is held together by the concord of opposites.

Let us try to understand this in our own terms. As the Tower comes to seem the center of the world, the conscious, more or less rational ego comes to seem the center of the self. Yet all the while there is a nagging feeling that other things—wild things—are being shut out, or shunted off to the side. The ego feels lighter and flimsier than a real center should; the Tower feels gimcrack and unfounded, as if the snapping of a single strut would make it topple over. Perhaps the real center of gravity of the self, as of the world, lies elsewhere.

The garden tries to find the true center, joining opposites in an ordered whole. It is at once an ensouled world and an embodied soul.

What center does the paradise garden find? At first glance, it would seem to be the Mountain. But while this does seem to be the thing symbolized, the form of the symbol makes a difference. The wild mountain, the source of untamed torrents, has become an even-tempered fountain, or simply the junction of straight channels. Though the paradise garden may be an *image* of the Mountain and its rivers, it is an *instance* of the Tower and its proud canals. This becomes clearer in those Persian and Mogul royal gardens where the ruler's pavilion or throne room is perched at the junction of the waters.

Arcadia tries to heal the breach between Mountain and Tower by finding the literal midpoint between them. The Temple on Mount Zion, it might be said, tried to fuse Mountain and Tower in a fairly literal way. The pleasure garden fuses Mountain and Tower, too, but does so by a subtle mingling of the literal and the figurative.

The need for a unified self and a unified world is probably a general human need, though it is often stronger in the West than elsewhere. To connect that need with monotheism, as the renegade Jungian psychologist James Hillman has done, is too pat; it would be more apt to say that monotheism is one of the many forms the same need for wholeness can assume. But Hillman is right, I think, to point out that a too-zealous pursuit of oneness can backfire. The harder we try to pack all the different elements of the self into a single neat box—as in the Jungian "integration of the personality"—the more likely it is that some of them are going to get bumped out. Bumped out with a vengeance, too: becoming "shadows," demons, enemies of the unitary self. What makes this even likelier is that a neat box wants to stay neat by avoiding change, which (at least where the self is concerned) can be accomplished only in a long, narrow box that tapers toward the feet.

Something like this happens in the garden. The Mountain has been brought back into the center of the picture, where it belongs; but its wildness is only token wildness. True wildness, troubling wildness, has been shorn away and walled out. Change, which is just as troubling, has been walled out, too. The exile from Eden is denied.

16

PATTING NATURE
ON THE HEAD

THE TWO GREAT peoples of classical antiquity deserve special notice when it comes to pleasure gardens: the Greeks because they had none to speak of, the Romans because they had, in some form or other, nearly every kind of garden the West would see until the present day. The difference is linked to different views of nature. The Greek view is matter-of-fact, the Roman view materialistic—which is not the same thing at all. Mixing sentiment for *mater terra* and contempt for her in equal proportions, the Romans take us straight into the fissures and false unities of the modern world.

WHEN GREEKS SUCH as Xenophon went to Persia to serve as mercenaries, they brought back tales of the paradises they had seen. But there was no immediate rush to copy them. To the Greeks, the Persians represented barbarism, despotism, and luxury, and the *paradeisos* smacked of all three.

The very idea of a pleasure garden was alien. When Homer musters all his descriptive power to render a splendid garden—the garden of Alcinous, the king of the Phaeacians, a pampered and faintly exotic people who spend much of their time dancing—the best he can come up with is an orchard, a vineyard, and a vegetable plot. Since the orchard is of four acres, fenced, and fed by a single spring, it bears some resemblance to a paradise garden; since some of its trees are blooming or fruiting no matter the time of year, it seems to have achieved the eternal spring for which the Persians hankered. Here Odysseus finds shelter against time and trouble, just as he does in Circe's barnyard and Calypso's forest grotto. But in none of these places is there any suggestion of a pleasure garden.

The classical Greeks were a public people—"urban animals." The idea of escaping from the city—inward to a private garden, or outward to a villa—

would have struck them as daffy. They shared Dr. Johnson's view: "When a man is tired of London"—or Athens—"he is tired of life." Talk, games, art, politics—all were to be found in the city. And the fields were only a few steps away. There was no need to bother with gardens. In private homes, the floor of the atrium was paved or beaten dirt, devoid of vegetation.

There was nothing to prevent the Greeks from making *public* gardens. Cimon planted plane trees in the Agora and the Academy. There were potted plants set in the rock on which the Temple of Hephaestus perches above the Agora. Gymnasia and playing fields had their rows of planes or other shade trees, temples their sacred groves. But all of this was on a modest scale. With the best will in the world, we can find nothing in the literary or archeological record that hints at gardens to match the splendor of the *Oresteia* or the subtlety of the Parthenon. The first city of democracy had no Central Park.

Apart from needlepoint, it is hard to think of an art the Greeks did not hone to that razor's-edge perfection we call classical. Why not the garden?

A HARDHEADED VIEW OF NATURE

In the smallish cities of Greece, with the fields and olive groves just a few steps away, the need for large public gardens, as for private gardens, may not have been so pressing. Also, great public gardens do not arise in any country until great private gardens have shown the way. (The temple gardens of Egypt and Mesopotamia are exceptions that prove the rule, for these were enjoyed by priests and kings and not by the people.) Another factor was the Greeks' matter-of-fact attitude toward nature—the lack of any strong urge to sentimentalize or romanticize nature, which is crucial to so much garden-making.

Look at the works of Homer, which (with those of Hesiod) are the Greek scriptures. Homer and the people he describes have a deep feeling for wild nature, especially the sea; but it hasn't yet become scenery. It hasn't been cloyingly peopled with fake gods, because it is still peopled with real ones. The seashore, for example, is not a place where one goes, like Keats, to be deafened by the roar of infinity. For Homer's people it is a place to launch a boat from, to pour libations, to give the gods a piece of your mind. A no-man's-land between man and nature, where mortals and immortals may powwow and palaver.

For the Homeric Greek, as for the American Indian, all nature is civilized. The land, sea, and air are reticulate with paths trod by purposeful feet, both human and divine; they echo with articulate speech. The ether from which the gods gather gusts of smoking thigh-fat is as much a part of the world's economy as a barley field. The very air is arable.

The same even-handedness can be seen in the most famous of all Homeric epithets: *oinopa ponton*, "wine-dark sea." On its surface this is a simple perception of color. But on a deeper level it links the wildest nature known to the Greeks with the pressed quintessence of their culture, the product of laboriously pruned vines and patient fermentation. Each of the symbols here yoked is itself two-faced. For all its wildness, the Mediterranean was still the freeway of the ancient world, a mirage of shimmering asphalt. Admittedly it was a freeway with lanes and exits unmarked, one whose shimmer might, in the blink of an eye, come to a boil. Yet without it (whatever that can mean) the Greeks would not have been Greeks at all but grunting, mutually unintelligible tribesmen, in manners and depth perception no better than Polyphemus.

As for wine, its smiling surface hides dangers of its own. The seal of civility, wine can make men wild. Although a currency of respectable ritual, it can at times engender the Dionysiac frenzies frowned upon by reasonable people. Wine contained in a krater can be the cause of *akrasia*, the failure to master or contain oneself. We joke about tempests in teapots—real storms can erupt in a bowl of wine. So a thoughtful person might well find the sea in his wine bowl—as he might see in the Mediterranean the communal wine bowl of the ancient world.

In Homer, nature has not yet become a sounding board for human cries and whimpers. What might at first be mistaken for the pathetic fallacy is nothing of the kind: nature has a mind of its own. Many minds, in fact. It is less a sounding board than a switchboard, with the messages of gods and men rocketing across the sky, often colliding or crossing wires. Superstition, pure and simple: and yet a far more accurate view of what nature is—a net of energy and information—than the Romantics ever dreamed up.

The need for wholeness, in the world and in the soul, was not absent from the Greek makeup, but the forms it took did not lend themselves to formal garden-making. Though the Greeks liked a well-rounded man (Odysseus being the textbook example), they did not insist that all his parts fit neatly together. To say they lacked the modern notion of the unitary self would be stretching things, but it is true that their self was a looser bag than ours, with the influences of various gods and goddesses bouncing about like kittens. Meanwhile, from the days of Thales in the sixth century B.C., their quest for a unified vision of the world had begun to shift from the realm of myth to that of philosophy and science. If the universe can be understood in terms of its fundamental laws or building blocks, the urge to find its mythic center may cool.

As navigators who ranged (at the very least) from the Indian Ocean to the Atlantic, the Greeks made real maps, and so had less need to keep a neat schematic map of the world in their minds or on the ground outside their houses. If they had never ventured beyond their own coastlines they would still have known that the world was irregular, more like a jigsaw puzzle than a chessboard. The center of the world might be the Mountain, but the Mountain might be Delphi, or Olympus, or Helicon, depending on what function you had in mind. The center of the world for human purposes might be the Tower, but this in turn might be Athens or Corinth or Thebes or some hardscrabble hamlet in Boeotia, depending on whom you asked.

The same hardheadedness shaped their view of change. Though gardens of a sort had given respite to Odysseus, the ancient Greeks were too clear-eyed to think that mortals could succeed for any length of time in walling out change, or evil. The slide from gold to iron—for people and nature alike—was inexorable. At best one could hope to find comfort and good eating in a well-tended orchard or vineyard.

After Socrates drank his hemlock, the prospects for gardens began to improve. By degrees, the Greek civic ideal grew cold. City-states ate each other up, and the survivors fattened into empire. The gap between rich and poor widened. Democracy came to seem unworkable. For the first time, the notion of walling out one's fellow citizens began to make sense. Philosophers retreated to private gardens, while the wealthy and powerful built villas. As the exploitation of nature and other people expanded, so too did guilt, regret, and the need to hide in idealizations.

The first great (that is, big and lavish) gardens in the Hellenic world were those of the tyrants of Syracuse, in Sicily, who were also the first Greeks to take a stab at establishing a real Eastern-style empire. (Their other notable work, still standing, is the Ear of Dionysius, a dungeon so constructed as to make the whispers of its prisoners clearly audible to the guard.) In the design of their gardens, they were undoubtedly moved by Near Eastern models. The same models inspired the host of gardens that sprang up in the Hellenistic world. Alexander the Great was a fervent admirer of the Persians he conquered. Hot on the trail of Darius III, he stopped to pay his respects at the tomb of Cyrus, and deplored the fact that the garden surrounding it had not been well kept up. As he marched eastward, he sent exotic plants back to his tutor Aristotle.

ROMAN GARDENS

As to specifics, we know little about Hellenistic gardens. It is convenient to guess that they were much like Roman gardens, about which we know more, partly thanks to the exuberance of their owners and partly to that of Vesuvius. Owners such as Horace, Cicero, Virgil, and Pliny the Younger wrote letters and poems boasting of their humble estates and the humble pleasures they had in them. Vesuvius did them one better by preserving in ash, in the year 79 A.D., two garden-rich cities and a number of outlying villas on the fashionable Bay of Naples.

By making casts of the impressions left by the roots of trees and shrubs in the soil—impressions that filled with volcanic debris after the eruption—archeologists are getting a fairly good idea of what and where the larger plants were. Samples of pollen and fragments of pots help fill out the picture. So do the murals found on garden walls, which sometimes mirrored the gardens they hemmed and sometimes improved on them. Taking all the evidence together, we can picture almost the full range of Roman gardens, from the tiny peristyle garden of a fish-sauce shop at Pompeii to the vast man-made landscapes of Hadrian's villa at Tivoli. They deserve our attention because they are the first pleasure gardens that are recognizably Western. In them, the basic vocabulary of the Western garden is set.

Every ingredient needed to make a lavish culture of gardens the Romans had in abundance. To begin with, they loved the soil. The Greeks we know best were seafarers first, farmers second, and though the stolid voice of the hinterland can be heard in Hesiod and Theognis, mostly it is in the background of Greek letters, a black-cloaked chorus. The Romans, on the other hand, were farmers first, taking to the sea late and reluctantly (although then with their usual thoroughness, so that the Mediterranean became *mare nostrum,* "our sea"). The how-to manuals of Virgil, Cato, Varro, Pliny the Elder, and Columella drew on Greek and Punic book learning, but also on traditions of farming that had been common among the peasants of central Italy long before Romulus met the wolf. The landowners who read such manuals wanted to have a hand in managing their own estates. Some were even willing to get dirt under their fingernails. Republican ideas clung to the Roman mind even as Rome broadened and fattened into empire. Lawyers and senators liked to see themselves as honest farmers called from the plow, like Cincinnatus, by the urgent demands of the day.

For all that, Rome *was* an empire. Its lawyers and senators lived off the fat of a nation that lived off the fat of the known world. A few people, relatively speaking, but a lot in absolute numbers, had a lot of money to spend on gardens. A taste for things rich and recherché was easily acquired, easily fed, and never satisfied. Grafted upon the stock of Italian farm lore were the lavish formalisms and fabulous species of the Hellenized East. As Europe, Asia, and Africa were scoured for lions, boars, and hyenas to feed the stymied fury of the poor, so they were combed for pomegranates, octopus, and peacocks to stock the orchards, fishponds, and aviaries of the rich.

We can begin, though, with the more modest gardens of the urban bourgeoisie, as we find them in Pompeii. Pompeii was a thriving commercial town in an old farming region that was now a trendy watering hole as well. To walk the streets of Pompeii was to walk a maze (an orderly, commodious maze) of tufa. Except for the public parks, nature was all inside.

The courtyard that in Greece was paved here becomes a garden, with living rooms behind the porticoes. But the garden is—as often in sunny climes—itself a room. And the indoor rooms are made to look like gardens, with wall paintings of garden scenes. These cover the walls of the garden, too. If everything worked just right, walking into a Pompeian house was like being in one of those children's stories where a gap in a brick wall on a back alley leads to an enchanted world of green.

The painted gardens are alarmingly cute. Their centaur fountains, rose-covered pergolas, lattices, knotted ivy, alcoved statuettes, and fluttering birds seem a kind of ancient kitsch. (They differ only in detail from the gardens kept by modern Italians in sections of Queens, New York.) The actual gardens seem to have been just as cute, if somewhat more modest. Most of the color was green, year-round: ivy, viburnum, acanthus. Scalloped flower beds trimmed in box might have held violets, poppies, Madonna lilies, and daisy chrysanthemums.

Wall paintings might mirror the actual garden myrtle for myrtle, pergola for pergola. (In the peristyle of the Villa of Diomede at Herculaneum, a fish pool dense with fish, eel, and squid shimmies up the wall beside the real fish pool.) But they might also tease the eye outward into a hazy, sylvan infinity . . . in which case the role of the wall was to deny not only itself but all walls, anywhere. Often one or more walls would show a fierce wilderness—jagged rocks, gnarled trees—where lions, bulls, gazelles, and boars chased each other around. This painting was the poor man's (or, rather, the moderately rich man's) hunting park. In fact, it was called a *paradeisos*. The effect was not exactly *trompe l'oeil*, for the scene of wild carnage was bordered with painted or sculptured pomegranates, myrtle boughs, grape clusters, medallions, robed musicians, and other tokens of domesticity.

The walls denied that walls existed; yet one did not feel exposed. The cozily filigreed borders, and the living rooms all around, and the knowledge that Rome ruled the world, both man and beast—all these conspired to make one feel safe. The domestic realm stretched, yawning, to the ends of the known world.

The same mind-set can be seen in villa gardens. The Roman drew the nearby plowland, pasture, mountains, and sea into the orbit of his garden, admiring them all in a proprietary way. Was not the sea *mare nostrum*? Was not the land *mater terra*, dutifully serving her chosen sons?

NATURE PATRONIZED

The houses of Pompeii, which turned their backs on the street and hid their gardens inside, were typical of older Italian houses, but the new villas were another matter, carefully sited where the owners could see and be seen. Wall paintings (confirmed by the spade) show the Bay of Naples strung with villas like a necklace whose pearls jostle each other for notice. Rich men and women vied to show off the most striking views, the most phantasmagorically clipped hedges, the most thalassic natatoria.

The truly rich did not have to rely on painted paradises. Their game parks and landscaped grounds reached outward into nature, at once embracing her, manipulating her, shivering before her grandeur, and patting her on the head. The domestication of nature one finds in the city gardens—making nature cozy and homey—here attains a mind-boggling scale. You can look at the use of animals, both wild and domestic: a distinction the Romans often muddled. Around the middle of the first century B.C., Persian-style hunting parks came into vogue. The lawyer Quintus Hortensius walled in an area of thirty acres and set up a lodge on high ground in the midst of it. There, at meals preceding a hunt, guests would hear a slave dressed up as Orpheus sing and then sound a horn. Promptly the earth would rumble and a crowd of stags, boars, and other game would come running, which not only confirmed a pretty legend but made the hunt easier, too.

The same attorney was an early convert to the craze for fishponds. (He was too goodhearted, though, to eat his own fish, and even wept at the death of a favorite eel.) Then there were the aviaries. Lucullus had one under the same roof as a dining room, so his guests could take in with the same glance the cooked birds on their plates and the ones flying around uncooked. Varro says the idea was a flop, since the eye's gain was the nose's loss.

Talking birds were especially prized—the Elder Pliny writes of starlings and nightingales who declaim long sentences in Greek as well as Latin, of a talking raven given a hero's funeral by the Roman people, of magpies who "not only learn words but love them," even meditate upon them, and die of heartbreak if they come upon one they cannot pronounce—but this did not stop the tragic actor Clodius Aesop from eating them, at a cost of 6,000 sesterces apiece, "impelled by the desire to engage in a sort of cannibalism."

The Romans loved nature, even (up to a point) wild nature. In stories they told about themselves, they traced their descent from wild boys suckled by wolves, or from men who grew like trees in the forests of Latium, or from other men (the Arcadians) who had sprung from acorns in the mountains of the Peloponnese. The love of wild nature showed up not only in the content of Roman stories, but in their style: Virgil's sketches of crags and forests have no real precedents in Greek.

But shadowing the Romans' love of nature was a kind of contempt. They loved to manipulate nature, to twist and master and mimic nature—to one-up nature. Of course, it was sometimes different people who had these different urges, but sometimes it was the same people. Maybe the Romans could afford to love nature precisely because they had mastered her. In this, as in so many things, they stated a theme on which the West would play variations for the next two thousand years.

The Romans were materialists, in the deepest sense of the word: a word that contains the Latin word *mater*, "mother." As James Hillman has noted, the Romans were on very cozy terms with Mother Earth. Their earth goddess, Tellus, was a cozy figure, a patroness of grain and cattle, of marriage, and of all the bodily functions the Romans performed with such relish. The Romans saw no contradiction between their love of this earth goddess and such other expressions of their materialism as sewers, highways, lead smelters, and strenuously artificial gardens. Familiarity bred contempt.

THE FOUNTAINS OF ROME

Nowhere was the Romans' Janus-faced view of nature manifested more clearly than in their use of water. In the Near East, water was at once a bare necessity and the headiest of luxuries. A steady trickle could make life spring up in the teeth of a wasteland. An ankle-deep pool or handsbreadth-wide channel could keep a poet in happy reverie for years. Translated to Rome, where Father Tiber so spoiled his children, this quiet awe of water became a self-indulgent splashing about. No villa was complete without its three baths—*caldarium, tepidarium,* and *frigidarium*—to say nothing of its fountains, channels, and fishponds.

But this splashing about had its steely side. Since Sumer at least, many of the nerviest efforts to control nature have involved the control of water.

This is partly a practical matter and partly symbolic. To control the waters of life, the rivers of life, is to take on the role of the sky god. Gardens let Romans play this role fancifully, as aqueducts let them play it in earnest. Since water is also the principle of connection to the outside world, their virtuoso control of water inside the garden hinted at control of nearly everything outside.

The abundance of water in central Italy had another effect. In arid lands a garden is built on too little, in wet lands it is carved from too much. In the first case addition, in the second subtraction is key. As in Rome the formal garden first touched the tangled bank of damp Europe, so in Rome *opus topiarii*, "the work of the gardener," came to mean shapes hewn and prodded from profusions of boxwood. Overeager nature had to be restrained. Restraint, though, was not the *topiarius's* strong suit. He tickled his charges into forms more fanciful than they would ever have taken on their own: lions and boars, legionnaires in armor, naval battles—in short, just the sort of thing the Romans liked to see in their theaters. In case the triumph of mind over nature should somehow be missed, other bushes were carved into letters that spelled out the name of the gardener or his master.

Compared with aqueducts and artificial lakes, topiary seems a gentle sort of toying with nature. The conceptual daring involved is no trifle, though. "Everything is what it is, and not another thing," said Bishop Butler; but the topiarist denies this. A box tree can be a lion, a galley, a gladiator, even a word.

Some of these tendencies are best illustrated by the later Roman emperors, who toyed with nature as wantonly as with their people's lives. (The two toyings were of course related, both in theory and in practice.) Caligula had built in a vast plane tree a dining room that would fit fifteen guests and their attendants (much of the shade, the Elder Pliny says, was cast by the trunk of the emperor himself). Elagabalus is said to have had a special ceiling in his banqueting chamber which, at a signal, inverted itself and smothered his guests—some mortally—in violets. Gibbon says of him that he loved "to confound the order of seasons and climates," and "never would eat sea-fish except at a great distance from the sea."

Perversity was not *always* the point. Sometimes the point was to mingle nature and culture in perfect measure. Arcadia, the invention of a Latin poet, did this by seeking a golden mean, which entailed some compromise. But the Roman garden tried to get the best of both worlds—no compromise necessary. Nor did it settle for doing this in the realm of symbolism, like the paradise garden. Being good materialists, the Romans wanted to get the best of both worlds in a form they could see, touch, smell, taste, and

vomit behind. They broke down the wall that separated the pleasure garden from the outside world. In so doing, they laid the groundwork for the
great gardens that would emerge in the Renaissance and later.

If the ideal country house held all the luxuries of the town, the ideal
town house condensed all the joys of the country. The last ideal found its
pint-sized expression in the painted wildernesses on the garden walls of
Pompeii. But it was Nero who took this ideal to its limit—dispensing even
with condensation—in his Golden House, which took up 125 acres of what
had been the most densely populated part of Rome. Tacitus says that Nero
"took full advantage of the devastation of his native city" in the Great Fire
of 64 A.D.—during which he so famously fiddled—to build himself "a
house in which the novelties were not the precious stones and gold, now
old hat and debased by the long practice of luxury, but the arable fields and
lakes and artificial wild country with woodland and open spaces or vistas
mixed in." The landscape teemed "with every kind of domestic and wild
animal." Later, the Flavian emperors would build the Coliseum on the spot
where Nero's lake had lain. Re-creating the Nile or the Alps beneath its
bleachers, they would turn his private madness into mass entertainment.

Romans were not fond of the Domus Aurea. "Rome has become a house.
Citizens, emigrate to Veii!"—twelve miles away—"but be sure that the
House does not reach that far." Though the house inconvenienced a few
hundred thousand persons, it allowed one man to achieve full personhood.
Here Nero reached his goal of creating "a semblance of what nature had
refused." Here he could sum up human experience—urban and rural, savage and civilized—and, as he said, "begin to live like a human being."

The project of making a complete world and a complete self, which the
paradise garden could only hint at figuratively, was now carried out in the
most literal way. If Nero could not simultaneously go to the Mountain and
the Tower, the Mountain and the Tower would come to Nero.

NATURE FRAMED BY CULTURE

While many educated Romans shared Nero's ideal of a fully human life, not
all agreed that it could only be attained by razing half a city. The din of
Roman villa-building and landscape-making, loud as it was, was nearly
drowned out by the clucking of the philosophers. From the strictures of
Cicero, Plutarch, and Seneca we get, in fact, much of our information, for
what they skewered they also preserved. The same goes for the agricultural
writers. Cato, Varro, Columella, and the Elder Pliny spent much of their
time telling landowners what not to do: mainly, not to waste good farmland
on frivolities.

Philosophers such as Cicero liked to praise nature. They liked to think that in their gardens and the grounds of their villas they encountered nature herself. They claimed to find escape in their gardens from the lies and backstabbings of human society. What they really found escape in, though, was not nature in the raw, but nature framed by another culture.

How could it have been otherwise? Even an unwalled garden is walled in the sense that it offers escape from unpleasant realities. But that escape is never simply into nature. For you can no more escape culture than you can escape your own skin. The best you can do is hide in the folds of another culture: preferably a culture of the past, and necessarily idealized.

If we think of the garden as an ideal, unified image of the world and of the self, it is easy to see how the messy facts of the present might get in the way. The facts of past and distant cultures may be just as messy but are either unknown or easy to forget.

The culture that the Romans generally chose—not just the philosophers, but normal people, too—was that of classical Greece. This meant copying classical Greek gardens, which is rather like taking the Bushmen as models of haberdashery. Greek porticoes, Greek sacred groves, Greek garden sculpture, Greek grottoes and shrines, even (as at Hadrian's villa) whole Greek landscapes were dutifully reproduced. While the scale and lavishness and much of the contents of Roman gardens came from cultures farther east (as the phrase "Xerxes in a toga," used of the wealthy Lucullus, suggests), the most explicit influence was Greek.

Although fourth-century Greece did not really offer any great models of gardens, it did offer some great models of what to do in gardens. At each of his estates, Cicero had a portico and park that he called the Academy, and in Tusculum he had an extra set called the Lyceum. By imitating what little there was to imitate in the gardens of the Greek philosophers, Cicero and other wealthy Romans could strike a philosophical pose. And so whatever contradictions they darkly felt in their real relations with nature and their fellow men—orating about the public good while the public lived in squalor; enjoying a lifestyle that ate up the nature they professed to love; owning seven pleasure estates while praising the simple life—would be annihilated in the green shade of "the Academy" or "the Lyceum."

By late-Roman times, the basic vocabulary of the Western garden was set. So was the vocabulary of talk about the garden, and the terms of all future debate until the nineteenth century at least. Nature versus artifice, image versus reality, country versus city, pleasure versus utility, symmetry versus asymmetry, luxury versus simplicity: the lines were drawn, and though they have often shifted—what Cicero thought simple and natural would have appalled Pope—they have not been erased.

Such tensions make for good gardens. If we cared only for unmixed nature, we would not make gardens at all. If we cared nothing for nature, we would not make them either. A purely rustic garden would include no purely ornamental plants; a purely urban garden would perhaps include no plants. The will to meld Mountain and Tower in actual fact sends the garden spiraling outward, grasping the whole landscape in its sensuous embrace; the urge to join them in a symbolic whole turns the garden inward, deepening the hue of each petal.

17

THE CLOISTER
AND THE PLOW

IN THE MIDDLE AGES, the garden retreated behind closer walls. Instead of reaching outward to conquer this world, it turned inward to contemplate the next. The retreat was somewhat deceptive, though, since all the while people were mastering nature more thoroughly than the Romans had ever managed to do.

The barbarians who carved up the Western half of the Roman Empire after 476 A.D. were hungry for Roman luxuries. They must have employed Roman gardeners just as they did Roman ministers and scribes. That some rudiments of the physical form of the Roman garden were preserved under the Vandals and Visigoths and Ostrogoths is clear from the testimony of writers in places as far apart as Burgundy, Calabria, and North Africa. An art so dependent, in its higher forms, on the precise manhandling of water and greenery could hardly have thrived in such straits. Yet the spirit of the garden, the reverencing of it as a place of song, and talk, and reverie, was kept alive.

Much of the credit must go to the monks, who knew a good thing when they saw it, pagan or not, and so took up the garden lore of the old poets and philosophers. In Byzantium, monks commonly lived in cells around a central garden. Instead of looking outward upon the fallen earth and its temptations, they looked inward on a small parcel of earth arranged in such a way as to prefigure heaven. Here they walked or sat or worked, meditating all the while. The Byzantine name for a Roman peristyle garden, *paradeisos,* took on a new meaning. The Septuagint and Vulgate rendered the Hebrew *gan,* in *gan Eden,* as *paradeisos* and *paradisus.* Two great streams of earthly felicity, one flowing from Persia and the other from Canaan, now converged in a tiny square of earth whose point was not earthly at all.

For much of his early life, Augustine swung between bouts of sin and periods of rural seclusion in which—inspired as much by Plato and Cicero as by the desert fathers—he cultivated his mind to receive "a divine plant-ing." Finally, in Tagaste, in North Africa, about 387 A.D., he assembled, "in a villa garden that Valerius had given me, certain brethren of like intentions with my own, who possessed nothing, even as I possessed nothing, and who followed after me." A few years later, he built a church and monastery, with its cloister modeled on the museum, exedra, and portico of the pagan philosopher's garden.

But in the West it was above all St. Benedict who made gardening a staple of monastic life. Taking seriously the pagan philosophers' ideal of self-sufficiency, in about 529 he set up at Monte Cassino (an old pagan shrine) a community that grew its own food, which of course required more than a courtyard. A detailed plan for a monastery based on the Rule of St. Benedict has been found in the library of the Abbey of St. Gall, in Switzerland, the cultural heart of Charlemagne's empire. Drawn up about 832, it amply meets the rule's demand that "all the necessaries" for monkish life be found within the cloister walls. Inside this vast rectangle are a vege-table garden, a physic garden, a graveyard planted with fruit trees, and—at the very center—a cloister garden.

The standard cloister garden was a square divided into four squares by intersecting paths, with a well, fountain, lustral basin, or tree in the center: in short, a modest paradise garden. Though traces of this pattern can be found in Greek and Roman gardens, its full-blown form probably came to the medieval West from Persia by way of Byzantium. It got a stamp of approval from the Bible (just as it would, indirectly, in Islam). The four paths were the four Rivers of Eden. The fountain was the spring (*fons* in the Vulgate) from which they sprang; or the tree was "the tree in the midst of the garden." The garden as a whole could be seen as an echo of Eden, or as a pre-echo of a world redeemed.

The cloister garden was a paltry thing, yet it bore a great symbolic weight. The glories of heaven and earth could be stamped in a plot of grass a few yards square. A single pear tree—a single pear—gave a foretaste of paradise.

From the beginning, the world-pole figured in the paradise garden had mythically joined heaven and earth. In ancient times, the point had been to draw the bounty of heaven down to earth. Now the point was different: by contemplating the earthly garden, you could hoist your soul to heaven. The trick was not to forget the end in your enthusiasm for the means. In the *Hortus Deliciarum* or *Garden of Delights* by Herrad of Landsperg, a hermit who has climbed to the top rung of the ladder of Virtue makes the last-minute mistake of looking down. He sees his flowery garden, is seized by yearning, and plunges down—losing, for the sake of this earthly paradise, the heavenly one that was just inches away. In a famous woodcut illustrating the story, in which demonic archers swarm around the ladder and sword-wielding angels fend them off, the garden is represented by a rocky crag with a single plant and a total of two blossoms. Yet the poor hermit— his unkempt beard dangling vertiginously as he peers downward—looks as wide-eyed and besotted as if he were watching the Folies-Bergère.

THE MOUNTAIN GEOMETRIZED

For the Christian thinkers who played the ancient game of seeking cosmic significance in numbers, the number four had a special cachet. So did the figure of four equal sides. The square meant *aequalitas*, which meant perfection of being. Hugh of St. Victor said the cloister should be square—like the courtyard of the Temple in Ezekiel's vision—betokening both the even temper of the monkish life and the perfection of Godhead. Ezekiel's Temple was a geometer's remake of Eden, as was the city "that lieth foursquare" (*tetragonos*), the New Jerusalem of John.

The St. Gall plan is the first medieval example of design *ad quadratum*, in which all the buildings are generated, as in some metaphysical domino set, from multiples of a square module—in this case, the forty-by-forty-foot square of the crossing of transepts and nave. The square is not used simply as handy shorthand for Eden; it is the actual shape of Eden. Geometry is God's trademark. If the things we see around us do not bear that trademark, the reason must be that in their fallen state it has gotten rubbed off. But what that means in practice is that godliness can be found only in the work of man, or in nature remade by man. Even heaven itself is pictured in most Christian sources, from John onward, as a city.

In fact, the turning inward that one finds in medieval gardens and in other facets of medieval life is somewhat deceptive. As historians have been telling us for decades, the Middle Ages were convulsed by social, technological, and ecological changes without which "the Renaissance," "the Scientific Revolution," and "the Industrial Revolution" could never have happened. The most crucial of these changes came at the very start of the Middle Ages, and they had to do with farming.

The human allies that had become so well established in the Mediterranean basin in ancient times had not gotten much more than a sporadic toehold in Northern Europe. The annual grasses known as grains could compete with native flora only where the soil was disturbed—a service which, in the South, was rendered easily enough by humans and their four-legged allies. But the light scratch plow that was adequate for the light soils of the South was no match for the damp, heavy clay soils of the North. Also, most of these grasses came from dry parts of the world and did not like getting their feet wet.

The answer was the heavy wheeled plow. Pioneered by the Slavs in the sixth century and in use through much of Europe by the eighth century, in its classic form it involved a vertical coulter to unzip the turf, a horizontal plowshare to peel it up, and an angled moldboard to rip it loose, turn it over, and dump it off to the side. Instead of cross-plowing in a check pattern as in ancient times, one could now plow in long strips. As a bonus, the moldboard's sidewise action naturally created a drainage ditch between strips.

The deep, fertile soils which were now opened up to farming—soils on which rain often fell in all four seasons—were not fully exploited by the old system of biennial fallow, in which half the land lay idle at any given time. As early as the eighth century, peasants in northern France began dividing their land into three fields. One was sown in autumn with wheat, barley, or rye; a second was sown in spring with oats or legumes; a third was left fallow.

The new harvest of beans, lentils, peas, and chickpeas meant more protein for people, but the biggest advantage of the new system came to people by way of a four-legged ally. Oats could be fed to horses. The heavy, high-friction plow had at first been pulled by teams of oxen; horses, which were faster and potentially more efficient, could not be used because the old-fashioned yoke tended to choke them. A pair of yoked horses could pull a chariot or light cart, but not much more. Around 800, though, the stiff collar harness appeared, allowing a horse to do four to five times as much work as it had before and twice as much as an ox. The problem of the horse's relatively tender feet was solved by the iron shoe, and the problem of its need for high-energy fodder was solved by oats.

From this time forward, the central symbiosis in European husbandry was between horses and oats. Wheat, barley, rye, beans, peas, and people merely came along for the ride. But what a ride! With the invention, by the eleventh century, of the whippletree—a hinged rod that kept horses' abrupt movements from breaking the harness or toppling the wagon—people, produce, and other heavy things could be hauled efficiently, and the possibilities for a market economy broadened.

The new farming spawned a new landscape. The small, square, fenced fields that cross-plowing had favored gave way to vast open fields in which each villager farmed a strip (this contiguity, and the need to pool resources for plowing, meant that people had to work together). Above all, the fields spread, and spread, and spread. In the space of a few centuries, the forests of Northern Europe were largely cleared and the wetlands drained. By 1150, the process of reclaiming land from the sea by means of dikes (and, later, windmills) had begun in the Low Countries—a godlike act of creation hardly matched in our own day. The harnessing of wind and water power sped up the milling of all this new grain and new timber, and the making of other products as well. Mining, metallurgy, and glassmaking were revolutionized. Mechanical clocks started to wrench people loose from the cycles of nature and subject them to the often sterner regimens of landlords, managers, and industrialists. For the first time, humankind's living allies were joined by mechanical ones, which in the long run proved even more high-strung and demanding.

The effects of these innovations (some of which, admittedly, were borrowed from China) were far greater than those of anything the Greeks or Romans had done. As their allied grasses and quadrupeds drove out the native trees and peat mosses, the human population of Europe boomed. It is estimated that, between 1000 and 1340, the population of Northern and Western Europe nearly trebled, from 12 million to 35.5 million. At its peak, the rate of growth was comparable to that of the nineteenth century. Thousands of new towns cropped up, many of them free cities where manufactures, trade, and young capitalism could grow almost untrammeled. "In the course of three centuries," writes Mumford, "the Europe we know today was opened or re-opened for settlement. This feat compares exactly with the opening of the North American continent between the seventeenth and the twentieth centuries."

In the performance of this feat the monasteries played a central role. They took the lead in the clearing and draining of land. Far from turning up their noses at labor-saving machines, the monks actually invented many of them in order to have more time for prayer and study. The Cistercian Rule, which forbade the use of other people's labor, went so far as to advise that monasteries be sited next to streams so they could make use of water power. At Clairvaux Abbey in the time of St. Bernard—the same St. Bernard who praised the edifying power of a shady lawn—the neighboring stream ran a grain mill, machines for making textiles and leather goods, and a slew of other gadgets. The same Hugh of St. Victor who explained why a cloister garden should be square was the first Western thinker to give the mechanical arts a place in the pantheon of arts and sciences. In 1248, the archbishop of Mainz praised a group of monks: "I have found men after my own heart. . . . Not only do they give witness of unblemished religion and a

holy life, but also they are very active and skilled in building roads, in raising aqueducts, in draining swamps . . . and generally in the mechanic arts."

Even where the monasteries were not directly involved, their ideology
was. The passion to reclaim a fallen world—to reinfuse it with the neatness,
efficiency, and attentiveness to human needs that the Church saw in
Eden—was incubated in the cloister. So the cloister garden, though osten-
sibly set up to concentrate the mind on the next world, may have done
more to spur the conquest of this one.

THE JOYFUL SCIENCE

The cloister garden was the model of all medieval gardens, even those
devoted to more frivolous pleasures. As if deployed by some crazed clerk,
the square stamp of Eden came down again and again on the varied soils of
Tuscany, Brittany, Saxony, Ireland. Not all these squares had buildings
arrayed around them, but all had walls. Part of Eden's holiness, and part of
its charm, was that it was "a place apart," "a happy retreat," a *hortus con-
clusus* like the one in Canticles. Eden was cozy—"the nest of human nature," as Dante said. In an age when most people spent much of the winter
huddling for warmth, coziness was nothing to sneeze at.

Most medieval pictures of gardens are risibly skimpy. While conventions
of representation are part of the reason, the fact is that, on earth as well as
on paper, the imagining of an earthly paradise was impoverished. Artists
cooped up our first parents on a tiny plot of grass engirt with a dry-wattle
fence. But the vision of a skyborne paradise made up for that. A blade of
grass, a curl of rosemary could give off an otherworldly glow.

Or a this-worldly glow, for that matter. In the eleventh century, the same
winds that blew the chaste robes off the Virgin's back and replaced them
with the finery of the Domna—that turned Mariolatry into chivalry and
romantic love—ruffled the garden, too. The place of meditation became a
place of assignation. Those winds may have blown mainly from Spain,
where Arab poetry mingled sacred and profane, male and female, botany
and anatomy in elegant disarray.

The troubadours give no plans of Provençal gardens. A garden is a
simple thing; a few quick strokes assemble it before the hearer's eye. The
ingredients are well known: grass to sit on, a fruit tree or pine to sit under,
a rosebush, a nightingale. Not because the garden is a symbol but because
it is a real place—a place of assignation—only the most basic directions are
needed. In one of Marcabru's poems a lady instructs, in beak-sized phrases,
a starling who acts as go-between: *"Vai e.l di / qu'el mati / si.aisi, / que sotz*

pi / farem fi, / sotz lui mi, / d'esta malvolensa." ("Go, tell him, be here in the morning; under the pine tree, we will end—I beneath him—all the bad blood between us.") Or the poet finds a lady "by the fountain in the orchard, where the grass is green down to the sandy banks, in the shade of a planted tree, in a pleasant setting of white flowers, and the ancient song of the new season."

The very spareness of the medieval garden must have given each leaf, each petal—each thorn—a piercing beauty. In the songs of the *Gai Saber* ("the joyful science," as the trade of the troubadour was called) they pierce us still. Gertrude Stein once said that things you are cooped up with (in a room, on an island) will, in the fullness of time, start to sing. People in medieval Europe often liked to coop themselves up and wall themselves in. In an age of plague and famine, of marauding Vikings and Moors, this made them feel a little safer. Yet the High Middle Ages were also a time of widening horizons, both figurative and literal. The way these two tendencies played off each other can be seen by looking briefly at the best-known long poem of the day.

INWARD, OUTWARD, NATUREWARD

The *Romance of the Rose* was begun about 1237 by Guillaume de Lorris, Lorris being a small town in the Loire Valley, where the great châteaux and gardens of France would later rise, and which was now close to the center of high-medieval culture. The hero and narrator dreams that he has awakened early on a May morning—when Earth "exults to have a new-spun, gorgeous dress"—and set off for a ramble, taking a needle threaded with silk so that he can baste his sleeves with zigzags as he goes. Following the path of a rippling stream, he comes to a garden enclosed like a fortress with a crenellated wall. Horrid gargoyles stud the wall—Hate, Felony, Villainy, Covetousness, Avarice, Envy, Sorrow, Old Age, Hypocrisy, and Poverty—these being the traits that exclude one from the garden, "to which no low-born man had ever come." The dreamer is keen to get in, lured by the sound of birds— "three times as many as there can be in all the rest of France." Admitted by a dimpled blonde named Idleness, he learns that the garden is owned by Mirth, who has had it planted with trees from the land of the Saracens. "It seemed to me a better place than Eden for delight." The dreamer meets Mirth and his friends—Gladness, Love, Sweet Looks, Beauty, Wealth, Largesse, Freedom, and Courtesy—and joins in their dancing and caroling. When they pair off and disperse to make love, he (the odd man out) wanders in the garden. It is "a perfect measured square, as long as it was broad," and contains every kind of fruitful tree—pomegranate, date, fig, almond, nutmeg, licorice, gillyflower, malagueta pepper, zodary, anise, cinnamon, quince, peach, pear, chestnut, medlar, apple, plum, cherry, olive—as well as laurel, hazel, cedar, cypress, hornbeam, beech, aspen, ash, maple, oak, spruce.... "Why mention more?" They are planted regularly, five to six fathoms apart, and their crowns merge in a single shady canopy. Among them frolic roebuck, deer, squirrels, and thirty kinds of rabbit. There are clear springs and brooks on whose grassy banks "one might lie beside his sweetheart as upon a couch."

If this marvelous place is cut on the last of the cloister garden, it is ruled by a different set of gods; and stretched to match the new horizons of the High Middle Ages. In fact, it verges on a Persian-style paradise park. It is inspired by, and will inspire, the larger walled orchards and hunting preserves now coming into fashion among the nobility. The Garden of Mirth seems to mingle wildness and civilization in perfect measure. Like the medieval image of Eden, it seems to hold all the world's nicest species and none of the nasty ones. Like Arcadia, it is a stage for the best civilized graces. Unlike the mostly symbolic cloister garden, it shows a good-faith effort to imagine what paradise might be like in the flesh. In this sense it marks the beginning of the movement—in part, a return to antiquity—that will hit its stride in the actual gardens of the Renaissance.

This garden's wider horizons excite the dreamer, but dizzy him, too. Although the human traits excluded and included seem to promise an image of the ideal self, the garden is too big, too various, and (in its odd way) too real to let that image come into focus. The dreamer seeks the still center, the mirror of the self.

He finds it at last in "the fairest spot of all," under a great spreading pine. ("Not since King Pepin's time, or Charlemagne's, has such a tree been seen"—further testimony to the land boom's effects.) Under the tree, Nature herself has set a spring on whose marble verge are carved the words: "Here Fair Narcissus wept himself to death." On the pebbled floor of the spring sit two crystal stones in which, when the sun hits them, one can see reflected everything in the whole garden, no matter how small or hidden. Clearly the tree is the world-pole on which the world turns, the spring the spring of Eden.

One thing catches the dreamer's eye: a rosebush surrounded by a hedge. He rushes off to find it. As he followed the stream to the garden, so now he follows water's urgings to the garden within the garden. To make a long story short, he falls in love with a single rose—signifying, of course, a woman: his soulmate, we might say—and his efforts to get past the thorny hedge and pluck it bring about the erection by Jealousy of a fortress around the rosebush ("foursquare in shape, six hundred feet each way") complete with towers, turrets, battlements, moat, portcullis, ballistas, mangonels, arbolast, and garrison.

The lovely poem of Guillaume de Lorris breaks off here. In it, the medieval paradise has come down to earth. It moves outward to embrace more of the world, but also inward to the true self. Each is a rediscovery of nature.

The process is not trouble-free, though. The gargoyles on the wall are meant, in the manner of apotropaic magic, to drive out the vices they represent—vices of the "low-born" that are so threatening to the well-being of nature and her favored children. Although the wall is allegorical, it had its counterparts in the real world. As the European land boom devastated forests and game, the ruling class, like the Persian ruling class before them, found it necessary to wall nature in and wall other people out. Nor did this happen only in pleasure gardens and hunting parks.

The conquest of Europe by humans and their grassy, four-legged, and two-or-more-wheeled allies was not an unmixed triumph—not even for humans. By the mid-thirteenth century, when the *Romance* was written, land had pretty much run out. Though no one doubted that land clearance was a good thing, it became clear that it was a thing one could have too much of. Timber was scarce, and demand was rising as the towns grew. Some of the lately cleared land turned out to be poor and was abandoned. Hillsides that might have fared well enough under the old scratch plow were gouged too deeply by the new heavy plow, and their topsoil washed away. Sandy loams that might have muddled through under the two-field system were worn out by the three-field system. (Though legumes added some nitrogen to the soil, and more was added when livestock grazed in the stubble, in practice the latter happened less often than we might suppose, and manure was often in short supply.) Yields fell off.

The growing urban population showed a growing appetite for wine, meat, cheese, fruits, herbs, and vegetables, as well as for leather, wool, linen, hemp, dyestuffs, and the ever-scarcer firewood and timber. To meet these demands, landowners began to replace their grainfields with vineyards, gardens, plantations, pastures, hayfields, and woodlots. The open fields and common grazing lands of the middle Middle Ages were gradually crisscrossed by hedges, tending toward a system the French would call *bocage* and the English would call enclosure. As a growing population tried to get what it wanted from a stagnant or shrinking land base, land prices rose. In many places, old forms of feudal tenancy broke down. Many peasants became landless laborers; others—in many regions, a majority of the peasantry—were left with barely enough land to survive on.

At the moment when the *Romance* was written, the upper classes and urban middle classes were still flourishing. By the first decades of the fourteenth century, though, things would start to get out of hand. Strained to the limit, the food system would need only a series of bad harvests—duly supplied by the cool, soggy weather of the years 1315 to 1317—to collapse into famine. The population of Europe would begin to decline. And hunger, overcrowding, and ecological stress would help pave the way, a few years later, for the Black Death, which would have only slight respect for social rank.

In the *Romance of the Rose*, the outer wall of the paradise park is supposed to exclude Hate, Envy, and evil in general by applying a simple class test. Whatever interferes with the ideal world and the ideal self is walled out. But as we noted earlier, you cannot wall evil out of paradise unless you wall yourself out, too. And in fact the dreamer, though admitted to his paradise, finds himself walled out of the part that matters most. Though the poem is intended as an allegory of the love between man and woman, it applies just as well to other forms of the heart's desire: not least the desire for a privileged place in the bosom of nature.

Those who seek paradise in a walled garden may have the same problem as those who seek it in Arcadia. As they advance, paradise recedes—or throws up new walls.

18

BRINGING A STATUE
TO LIFE

I**N HIS** *Oration on the Dignity of Man*, the Humanist Giovanni
Pico della Mirandola wrote: "The saying 'Know thyself' urges and
encourages us to the investigation of all nature, of which the nature
of man is both the connecting link and, so to speak, the 'mixed bowl.'
For he who knows himself, in himself knows all things."

The Humanists—the poet-scholar-courtier-pundits of Renaissance
Italy—wanted to recover their own nature, and knew that this entailed
recovering all of nature. It also meant recovering earlier cultures. For them,
antiquity was a paradise of the most poignant kind, a dream from which
they had awakened too soon and which they struggled to piece together
from the fragments that floated past. Amazingly, they succeeded. Their
great good luck was that the dream was set in the landscape of their waking life. They could close their eyes and dredge texts from the darkness; they
could open their eyes and match the texts against the world.

The landscape of antiquity was remarkably like the landscape of Italy.
Not only were there ruins on every hill—there was a hill under every ruin,
substantially the same hill that had been under it a thousand years before. The trees might be sparser, but they were substantially the same trees,
laurel and oak and pine. Wheat, the olive, and the vine were still cultivated
much as they had been on Virgil's estate. Place names had been Itali-
cized, but a little rubbing exposed the hard originals. Best of all, the human
body, when exhumed from centuries of shame, bathed, and let dry in sunlight, was substantially—could it be exactly?—the same human body that
Praxiteles and Phidias had gloried in. (If the sexual organs were bigger now,
so much the better.) This eager, too-good-to-be-true dancing back and
forth between dream and waking, between letter and landscape, between
remembered ideal and flesh-and-blood reality, is perhaps what gives the
Italian Renaissance its unrivaled vividness, its sensuous fullness, its red-
blooded perfection—what makes Lorenzo and Leonardo nearly as alive for
us as Hadrian and Virgil were for them.

The Humanists unearthed the statue of a beautiful woman, brushed it off, put their lips to it—and felt it grow warm and tremulous under their hands.

More literally than books, more sturdily than buildings, gardening brought the ancient world to life. Gardens were the living image of what the Humanists aspired to do: to make Italy bloom again after a millennial winter. Though the word "Renaissance" begs for debunking, it does give a sense of what these people were after. They had no interest in mummified antiquity. What they wanted was rebirth. The spirits of the pagan poets and sages were called up from the cold storage of Limbo and invited into the lithe bodies of Italian youths. How should they feel at home if they could not stroll and sprawl in familiar gardens?

The process began as early as the 1320s with Petrarch, perhaps the West's first amateur gardener of note. Although he gardened wherever he went, his favorite garden was the one at Vaucluse ("hidden valley") in Provence, where the river Sorgue bursts from a grotto beneath walls of rock 350 feet high. This, he said, was his Rome, his Athens, and his Helicon—the seat of his Muses. The garden was sacred both to Apollo and to the poet's beloved, a girl from Avignon named Laura; the laurels featured in the garden were tokens of both. As Cicero in his gardens remembered Plato, so Petrarch remembered Cicero. An even greater inspiration was Virgil: the *Georgics* guiding Petrarch's gardening, the *Eclogues* his rambles through the woods. It was Petrarch who brought the pastoral back into the mainstream of Western literature, where it would float majestically for the next four centuries. And though Petrarch's ascent of Mount Ventoux is often overplayed (he was by no means the first person in history to climb a mountain for spiritual or aesthetic reasons), it does have a certain symbolic weight. The great men of the Renaissance, and their gardens, alternated between a need for hidden valleys and walled places where they could seek unimpeded the ideal past and their ideal selves, and a need for high places from which they could see the world. Antiquity was at once bower and belvedere—an escape from the real world and a vantage point from which to plot its conquest.

Consider the first city of the Renaissance, Florence, and its first family, the Medici. In the mid-fifteenth century, Cosimo de' Medici, inspired by the example of Cicero, hoped to ensnare the spirit of Greek philosophy by preparing for it a place that would feel like home. At his villa at Careggi he installed the young Platonist Ficino and his circle of Humanists in a wreath of groves and gardens modeled on the Academy. Later when Lorenzo de' Medici came to be its master, the circle at Careggi would spend much of its time playing with the ancient imagery of Arcadia and the Golden Age. Lorenzo himself wrote rustic, often raunchy lyrics that joined a fresh feeling for nature with a classical sense of form. Among the acquisitions that helped win him the name "Magnificent" were not only objects of art both ancient and modern, but many rare and exotic plants in his garden at Careggi, among them the turkey oak, cork oak, and incense juniper.

The villas and gardens described by Roman authors were studied word for word, reimagined, and rebuilt, with equal attention given to stone and leaf. If Roman topiary was meant to force life into the mold of art, the sculptured hedges of the Renaissance tried to bring art to green and whispering life. Meanwhile, in grottoes such as that of the Boboli Garden in Florence, figures seem to emerge from rock or crushed seashells, then melt back in.

The Renaissance coupling of nature and culture reaches its promiscuous height in an astonishing book written in Lorenzo's time, the *Hypnerotomachia Poliphili*, or *Sleep-Sex-Struggle of the Lover of Polia*. Authored by a monk, the story is an extended wet dream in which the dreamer is inducted into the pagan mysteries of nature. He wanders or is led among naked bathing maidens, naked dancing nymphs, the goddess Venus, and Nature herself—"nursemaid of all things"—spouting a river of milk from each breast. Diluting these images, though, are maddeningly minute accounts of the architectural surroundings. Capitals, pediments, arches, and entablatures, inspired by a mix of antiquarianism and geometrical mysticism, are catalogued for page after page in detail worthy of an archeologist's dig. The mingling of nature and culture is summed up in a scene early on: the dreamer wanders among classical ruins on a hillside, enters a cave, and comes out the other side—on the flank of a wild and craggy mountain.

As I noted earlier, one has to distinguish between Arcadia, to which the garden often alludes, and the garden itself. Arcadia is not meant to be a world-pole, but a mean point between world-poles. True, it too is supposed to combine the best of both worlds. But it does so by fine balance and gentle mingling. The garden, by contrast, has no qualms about ramming one world-pole down the throat of the other. Greenery and masonry are conjoined, or metamorphose the one into the other. Shrubs become statues and battlements, stonework erupts in rosettes. All the pleasures of civilized life are to hand, but so is the (no doubt idealized) calm of wilderness.

RATS IN FLORENCE

In the comic-strip view of history, the Renaissance is drawn as a time when windows were thrown open and walls knocked down. Yet the Renaissance was also the age in which the use of the garden as a wall, or frame, or screen attained the level of high art. On the simplest level, the Renaissance garden was a shield against ugly facts, both social and biological. Think of the premise of the *Decameron*, the grab bag of comic, tragic, and bawdy stories for which the poet Boccaccio is best known. In the spring of 1348, the great wave of bubonic plague that had swept the Mediterranean reached Florence. "Such was the cruelty of Heaven," Boccaccio says, "and perhaps in part of men, that between March and July more than one hundred thousand persons died within the walls of Florence, what between the violence of the plague and the abandonment in which the sick were left by the cowardice of the healthy. And before the plague it was not thought that the whole city held so many people."

Only when struck by plague does the city come to know itself: how big it is, how thick with people, how thin its veneer of order and morality. The plague is a hammer that smashes the city wide open. Out spill lechery, gluttony, selfishness, and thousands of dead bodies. The poor, especially, who had squirreled themselves away in holes and corners while alive, now litter the streets in numbers far greater than anyone had guessed. In more ways than one, the cruelty of heaven exposes the cruelty and folly of men.

In Florence, gentlefolk walk the streets with nosegays pressed to their faces, to ward off the stench and evil humors. Better is a house with a garden, and best of all a country villa engirt with gardens. To a chain of such villas, Boccaccio's young nobles make their escape. Although the plague has made inroads into the countryside, the villas and their gardens remain inviolate. By waiting out the plague in these places, the young people avoid both dying and watching other people die.

When Boccaccio speaks of "the cruelty of Heaven, and perhaps in part of men," it seems that he blames men not for the plague itself, but only for failing to care for each other once it had broken out. A look at the plague's history suggests that he may be too lenient.

Before *Pasteurella pestis* visited human cities, it had long been a resident of the underground cities of burrowing rodents in northeastern India (or perhaps central Africa). The bacillus and its host had reached an understanding; the disease was not acute, did not kill off too many of its hosts, and felt no great need to spread. But when black rats—which, as camp followers of humankind, had greatly expanded their range—came into contact with the parasite, it fell back into a pattern of virulence, swiftly killing off one host and leaping to the next. The same pattern applied to humans. The rats themselves, nimbly climbing mooring ropes and stowing away in the holds of ships, spread the disease to port cities throughout the Mediterranean. In 542 A.D. it announced itself in Rome as the Plague of Justinian.

In Constantinople, it killed ten thousand people a day. It sealed both the fate of the Roman Empire and that of Christianity, which battened on death just as Buddhism did, about the same time, in plague-stricken China.

By about 750, the plague seemed to have burned itself out. In the fourteenth century, though, the northward expansion of Mongol caravans introduced the bacillus to the vast network of marmot burrows that stretched beneath the Eurasian steppe. This new reservoir was dipped into repeatedly and unwittingly by the Mongols and other nomadic tribes, who spread the plague among themselves when they gathered for the hunt or for war, and by war and trade transmitted it to the settled peoples of Asia. Reaching the Crimea in 1346, the bacillus quickly resumed its ancient habit of travel by ship throughout the Mediterranean, a habit facilitated by the very trade that was making Italy great. From the ports it moved inward through Europe, killing—between 1346 and 1350— one-third of that continent's population.

Europe was a sitting duck. Since the land boom of the tenth century, Europe had gradually filled up with people, and by the fourteenth century farmland had run out, or was worn out. Malnutrition, overcrowding, and ecological damage all conspired to make life easier for black rats and the bacteria they carried.

Those invaders may have found Florence especially inviting, for Florence in the fourteenth century had perhaps the first large industrial proletariat in the world. Of a population of some one hundred thousand before the plague struck, roughly a third were employed in the textile industry. Of these, the great majority were beaters, carders, washers, combers, and other wage laborers who scraped out a bare living. When you consider that another third of the population were classed as paupers—either unemployed or employed as servants, and too poor to pay taxes—you get some idea of the haven *Pasteurella pestis* would have found in the cramped warren that was Florence.

The powerful merchants of the Lana and Calimala guilds controlled every step of the business, from the purchase of raw wool to its marketing abroad. They had their fingers in just about every pie in Florence—and many elsewhere—from real estate to international banking and trade. The banking operation of the Medici in the time of their founding father, Giovanni, shortly after the first plague years, had offices in sixteen European capitals, and soon became the largest in the world. Pioneers of double-entry bookkeeping, letters of credit, bills of indebtedness, and other techniques, the merchants of Florence were the first modern capitalists.

Workers who offended the wool guild could be tried in the guild's own

courts, jailed in its own jail, blacklisted, fined, docked pay, flogged, muti-
lated, or hanged. In 1345, a carder named Brandini was tried on charges of
urging workers to join a union, "in order that they might more strongly
resist their masters," and—despite, or perhaps because of, a protest strike
by the combers and carders—was hanged. In 1378, the Ciompi (as the
menial wage laborers were called) joined forces with some of the less privileged artisans' guilds in a revolt that actually succeeded in toppling the government. But though the demands of the Ciompi were almost laughably
modest—a voice in government, tax equity, the right to a guild of their
own—they were enough to frighten the artisans, who broke ranks. In a
matter of weeks, the uprising was bloodily crushed.

The Ciompi episode, which Marxist historians would seize upon as the
first (and somewhat premature) modern proletarian revolution, would
recur on various scales for the next few centuries, and would again be
suppressed—in one notable case, by roasting at the stake the rabble-
rousing monk Savonarola. The Black Death, too, would recur: after its first
visit, it would return roughly once every decade for the next eighty years.

GARDENS AND TALES

On the simplest level, then, the Renaissance garden offered a physical
refuge. But note that the young aristocrats of the *Decameron* seek refuge in
two things, gardens and tales. And each garden is itself a tale: a tale about
people and nature, about men and women, about the past and the present
and a certain slant of light that is perhaps the future. If the tale-teller offers
his fictions as fact, so does the garden. This is what the world is, the garden
says: not a howling waste where worms of sickness fly in the night, but a
bower where roses bloom. Yet beneath this delicious pretense is the still
more delicious knowledge that the world *is* a place of sickness and death,
and that here in the garden one is safe. The young people's tales are not
purely escapist; they dance between danger and bliss, stealing voluptuary
glances into the pit of horror that gapes below.

In the same way, the typical garden of the Italian Renaissance is walled
off from the outside world but has a good view of it, too. At times the
viewer relishes that world and plans his conquest of it; at times he con-
gratulates himself on getting safely away.

Though the Villa Medici at Fiesole abandoned the closed, fortified plan
of the medieval castle and garden in favor of the more open layout of the
Roman villa—terraced on a hillside, with a spectacular view of the surrounding country—it was careful to keep the lower kitchen garden and
service area low enough to be out of sight.

Even an unwalled garden, I said, is walled in the sense that it offers physical and psychical shelter—shelter from the waves of change, and from the contradictions you feel in your relation to nature and other people. Nature in the raw would not offer such shelter. For no matter how much you might wish just to stand and look, you know in your bones that you must stand in some relation to nature and other people. And your actual relation—the relation that stretches beyond the walls, and that provides you with food, clothing, and literal shelter, as well as such bonuses as sex and power—would haunt you. So you pretend to stand in another relation: a relation defined, unavoidably, by another culture, the culture of another time or place. It has to be an idealized culture, because a real one would probably not be much better, all in all, than your own. In short, there can be no escape into nature, but only into another culture.

The cultures into which the makers of the Italian Renaissance mainly escaped were those of ancient Greece and Rome. To say that the revival of antiquity was also a quest—a daring effort to recover the wholeness of the world and of the self—is no contradiction. It is very hard for the human mind to achieve wholeness except by throwing out or covering up unwanted parts. The elites of the Renaissance may have done their throwing and covering more deftly than those of the Middle Ages, but they did them all the same.

You might ask what parts of the world the Florentine elite needed to suppress: were they not sitting on top of it? They were, if you ignore the Black Death; if you ignore the dangers posed by hungry Milan and imperious Rome, by the Ottoman Turks and gigantic France; if you ignore the factional strife and foreign intrigues that could send hundreds of Florentines, including Dante and the father of Petrarch, into exile in a single year; that sent hundreds of the Medici's rivals into exile during the fifteenth century, and sent the Medici themselves into exile from 1494 until 1512; that got Lorenzo's brother hacked to death in a cathedral, a fate which Il Magnifico himself would have shared had he not drawn his sword and slashed his way out. As Lorenzo said, Florence was not a good place to be rich unless you controlled the government—which was not so easy to do. Nor, despite the soothing hum of the Humanists, could the ruling class be completely deaf to the screech of its own hypocrisy. Florence had evolved from the medieval commune and was still, in theory, a republic. The sacred name of the *popolo* was still invoked by those eagerest to squeeze them dry. Moreover, by the late fifteenth century the ruling class had itself ceased to rule; the logic of oligarchy had reached its conclusion, leaving a single prince in power and reducing everyone else to courtier or bystander. By this time, too, the economic picture had lost (as we shall see) much of its luster. Finally, there was the matter of the Renaissance stance toward nature, which combined love and the urge for mastery in a suave but somewhat uneasy embrace.

From all these irritants the Florentines fled into dreams of Greece and Rome. As if that were not far enough, they fled into the very dreams into which the Greeks and Romans had fled: dreams of Arcadia and of the Golden Age.

THE AGE OF GOLD

To find a phrase more tersely eloquent than "the Golden Age" is no easy task. The color is a kind of concentric symbol, leading us deeper and deeper into the sources of all wealth: metal, grain, the sun. We might include beer in its compass. We might include blondes. Imagine a Budweiser commercial filmed in a wheat field in which a Brinks truck has stopped for a picnic, attracting flaxen beauties from a nearby beach. . . . But some of gold's hues are darker.

The original El Dorado was not a city but a man, a native priest-king who once a year was ritually powdered from head to toe in gold dust. That image takes on a Midas-like horror in the story Vasari tells of a parade held in Florence in 1513 to celebrate the election of Lorenzo's son to the papacy as Leo X. The main sequence of "triumphs"—what we would call floats—portrayed a string of golden ages. The first float, drawn by oxen decked out in grass, featured Saturn with his scythe, escorted by six pairs of shepherds on horseback. The shepherds were dressed in sable and marten's fur, the horses in the skins of lions, tigers, and wolves; the cruppers had gold cord, the bridles silver cord. Each shepherd had four footmen also dressed as shepherds, but in less costly skins, and carrying torches. After this came floats representing the golden ages of Numa Pompilius, Titus Manlius Torquatus, Julius Caesar, Augustus, and Trajan. And then the grand finale:

> In the midst of the car arose a great globe, on which lay face-down a man who seemed dead, armed with rusty weapons, his back gaping open and from it emerging a boy entirely nude and gilded, representing the Age of Gold resurgent and the end of the Age of Iron; the former being revived by the creation of this Pope. . . . I will not omit to mention that the gilded child, who was the son of a baker, died shortly after from the pains he underwent to earn ten crowns.

In his poetry, Lorenzo praised the mythical Golden Age in which people lived *communamente* and acknowledged no difference between *tuo* and *mio*. His courtiers and hired poets declared that in his reign the Golden Age was come again. In the reign of Leo X it would be extended to the rest of the world.

In fact, Florence had ushered Europe into a golden age, but it was not exactly the kind Lorenzo poetized about. In 1252 the city had begun minting a gold coin that bore the image of its patron saint, John the Baptist. Known as the florin, it was among the first gold coins of reliably uniform weight and fineness. Within a century, gold was displacing silver as the standard European currency, and the florin was the standard on which most gold coins were based. Less literally but more to the point, Florence had helped usher Europe into a real age of gold: the age of capital.

In the Golden Age of myth, nature's fruits are there for the plucking and there is plenty for everyone. You take what you want when you want it. In real life, nature's flow is not so copious or so steady. It gushes one day, trickles the next—a fickleness which is not merely annoying, but life-threatening. Agriculture ironed out nature's irregularities on various time scales. By sowing seed, people could protect themselves to some degree from the year-to-year vagaries of a wild harvest. By storing grain, they could pile up fat years and fat seasons against the knifing wind of lean years and lean seasons. By doing agriculture in the first place, they could even out the great fluctuation that had plagued hominids for a million and a half years: that between the grass-rich, game-rich glaciations and the tree-infested interglacials.

But this effort to flatten out inequalities in time produces inequalities in space—or, rather, between people. As soon as nature's bounty can be stored, it can also be stolen and hoarded. Both the means and the reasons for one set of people to lord it over another are greatly increased. While this is by no means the origin of inequality—differences in status and in access to food and mates exist among most hunter-gatherer peoples, as among animals—it does allow inequality to become fixed in a class system. That is something found rarely in nature except among the social insects.

Hoping to make the Golden Age permanent, people turn to gold: the gold of grain, the gold of gold. In so doing they generally put an end to some of the very qualities, such as freedom and equality, that made the Golden Age golden in the first place. The irony was not lost on Ovid, who wrote of his own plutocratic age, *"Aurea sunt vere nunc saecula..."*—"This is the Golden Age, all right...."

As the medieval economy of northern Italy gave way to capitalism, the poet Cecco Angiolieri wrote:

> Preach what you will,
> Florins are the best of kin:

Blood brothers and cousins true,
Father, mother, sons, and daughters too. . . .

Of this transformation, the gilded boy on the float is the perfect image:
the son who becomes a florin.

Oddly enough, the great age of the Medici was not a golden age at all,
economically speaking. By the early fifteenth century the Florentine wool
industry was taking a beating from Spanish and British competition.
Though the silk industry took up some of the slack, the city's economy was
in decline. The glory of the Florentine Renaissance was possible only
because the patrons of art were taking and spending a larger share of a
shrinking pie. And by the time of Leo's parade, Italy as a whole had begun
a long slide into a dark age that would last for centuries. Europe's center of
gravity was moving north. From now on, the great powers would think of
Italy mainly as a picturesque chessboard on which to fight their wars.

COURTLY RUSTICITY

That the prize offered the singer in the First Idyll of Theocritus should be
a cup is apt: apart from flutes, cups are almost the only technology needed
in the poet's Sicily. Nature overflows; the trick is to catch her bounty. Pastoral says: once we feel the need to put a wall around paradise, or lock it up
in a chest, it is already lost. When we felt so close to nature that we needed
no wall, so secure that we needed no granary, so rooted that we never
craved a piece of real estate to call our own—that was paradise. Paradise
was no place, and every place.

It may have been so, though when it was so is hard to say. In any case,
pastoral imagery has more often served to disguise the lust for wealth than
to uproot it.

The "Triumph of Bacchus" is one of the carnival songs Lorenzo wrote
for the pageants with which he kept the people amused. As Bacchus, Ari-
adne, and sundry nymphs, satyrs, and sileni parade by, Lorenzo holds them
up as models of conduct. "Banish every sad thought / Let's make every day
a holiday!" Lorenzo and the poets of his circle wrote rustic verse of great
charm, and these and other Italian poets would shortly make pastoral the
rage throughout Europe.

What was the real relation of the Florentine elite to the "pastoral" landscape? Although northern Italy escaped most of the famines that struck
Europe in the early fourteenth century, its peasants shared in the disloca-
tions that had started decades before. Grains gave way to olives, vines, cat-

tle, and other items of interest to city tastes. (Much of the grain Florence ate came from outside Tuscany.) Old feudal holdings disintegrated. Rich townsfolk bought up country estates for pleasure and profit, so that by 1300 nearly the whole countryside of northern Italy was in their hands. A century later, some eight hundred villas studded the hills around Florence. The new landowners ran their farms on strictly capitalist principles—devil take the hindmost. Many peasants became sharecroppers or migrant laborers; many were absorbed by the city's cloth industry, or took work farmed out by the industrialists. In either case they lived hand-to-mouth. Even as the burghers flourished, the rural population of Tuscany had started to decline a century before the plague arrived.

In Lorenzo's time, the lot of the peasants was no better. They were neither well off nor well liked by city people. The friendly rustic songs of Lorenzo and his circle may show a closer observation of country life than any poet of good blood would have stooped to in earlier days, but they show an edge of mockery, too.

Anyway, nothing remotely like the landscape of Arcadia (either the real or the literary) survived in Tuscany. Pastoralism was long gone; the land was densely settled and given over to more intensive uses. That is one reason why, for all their pastoral verse, the Florentines never invented pastoral landscape painting. The landscapes of Mantegna and Leonardo are cramped, corseted, and often loomed over by towns. Modern landscape painting has its origin not in Florence but in Venice, which was close to the Alpine foothills in which old-fashioned pastoralism was still practiced.

The real pastoral landscape of that age, however, was much farther away. In Spain, the transhumant pastoralism of the ancient world had become big business—largely at Florence's behest. The fleece of the merino sheep, a cross between Spanish and North African breeds, was much in demand in Florence and elsewhere. To feed the demand, huge flocks owned by ranchers in northern Castile would summer in the mountains, then head south in the fall to the vast plains of La Mancha, Extremadura, and Andalusia. Millions of sheep, "their precious fleeces smeared with red clay," would trek hundreds of miles across the Iberian Peninsula, driven by shepherds with slings and long crooks. Their picturesqueness did not endear them to the peasants whose crops they nibbled and trampled and whose common pastures they freeloaded on. But in the "range wars"—not unlike those of the American West—that went on for several centuries, the farmers' attacks on the fleecy host were as hopeless as Don Quixote's. The Mesta, the sheep-ranching syndicate whose letters of privilege dated to 1273, was immensely rich—not least, in florins—and immensely powerful. Since the crown, the

Church, the nobles, and the towns all got a piece (for which they squabbled fiercely) of the tolls that were charged along the route, all had an interest in keeping the vast flocks moving. The end result was the dusty North African landscape you can see in much of southern Spain today. (You could see it already in the time of Cervantes, one of whose running jokes is just this: after feeding his fevered mind on the bosky landscapes of courtly pulp fiction, the Don sallies forth into a landscape almost devoid of trees.)

The tragedy of Arcadia—or any ecological tragedy—is not always a simple matter of waves moving outward in serried ranks, of one molecule jostling the next, of local cause and local effect. Already in ancient times, trade and empires produced action at a distance. From the Renaissance onward, the action came faster and jumped farther, often from one side of the globe to the other. Walls were not needed to hide the effects of one's actions when oceans and mountain ranges did the job better.

MASTER OF ALL HE SURVEYED

Despite belvederes, gazebos, and mounds, most gardens do not offer views of the outside world. They aim to be worldlets, walled in and self-sufficient. Even when we picture Eden perched on a peak, we do not think of Adam and Eve leaning against the railing and gazing out at the plains beyond. We assume that they were utterly content where they were (well, almost utterly). The notion that there might be something outside never crossed their minds. As the cherub picked them up by the scruff of the neck, they must have thought they were about to be tossed off the edge of the world. When, outside the gate, they picked themselves up and brushed themselves off, they must have blinked in astonishment at the vast prospect before them—daunting, but at the same time a great relief.

In Renaissance Italy, many garden walls become hedges, or fade from sight. Of his three main villas, Cosimo de' Medici (the first of that name, a down-to-earth businessman) said he liked the one at Cafaggiolo best, because everything he could see from its windows he owned.

In many cases, what the villa owner owned included (for all intents and purposes) the village that huddled beneath the front door, like a modern mill town groveling at the foot of the owner's mansion. Behind the house, the gardens climbed up the hillside and into what passed for woods. The town at the front door, nature at the back: anyone who has fantasized about having a house with a front door opening on Broadway and a back door opening on the Canadian Rockies will (if his social conscience can be kept quiet) find this arrangement appealing. Though it may be that no one but

Nero has ever come close to realizing the fantasy in question, some of these villas achieved a rough approximation.

In the Renaissance, the square paradise garden gets stretched into a rectangle. Most often it gets stretched up the side of a hill, as if to dry—a departure from ancient gardens, which were built in sections on large, flat terraces. As if warped by the sun, it flares outward at the top. The long axis becomes the crucial one. Not only is it the axis of symmetry, on either side of which the garden roughly mirrors itself; it is also the main, often the only, channel of water, which forgivably flows downhill.

Once the garden is stretched across a hillside, walls can no longer contain it. Visually, it cannot help embracing the wider world. No longer is it a pure geometrical form; no longer does pure symbolism suffice to mark it as paradise. Now what you see is what you get. Perspective—both within the garden and beyond—takes the place of the magic rectangles plotted by the architect of St. Gall.

THE MASTERY OF WATER

As walls decline in importance, water—the principle of union—surges. Sixteenth-century gardens, not only in Italy but throughout Europe, vie with ancient Rome to display their mastery of nature in the mastery of water. The streams that rise up in Eden have lost their mystery. Subject to human control, they become an expression of human strength, exuberance, and fancy, and maybe some less savory traits as well.

Two things in which Grand Duke Cosimo I de' Medici, great-great-grand-nephew of the first Cosimo, took special pride—the aqueducts bringing water to Florence; and the Pitti Palace, his suburban home, with the vast Boboli Garden rambling up the hillside behind—were closely related, for the water from springs in the hills beyond the garden ran first to the garden and its fountains, then to fountains throughout Florence from which the people drew their water. The symbolism is hard to miss: the Boboli Garden is Eden, Cosimo the god who dispenses its frothing largesse to the rest of the world. Almost a century later, Ferdinando II completed another aqueduct that likewise brought water first to the garden, then to the city. To make sure no one missed the point, he set up a giant figure of Moses, with a legend likening the works of the Medici dukes to God's provision of water in the wilderness.

All of this was fairly typical of the water imagery that arose in the late Renaissance and coursed through the age of absolutism. But the play of water's meanings was often more delicate than that. The keen interest which Montaigne, in his travel journal, shows in bravura uses of water is typical of his age. He comments on water wheels, locks, canals, viaducts, and a remarkable invention called a *doccia,* or shower. In the gardens of Pratolino, Castello, Bagnaia, Caprarola, and Tivoli he finds water producing organ music, birdsong, harquebus fire, and rainbows; water moving mechanical figures of every description; water flowing through every possible orifice of every imaginable sculpture, from a washerwoman wringing out a marble tablecloth to a twenty-foot Antaeus squeezed so mercilessly by Hercules that water spurts from his mouth thirty-seven fathoms in the air. On top of all this, or rather beneath, he gets unexpected douches from hidden jets—a trick more often played on women. ("While the ladies are busy watching the fish play, you have only to release some spring: immediately all these jets spurt out thin, hard streams of water to the height of a man's head, and fill the petticoats and thighs of the ladies with this coolness.")

Funny things happen when men try on the sky god's robes. In the eighteenth century, princelings will compete to build the longest cascades—eight hundred feet at Kassel, a full two miles at Caserta—like schoolboys trying who can pee farthest. What urges (beyond sheer fun) impel all this spraying, spurting, and novelty-shop spritzing? Do men really believe they can control the waters of life, and therefore life itself? Or do they, like the author of "Enki and Ninhursag," fear a clogging up, a salting or silting up, a final sterility?

Consider Montaigne. He faults the Villa d'Este because its marvelous fountains flow with the turbid water of the Teverone, while those of Pratolino run spanking clean with springwater. Yet he must have felt a secret kinship with the Villa d'Este. For as he goes from villa to villa admiring the fountains, his own plumbing is clogged up. He suffers from kidney stones. It is no laughing matter; his father died of the same complaint. With offhand stoicism, in the space of a few pages he shifts his gaze from marvels of architecture to the stone he has just passed (one is "as big and long as a pine nut, but as thick as a bean at one end, and having, to tell the truth, exactly the shape of a prick"). Marble and kidney stones are limned with the same exactness. In hope of relief he makes the rounds of European spas, dunking in or drinking from every body of water with the faintest medicinal reputation—no matter how sulfurous, ferrigenous, or flat-out foul it smells. Nor is he alone in taking the waters. It turns out that the waters of life have not quite lost their mystery—not yet. People in search of healing still turn to the earth's own secretions, its hidden springs and seeps, rather than to the waters men have channeled.

In the Renaissance love affair with water, the Villa d'Este is the climax. But when I visited it a few years ago, making the rounds of the villas as Montaigne had done, I left unsatisfied. The fountains had been turned off for reasons of public health. The waters of the Teverone were no longer just turbid, they were badly polluted, and the fountains had been acting as a giant atomizer, filling the air of Tivoli and the lungs of its citizens with tiny evils. The irony would not have been lost on Montaigne, who wrote in his essay "On the Cannibals": "We have so much by our inventions surcharged the beauties and riches of Nature's works, that we have altogether over-choked her."

Montaigne's language of surcharging and choking brings me back to the image of the gilded, suffocated boy in the Florentine "triumph"—but something would have had to bring me back. For me, that image stands at the focal point of the Renaissance, and so of the modern world. "In many of their chief merits," Burckhardt wrote, "the Florentines are the pattern and the earliest type of Italians and modern Europeans generally; they are also in many of their defects." The artists and thinkers of the Renaissance could bring a statue to life. They could also take a living, breathing, sweating boy and turn him into a statue, rigid and (presently) dead—roughly what the modern West is doing to nature in its human and nonhuman forms. Possibly the organizers of the parade did not anticipate the effect of sealing the body's pores; but, then, we have often failed to guess the effects of sealing nature's.

19

LEAPING THE FENCE

W HEN CHARLES VIII of France conquered the Kingdom of
Naples in 1495, he wrote home: "You would not believe the
beautiful gardens I have in this city, for, upon my word, it
seems that only Adam and Even are lacking to make it a terrestrial paradise,
it is so lovely, and so full of all good and singular things." Among the booty
Charles took home were orange cuttings and a cluster of Italian craftsmen
and artists, not least of them the gardener Pasella de Mercigliano. Their
arrival is generally thought to mark the start of the Renaissance in France.

If you visit some of the sixteenth-century châteaux along the Loire, you
can see the results of this infusion. You can see the various stages by which
fortress yields to villa, walled garden to open garden, the Middle Ages to the
Renaissance—or, you might as well say, to antiquity; for the French were
learning from live Italians how to learn from dead Greeks and Romans.

THE PLAYGROUND OF REASON

In time, the French learned to learn in their own way. They really hit their
stride in the seventeenth century, in the gardens designed by André Le
Nôtre. At Vaux-le-Vicomte, classical forms are abstracted and tossed about
with what seems, to modern eyes, an almost postmodern cheek. You seem
to have stumbled into a god's board game in which boxwood spheres and
cones are the pieces. Or maybe an Arcadia of reason, where geometric
shepherds tend Platonic solids.

Although, like most French gardens, Vaux occupies much flatter terrain
than its Italian or Roman models, it too makes walls seem beside the point.
For the first time in Western history, the garden is so vast that the question
of what is inside and what is outside hardly arises. The wall between nature
and culture, too, seems a childish construct. On reason's playing board, all
pieces are movable.

Vaux-le-Vicomte was owned by Fouquet, a finance minister of Louis XIV. The great attraction of this job was that you could siphon off vast amounts of tax money for your own effervescent ends. Everyone did this, but Fouquet was just a touch too brilliant, too handsome, too deft at collecting great artists and sparkling women. An industrious functionary named Colbert had his eye on Fouquet's job and was quietly building a case against him. When Fouquet showed off his spanking new garden to the king, each display of magnificence—the whale spouting fireworks, the walking statues, the new play by Molière introduced by a naiad who popped up suddenly out of a shell—only dug that much deeper the pit into which he would fall. Three weeks later, Fouquet was arrested. He spent the rest of his brief life in jail, while Louis systematically looted Vaux and his other estates of plants (including a thousand orange trees), statues, and other booty.

Louis then hired Le Nôtre and the rest of Fouquet's team to build him a bigger and better garden at Versailles. Whatever there was at Vaux, there would be six of at Versailles. For good measure, the abstracted classical forms would be joined by literal classical figures employed in the most heavy-handed way to celebrate the glory of the Sun King. A great fountain from which Apollo's chariot bursts and a tableau of Apollo tended by nymphs, set in a grotto, are two of many items suggesting that Euro-Disney opened in France long before the official date.

The way we look at historic gardens is pitifully blinkered. It is as if we had stumbled on a warehouse full of theatrical sets and assumed that they were simply paintings, meant to be hung on very large walls and wordlessly gazed upon. We know, of course, that gardens are unlike paintings in that they can be walked in, sat in, and regarded from various angles. But we tend to forget that they were stage sets for real-life dramas of all kinds—festive dinners, philosophical disputations, religious rites, affairs of the heart and affairs of state—as well as dramas in the narrow sense, pageants, masques, concerts, recitations, and so forth. This was true in Renaissance Italy and even truer at Versailles, where many of the statuary groups and other features were created as backdrops for particular events: above all, for the *ballets de cour* that choreographed, in heliocentric curves, the new order of absolutism. In the early years, the grounds were more like the back lot at MGM than like a finished work of art.

Versailles was not just a country retreat; it was the seat of government, holding in its heyday some twenty-five thousand bureaucrats, courtiers, and servants. Together they ate, drank, and otherwise consumed six tenths of all the revenues collected in France.

The basic principle of the paradise garden can be used to divide or to multiply; Le Nôtre used it to multiply, slapping down new parterres like dominoes as the king's power and vanity demanded. As Mumford has said, this was an age in which power expressed itself horizontally. In part, this was a matter of military necessity: the towers of the Middle Ages were not much use against cannon fire, so fortifications sprawled outward in star shapes that let defenders catch attackers in crossfire. Louis's Field Marshal Vauban ringed France with an "iron barrier," forerunner of the Maginot Line, that included numerous star-shaped forts. The star shape, in various forms, spangled the landscape of the age. It can be seen in the radiating paths of French hunting parks; in the asterisk shape of Baroque towns such as Rocroi; in the large and small paths radiating from various pools and plazas at Versailles; in the avenues of the new city of Versailles, extending over eight miles from their origin in the king's apartment; in the radiant logo of the Sun King himself. Though it would be two centuries before Baron Haussmann would impose this shape on Paris itself—making the city safe for its rulers, who like hunters would have a clear shot at their human game—the basic idea was already established.

To see, to control, to project power: the Tower had been doing these things since ancient times. But now the effective base of the Tower—the area it affected—was for the first time fully global. The age of colonialism that had begun in 1492, the year of the death of Lorenzo de' Medici, was now in full swing. France had possessions in Canada, Louisiana, the West Indies, and West Africa. An absolute monarch really was a kind of sun whose rays touched the ends of the earth.

THE SUN KING AND THE ICE AGE

Waves of biological change now moved across the whole globe at unprecedented speed. Although this was partly due to improved ships and navigational aids, it had more to do with the state of Europe's ecosystem. As Henri Pirenne said of the Norse voyages of the eleventh century, "America was lost as soon as it was discovered, because Europe did not yet need it." By the sixteenth century, if there had not been an America, Europe would have had to invent one. Having rebounded from the Black Death, its population was once more straining the continent's seams. The alliance of humans, quadrupeds, and annual grasses was again feeling the instability that came of its own success. It had to expand. As half a millennium before it had displaced the forests of Northern Europe, so now it began to displace the forests and grasslands of the New World.

The drift toward a New Pangaea was just beginning. So, on a smaller scale, was the move toward a centralized, homogeneous nation-state; and France led the way. Versailles was a fit symbol of this projection of reason and power. It was, in the words of a later (and rather chauvinistic) French writer, a *jardin de l'intelligence.* Apollo, whose imagery commands the gardens, was of course not only the sun god but the god of reason and order. The radiant avenues passed indifferently through forest and cities. Wild or civilized, whatever the sun looked upon was subject to reason.

When you look back on Louis's reign, though, what you see most clearly is the failure of reason—at least, of reason preconceived and imposed from the center on everything it can reach. By the start of Louis's reign, France held some twenty million souls. Given that farming methods had not changed much since the High Middle Ages, it is no wonder that the system was strained to the breaking point—or that it did break down repeatedly, throughout Louis's reign, whenever bad weather or some other pretext offered. There was plenty of bad weather, for the reign of the Sun King was also that of the Little Ice Age, when glaciers crept down the slopes and wine froze on the royal tables. The mercantilist stress on manufactures meant that farming was often neglected—though a thoroughgoing benign neglect would have been better than the malign attention it did get. The elaborate system of taxation and other controls that Louis imposed on French peasants only tied their hands, so that they could not respond flexibly to changes in climate and other conditions. Internal tariffs made it hard for famine in one province to be eased by plenty in another; often, farmers let their crops rot rather than go to the trouble and expense of shipping them. Since a tidy, well-appointed farm was an invitation to the tax collector, many peasants let their farms go to seed. Nor were matters helped by the legally sanctioned process of *triage,* whereby a third of the peasants' common land in a district could be seized and enclosed by a large landowner. Major famines took place in 1662, 1663, 1694, 1709, and 1713, often causing desperate riots and revolts that were redly suppressed.

If Versailles was an image of the age's pretensions, the way it was made and maintained was an image of the age's realities. The hunting lodge built by Louis XIII sat in the middle of a swamp. Making this wasteland bow to the demands of reason cost countless *livres* and countless lives. A great rectangular lake took several regiments of Swiss soldiers nine years to dig; the bodies of those killed by marsh gas, Madame de Sévigné noted, were trucked away by the cartload every night. Supplying water to the fountains of Versailles, the Trianon, and nearby Marly—at final count, some fourteen hundred of them—was an even bigger job. Full-dress water shows at Versailles used more water in a day than the pumps of Samaritaine sent to the 600,000 citizens of Paris. An attempt to divert the river Eure through forty miles of aqueducts and canals—second in ambition, among Louis's water projects, only to the canal joining the Atlantic to the Mediterranean—was given up after ten years, during which time thousands of soldiers sent to work on it died from illness or mishap. More successful was the "Machine of Marly," a vast contraption of fourteen water wheels that raised water from the Seine to reservoirs five hundred feet above.

Even so, the fountains could be kept working only on special occasions. Within a few years of Louis's death, the whole system began to fall apart. "What has not been spent to bring water to Versailles?" asked a critic in 1755. "After devoting innumerable sums to this work, all is reduced to being in working order two or three times a year, to produce . . . water which miraculously shoots into the air for a few minutes. . . . The remainder of the time one sees not a drop of running water; one meets only dry fountains, and half-filled basins, stagnant and ill-smelling."

THE ENGLISH LANDSCAPE GARDEN

The great reaction to the French style, and to its Dutch and late Italian variants, took place in England. For much of their history, the English had been content to give the latest continental fashions a native twist—rigging the parterre into the more intricate "knot garden," giving topiary a foggily looming, faintly monstrous air. In the late seventeenth century, when the first really vast estates had been assembled, many of them aped Versailles, with avenues and allées radiating from the house into the countryside, the farther the better. The duke of Montague is said to have pondered running an avenue from his house in Northampton all the way to London, about seventy miles. The competition was fierce: Defoe says that when Chiswick, "the flower of all the private gentlemen's palaces in England," was honored with a visit from King William, the king—who was recognized as a connoisseur—"stood still on the terras for near half a quarter of an hour without speaking one word, when turning at last to the Earl of Portland, the King said, This place is perfectly fine, I could live here five days."

Just a few decades into the eighteenth century, though, one great house after another was shorn of its formal gardens and found itself floating in a sea of grass, amid archipelagoes of wood.

At first glance, you might take this for a genuine return to nature. The love of nature that had sprouted in Renaissance Italy had finally fought its way free of the ruins and the flaking parchment and come up in the fresh air. People were fed up with artifice and ready to see nature face to face.

You would, of course, put this in some kind of social perspective. The rising middle classes put their faith in nature, you would say, as revealed by science and the free market. Formal gardens were emblems of discredited custom, the folderol of a failed aristocracy. The green expanses of the new style were a *tabula rasa*. They were a level (more or less) playing field on which virtue and industry would triumph and find their reward.

No doubt the lives of these rising men and women were riddled with contradictions, as all lives are. No doubt they, too, needed some kind of escape. But they looked to the future, not the past. So they found their paradise in nature.

A stroll through any of the great English landscape gardens—Rousham, Stowe, Blenheim, Stourhead—will quickly jumble this neat picture. To begin with, they are infested with ruins. Everywhere you look there are temples, arcades, grottoes, obelisks. Some are built intact, others predilapidated. Some are meant to be walked through, others to be seen from a distance. (The latter, known as "eyecatchers," are often just façades or shells.) Most are in Greek or Roman style.

At Stowe, which is certified by the National Trust as having no fewer than twenty-one Class A monuments, a certain stretch of stream bank is called the Elysian Fields. It was designed by William Kent, perhaps the first of the great landscape gardeners. Ranged parallel to the stream is a pediment holding busts of Newton, Locke, and other "English Worthies"—Whig counterparts of the Greek heroes who while away eternity on the darkling plain of Elysium. Also at Stowe is one of the first landscapes to bear the imprint of Lancelot Brown, who would shortly become the dominant figure in the landscape movement. An ordinary-looking expanse of English grass between two dollops of English forest, it is called the Grecian Meadow.

The landscape at Stourhead, with its various temples and galleries garnishing the lake, was inspired by the story of Aeneas, and in particular by a painting of Claude Lorrain's called *Aeneas at Delos*. Dozens of other English properties were reshaped to match paintings by Claude or Poussin or Rosa of Greek or Italian scenes, or memories of those same scenes brought home by travelers. The sinuous curve of water, the windswept pine placed just so, the distant not-too-ruined ruin were zealously reproduced. The landscape garden was supposed to represent a coming home to nature from the alien artificialities of French, Dutch, and Italian gardens. Yet the landscapes were Italian, the landscape paintings they were copied from were French, and the word itself was borrowed from the Dutch *landschap*.

Lancelot Brown probably ripped out and turfed over more formal gardens than any man before or since. He was called Capability Brown not on account of his competence, but because he was always talking to clients about the "capabilities" of a piece of land. This was something like Pope's "genius of the place"—the inherent spirit that dictated, to the attentive eye, how a given landscape should be gardened. The odd thing was how many geniuses of English places seemed to wish they were living a few hundred miles farther south.

For Pope, it was not enough to praise "the amiable Simplicity of un-adorned Nature." He had to prove that "this was the Taste of the Ancients in their Gardens." If Francophiles found classical models for their lavish topiary, Pope could quote Martial's ridicule of "tonsured boxwood." Of course, there is a fine line between classical satire and homegrown common sense, as in Pope's famous catalogue of a topiary sale:

> ADAM and *Eve* in Yew; *Adam* a little shatter'd by the fall of the Tree of Knowledge in the great Storm; *Eve* and the Serpent very flourishing.
> THE Tower of *Babel,* not yet finished.
> St. GEORGE in Box; his Arm scarce long enough, but will be in a Condition to stick the Dragon by next *April.*
> A *green Dragon* of the same, with a Tail of Ground-Ivy for the present.
> N.B. *These two not to be Sold separately.* . . .
> A Quick-set Hog shot up into a Porcupine, by its being forgot a Week in rainy Weather. . . .

According to Walpole, it was Kent who "first leaped the fence, and saw that all nature was a garden." Usually this is read to mean that Kent saw the beauty of nature unmonkeyed with by man. But you might just as well say he saw that all of nature—at least, all of nature that he could see—had been monkeyed with by man, and one might as well do it in a professional man-ner. Before he went into landscape gardening, Kent designed sets for the theater. The "eyecatchers" scattered about Rousham and Stowe—among them Gothic churches and English cottages as well as classical structures—reflect this background, but so does the whole idea of arranging landscapes as if they were landscape paintings.

You might even say that the fence itself leaped outward and began to enclose all of nature. Enclosure, after all, was the order of the day.

ENCLOSURE AND THE HA-HA

The pastoral ideal has often furnished a mask for bashful capitalists. Some fine examples can be found in England, a nation that, as Raymond Williams has noted, was the first in the world to be urbanized and industrialized and the last to let go (if indeed it has yet let go) of its pastoral self-image.

The pastoral ideal could establish itself in England only by means of a process of expropriation. That process did not begin, as is sometimes assumed, with the four thousand Acts of Enclosure passed by Parliament in the late eighteenth and early nineteenth centuries. These only speeded up what had been going on, by force and cunning and the power of money, since the thirteenth century at least. The common land of a village—whether woodland, pasture, or arable farmed under the old open-field system—would become private property. At first this was done piecemeal, by buying and swapping of rights. Later, it might be done "by agreement" (in theory, free agreement) among all the landowners, with the lord of the manor getting one-twelfth of the land and the rest being doled out to the freeholders in proportion to their prior holdings and rights. From the mid-eighteenth century onward, though, unanimous agreement was no longer needed: if the owners of four-fifths of the land petitioned Parliament for an Act of Enclosure, they would get one. Obviously, this was very different from requiring that four-fifths of the landowners consent. The result was the enclosure of almost all the common land left in England.

Historians are not unanimous on just how dastardly all this was. Peasants often welcomed enclosure. Often they got their fair share out of the deal. But then they might find themselves unable to afford fences, hedges, and drains, and so be forced to sell out. They were not always dupes; often they got a good price. Yet the net effect was that small landowners and copyholders (more or less permanent tenants) were dying out. Far more than elsewhere in Europe, the rural social order was becoming a pyramid with three distinct steps: large landowner, tenant farmer, and landless laborer.

Most of the estates born of the later wave of enclosures were based on, and justified by, the "scientific" agriculture advanced during the seventeenth and eighteenth centuries by such men as Weston, Worlidge, Townshend, Jethro Tull, and Arthur Young. The old open-field system, unchanged at bottom since the Middle Ages, relied on such practices as strip cropping, fallowing, and common pasturage. It was, as the reformers claimed, inefficient and far from kind to the soil. In its place, the great landowners of southern and eastern England used their capital to erect a system of "improvement" in which four- or six-year crop rotations, new crops (especially root crops and legumes), stock breeding, and the use of natural fertilizers all played a part.

In other words, the alliance of humans, annual grasses, and quadrupeds, which had ravaged the soil in some parts of Europe and had done some damage in England, was now changing in important ways. It was expanding to include legumes (such as clover) and their symbiotic bacteria, which restored to the soil some of the nitrogen the grasses used up; and root crops whose nutrient demands were less than, or at least different from, those of grasses. But even as the alliance expanded, it tightened into a more integrated system, in which plants and animals fed each other.

Improvement was all the rage. Before he went mad, King George III was thought only mildly eccentric for his habit of riding about in fair weather and foul, in top boots and greatcoat, to inspect the farms on his Windsor estate. "One of the most scientific botanists in Europe"—in the view of one contemporary—"Farmer George" brought in merino sheep, spruced up the management of the royal lands, and even contributed (under a pseudonym, of course) to the *Annals of Agriculture*.

If the landowners were proud of their improvements—and they were— you would not have known it to look at the parts of their estates that were visible from their houses. The landscape they favored, which reached its grassy summit in the work of Capability Brown, was that of Arcadia. On these endless undulations of turf one would expect to find dallying shepherds, not diligent scientific farmers. And whereas the Arcadias of Italian and French gardens relied heavily on symbols—a laurel, a statue of Faunus—the English version was literal. The old gardens would have been devastated if sheep or goats got loose in them. These new ones really were pastures or deer parks, at least in part.

Literalness, however, makes strict demands on the set designer. All the scientific agriculture that was going on (and helping to pay the designer's bills) had to be hidden, along with the fences that scientific agriculture required.

The new husbandry gave landlords a reason to fence or hedge their fields, breaking up the old open-field system and uprooting many tenants in the process. The vogue of landscape gardens, with their vast deer parks and sheep pastures, provided another reason for walls and wardens. But the ethos of the landscape garden denied the need for walls. In fact, it flat-out rejected them. The whole point of such a garden was, in Pope's words, to "call in the country."

We have seen how garden walls, in forms material and immaterial, shield people from the contradictions of their way of life. In this case, the wall itself was a contradiction.

The answer was the invisible wall, or ha-ha: a stone embankment sunk in a ditch, which got its name from the practical joke it played on unwary walkers. Now one could call in the country without any danger of calling in unwanted creatures, or letting wanted ones out. One could be master of all one surveyed, yet never have to confront the mechanics of mastery. One could run one's estate on the most efficient modern principles and still feel, for much of the day, as guileless as a shepherd in Arcadia.

The usual explanation of the ha-ha—that walls interfered with the view—is nonsense. What the landowner was doing was denying the fact of enclosure, the basis of his fortune. With the help of the ha-ha, the joining of field to field which Cobbett, like Isaiah, condemned came to seem the most natural thing in the world. If all nature was a garden, why not own it all?

In a subtler way, the hedgerow turned the same trick. It made the fence seem part of nature—a tendril of forest making its way across the field. If this really was true, in a way, of the hedgerows of medieval enclosures, which meandered and were composed of diverse plant species—maple, elm, hazel, oak, dogwood, hawthorn, bramble, sloe—and therefore played host to diverse animal species, it was much less true of the hedgerows of modern enclosures, which were monocultures of hawthorn.

If ha-has could hide the need for fences, they could not hide such other artifacts of agriculture as plowland or the men who plowed it. For that purpose, Brown would plant overlapping belts of trees, cutting off from view the working landscape beyond the landscape garden.

A TALKING DUCK

Unfortunately, there were times when neither ha-has nor trees could hide the offending objects. When landscape gardens were created, whole villages often had to be razed because they spoiled the clean Arcadian view. They might be rebuilt elsewhere, or they might not. They might be rebuilt elsewhere, razed again, and rebuilt in a third spot. In 1737, the village of Shugborough was moved so Thomas Anson could make a big park around his house; in the nineteenth century it was moved again so Thomas Lord Anson could make the park bigger. Sometimes the parish church was left standing when the rest of the village was shaved away; even if it stayed in use and in good repair, it became in effect a quaint ruin. Just how common these practices were is suggested by a survey of parks and gardens in Hampshire, which found old village foundations or orphaned churches in over half of them.

The landscape garden of Castle Howard, one of the first and grandest essays in the form, was intended by Charles, Earl of Carlisle, as an "elysium" that would recover the Golden Age and make flesh the words of Virgil; a place of "Woods and verdant Plains / Such as possess'd of old th' Arcadian Swains." In the words of a poem apparently by the earl's daughter, the garden would "make happy those who on you depend." Either happiness was construed broadly or dependency was construed narrowly, for the park was made by evicting the villagers of Henderskelf without provision for their resettlement.

In a nation as hard-nosed as England, the clanking of pastoral tropes against the facts of the countryman's life could not go on forever without exciting comment. An early comment came in 1736 from a real-live countryman, Stephen Duck.

> No Fountains murmur here, no Lambkins play,
> No Linnets warble, and no Fields look gay;
> 'Tis all a gloomy, melancholy scene,
> Fit only to provoke the Muse's spleen.
> When sooty Pease we thresh, you scarce can know
> Our native Colour, as from Work, we go:
> The Sweat, the Dust, and suffocating Smoke
> Make us so much like Ethiopians look.

The "thresher poet," who was as much a curiosity as a talking duck would have been in those pre-Disney days, was adopted as a pet by Queen Caroline and within a few years was writing standard linnets-and-lambkins pastoral verse. But anti-pastoral was born, to be adopted in the next ten decades or so by such poets as Crabbe, Goldsmith, and Clare. Ironically, many of them quickly fell into one of the most well-worn of all pastoral ruts, the lament for a lost golden age. A generation before, the English peasant had been a happy, self-reliant yeoman; now, thanks to enclosure and other evils, he was a landless and often foodless fieldhand. The chorus was taken up by political writers such as William Cobbett. For the poor, Cobbett said, the hedgeless common had been a vital hedge against want.

> I used to go around a little common, called Horton Heath, on a Sunday. I found the husbands at home. The common contained about 150 acres; and I found round the skirts of it, and near to the skirts, about thirty cottages and gardens, the latter chiefly encroachments on the common, which was waste (as it was called) in a manor of which the Bishop was the lord. . . . I remember one hundred and twenty-five or thirty-five stalls of bees, worth at that time ten shillings a stall, at least. Cows there were about fifteen, besides heifers and calves; about sixty

pigs great and small; and not less than five hundred heads of poultry! The cattle and sheep of the neighbouring farmers grazed the common all the while beside. The bees alone were worth more annually than the common, if it had been enclosed, would have let for deducting the expense of fences. . . . Waste indeed! Give a dog an ill name. Was Horton Heath a waste? Was it a "waste" when a hundred, perhaps, of healthy boys and girls were playing there of a Sunday, instead of creeping about covered with filth in the alleys of a town?

If many of the lamenters tended to overstate both the goldenness of the prior age and the abruptness of the change—each successive writer placing it just after his own childhood—they were not wholly wrong about the way things were going. Even Arthur Young, the foremost proponent of scientific farming and the enclosure it called for, had second thoughts. "I had rather that all the commons of England were sunk in the sea," he wrote, "than that the poor should in future be treated on enclosing as they have been hitherto."

The net effect of enclosure on nonhuman nature is harder to assess. One of the most widely cited articles ever written, Garret Hardin's "The Tragedy of the Commons," uses the medieval commons to illustrate a common ecological dilemma. When every villager has a right to use the common pasture, Hardin says, every villager has an incentive to graze as much livestock on it as he possibly can—for, if he doesn't, someone else will. The logical conclusion is a denuded pasture.

As social science this model is very helpful: it can be applied to the use and abuse of almost any common resource or sink, from ocean fisheries to air. As history, though, it is "hogwash" (to quote the ecological historian Donald Worster). In most medieval commons, the rights of use were regulated by the village as a whole. Many common lands were subject as well to forest law, designed to encourage the proliferation of deer and other game for royal hunts. It is true that regulation was far from airtight, and that woodlands in particular often fared ill when subject to common use. In fact, enclosure often led to reforestation. But it could also lead to the rapid felling of timber for profit. Plowland worn thin under the open-field system was often improved by private owners, and some plowland that should never have been plowed went back to grass as the price of wool and dairy products rose. At the same time, though, thousands of acres of wild or nearly wild moor, heath, and fen were brought under the plow.

The balance sheet is very hard to draw up. But there is no doubt that the main thrust of enclosure was to take "wastes" and "improve" them. While gardens were going "natural," shimmying with S-curves right and left, the more natural lands beyond were being drained, fenced, and ruled out in squares. While in parks trees were scattered in the style of Claude and Rosa—as carefully tousled as a politician's hair—in plantations meant for profit they were plunked down in rows that gave Wordsworth the impression of a "vegetable manufactory." "Nature abhors a straight line," said Kent—a patent falsehood, of course—but his clients did not abhor straight lines that helped maximize the profit from their estates.

In fact, it has been argued—by no less an authority than the great Victorian gardener J. C. Loudon—that the landscape garden arose in eighteenth-century England precisely as a reaction to what was happening in the larger landscape. When the landscape was still curvy and patchy, great landowners distinguished their gardens by making them geometrical. But when the landscape started getting geometrical, distinction had to be sought in curves and patches. Like the paradise park, the landscape garden was an island of pastness, walled off against the very changes that paid the gardener's bills.

COAL, CAPITAL, AND COLONIES

The changes in eighteenth-century England were not, of course, limited to farmland. The heyday of the landscape garden was also the start of the Industrial Revolution, which in the long run would change the planet as radically as anything else in human history. Three landscape gardens— Himley Hall, Stourhead, and Stowe—can be seen as symbols of the three conditions that made England the nursery of industrialization: coal, capital, and colonies.

Coal rightly takes pride of place. Without coal, England's fleshy lungs might have breathed more freely, but its economic lungs would have been choked off. It would have remained what it was in the Middle Ages—a sparsely settled backwater—with the minor difference that it would have been utterly bare of trees.

At the start of the sixteenth century, England was one of the most densely forested countries in Europe. By the end of the century, shipbuilding, iron-smelting, glassmaking, and the fuel hunger of a growing population had brought about a crisis. But if Britain's historic forests were mostly gone, its prehistoric forests were still present; they had just gone underground. The conversion to coal that, in the case of contemporary oaks and beeches, was performed in kilns had, in the case of Carboniferous cycads and ginkgos, been managed more cheaply by a long geological squeeze.

By the mid-1600s, coal or its refined cousin coke was in general use not only at home but in shops and works producing salt, soap, alum, yarn, glass, metal products, spirits, and beer. Without coal, Britain's population growth would have been stymied, no matter how much its farming was "improved." Without coal, a city such as London could not have come to be: neither a city so big—twice as big as any other in Europe—nor a city so foul. London's air had drawn unfavorable comment since the thirteenth century, but now the comments grew to a hoarse, gasping roar. Smoke and soot from coal that was twice as high in sulfur as modern types hovered above the city like a pillar, guiding travelers from afar. At close range, the smog blackened statues, gnawed at buildings, killed trees and flowers, ruined clothes and furniture, and turned long-term pedestrians as dark as Stephen Duck's threshers. In John Evelyn's view, London had become "the suburbs of Hell."

For all its uses and overuses, there was one thing—one crucial thing— coal could not do. It could not make respectable iron. This was not for lack of trying. Many experiments were made, of which some may have been qualified successes. In 1619 Dud Dudley, bastard of a family whose estate in the West Midlands included coal mines, iron mines, and ironworks which they had exploited since the thirteenth century, came down from Oxford to manage his father's furnace and forges. "Wood and Charcoale growing then scant, and Pit-coles, in great quantities abounding near the Furnace," he was moved to invention, and by his own account "made Iron to profit with Pit-cole." James I gave him a patent, but Charles II refused to renew it, and Dudley's secret process sank into darkness. In 1709, Abraham Darby, a native of the town of Dudley, pioneered a new process which his son of the same name had perfected by 1735. The pig-iron industry, freed from the need for trees, swelled like a prize sow. Ironworks moved from the forests to the coalfields. They were chockablock in places like the West Midlands, where rich deposits of coal and iron ore lay side by side. And this became even truer after 1783, when Henry Cort figured out how to make pig iron into wrought iron with coal as the sole fuel. "All the activity and industry of this kingdom," said Arthur Young in 1791, "is fast concentrating where there are coal pits."

Once this buried energy was exhumed, machines flocked to it like buzzards. Their first job was to pump out the coal mines, which, as demand soared, got deeper and deeper and therefore damper and damper. The first steam engine was built by Thomas Newcomen in 1712 to pump water from a Dudley coal pit. Watt's improved steam engine was likewise first used to drain mines. Another early application of the engine was to pump air into a blast furnace, which boosted the quality of coal-fired pig iron. Without coal-fired smelting, foundries could not have made the larger and more precise castings needed to make steam engines; without steam engines, the coal and iron industries would not have been able to make more steam engines.

Railroads, too, began by serving the mines and the ironworks. Rails had been used by horse-drawn carts in German mines before 1500, and in English mines by 1600; once Watt had produced a version of his steam engine that could turn a wheel instead of pumping, it was only a matter of time before the first steam-powered locomotive appeared, hauling iron ore from the Pen-y-Darran mines in 1804.

As railroads helped increase the supply of coal and iron, they also increased the demand by using coal and iron themselves. The use of iron for everything from railroad cars to men's collars made the material world heavier, so that more coal was required to move things around. In short, these two substances sprung from the earth joined in a breathless symbiosis, each leapfrogging clumsily over the other in a frenzy of new applications.

Many landed families were active in industry. The earl of Derby had coal and lead mines in Lancashire, a cotton factory in Preston, and interests in canals and toll roads. Defoe in his travel letters mentions the alum mines near Musgrave, in Yorkshire, "from whence the Lord Musgrave, now Duke of Buckinghamshire, has his title, as he has also a great part of his estate from the allom works not far off." Lead mines on the Powis estate provided both the funds and the lead ornaments for the lordly gardens of Powis Castle.

Such people were helped by a peculiarity of English law: in England, unlike the rest of Europe, rights to all minerals except silver and gold belonged to the owner of the land. They were also helped by enclosure, which allowed them to become owners of common land that had valuable minerals under it. A case in point is the Dudleys: beginning in 1776, they pushed through a series of Acts of Enclosure that gave them an extraordinary 40 percent of the district's common land, with full rights to its iron, coal, timber, and other natural wealth. The Acts freed them from "paying or making Satisfaction to any Person . . . for the Damage to be done . . . in the said Lands." This was a wise move, for the scarring of the landscape was severe, to say nothing of the pall of smoke and coal dust that settled over everything and would shortly earn the region the name of "the Black Country." By 1830, the native oaks that had not yet been cut were stunted and brittle from the smoke, and an expert advised planting beeches instead, as "the smoothness of the Bark and Leaves prevents the soot resting upon them to any great extent and the first rain washes it off."

As the Industrial Revolution gathered steam, it became harder for the rich to wall out from their private grounds the changes that were making them rich. Often they had to make a run for it. When soot from the mines and ironworks made the Dudleys' ancient seat, Dudley Castle, unsittable, they moved six miles west and built Himley Hall and its fashionable park. But the logic of extraction followed them. By 1788, the second viscount had started selling timber from the park—not sparing the "vistas, walks and rides," and prompting a lawsuit by his half-brother. In 1836, another ironworks, not their own, was opened nearby and Himley was made "uninhabitable." The Dudleys bought a new manor some twenty miles to the southwest, well out of the Black Country.

BANKERS AND SWAINS

What was revolutionary in the Industrial Revolution was not the use of machines. The innovations that made the cotton industry grow tenfold between 1760 and 1785—Hargreaves's spinning jenny, Arkwright's water frame, Crompton's mule—were technically piddling. The water frame, for instance, was just a late-medieval machine, the spinning wheel, adapted for a late-medieval power source, water. What was revolutionary in the Industrial Revolution was the wide application of fossil fuel.

The human alliance with machines had gotten rolling in the Middle Ages. Human saprophagy—drawing energy from decomposed organic matter—had begun in earnest in the seventeenth century. What happened in eighteenth-century England was the confluence of these two trends: that is, the advent of machines that could feed on coal, just as oxen had fed on grass.

Living allies remained crucial, of course. Even in our own age of synthetic fertilizers, no one has yet come up with a palatable way of eating fossil fuels. The Industrial Revolution could not have gotten off the ground without the agricultural revolution—or evolution—that came first, swelling the population and freeing hands from the plow so they might take up the shuttle. The fledgling agribusiness of the age was also the source of much of the wealth that came to be invested in other industries. Corn produced capital. But corn alone could not have produced enough capital, or capital that would move freely enough. For that you needed the whole modern system of equity and credit that was pioneered by bankers like Henry Hoare, who at Stourhead created one of the greatest of all landscape gardens.

The Hoares had started out as goldsmiths. In the Tudor period, they and some of their fellow goldsmiths had become England's first real bankers. As goldsmiths necessarily had vaults, they often held gold and other treasure in safekeeping for the crown and other well-heeled clients. The receipts they gave out came to be used as money and so became the first banknotes. At this point, the goldsmiths were functioning as glorified pawnbrokers; but it was not long before some bright smith realized that not all depositors would claim their gold at the same time, so he was free to issue, as loans, banknotes exceeding the value of what was in the vaults. By taking the next step in the sequence we have spoken of—sun, grain, gold—the goldsmiths gave England the big, fluid pool of money it would need to launch its Industrial Revolution. Strange but true: it was the goldsmiths who left the clutchiness of mercantilism behind and led the way from the age of gold into the age of credit.

The new form of wealth, it turned out, did not have gold's immutability. It was more like grain: if you planted ten seeds, you might get fifty back—or you might lose them all. This was proved in the world's first stock-market crash, the notorious South Sea Bubble of 1720. In an effort to consolidate its huge war debt, the government—following the lead of France, which itself was advised by a Scot—issued shares in a company that was chartered to trade in "the South Seas and other Parts of America." Though the South Sea Company hardly did any actual trading, the shares became the object of a speculative frenzy that launched a general stock-market boom. A Dutch attorney visiting London said it was "as if all the Lunatics had escaped out of the madhouse at once." Within months the French shares crashed, and in short order the British shares did, too—from a high of 1,050 to a low of 180—taking most other stocks with them.

Henry Hoare was one of the canny few who got out when the getting was good. He seems to have invested much of his winnings in land—including the land that would become Stourhead.

As I mentioned earlier, the landscape at Stourhead appears to have been patterned on Claude's painting *Aeneas at Delos*. The scene is one in which the hero asks Apollo to grant his weary band of Trojan refugees a home of their own—"a walled city that shall endure." The path around the lake at Stourhead is based on Aeneas's journeys. And the statue of a river god is thought to refer to Rosa's etching, *The Dream of Aeneas*, in which Father Tiber assures the hero of victory, tells him this place (the future Rome) will be his home, and advises him to seek out the Arcadians as allies.

Of Hoare it has been said, "Like Aeneas, he was establishing his family in a place." The path at Stourhead is the path of the hero, the builder. But his allies are the Arcadians: the old landed classes, perhaps, whose money he takes and multiplies and into whose ranks he ascends; or simply the English pastoral self-image, the plush grassy lawn on which rough-and-tumble games of industry can safely be played. Hoare spoke gloatingly of the "fruits of industry . . . the envy of the indolent who have no claim to temples, grottos, bridges, rocks, exotic pines and ice in summer." A pungent irony: the idyll of the leisured shepherd becomes the preserve and reward of the industrious banker. But that is more or less what most great gardens in modern times have been.

AN UNSTABLE UNITY

The Industrial Revolution could not have happened as it did without the quickening drift toward a New Pangaea that had begun three centuries before. In more conventional terms, it is clear that England could become the "workshop of the world" only because men like William Pitt the elder, striding through the meadows of Stowe, determined that England would do whatever it took to get markets and raw materials. To supply the British cotton mills, a plant long associated with the Indian subcontinent was grown in America using labor from Africa.

This, of course, is where colonies come in. Stowe was the unofficial headquarters of the Whig political elite. Pitt himself married into the family. With Pitt at the helm, England took Canada and India from the French, firmly establishing the empire on which the sun never set.

In the Temple of English Worthies at Stowe, the busts of King Alfred ("who secured the seas"), Queen Elizabeth, Raleigh, and Drake all celebrate English naval power and the riches it could bring. But other busts are equally revealing: Sir Thomas Gresham, who (the inscription says) "followed the honorable Profession of a Merchant"; Hampden, who helped bring about a form of government favorable to commercial interests; Bacon, Newton, and Locke, who helped shape the modern view of nature and what to do with it.

William Blake would have been disgusted to find the last three in a garden, of all places. His redeemed Milton vows to "cast off Bacon, Locke & Newton from Albion's covering"; and in *Jerusalem*, Blake sees

> the Loom of Locke, whose Woof rages dire,
> Wash'd by the Water-wheels of Newton
> . . .
>
> cruel Works
> Of many Wheels I view, wheel without wheel, with cogs tyrannic

Moving by compulsion each other, not as those in Eden, which,
Wheel within Wheel, in freedom revolve in harmony & peace.

Between England's "mountains green" and her "dark Satanic mills" no
truce was thinkable. But in 1735, when the Temple was born—and Blake
was twenty-three years short of his first birthday—the love of nature and
the urge to control nature were still one.

Since ancient times, the garden had tried to bring Mountain and Tower
together in a symbolic whole. As the Arcadian ideal came to the fore, gardens tried to make this mingling more and more real, so that the chief
delights of wilderness and civilization really could be tasted in a single
place. The symbolic wholeness of the medieval garden gave way to the
imagistic and magical wholeness of the Renaissance garden, which in turn
gave way to the real (or putatively real) wholeness of the English landscape
garden. This meant that the garden became part of the larger landscape, or
vice versa: not a symbolic Arcadia but a real one, in which sheep might
safely graze.

Behind this new kind of garden lay a new way of looking at the world
and of dealing with the world: a way in which looking and dealing were
closely allied. Already in the Renaissance the joyful discovery of nature and
the urge to conquer nature were hard to disentangle. Now they were bound
even tighter. Unlike the Renaissance mages with their ancient tomes and
their alchemy, Newton had revealed a world that obeyed mathematical laws
and answered predictably to human action. His calculus and his trajecto-
ries made curves and other features of wild nature seem less wild, more
subject to reason. Scientific agriculture, smiled upon by a somewhat
warmer global climate, made widespread famine a thing of the past. The
sceptred isle had managed to purge itself of unfriendly wild animals more
effectively than its continental neighbors; wolves, for example, were a dark
memory. For the upper classes, at least, it was possible to find nature as
a whole fundamentally good—admirable in itself, and yet capable of
improvement by the gentle hand of man.

True, this unified world was lived in mainly by the upper classes. Its
unity depended, as often as not, on the exclusion or eviction of what did
not fit. And this exclusion was unusually cruel in that it excluded the very
idea of exclusion: the very walls had to be unseen. Still, as unities go, this
one was fairly well knit.

Before the century was out, it was flying apart.

Coal and iron had their own logic. London got bigger and bigger, blacker
and blacker, giving poets like Wordsworth the world's first taste of the

grimy, alienating modern city. (Locke himself, Blake's bugbear, had already learned to steer clear of London: a chronic invalid, he couldn't stand the smog.) Factory and mining towns sprang up, of the sort that would serve as models for Dickens's Coketown: "a town of red brick, or of brick that would have been red if the smoke and ashes had allowed it," in which the streets were all alike, the people were all alike, and every day was like the last. The scars on the landscape started to be apparent even to the casual observer. The extirpation of wild animals and wild places started to seem a bit too complete. As the population grew, so did the molar-grinding of people like Malthus—and the need of people in general to escape the crowd, which they proceeded to do in nearly wild nature. Relief from the crude utilitarianism of the new industrial landscape could not be found in the square fields of the "improved" countryside. One had to go farther afield.

But the search for wild nature was not just a reaction or an escape. As the mastery of nature had its own logic, so too did the love of nature. Once take nature as your model, and each successive compromise comes to seem too tame. The mixed formalities and informalities of Kent must give way to the ruthless green of Brown. Brown in turn must be accused of smoothness, and must yield to the rugged crags of the Picturesque. Finally even the Picturesque comes to seem too—well, picturesque. Why bother with a garden at all? Why not head for the mountains?

Well, why not? It was now fairly safe. One could even take the railroad! A sense of safety and well-being had allowed the flirtation with wild nature in the first place. As that sense deepened—thanks to "progress"—the flirtation could become a full-blown affair. And so, in 1844, we find Wordsworth railing against a railroad that would bring "the whole of Lancashire and no small part of Yorkshire" into his Lake District.

A common account of the rise of the modern sensibility goes like this: First came science, technology, and the urge to dominate nature. Then, in reaction, came the urge to cherish and protect nature. What I am saying is that it was not simply a matter of action and reaction. At one point, the two urges were one.

And not just at one point. The love of nature and the urge to master nature have always, I am sure, been basic to the human mind. And they have always gone hand in hand, as they seem to do in the cave paintings of Lascaux. Yet there has always been a tension between them—a tension expressed, for example, in the rituals by which some indigenous hunters placate the spirit of the animal they have just killed.

What is remarkable in the modern case is that each instinct has had the luxury of following its own inner logic. For a moment, this produced a more glorious union: in figures such as Conrad Gesner, Leonardo, Newton, Gilbert White, Joseph Wright of Derby, Goethe, and dozens of others, science and art, love and mastery blazed up like two candles touched together. Soon, though, the tensions began to be felt. Once command of nature was freed of awe and magic; once love of nature was freed of fear and need, a parting of the ways was inevitable.

Of course, this parting did not mean that the two urges could no longer coexist in the same brain. In fact, the typical modern person profits from the exploitation of nature while seeking out, and seeking to protect, wild things and wild places. In other words, he lives in the Tower and flees, or fantasizes about fleeing, to the Mountain.

By some lights this is sheer hypocrisy, or at best self-deception. The historian Keith Thomas, after charting in rich detail the growth of "the modern sensibility" in eighteenth-century England, concludes: "For adults, nature parks and conservation areas serve a function not unlike that which toy animals have for children; they are fantasies which enshrine the values by which society as a whole cannot afford to live."

By using the word "fantasies," Thomas reminds the reader of the epigraph of his final chapter, which comes from Freud:

> The creation of the mental realm of phantasy finds a perfect parallel in the establishment of 'reservations' or 'nature-reserves' in places where the requirements of agriculture, communications and industry threaten to bring about changes in the original face of the earth which will quickly make it unrecognizable. A nature reserve preserves its original state which everywhere else has to our regret been sacrificed to necessity. Everything, including what is useless and even what is noxious, can grow and proliferate there as it pleases.

But Freud's observation can be given a very different spin than it gets from Thomas. If nature reserves are parallel to the realm of fantasy, this does not mean they *are* fantasies. Certainly it does not mean that they are *mere* fantasies, with no other function than to salve our childish qualms. Fantasy plays a vital role in our psychic lives. Without its dark upwellings, the ego would dry up and die. Nature reserves—and wilderness generally—play a vital role in the life of the planet; without them, civilization would dry up and die, though it might take some decades or even centuries to do so. And the cause of the death of civilization would not be mere psychic impoverishment, though that might lend a hand; it would be the loss of breathable air, bearable climate, temperate hydrology, and all the other amenities that wildness provides.

In exploiting one part of the planet while protecting another, recent humans are simply carrying out the command given to the first humans—"to work it and protect it." To be sure, this is not possible unless the part that is exploited is also protected to some degree. But neither is it likely to work unless some kind of division is accepted.

For the psyche to function adequately, there has to be a barrier—though not an impermeable one—between fantasy and reality. When that barrier breaks down, we enter the realm of delusions and other psychoses. To extend Freud's parallel a little further, we might say that the real self-deception lies in the construction of a false unity: in thinking that Mountain and Tower can be juxtaposed with perfect comfort, or that a perfect mean can be found between them.

To deny the need for walls is to deny something deeply entrenched in nature and in the human mind. I myself came to England with a prejudice against formality and walls, fully intending to fall in love with landscape gardens. To my surprise, I found that I was happiest in the old-fashioned, left-over walled garden at Rousham, with its turretlike dovecote, its knot beds, and its high brick walls overgrown with vines and crowned with peacocks.

In the eighteenth and nineteenth centuries, many such delectable spots in England and elsewhere in Europe were erased in the name of nature. Every estate had to have its *parco inglese* or *jardin anglais* (even Versailles got one, which, oddly enough, like the artificial village where Marie-Antoinette played milkmaid, fit just dandily into the Sun King's theme park). If the rational basis of the eighteenth-century faith in a unified world was breaking down by the century's end, the effort to impose unity on the landscape just kept rolling along. The park gobbled up the garden within and the wilderness without, masticating both to a green sameness. The chewing would be even more vigorous in America. Olmsted's tour of English landscape gardens would have a fateful effect not only on America's estates and parks, but—more ominously—on her suburbs.

AND THE WORD WAS MADE GREENERY

Gardens try to heal the rift in the world and in the self that began when we were kicked out of Eden—that is, when we became human. They try to join Mountain and Tower, wildness and civilization, in an orderly whole. They may try to do this by collage or by compromise, in symbols or in the flesh. They may—perhaps they must—unify by dividing, by walling out what does not fit. To unify, gardens have to mystify, for the cozy wholeness they are after is not to be found in the real world.

Gardens wall out change, including some change that has already happened. They wall out the facts of their own time and place. The here and now is a thicket of conflicts and contradictions; if wholeness is to be found anywhere, we are fairly sure it is to be found elsewhere and elsewhen.

The story of the Western pleasure garden looks like a visual form of "telephone," the party game in which a whispered message is passed along a line of people until it loses all resemblance to itself. So the secret of paradise was passed from Greece to Rome, from Rome to Italy, from Italy to England, from England to America. This was not a simple process of signal degradation, though. At each point information was lost, but vital new information was gained. Besides, each player did not just rely on his predecessor, but referred back to certain common originals.

In a way it is amazing that Western civilization has produced any decent gardens at all. For its gardens have been patterned on the texts of two peoples who had no pleasure gardens to speak of. While the great examples of Egypt, Persia, China, and Kashmir have had some effect over the years, especially on the choice of plants, they have had less effect than the patterns cut in black and white by the Greeks and the Hebrews. Looking at Western gardens, one is struck by the power of the written word to shape our vision of nature. The pen has been mightier than the eye.

As Cadmus (the mythic inventor of Greek letters) sowed dragon's teeth that sprang up as armed men, and as Ezekiel sang the dry bones to jaunty life, so the Greeks and Hebrews sowed letters—bare and dry as bones—that sprang up centuries later as roses, tulips, and boxwood dragons.

In the beginning was the word, and the word was made greenery.

The love of nature is always, I think, mixed up with the love of some other or earlier culture. For the role of spectator gets old very quickly. We want to live in nature, not just look at it. We want, or think we want, a mode of subsistence that matches our love. Since no human can invent a way of life from scratch, we turn to the ways of life that are, or were, practiced in places where there is, or was, more nature than we find here and now. The past tense fits best, for obvious reasons. Ironically, the cultures that had a lot of nature to work with were not always cultures that used nature well.

Even when a garden is modeled on Eden itself, some cultural relation to nature is often assumed. What the maker has in mind is hardly a real bare-knuckle bout with wilderness. Since Adam's labors never made him break a sweat, the hard work must have been done by someone else. Whether that someone else was God, nature, or a lot of peasants would not matter much to the owner of the facsimile garden. In this sense, Eden was as good a screen as Arcadia.

Perhaps this state is seen as precultural by those who pine for it—seen, in fact, as a state of nature. But such a view is possible only for those who have idealized nature beyond all resemblance to its present state, making it far friendlier to unaccommodated man. Either they do not look at nature at all, or they look at her and think, as the Church Fathers did, that as surely as Eve she is a fallen woman.

In the New World, however, the dream of Eden has taken a new form. The fantasy of a direct relation to nature has been indulged on a grand scale, with results that we will now consider.

20

WESTWARD IN EDEN

ONE OF THE ticklish things about gardens is that they not only embody but also affect our relation with nature. In the first place, they are concrete examples of the way we use soil and water and plants: of our methods of irrigation, tillage, plant breeding, propagation, fertilization, pollination, pruning, and pest control. At the same time, they are stylized images of our relation to nature. We have to be careful what message those images convey.

The garden is at once the sign and the thing signified, and the two roles are often at odds. It is rather like a marzipan fish: considered as an image of food, it is one thing; as an example of food, it is something else again. As you move out into the larger landscape, the symbolic element dwindles and the practical element grows; but not so much as you might expect.

The English landscape gardeners, we are told, leaped the fence and saw that all of nature was a garden. Well, all of nature *was* a garden, in the sense that it had been shaped by human hands. At least, nearly all of Western Europe was a garden. By the eighteenth century, nearly all its wilderness was gone. Densely settled and intensively cultivated, the continent might as well have had a fence around it.

A garden, I said, can be an island of pastness or an island of futurity. As befits a very large garden, Europe was both. It was crammed with the symbols of past and passing cultures, which served both to disguise and to enforce certain relations between people and nature and between people and other people. This made Europe seem at times an island of pastness which the waves of change could hardly affect. But Europe was also the eye of the storm—the point from which many of the waves went forth. Like a vegetable garden or grainfield hewn from wilderness, Europe was an earnest of what was to come: what in time would fill almost the whole world.

It was inevitable that Europe would expand; as I said earlier, nothing succeeds like failure. Because Western agriculture boosts populations as it depletes soil and other ecological capital, it has no choice but to expand. Western industrialism behaves in much the same way. The tide of white humans and their allies that had flooded Europe in the Middle Ages had ebbed in the face of the plague, but then had gathered force again. Reinforced by new recruits—including machines and, ironically, a number of crops from overseas—it now burst the confines of the continent.

True, in parts of Europe farming showed signs of becoming sustainable. Favored with a more forgiving climate than their cousins on the dry limestone hills of the Mediterranean, the farmers of northwestern Europe had some margin for trial and error, and they made good use of it. Europe was a humanized landscape, yet had some of the diversity and stability of a natural landscape. It was this achievement that René Dubos, who grew up in the Ile de France, would celebrate, in a phrase borrowed from Tagore, as "the wooing of earth."

But the garden of Europe was far from snakeless. It was crowded, rigid, rife with injustice and pettiness and suspicion. To maintain a tidy ecosystem so unlike any natural ecosystem, humans paid a stiff price. Often, that price was imposed in the coin of symbolism (in Greek, *symbolon* meant, among other things, half of a coin split between contracting parties). Cultures of other times and places were invoked by various parties to keep the garden ordered the way they liked. This was one of the things that drove other parties to seek a wilder Eden across the sea, which in turn furthered Europe's expansion.

So Europe's futurity fed its pastness, and its pastness fed its futurity.

A SMARTER ADAM

From the time of its "discovery," America was plastered with every label the myths of the West could inscribe. It was Arcadia, the garden of the Hesperides, the Isles of the Blessed, the place where the Golden Age still reigned. It was Eden, the Earthly Paradise, the Promised Land, the New Jerusalem. It was the Wilderness of Zin, Gehenna, a howling waste, a place of Azazel.

Whether the settlers of America saw a paradise or a howling waste depended partly on their prior beliefs and partly on what actually happened when they arrived. Much has been made recently of the "fertile, humming, blooming paradise" that the Philadelphia naturalist William Bartram found in the Southeast—the wildflowers, the strawberries, the teeming trout served up in orange juice. Winter in the Northeast was not, perhaps, quite so enthralling a wilderness experience as Bartram's rambles in Florida. The word "Adirondack" means "bark-eaters"—the name with which the Iroquois baited the Algonquins, implying that this was what they were reduced to when their feeble hunting skills failed them. Most of Bartram's American contemporaries found his report obnoxiously rosy, not because they lacked his innate receptivity to nature (though they may have) but because the wild nature beyond their own fences was so obviously hostile. It was in Europe that Bartram's sparks found tinder—in the volatile minds of Wordsworth and Coleridge, for example. When European savants of the eighteenth century met the American wilderness face to face, they blinked. Linnaeus's protégé Pehr Kalm, although duly enthused about this vast vacuum waiting to be filled with Latin, was horrified by the wood lice, gnats, mosquitoes, and rattlesnakes that ambushed him as he traveled through the woods from Albany north to Quebec. ("Woodlice [Acarus Americanus L.] abound here. . . . Scarcely had a person sat down before a whole swarm of them crept upon his clothes. . . . There are examples of people whose ears were swelled to the size of a fist on account of one of these insects creeping into them and biting.") Even Crèvecoeur, whose *Letters from an American Farmer* was the best advertisement for the continent ever written, saw the wilderness as a beast to be tamed. And Bartram himself, although the kind of holy fool whom God protects, had moments of horror in the "dreary wilderness" he explored. The trout he had for dinner one evening, with orange juice as vinegar, just missed being the dinner of an alligator—as did Bartram, who was cleaning the fish when the monster lunged. An hour or two later, his postprandial musings were interrupted by two large bears.

Sometimes, America must have seemed heaven and hell rolled into one. That, of course, is exactly what the World Mountain is: paradise, but not for us. A wild place, a place on which human culture has not set its seal (or a place that seems wild because the seal is unintelligible to us) may fill us with awe or even joy, but most of us will never feel at home there. What is stunning about America is that people were determined to make Eden their home, and still keep it Eden.

What does this mean? Not that they meant to keep it wild—not at all. What it means is that they were determined to have a direct relation to nature. In America, the dream of Eden was put to the test.

They knew they would have to work. A utilitarian Eden, then. But this posed a problem. They could not get at nature's bounty without the help of culture. But culture was oppressive, hierarchical, stifling, old-fashioned: in a word, European. The answer was to use only the useful part of culture:

in a word, technology. As for your spiritual relation to nature, that was your own lookout. If you wanted to have one, there was no reason to get dead people mixed up in it.

The Puritan's direct relation to God became, in America, a direct relation to nature. Every person could relate to nature directly, without worrying about other people—living, dead, or unborn. There was so much nature here that no fancy rules were needed for dividing or sharing it. There would be plenty for everyone, forever. It would never run out.

The Indian proved there was such a thing as Adam. The task now at hand was to be a smarter Adam, more efficient, more productive—to eat of the Tree of Knowledge yet stay in the garden. This was the special task and the special privilege granted to Americans, though withheld from all humankind since the Fall. America was paradise regained, but with a difference. No longer were God's grace and nature's bounty to be purchased at the cost of ignorance and bare feet.

A privilege and a task. Not an easy task, either. Beneath the smooth flow of American optimism lie treacherous eddies of doubt. Hawthorne, Thoreau, Melville, Mark Twain, James, Wharton, Lewis, Fitzgerald, Mailer: one American writer after another pokes a foot in. Is innocence lost? Can Adam be a Connecticut Yankee—wily, pragmatic, wealthy but pennywise, with a whole barn-full of the ripe apples of knowledge—and still be Adam? Without genuine (that is, European) culture, is he any better than a savage? With genuine culture, is he any better than a European?

In *The House of the Seven Gables*, Holgrave asks, "Shall we never, never get rid of this Past? It lies upon the present like a giant's dead body!" In "Earth's Holocaust," a "modern philosopher" urges the burning of all books: "Now we shall get rid of the weight of dead men's thoughts. . . ." The irony is that what oppressed Hawthorne's people was the weight of New England's brief history—a history invented as a counterweight to the history of Europe. A history meant to be redemptive, Edenic. What oppressed him was not the culture of Europe but the Puritan anticulture of America.

Europeans had sought Arcadia in America; had become Americans; now turned around and sought Arcadia in Europe. The newness of America, once so alluring, palled. Americans built ruins along the Hudson—just as the English had built them along the Stour the century before. And they went to Italy for ruins. Absent ruins, there could be no Arcadia. Why? Because the mood of Arcadia must be elegiac? Because the Golden Age can only be remembered, not known? Or because the dream of a direct relation to nature was wearing thin?

The dream had to wear thin. For better or worse, our relation to nature can only be cultural. Nature is never just nature. A landscape—even a place that is utterly wild—is admired not only for itself, but for its links to previous human experience. Whig bankers tried to re-create classical landscapes not merely because they were pretty but because they were the proper backdrops for a heroic life. The vast emptinesses of the American West shimmer with the ghosts of cowboys and murdered Indians. Africa's savanna gains hardly more glamour from its lions and elephants than from the legendary white hunters who stalked them. Even an explorer who treads where no other human has trod inevitably sees the place as the kind of place where other explorers—his heroes—have gone. If one day humans scale the peaks of Mars, they will probably see the view from the top through the borrowed eyes of Edmund Hillary.

Every longing for nature is also a longing for some other culture. Wordsworth would lief be "a pagan suckled in a creed outworn." Nature was not the only book Thoreau read at Walden. Without the help of Homer, the Bible, and the Upanishads, the book of nature would have been unreadable. And there was the unwritten culture of the Indians to guide him—the mute arrowheads that "sprang from the ground when he touched it." Perched on the cliffs of Yosemite, Muir had his dog-eared Emerson; but exactly what culture did he admire? Or was he the harbinger of a truly cultureless love of nature—the love of pure wilderness, excluding all humans? The love that has no consequences for our way of life, that is therefore a dead end?

INNOCENCE AND POWER

Set an American down face to face with a tree, a flower, a flock of birds, an expanse of grassland, and he is at a loss. But give him a chain saw, a weedeater, an assault rifle, or a dirt bike, and he is happy as a clam. Nor is it only males who feel this way. I heard a woman on a radio swap show asking, in a Yankee voice as dry as aged cordwood, for some of those old-fashioned kerosene cans, "the kind your mother would send you down to the store with, and put a potato on the spout so it wouldn't spill." Ah, the good old days. Then she explained that she had bought "a new toy," a gasoline-powered leaf-blower, and you had to mix kerosene in with the gas. In the name of a direct relation to nature, we have made technology a wall between ourselves and nature.

We have come, it seems, from the Mountain back to the Tower. We feel as if we have taken the path of Poliphilus, only backward: entering a cave in a wild mountainside, and coming out in the midst of some colossal structure.

But this is not the old Tower, not exactly. The Mesopotamian relation to nature was cultural through and through. So was the Roman. Their confidence was in the power of culture as a whole—politics, art, the cult of the temple and of the hearth—to subdue or soften nature. Only with seventeenth-century empiricism do people start looking in earnest for a direct relation to nature, one that does away with the accumulated nonsense of culture. (People had started looking in ancient Greece and Rome, but—except for such odd cases as the street philosopher Diogenes, whose shocking habit of eating in public won him the name "dog," or *kynos*—hence "Cynic"—their ideas never took on much practical force.) Locke the mentor of industry and Rousseau the tree-and-nanny-hugger are not so far apart after all. Both find their true home in America, often within the same set of ribs. Only in America does the idea of a direct relation to nature have the open field it needs in order to thrive.

And yet the feeling that we have gone from Mountain to Tower by a back door or a hidden passage is not entirely wrong. This part of the book has been largely about efforts to bridge the gap between wildness and civilization, in the world and in the self. Pastoral tries to bridge the gap by finding the perfect spot in between. Pleasure gardens pile Mountain upon Tower, or vice versa, in a single iconic scheme. And now, in a new world, a new gambit is tried (one that, like the others, is largely unconscious). Drop all the ranigazoo of culture, and you can have the innocence of the Mountain and the power of the Tower at one and the same time.

MINING THE BREAST

The line from Virgil's Fourth Eclogue on the dollar bill, *Novus ordo seclorum*, is one more occasion for irony about the age of gold—the new world order in which gold rules. That paradox has been common since Ovid, but America adds another with its "greenbacks." By linking gold and green, we come full-circle, back to the source of wealth in nature's bounty—amply evident in America.

One of the charges lately brought against Columbus is that he viewed the new world he had found as if it were a rack of merchandise—spices, jewels, gold, slaves. A like charge has been leveled against the bulk of his successors, the white explorers and settlers of the New World. Even at the time, some observers found it odd that "Wilderness should turn a mart." Yet Columbus himself thought the world he had found contained the

Earthly Paradise—the tip of a breast-shaped swelling of the aqueous globe, from which abundant fresh water flowed—and saw no contradiction in mining it for all it was worth.

The richness of paradise is an ancient theme. There are jewels in Ezekiel's Eden, golden apples in the garden of the Hesperides, bejeweled trees in the Mesopotamian garden of Siduri and in Persian pleasure gardens. For that matter, the bounty of any natural paradise or golden age is itself a form of wealth. It's not as if the Indians had no concept of merchandise or exchange value; theirs was just different from ours, and differed from tribe to tribe. The Indians of the Pacific Northwest, perhaps the wealthiest hunters and gatherers we know of, were particularly modern in this regard. Think of the Karuks, who lived along the Klamath River, "inventing money in the rainy woods and then counting every single thing in it and giving it a price: They kept neat oral lists of equivalence and discount. They ran their happy anarchic cornucopia as though it were a giant mall."

Gold and jewels can be just another way of saying, "No need to work—not now, not ever again." Crystallizing the bounty of paradise in stone and metal is a way of making it imperishable. The Greek Isle of the Blessed, like the age of Cronus, knew perpetual spring. In Persia, where spring is even briefer than in Greece, the Emperor Khosrow made spring eternal by having it spread-eagled on a carpet studded with emeralds, gold, and pearls. The need to see nature's bounty in imperishable (and negotiable!) form betrays a distrust of paradise. So the anxieties of real life bleed back into myth—even the rosiest myth.

All paradises want to be immortal, but they go about it in different ways. One kind of paradise seeks immortality by making itself proof against change. This is the alabaster city—Dilmun, Babylon, El Dorado—which some will recognize as our old bugbear the Tower. It owes its immortality to its distance from the Mountain, whose on-again, off-again flow of good things it channels, stores, and doles out at its pleasure.

Oddly enough, the other kind of paradise owes its immortality to just the opposite condition: its closeness to the Mountain and the ever-renewed, ever-renewing flow of wildness. Such a paradise foils change by accepting change: that is, rejuvenation. Instead of setting yourself above nature, you live closer to nature. You live more simply. You move around a lot. You jettison the impedimenta of culture. Rejuvenation is easy, the thinking goes, if you stay simple and don't stray far from the basics, the sources of life. If this paradise is less secure, it also tends to be less unjust—there being nothing to hoard and no one to exploit.

Each kind of paradise seeks a kind of permanence: agelessness on the one hand, frequent rejuvenation on the other. Each seeks a golden age: the reign of Cronus, in which nature always provides, or the more ironic golden age of bullion and gilt-edged securities. The choice is simple: do you consider the lilies of the field, or do you gild them?

On the surface, it looks as if Americans have put their faith in rebirth. In *The House of the Seven Gables*, Holgrave declares that capitals, courthouses, churches, and the like should not be built of brick or stone but "should crumble to ruin once in twenty years or thereabouts, as a hint to people to examine and reform the institutions which they symbolize." The desire for a shifting, primitive life—the cabin on Walden Pond, the grass huts of Melville's South Sea islanders—runs deep in the American mind.

But the reality of such a life has rarely satisfied Americans for long. What they really want is the best of both paradises. They want to stay close to the source of nature's bounty, yes; but they want to extract, exploit, and control that bounty with all the means at their disposal. They are not content to sip from the spring; they want to pump it out as fast as they can. They want to mine the breast.

It is true that Americans move around a lot. (They always have: in the nineteenth century, it was rare for a family to live at the same address for ten years, and de Tocqueville noted that "an American will build a house in which to pass his old age and sell it before the roof is on.") But being nomadic does not keep them from being acquisitive. They shoot the ballast of culture, only to weigh themselves down with technology. Their buildings, cars, and appliances may crumble to ruin every few years; but, far from being prompted to examination or reform, they rush to replace them with more and shorter-lived junk. In lieu of renewal, they have this year's model.

Trying to be a smarter Adam—which is what this boils down to—is an American specialty, but also an extreme case of an ancient human trait. Unluckily, the effort to make earth's plenty permanent in this way—mining her for grain and minerals, plucking her plumes and furs—often has, in the long run, just the opposite effect. Afraid that the bonanza can't last, we grab what we can as fast as we can. Unsurprisingly, the bonanza doesn't last. Yet we *are* surprised, for, while we fear that nature is fickle in the short term, we feel deep down that in the long term she is inexhaustible, no matter what we do.

What about the hope that this simpler way of life would be less unjust? There was something deeply democratic, no doubt, in the wish for a new world where the Golden Age was fact, not wallpaper—where nature's bounty flowed for everyone, not only for the few, and where it did not have to be surreptitiously yanked out of her by invisible laborers. Encouraged by Spanish accounts of the West Indies, some Europeans, at least, hoped to find just such a new world. In Marvell's "Bermudas," the fugitive Puritans sing of a refuge where their God

> sends the Fowl's to us in care,
> On daily Visits through the Air.
> He hangs in shades the Orange bright,
> Like golden Lamps in a green Night. . . .
> He makes the Figs our mouths to meet;
> And throws the melons at our feet.

The tropes here are the same ones poets used to flatter the lords of English manors, except that now the mouths in question are those of the whole community. Unfortunately, the first white immigrants did not find the Golden Age, any more than later ones found the streets paved with gold. The new world *was* rich, but to pry its riches loose took work—if not your own, then someone else's. Like other real-world paradises, this one depended on exploitation, expropriation, even extermination. The half-visible victims might be Indians, or Africans, or Irishmen, or coolies. Or they might be members of other species entirely.

The abstract equality of opportunity on which America prides itself—achieved, it likes to think, by clearing the playing field of the stumbling blocks of class, religion, family, ethnicity, and culture generally—has given it the largest gap between rich and poor of any industrial nation. Abstract freedom is often the freedom to abstract someone else's substance.

The gridiron plan imposed on most American cities and much of America's countryside expresses this neatly. Whereas the radial plans and warrenlike layouts common in European towns privilege some blocks over others, in the gridiron all lots are equal. It sounds like a good idea, democratically speaking. But the upshot—and, in general, the intention—is to make land more fully and flatly a commodity, so that each square inch can yield the highest price and huge fortunes can be made in real estate.

DIVIDERS OR MIXERS?

The love of pure nature in America drew some of its steam from European Romanticism, which in turn was powered by a reaction—at times dithyrambic, more often bitter—to industrialism. (Though the love of nature and the urge to master nature were closely joined in early-modern times, their parting, when it came, was violent.) What industrialism most visibly threatened was not pure nature, but the garden landscape of Europe. Yet the Romantic response, as machines rumbled in the garden, was to rediscover wilderness.

Arguably, this new fervor for pure and untrodden nature only ratified the exploitation of whatever part of nature humans did tread on. Part of the earth would be stripped for food, clothing, and suspension bridges, and part would be reserved for the contemplation of poets. On the other hand, if the stripping was going to go on anyway, it was well that some part be set aside where wild nature could go about its business, whether under the poets' benign gaze or not. In the long run, whether this division of the world was a good thing or a bad thing would depend on how the pie was sliced.

Anyway, it is clear that the Romantics took this division only so far. It is tempting to say that Americans took it further. When I began work on this book, I thought the problem with America was a too-strict division between nature and culture. Having nature to burn, America let culture have its way. When voices such as Muir's convinced Americans that nature was running out, they put some of it in boxes called parks or preserves. That way, they could exploit the rest of it to their heart's content. Inside the box was pure, untampered-with nature; outside the box was pure, untrammeled culture.

A similar view is expressed by a writer on gardens, Michael Pollan, who claims that the two main contributions of this nation to the world's landscape are the wilderness preserve and the front lawn. Polar opposites: the one pure nature, the other almost pure culture. Gardens, on the third hand, are a way of mediating between nature and culture, which is why Americans have given them such short shrift.

Plausible enough. Yet the critic Leo Marx made a good case years ago that what Americans wanted, from the outset, was the middle landscape— the new Arcadia—the perfect mean between wildness and civilization. This, too, is plausible, especially when you look at what Americans have gotten: namely, suburbia.

The ideal of the "middle landscape"—the mean between city and wilderness—first expressed itself in a vision of rural life: of a whole continent made fruitful, quilted with the fields of yeoman farmers, spangled with church steeples. Although this vision is often called Jeffersonian, the Puritans had a version of it, too. To a stickler it might not be pastoral, since it has to do mainly with settled farming. But it is pastoral in spirit—in its quest of the blessed middle.

How can these two claims be reconciled? How can Americans have wanted simultaneously a perfect blend of nature and culture, and a sharp line between them?

As far as I can see, the only way the two claims can be reconciled is by means of a third claim which I have already made. What Americans wanted was Eden: that is, a direct relation to nature.

Let us go back to the specifics.

For starters, I suspect most Americans think of lawns as *more* natural than gardens. Gardens are fussy, overelaborate, contrived—a remnant of the Old Regime. Lawns are simple, honest, straightforward. One color, one height. It grows up, you mow it down. Gardens are hierarchical, lawns are democratic.

America's lawns are descended from the lawns and pastures of eighteenth-century English estates, which were supposed to be more natural than gardens. Ubiquitized, lawns affirm the democratic self-image: every man a king, or at least a squire. Lawns are Whig, gardens are Tory. The lordly formal garden is too lordly for democrats; the humble cottage garden is too humble—a cringing, down-on-your knees affair.

As the English park gobbled up both garden and wilderness, so has the American lawn. Only the American lawn has had a lot more wilderness to gobble.

Lawns are not an extreme of culture, but a way of escaping culture. Obviously they are a way of escaping the culture of the soil, since they replace the patient, complex, and unpredictable work of gardening with the zombie work of mowing. (Such refinements as spreading lime, herbicide, and fertilizer are done only once or twice a year.) In a larger sense, too, lawns represent not culture's triumph, but its trouncing. For what is remarkable about this new Arcadia is that, instead of mingling nature and culture, it does its best to exclude both.

In the suburbs, nature and culture are mingled only by reducing both to a lowest common denominator. In its geometrically perfect form, that denominator is the flat surface, whether black or green. The space that might have been devoted to the intricate play of nature and culture in their higher forms—to woods and paths, duck ponds and produce markets, gardens and cathedrals—is taken up by lawns and parking lots. Together, these make up the flat and neutral playing field on which happiness is pursued.

What about America's wilderness preserves? The first national parks were meant to preserve places like Yosemite, with their open, parklike— indeed, lawnlike—vistas. One of the most prominent men calling for such preservation was Frederick Law Olmsted, who helped bring the English lawn to America and was an early booster of suburbs. (To be fair, neither his lawns nor his suburbs were much like what we have today.) So the front lawn and the national park started out carrying the same banner, the banner of the wide-open, democratic middle landscape. The solid green banner of a direct relation to nature.

But that is only half the story. While America's wilderness preserves are partly an expression of the nation's wish for a direct relation to nature, they are also a tardy response to its consequences. The impulse to just take nature and use it, culture be damned, led to a gobbling up of nature such as Europe had never seen. When nature did start to run out, Americans might have been driven to re-examine their culture, or lack of culture. They might have looked for subtler ways of relating to nature and other people. They might have wondered whether technology, individualism, and the market economy really had all the answers.

With a few exceptions, they did not. Instead, they invented wilderness preserves. That way, they could retain their direct spiritual relation to certain parts of nature, and their direct utilitarian relation to all the rest.

The harder you try to have a direct relation to nature, the more problematic that relation is likely to be. Wendell Berry expressed a similar notion when he wrote, "The only thing we have to preserve nature with is culture."

GREEN STERILITY

America's biggest crop is not corn, wheat, or soybeans. It is "turfgrass"—the stuff lawns are made of. The crop in question covers some 25 million acres, or forty thousand square miles, which is just a shade less than the area of Pennsylvania. Of this, some 81 percent, or about 20 million acres, consists of home lawns. Some 58 million American households participate in "lawn care," as against 39 million who grow flowers and 29 million who grow vegetables. For every man, woman, and child, an average of thirty hours a year is spent mowing lawns. Along with time, Americans invest money: the turfgrass industry has sales of $25 billion a year, which works out to a plush $1,000 per acre of lawn.

All this is to the good, according to the Lawn Care Institute and the Professional Lawn Care Association of America. A fifty-by-fifty-foot lawn, we are told, gives out enough oxygen to meet the needs of a family of four. Unfortunately, the institute and the association have forgotten to figure in the oxygen imbibed by microbes decomposing the clippings, sucked up by lawn-mower engines, or used in the production and trucking of fertilizers and pesticides. When these details are figured in, it turns out that there is a large net loss of oxygen from the atmosphere—and a large net gain of carbon dioxide, especially where trees have been felled to make room for the lawn.

But these are the least of the lawn's environmental sins. In an hour of pushing or riding your power mower, you release as much pollution as if you drove your car 350 miles. In California alone, lawn-care machines give off as much pollution each year as 3.5 million 1991-model cars driven sixteen thousand miles apiece. On the East Coast, up to 30 percent of all urban water is used to water lawns; in the West, the figure is closer to 60 percent. In 1988, about 67 million pounds of pesticide—some $700 million worth—were sprayed (or sold to be sprayed) on American lawns. The average acre of American lawn gets four times as much pesticide as the average acre of farmland.

One reason for this is that the definition of a pest is exceptionally broad. Clover, for instance, was an acceptable member of the lawn community until a major advertising campaign by a large seed-and-chemical company drummed it out. Once the principle of a sterile monoculture is established, a cycle is set up which is very healthy for the industry. Because the pesticides kill the microbes that eat the clippings, the clippings have to be removed; because the clippings are removed, chemical fertilizer is needed.

The man on the riding mower, alone in a sea of green, is a good emblem of Americans' direct relation to nature. An even better emblem may be that staple of advertising, the man in a car on a road that runs through wilderness.

ENTROPY ON WHEELS

A sudden rush of free energy can destroy structure. Where energy is dear, cunning strategies for catching and using it are devised. As different creatures' strategies work best in different places, and as it takes energy to move around, species tend to be localized on various scales. Genetic information translates, in effect, into complex structures not only in the body, but on the ground: structures that we have only begun to explore in our maps of ecosystems, communities, ranges, and distributions. On the human landscape, the interplay of genetic and cultural information has shaped the radial patterns of cities and villages and the linear patterns of settlement along seacoasts, river valleys, and rail lines.

In the last century, however, a rush of nearly free energy in the form of fossil fuel has disturbed those patterns, replacing them in some places (in

North America especially) with something much closer to entropy. We no longer have to settle next to food, water, or fuel, for they can be brought to us at small cost. We no longer need shun places that are too hot or too cold for our comfort. We no longer need to live near the place where we work, or near the place where we shop, or near any form of public transportation that goes to those places. The automobile sprinkles people across the face of the land like grass seed dispensed by a spreader.

In Part One, I spoke of the way genetic and cultural information has been blown away by a blast of ancient energy from beneath the earth's surface. The spatial information I am now talking about, shaped by nature and culture, has been caught up in the same upheaval. It is something like what would happen if you turned an electric fan on a Tibetan mandala made of colored sand.

In nineteenth-century England, the coal and iron mines of the Midlands exuded the centerless, pseudo-urban forms that Patrick Geddes (refusing to call them cities) called "conurbations." At the turn of the century, oil wells near the Los Angeles area oozed out loose industrial suburbs that set the pattern for settlement in that region. Oil-fed cars now claim more than one-third of the area of metropolitan Los Angeles as their own, in the form of highways, parking lots, and interchanges. Downtown, the figure rises to two-thirds. And Los Angeles is merely the Platonic form on which the rest of America has increasingly been patterned.

How exactly did this happen?

At the turn of this century, America was the paradise of public transportation, with the best and best-ridden network in the world. Each railroad suburb was effectively a village, clustered within walking distance of its station, with plenty of healthy countryside between one suburb and the next—a happy artifact of the steam engine, which was hard to start and stop. The electric streetcar made a closer spacing possible, and (as Sam Bass Warner has shown) began the process of sprawl; yet if cities trebled in size, they remained recognizably cities. At the nexus of the web, the city's business district remained healthy.

Two decades later, the web was ripped to shreds, largely by one man. Henry Ford had promised to "build a motor car for the great multitude," so that every man might "enjoy with his family the blessings of hours of pleasure in God's great open spaces." In 1908, he made good on that promise by launching the Model T, the first mass-produced automobile.

If in 1905 you had tried to fit all Americans into the existing fleet of registered passenger cars, 1,078 would have had to squeeze into each. In 1920, only 13 would have had to share each car—unsafe, but not inconceivable.

In the same year, the figure for the United Kingdom was 228; for France, 247; for Germany, 1,017.

All over America, a black carpet was rolled out for the newcomer. Automakers, tire-makers, oil companies, road contractors, and developers joined in lobbying for new roads, which were promptly built at public expense. While streets and cars were heavily subsidized, streetcars were left to the mercies of the "free market." Most streetcar companies had signed agreements in the 1890s guaranteeing a five-cent fare; though unlooked-for inflation (fed by new Alaskan gold) soon made that fare untenable, neither city officials nor voters were willing to grant increases. Cash-strapped and unable to modernize, streetcar companies started going under.

When they did, General Motors was waiting. In 1926, it opened a subsidiary whose job was to buy up teetering streetcar companies and replace the cars with buses. During the next thirty years, GM ripped up thousands of miles of track in New York, Los Angeles, St. Louis, Philadelphia, Baltimore, Salt Lake City—over a hundred streetcar companies in all. At length, a federal jury found General Motors guilty of criminal conspiracy. The fine was $5,000, which was less than the company made by converting a single streetcar.

General Motors was also the largest contributor to the American Road Builders Association—a lobby second in power only to the arms industry—which helped push through the Interstate Highway Act of 1956. By the authority of that act, 42,500 miles of highway were built, with the federal government picking up nine-tenths of the tab. In the postwar generation, 75 percent of government transportation spending went to highways, 1 percent to urban mass transit.

Since then, priorities have changed only slightly. In the Boston area, a working mother from Roxbury taking the subway at rush hour pays 80 percent of the cost of her trip, while a stockbroker driving his BMW back to the suburbs pays only 20 percent of the true cost of his—an arrangement that has justly been called "car welfare." Congress whets its budget ax and mutters darkly about the subsidy of almost $1 billion that goes to Amtrak: an insufferable meddling with the free market! Meanwhile, the subsidy of over $20 billion awarded the automobile—the amount spent on road construction and maintenance, traffic-law enforcement, and the like, above and beyond what is raised by gas taxes, tolls, and other user fees—is ignored.

The figure of $20 billion is very conservative, since it takes no account of costs to health, the environment, and the social fabric of the inner cities; or of the large part of the military budget that is spent to defend foreign oil. Nor does it reckon in the immense subsidies to sprawl that are hidden in credit and tax policies. Beginning in 1933, the Home Owners Loan Corporation helped to finance long-term, low-interest mortgages for home-buyers. It also invented redlining. Its appraisal system ranked neighborhoods on a color scale: green, blue, yellow, red. Newly built suburbs lived in by white professionals were green; old urban neighborhoods inhabited by blacks, Jews, or other minorities were red. Even more far-reaching in its effects was the Federal Housing Administration, set up in 1934 to insure long-term mortgages. The favorable terms it underwrote often made buying cheaper than renting. This might have been a good thing, on balance, if the FHA had not blatantly favored suburbs over cities. FHA guidelines favored free-standing single-family homes whose lot sizes, setbacks, and house widths met certain minimum standards. They favored new construction over home improvement. They even recommended restrictive covenants, so that "properties shall continue to be occupied by the same social and racial classes." With the help of the FHA, the trickle of middle-class whites leaving the cities quickly became a hemorrhage.

Many other policies, at all levels of government, have nurtured sprawl. Tax laws have favored new construction over renovation. Costs for suburban streets and sewers have often been borne by a whole city. Revenues have been siphoned off from cities and sprayed about outlying areas, notably in the form of defense spending. Low taxes on gasoline have given the driver a free ride. The mortgage-interest deduction has given rich suburbanites a subsidy several times the size of that extended to welfare families. And so on and on.

In a real sense, all of these are subsidies to the automobile. But we should not assume that they were all brought about by the car-and-oil lobby, or by any other assemblage of crude interests. For most of this century, the American public as a whole has believed that cars were the answer to many of America's problems. For the most part, its political leaders and social thinkers have thought so, too. Cars would save the cities by cleansing them of horse manure and purging them of excess population. (As Ford put it with typical bluntness: "We shall solve the city problem by leaving the city.") Cars would save the countryside by taming its great spaces and salving its loneliness. Cars would bring people back to the land.

Floating above the crass commercial interests was the hopeful vision of a new order, of a whole continent made over as a *locus amoenus* for the masses. What held that vision aloft was a mixture of gases: the pastoral tradition, Puritan visions of Eden, Jeffersonian agrarianism, the landscape gardening of English country seats, European Romanticism, American Transcendentalism, Olmsted's ethos of nature for the many. A succession of movements—the Arts and Crafts movement, the Garden City movement, the House Beautiful movement, the Regional Planning movement—found their culmination in the Usonian vision of Frank Lloyd Wright: a vision of one-story homes on one-acre plots, each joined to all the others by the magic of the automobile. The "universal home"—two or three floor plans to fit every family—could even be built of prefabricated, machine-made parts. Wright's Broadacre City would be a well-ordered world of happy homeowners and cooperative farming: "no landlords, no 'housing,' no slums, no scum."

Broadacre City was never built, but Levittown was—built and built again. Though Wright despised Levittown, its founders claimed him as their spiritual guide; and they were right. As you look down at a model of Broadacre City, the details over which Wright labored blur and run together, and what is left is deadeningly familiar. For all his Druidic love of nature—his use of organic forms and native materials, his matching of the structure to the shape of the land, chiefly the prairie—his weightiest legacy was not the scores of buildings he designed, but the millions of ranch houses and thousands of strip malls he did not.

Philip Johnson's gibe that Wright was America's greatest nineteenth-century architect has a core of truth. A nation of self-reliant individuals living in sympathy with nature by the grace of machines: this bluff Emersonian vision failed to take full account of the way these nonliving allies of humankind can take on a life of their own. "Things are in the saddle, / And ride mankind," Emerson wrote in a disillusioned moment. Frank Lloyd Wright in a sedan was the World Spirit on wheels. But his vanity did not allow him to see in which part of this chimera the real power lay.

In the drama of American sprawl, ironic figures are a dime a dozen. One of the very few, though, to rival Wright in size is the man who started it all, Henry Ford. Devoted to the myth of small-town America, he did as much as any man to destroy the reality. While he built replicas of old-time villages and promoted square dancing, his product methodically ate the heart out of the heartland.

A CULTURELESS CULTURE

Why is man-made America so ugly? For four hundred years, people came here not so much to find nature as to flee culture: despotic culture, crowded culture, threadbare culture. When they got here they wanted to be American. But there was no American culture, except that of the native Americans. So everyone seized on the lowest common denominator, the dregs that were left when all the old cultures were washed away.

Lawns are the perfect product of the melting pot. They are not Irish, Italian, Chinese, or Jewish. In their suburban form they are not even English (most English suburbanites have dooryard gardens, not lawns). Calvin Trillin, who grew up in Kansas City as the son of an immigrant from Russia, says he felt like a normal American because his father mowed the lawn.

Starting with as stunning a range of natural landscapes as any nation on the face of the earth, Americans have managed to create a built landscape of stunning monotony. True, the bare idea of a direct relation to nature does not rule out sensitivity to nature in a particular place. In practice, though, such sensitivity calls for long acquaintance—longer than one lifetime. In other words, it requires culture.

If the white settlers of America were to have a direct relation with nature, you might think their culture (or whatever it was) would be cast like plaster on the contours of the land. But it oozed westward so fast that it had no time to set. Under the circumstances, the only way the culture could fit the land was by learning from the cultures that had known the land for tens of millennia already. Except in a few particulars, though, that was out of the question. Americans rarely stopped to see what the Indians had done in a particular place—or, for that matter, what nature had done.

What about the culture of a place like Vermont, with its picture-postcard sensitivity to the contours of the land? In the nineteenth century, those contours were shorn of their forests. They stayed green, but green of a less robust shade. As much as we admire the culture of early New England, we must keep in mind that it was not a stable culture. It didn't last long. For all its frugality it was, ecologically speaking, a culture of the fast buck.

In a culture in flight from culture, the most acceptable aesthetic is one based on utility. The greatest triumphs of American design—clipper ships, dungarees, skyscrapers—have come when form followed function. Some other triumphs, as well as many horrors, have come when form followed economic function, making beauty the gal Friday of advertising and marketing.

Of course, a motion as violent as this flight from culture has to cause a recoil in the opposite direction. Among the intellectuals—the Henry Adamses, the Jameses, the Eliots, even the Henry Millers—there is a fervent tropism toward fuller, richer culture, which usually means going to Europe. Other people feel a weaker and blinder version of this tropism; or flail about, clutching at scraps of foreign food, drink, and decor. So the most functional American architecture, that of the gas station, motel, and roadside diner, soon becomes the most flamboyant and naïvely foreign. The road leads finally to Las Vegas, where culture—unmoored from all ties to history and geography—floats free like a crazed zeppelin. But what seems an escape from function turns out, in a final twist, to be a cunning embrace of economic function.

PIONEERS OF CONSUMPTION

The history of the American frontier is a cartoon of human history, which in turn is a cartoon of natural history. Many times over the past three billion years, alliances of species have made their way across this planet, displacing other species and generally upsetting what was, from the alliance's point of view, "nature." Various human-led alliances have done the same thing faster and with less regard for keeping nature's machinery working. The European-human-led alliance that swept across North America in the space of a few hundred years was the fastest and most regardless of all. No wonder—despite the monkey trials—Darwinism finally struck such a deep chord in America. Cartooned, as in Social Darwinism, it mirrored American culture, confirming its naturalness.

The closing of the frontier put Americans in an awkward position, and not only for the reasons that Frederick Jackson Turner famously advanced. The Eden from which they were now expelled was not, like most imagined Edens, changeless. It was change itself—the paradise of the ax and the carbine. Most Edens are dreams of harmony between people and nature. This one was a dream of direct relation—and a memory of conquest. Yet in this memory, or dream, there was so much nature that it never ran out. The moose, the bison, the white pine, and the red Indian were cartoon characters that could be shot, chopped, or set aflame and would bounce right back. As you pushed the frontier back, you moved right along with it, and it was as if neither moved at all. In the thick of change, nothing ever changed: you were the edge of the blade.

All this assumed that nature would not run out. Ideally, the continent would be a globe unto itself, on which you could push westward forever. By the time you got back to where you had started, the wilderness would have grown back.

But nature did run out. Or came so close to running out that what was left had to be fenced off. The frontier was closed. This left Americans with two alternatives: either dream a new dream of harmony, or keep on conquering whatever shreds and tatters of nature were left. By and large, they chose the latter. The heirs of cowboys are building malls. And the rest have become pioneers of consumption.

Americans still move around a good deal, chasing the shimmering frontier of opportunity. But the pioneer spirit has been transmuted into a form usable by people in cities and suburbs. Americans are no longer (or not so much) pushing westward, eating nature up as they go; instead they sit in the living room and have nature delivered. Though they shake off the weight of culture, their nomadic nostalgia stops short of disdaining the weight of possessions.

They might instead have copied the rugged lifestyle of the pioneers, their ability to make do with less. That would have led to a very different outcome—a culture able to live in some kind of rough balance with nature.

THE CARTOON THAT CONQUERED THE WORLD

As Gertrude Stein said, America is the oldest country in the world. It is the country where the present world began. Tourists now come to America as once they went to Greece or Italy, to find the landscape that grew the myths by which they live. The Roman senator falling on his sword has given way to the cowboy, the leather-clad biker, the Texas billionaire.

The phenomenal success of American culture in so many parts of the world can be explained partly by American economic and media power, backed up by the discreet use of military power. But it also has a deep appeal that is inherent—"in the grooves," as record producers say. As a lowest-common-denominator culture, it fits almost anywhere. Though deeply subversive of traditional culture, it is quietly subversive, since it lends itself to instant syncretism. As a pioneer culture, it has the rough-and-ready opportunism of a pioneer species, ready to take root wherever the cultural soil has been disturbed. The key, though, is this: As a cartoon culture—a culture that caricatures nature—it offers cartoon fulfillment of a cartoon of our deepest needs: sex, material plenty, pleasure, power, status. Add privacy and freedom, and you have a cartoon of Eden.

For most people in most countries, these promises are not kept, even in cartoon form. Other, equally deep needs—for community, for a meaningful past, for intimacy with nature—are not met at all. But it takes a while to find that out. (Some of the many nationalist and religious movements now on the upsurge may have found it out, while others are just lashing out blindly.) Meanwhile, a dream of Eden destroys what is left of Eden, and a cartoon of nature destroys what is left of nature.

Of course, this whole chapter is a cartoon. I have drawn the American experiment in thick, harsh lines and lurid colors. If I really thought so ill of my country I would have emigrated long ago, or else would be walking in rags on Sixth Avenue, shouting that the end is near. So let me take a few steps back and try to correct the picture.

The deep, inherent appeal of American culture is not wholly meretricious. The dream of Eden—of a direct relation to nature—is lodged so deep in the human mind that even its cartoon realization brings a kind of joy. The weight of culture may be a necessary weight, but casting it off can be a great relief. Freedom from tyranny, from inquisitions, from meddling custom, from rigid class distinctions—these may be negative freedoms, but they are freedoms nonetheless. On its good days, America has come closer to achieving them than any other society in human history. More than any other society, it has let people take a crack at achieving a direct relation to their own nature. Some try and fail miserably, and are left to watch TV or howl in the streets; others seem to be inches away from catching happiness.

The dream of a direct relation to nature is a foolish one for society as a whole, yet when certain individuals act on it—when they go into the wilderness determined to meet it face to face—they may come back with wisdom that everyone can use. Thoreau, John Muir, Bob Marshall, Edward Abbey, Annie Dillard, Edward Hoagland, Gary Snyder: try to imagine them anywhere but in America. (You might be able to imagine them in another frontier country, like New Zealand—but France or Germany? Not a chance.) Whether they come back convinced that a direct relation to nature is impossible or that it is the only thing worth achieving on earth, they give us something we desperately need to chew on: the problem of understanding nature on its own terms. Their visions have helped bring about the preservation of vast tracts of wilderness and near wilderness. And though putting wilderness in boxes is not enough, it is a necessary start.

For American artists in general, getting to see a telescoped version of human history (and thus an ultratelescoped version of earth history) has been immensely useful. Seeing wilderness gobbled up right before their eyes has inspired them and enraged them, and they have passed their vision along.

Focusing on a handful of artists, however, can make the lousiest society look good. Let us turn back to the larger culture. The rejection of culture left an open field in which anything could happen. At times the vitality of the frontier, the free market, and democracy found real expression. Of course the melting pot was by no means as thorough as I have made it sound. While it succeeded in ladling out a material culture of exceptional ugliness, it failed to denature the British, Irish, French, Italian, German, Hispanic, Jewish, and (above all) African musics that went into it; and what came out was perhaps the most potent popular music the world has known.

The role of machines was likewise ambiguous. They helped produce an ugly material culture, but they also helped produce a stunning popular music. Jazz and rock are unthinkable without trains and cars, and without the juxtaposition of cities and wide-open spaces that these made possible. And they are almost as unthinkable without the phonograph as movies—that other great American form—are unthinkable without the camera.

In the last part of this book, I will have more to say about jazz: how that particular product of the American experiment might serve as a model for our dealings with nature. First, though, I want to look at one last attempt—at once mythic and scientific—to make the world whole.

21

A GODDESS QUANTIFIED

OUR LAST IDYLL seeks wholeness and stability not by walling things out, but by walling almost everything in: inside the tropospheric skin of a blue creature called Gaia. If gardens model the world's wholeness in greenery, the Gaia hypothesis does so in the fictive space behind computer screens (a space by no means safe from the infiltration of myth). Of all responses to the exile from Eden, the pop version of Gaia may be the boldest, for in a way it denies that the exile ever took place.

IN 1961, the British scientist James Lovelock was invited by NASA to help design instruments for export to Mars, where they would poke about for signs of life. Lovelock was just the man for the job. A medical chemist by training, he had long been interested in atmospheric chemistry, and early in his career had invented the electron-capture detector. Able to detect infinitesimal traces of almost anything in samples of almost anything else, the ECD is most familiar as the machine that sniffs out explosives at airports. But it has also been responsible for two of the most publicized environmental stories of our time. In the 1950s, it found traces of DDT and other pesticides in such unlikely places as the flesh of Arctic penguins and the breast milk of human mothers, thus giving birth to Rachel Carson's *Silent Spring* and the modern environmental movement. In the 1970s, it found that chlorofluorocarbons from fridges and hairspray cans were collecting in the atmosphere; though Lovelock, who made the discovery, at first thought this a harmless curiosity, it is now clear that CFCs have an insatiable hunger for ozone, which is our parasol against the sun's less kindly rays.

Lovelock, however, came to believe that his qualifications as gadgeteer were irrelevant to the job at hand. To find out if there was life on Mars, no gadgets were needed—none, that is, besides the infrared telescopes firmly planted on earth. The automated biochemistry laboratories that NASA wanted to land on Mars, their tastebuds finely tuned to amino acids and other signs of earth-flavored life, were beside the point. We already knew that there was at present no life of any kind—earth-flavored or otherwise—on Mars. We knew this because the telescopes told us that the Martian atmosphere was in a state of chemical equilibrium.

With the help of Dian Hitchcock, a philosopher employed by NASA as a sort of interplanetary logic consultant, Lovelock had turned the question on its head. If we were standing on Mars and looking at Earth, how would we know that it had life? By noting that its atmosphere was *not* in a state of chemical equilibrium.

If you took the elements that any planet starts out with, shook them up, and left them out in the sun (to be exact, in the light of a sun 1.5 astronomical units, or 225 million kilometers, away) for four billion years, what you would end up with is exactly what Mars has. Its atmosphere is 95 percent carbon dioxide, with a few percent each of nitrogen and argon and barely a whiff—less than 0.13 percent—of oxygen. Repeat the experiment on a higher burner (a sun 0.72 AUs away) and you get exactly what Venus has: an atmosphere that has even more carbon dioxide and even less oxygen, a barely measurable trace. Repeat the experiment at an intermediate heat (a sun exactly one AU away) and you ought to get an intermediate result. If Earth were sticking to the simple chemical recipe followed by the other planets, that is the atmosphere it would have.

As every breath should tell you, Earth's actual atmosphere is sparklingly different: 79 percent nitrogen, 21 percent oxygen, and only 0.03 percent carbon dioxide. Instead of the average surface temperature of 240 to 340 degrees C. it would have if it were swathed in CO_2, it breezes along at a pleasant 13 degrees C.

In the 1940s the physicist Erwin Schrödinger, trying to define life, said that living things have the ability to resist, even locally reverse, the downhill flow of entropy in the universe as a whole. In other words, they can suck high-grade energy from the environment and spit out low-grade energy.

Lovelock knew this definition. He knew, too, that it was too general, since it also fit hurricanes and refrigerators. But it did point "in the right direction." And he took it a step further. An environment—a planet, to take a random example—in which life was present in force would itself show a reversal of entropy. In the universal flow toward physical and chemical equilibrium, the planet would be an island, a shoal, an eddy of strangeness.

A LIVING PLANET?

If Lovelock had stopped here, the NASA scientists might have been a bit peeved at having their applecart upset, but scientists at large would have hailed his percipience. Of course the earth is not at chemical equilibrium. Of course the other planets are. Of course it must be life that makes the difference. Obvious once you see it.

But Lovelock took one more step. He noted that, while the earth's atmosphere was not in chemical equilibrium, it seemed to be in some other kind of equilibrium. It was not just that the mixture of gases had been constant for the century or so since people had started measuring them. There was reason to believe it had been roughly constant for the past two billion years. In the case of oxygen, for example, a strong piece of evidence was the frequent presence of charcoal in the geological record. If the concentration of oxygen had ever dropped below 15 percent, no forest fires could have occurred; if it had ever nosed above 25 percent, the fires would have been so intense and widespread as to keep forests from forming.

Even more remarkable was the apparent fact that the temperature of the earth had not changed much over the past few billion years. Since there are sedimentary rocks in every stratum of the geologic record, there must always have been liquid water to lay down the sediments; hence the average surface temperature could not have gone much above 50 degrees C. or much below 0 degrees C.—a wide range if you are planning a picnic, but very narrow in the context of neighboring planets (Venus at 459 degrees C., Mars at −53 degrees C.), and friendly to the basic processes of earthly life. What made this truly remarkable was that the sun, like any normal, healthy, growing star of the "main-sequence" type, had undoubtedly been getting brighter and hotter all that time, so that it was now some 30 percent brighter than it had been when the earth took shape. By rights, either the temperature today should be much higher than it is, or the temperature several billion years ago should have been far lower than it actually was. The likeliest explanation of the "faint young sun paradox" is that the earth's atmosphere used to be richer in carbon dioxide, a greenhouse gas that traps solar heat. Apparently the planet has very gradually been shedding carbon dioxide, like a cat shedding hair in summer. (Actually, it is tying up carbon, as we shall see.) In other words, it has maintained its internal equilibrium in the face of external change.

Most remarkable of all, perhaps, was the planet's recovery from major traumas. Over the past few billion years, the earth has been hit by about thirty planetesimals (small planets) up to ten miles in diameter. Some of these impacts were a thousand times as powerful as the detonation of all the nuclear weapons on earth; some left craters two hundred miles wide. One of them may have plunged 90 percent of all living species into extinction. But each time, life bounced back. It might take a million years, but it bounced back.

For someone with a medical background, the natural analogy was to the human body. Just as the body regulates its temperature, the composition of its blood, and a host of other vital signs, so the biota appears to regulate the temperature of the earth's surface, the composition of the atmosphere, the salinity of the oceans, and so on. For all intents and purposes, earth is an organism.

Where Lovelock had now ventured, few scientists would be hardy enough to follow. Geologists knew that life affected the planet; they might be willing to admit, even, that they had underrated its effect somewhat; but surely everything could be explained in terms of simple cause and effect, without recourse to cloudy notions of purpose. Biologists knew that organisms affected their environment; perhaps they had not paid enough attention to that fact in charting the evolution of species; but to suggest that the planet itself evolves—this was going too far.

Lovelock did find an early and vital ally, though, in Lynn Margulis, one of the most omnivorous of living biologists. Margulis, whose serial endosymbiotic theory of the origin of nucleated cells was described in chapter 2, appeared to be on a first-name (or at least binomial) basis with the whole range of living things, and above all with the very small ones most of us ignore except when they happen to be making us sneeze or throw up. (As the evolutionary biologist John Maynard Smith admitted, she knew a great deal about a lot of creatures most biologists knew nothing about.) While most ecologists tended to think of the world as made up of plants, herbivores, and carnivores, Margulis knew that the basic business of the biosphere was going on long before any of those classes emerged and would probably keep going if they all disappeared. She had a keen sense of the ancient and varied tricks life has found for squeezing energy from any handy cluster of molecular bonds, and of the way those tricks can interact. In short, she was just the person to help Lovelock flesh his hypothesis out. And having seen one heresy she had propounded—the endosymbiotic theory—elevated to the status of holy writ in the space of two decades or so, she was not afraid to propound another.

DAISYWORLD

A number of evolutionary biologists, notably W. Ford Doolittle and Richard Dawkins, thought Lovelock had simply blundered. For natural selection to maximize the fitness of a whole planet, that planet would have to be competing with a lot of other planets—planets with life on them, and with various kinds and degrees of self-regulation. Thousands or millions of Gaias or near Gaias would have to be competing for galactic space and energy. What is more, they would have to be reproducing themselves. They would have to be spawning lots of little baby Gaias, so that increments of fitness could spread through the population of orbs.

When Lovelock read these criticisms, he wondered if he really *had* blundered. Slowly, though, he became certain that Gaian mechanisms could evolve, if not in the Darwinian sense. They could be explained without recourse either to a galaxy red in tooth and claw or to any woolly teleology. All it took was a field of daisies.

With a third collaborator, Andrew Watson, Lovelock created Daisyworld. In its first, almost childishly simple form, this was a computer model of a world populated only by black and white daisies. The black daisies like cold weather, the white daisies like hot weather—no doubt because the black daisies absorb more sunlight than the white ones. If the strength of the sun is constant, the temperature on the planet's surface depends on the planet's albedo, or reflectivity. (The higher the albedo, the more energy bounces off, and the less is absorbed in the form of heat.) If black daisies begin to take over, the planet's albedo falls, and the temperature goes up; at which point the white daisies gain the advantage. But once the white daisies start to take over, the planet's albedo rises and the temperature falls, giving the black daisies the edge. The upshot is that a more or less constant temperature is reached. For all intents and purposes, the planet has a thermostat.

So far, we should not be terribly impressed; after all, a lifeless planet might well have a constant albedo and a constant temperature. But now suppose that, over the millennia, the sun that shines on this world slowly heats up, as suns tend to do. Without life, the planet would heat up, too. But on Daisyworld the ratio of white to black daisies will slowly increase, and the planet's temperature will remain roughly constant. At some point, the solar influx will get so strong that the thermostat will break down—just as it must on our own planet a few billion years down the road. In the meantime, though, life will have made itself comfortable.

Playing God in Daisyworld, Lovelock added trophic levels: rabbits that ate the daisies, foxes that ate the rabbits. He bombarded his planet with meteors. There were great die-offs, there were booms and busts. Almost always, though, Daisyworld bounced back. Like Gaia, it showed a striking resilience. And the more diversity he added—the more he approached, across a gulf of orders of magnitude, the vast diversity of life on earth—the more resilient his world became.

Although Lovelock seemed tickled with Daisyworld, not all scientists shared his pleasure. The physicist-turned-ecologist John Harte was curt: concocting a model with negative feedback was child's play. So was concocting a model with positive feedback. It took him a few minutes, he said, to concoct a world inhabited by lupines and microbes in which, when the sun gets hotter, the inhabitants make things hotter still. Because warmed-up microbes process soil carbon more efficiently, more carbon dioxide enters the atmosphere, trapping more heat. "Lupineworld," wrote Harte, "is no less real than Daisyworld, and no more evidence of anything at all."

When speaking of positive and negative feedbacks, it is easy to be misled by the positive and negative weights we place on the words "positive" and "negative." In scientific usage, a positive feedback is one in which the original signal is amplified, the original tendency reinforced. In some cases this can be a "positive"—that is, good—thing, as when a business makes profits, plows them back into its operations, and makes still greater profits; or when an artist's growing reputation swells the prices paid for his canvases, further swelling his reputation. But in most cases positive feedback is bad, because it can make a system spiral out of control. When you feel anxious, the smooth muscles of your stomach and viscera tighten; as you become aware of this sensation, you get even more anxious, and so on. Or take the case of an arms race: you build a big weapon, your enemy responds with a bigger one, and so on. Unless the cycle is broken, it is bound to end in a smash-up.

One way to break the cycle is to "damp" it by means of negative feedback, which responds to a signal not with more of that signal, but with its opposite. Most self-regulating systems use negative feedback. A thermostat uses it. Your inner ear uses it to help you keep your balance when you walk. You use it (consciously the first few times, unconsciously for the rest of your life) when you ride a bicycle. Negative feedback sometimes has "negative" effects, as when a pianist onstage becomes aware that he has not yet made a mistake, and proceeds to make one. But far more often its effects are "positive"—assuming we think it a good thing that most natural and artificial systems keep going from day to day, instead of crashing or going haywire. That life keeps going on earth may depend on the apparent fact—perhaps a contingent fact—that negative-feedback loops in the biosphere outweigh positive ones.

In Harte's view, the Gaia hypothesis adds up to no more and no less than this: the claim that, in the linkages of living things to their nonliving environment, stabilizing feedbacks outweigh destabilizing ones. While this is an admirably simple way of looking at the matter, it may be just a shade too simple. For one thing, it is not just a question of adding up all the positive feedbacks in one column and all the negative feedbacks in another. They are all minutely linked, so that the effect of a very tiny feedback loop can be multiplied a thousandfold by another feedback loop, in what is known as a "butterfly effect." A loop of gossamer thread can trip a loop of steel cable. Also, a negative-feedback loop often needs positive feedback at various points to amplify either the signal it is responding to or the corrective response.

Even if we correct for such complications, the Gaia hypothesis still cannot be reduced to the claim that negative feedbacks outweigh positive ones. For it is also the claim that the former have always outweighed the latter, that they have done so with some consistency for two billion years, and that this can hardly be an accident.

The Gaia hypothesis is not so much an answer as a question: Why is the earth such a nice place to live? Why has it stayed a nice place to live for nigh on three billion years? Barring divine intervention, the answer must have something to do with the action of life itself. But that in itself is not an answer; it is just a reasonable narrowing down of the question.

Every ecologist knows that living things affect their environment, and that the net effect of the action of all the living things in an ecosystem is to keep it functioning in a way that benefits most of them. Every ecologist knows that the net of cause and effect, of cost and benefit, is denser and finer than he ever *can* know. What the Gaia hypothesis does is remind him that that net entangles the whole planet—from the dance of ions three miles up to the heavings of magma three miles down.

A RUN OF LUCK?

According to some critics, Gaia has got the cart before the horse. The earth is a nice place to live, they say, because life has evolved to find it nice; life has evolved to fit the planet, not the other way around. But in fact life has not evolved to fit the planet it found, any more than a family moving into a new house evolves to sleep on the bare floor. Life has furnished the planet, air-conditioned it, laid down carpets, hung drapes.

Other, subtler skeptics dismiss Gaia as a slipup of epistemology. That Gaia has kept the planet hospitable to life for three billion years is no great wonder: only on such a planet could a life form have arisen that is able to wonder. The odds that any random planet should be such a planet may be millions to one, but the odds that a planet with intelligent life should be such a planet are virtually one to one.

There are holes in this argument you can drive a truck through. To begin with, the second set of odds does not change the first set of odds. The fact that a planet has remained hospitable to life for so long is still remarkable, even if the fact that it is our planet is not. For we are talking about a state that violates the statistical laws of chemical equilibrium: as if you sprayed air freshener in a room and it stayed in a little cloud instead of spreading through the room. That, too, is physically possible, but so vanishingly unlikely as to cry out for some other explanation. And it still makes sense to seek a mechanism for that unlikely outcome, rather than chalk it up to good luck.

Moreover, that "good luck," if such it was, would not be a one-shot deal, a single inspired roll of the dice. It would be a run of luck—a throwing of sevens and elevens, day in and day out, for three billion years. Now, intelligent life arose on this planet a good million and a half years ago, and brains almost exactly like ours have been around for some ninety thousand years. The question of why the planet is hospitable to life has been asked in mythic form, one suspects, for at least that long. It was posed in philosophical form at least twenty-three hundred years ago by the Greeks, and in modern scientific form eighty-five years ago by Lawrence Henderson in his book *The Fitness of the Environment*. And Lovelock posed the question in his own way two decades ago. At any point in this sequence, our streak of luck might have run out. Every day that goes by in which the proportion of oxygen in the air remains at 19 percent is a fresh miracle, a thing to gape at—unless, of course, some explanation is proposed.

It profits the skeptic nothing to reply (a reply unbefitting a skeptic anyway) that, if the planet has behaved a certain way for three billion years, it is not surprising that it keeps behaving that way. For this assumes there is some mechanism making it do what it does—which is exactly what the Gaians are saying. Remember that we are talking about behavior that bizarrely defies the statistical laws of nature. If a cloud of gas has stayed put for three hours by sheer chance, it is overwhelmingly likely that in the next few seconds it will disperse. If, on the other hand, there is some reason why it has stayed put—for example, an invisible balloon or a hidden magnetic field—then there is a good chance it will continue to stay put.

The Gaians do not claim to know exactly what the explanation or mechanism is, only that it must have something to do with life. Now that God has thrown in the towel, what other contender is there?

MECHANISMS

As the Gaia hypothesis works its way from guess to full-dress theory, its proponents have to spend much of their time working out mechanisms. The plural is important. Not even the most ardent Gaians (among scientists, at least) hope to find a single, overarching mechanism that explains all Gaian doings. If Gaia is a system, it is less like a mainframe computer than like the Internet.

Also important is the chilly word "mechanism." While some enthusiasts find in Gaia a new kind of science—holistic, not reductionist; organic, not mechanical—the Gaian scientists, including Lovelock, know perfectly well that they must take Gaia apart, at least in thought. They must analyze, they must isolate, and they must find good old-fashioned causes and effects.

Some Gaian mechanisms seem fairly obvious. The main temperature-control mechanism is a case in point. As the sun gets older and brighter, plants and plankton and some other microbes get busier and more numerous. Their photosynthesis takes carbon dioxide out of the atmosphere; the natural greenhouse effect is lessened and the temperature stays more or less constant, despite the waxing sun. But Gaia, like God, is in the details, and the details are not so obvious. The carbon dioxide pumped into the soil by the biota (breathed out by roots or released by decomposition) reacts with water and the calcium silicate of rocks to form calcium carbonate and silicic acid. These seep with the groundwater into streams and rivers and thence into the ocean, where they are used by plankton to make their shells. As the plankton die, the shells drift to the sea floor, where they are buried and finally folded into the earth's molten mantle by the action of plate tectonics. (According to one scenario, this buried limestone is what makes the lateral movement of tectonic plates possible, by acting as a fluxing agent that keeps the magma fluid.) Another putative thermostat is obscure even at first glance: as the sun heats up, phytoplankton blanket the ocean more thickly, excreting more dimethylsulphide. This oxidizes in the atmosphere, forming an aerosol on which water vapor condenses. Denser clouds shade the planet, and the feedback loop is closed.

GAIA AND CHAOS

If Gaia's progress in the world of science has been slow, she has done better in the world of myth. For this, much of the credit belongs to the late novelist William Golding, whose idea it was to fasten the name of a Greek earth goddess to his neighbor's hypothesis. (If instead Lovelock had called it Geochemical Homeorrhesis, scientists might have been reassured but the public would have yawned.) Some credit should go, as well, to Lovelock's own name—suggesting as it does the firm but caring embrace of some power larger than ourselves—and to his person. In fact, if you asked a brainy novelist to invent two parents for the Gaia hypothesis, you might get a pair very like Lovelock and Margulis. With his whispery singsong voice, kindly wizened face, and penumbra of white hair, he might be a gentle, donnish Druid, or a wizard of Middle Earth, and what he really is—an independent scientist living and working in an old barn on the border of Devon and Cornwall—may be the nearest modern equivalent. As for Margulis, her plumply languorous, almost Levantine looks, strangely joined to a fierce motherliness, make her the ideal priestess of a new earth cult. Protecting both her ideas and her students like a tigress astride her cubs, she has defied the scientific patriarchy with stunning energy—notably with the very "feminine" idea that symbiosis has played as great a role in evolution as have competition and predation. (Note that I speak here in the language of myth, and of the media, which Margulis herself deplores.)

Myth and science have always walked hand in hand. But they often tug in different directions. Already this is happening to the Gaia hypothesis. Science tugs it toward precise statements, quantitative models, and testable predictions. Myth pulls just as urgently (and with the strength of numbers) toward pendulous images of an Earth Mother who suckles all creatures.

In a sense, the myth of Gaia is the myth of the Mountain. In this new version, the Mountain reverts to an early, distinctly feminine form. The base of the Mountain undergirds the whole earth, and the apex vanishes: the Mountain is everywhere.

The Gaia of popular myth—like the garden, like Arcadia, like America—is one more attempt to recover a lost unity. If the Mountain is everywhere, then we *can* live on the Mountain. In fact, we can't help doing so. Instead of seeming fearsome and somewhat alien, wildness seems cozy. The Mountain becomes a plush breast, or perhaps an enveloping womb.

Even the womb, though, is not enveloping enough to meet the case. The fetus, after all, can be thought of as an organism distinct from the mother.

But in the strong form of the myth, Gaia is not just our mother, she is the organism of which we are cells. Mother and child may have a falling out; the cells of a single body (in popular thinking, at least) cannot. The organism is a functional whole.

As is usual in myths of unity, part of the idea is to deny change, instability, and disorder of all kinds. The pop Gaia is like New Age music: an endless tonic chord, with a lot of noodling but no real change. The fall of man is simply a failure of perception. If we could see clearly, we would see that we are still in Eden.

As is also usual in myths of unity, some things have to go by the boards. If the Mountain is everywhere, where is the Tower? What is the role of technological man, who thinks he runs the world, in a world that runs itself? This is a puzzlement. Is man a child gone mad, a matricide? Is he a cell gone berserk, a cancer that may finally consume the body? Is he a minor irritant which Gaia, turning in her sleep, will squash like a mosquito? Or is he just like any other cell, however different he may look—in the long run, a part of the body's functioning? Is the rebellious child doing just what Mother wanted all along? Each of these possibilities is advanced by one or another version of the Gaia myth.

It is not just man, though, that muddies the picture, but certain aspects of nature as well. Like so many attempts at unity, this one seems to have spawned its own shadow. Anyone who casts a cold eye on nature must at times find the image of a smiling, great-breasted mother fading or flickering, and behind it a darker image glimpsed: a hag, a Gorgon, a dragon. Gaia is not the only divinity of the earth. There is also Tanit, to whom infants were sacrificed; there is also Hades.

Happenstance *may* be the only reason the science of Gaia and the science of chaos should both have arisen over the past few decades. But the fact that chaos has followed Gaia onto the stage of popular myth is less easily chalked up to chance. For in the dark figure of chaos we find just that aspect of nature which seems to be excluded from the cozy oneness of Gaia.

Scientists have long known and loathed phenomena that resist their usual tool of prediction and control, the linear equation. Poets and painters have long known and loved and feared the same phenomena—clouds, snowflakes, hurricanes, heartbeats—for the erratic and the erotic are closely coupled. The great surprise of the new science is that this unpredictability often turns out to be oddly predictable, if only in general terms, by means of an arcane and highly computerized mathematics of period doubling, strange attractors, fractional dimensions, and the like. And the great surprise for poets, painters, and mythologers is that the tenuous order glimpsed beneath chaos only makes chaos that much more sinister and alluring. If, as Douglas Hofstader says, an eerie chaos lurks beneath order, and an eerier order lurks beneath chaos, how many more layers may there not be? A pit of pure randomness has no depth; an abyss of serried order and disorder leads us downward into infinity.

Once again, the cast of characters is well chosen to prime the pump of myth. Ranged against the gentle, white-haloed Lovelock and the tough but motherly Margulis, chaos has its own, equally typecast protagonists. It has the acerb, wild-haired Benoit Mandelbrot, mad scientist of fractal geometry. It has dark, intense Mitchel Feigenbaum—discoverer of a mathematical pattern common to many chaotic systems—chain-smoking as he paces the midnight streets of Los Alamos.

When we look at the science, it turns out that the two schools are not as opposed as they seem. But first let us consider Gaia on its own terms.

As we watch this worldview in the making, we have a rare opportunity to make some kind of peace between the needs of the heart and the strictures of the mind. Myth can point out to science aspects of the theory that may have profound effects, both emotional and practical, and so require scrutiny; science can point out aspects of the myth that do not stand up to scrutiny and that may mislead both heart and hand.

Start with the idea that the Mountain is everywhere. Science supports this, up to a point. Gaian science reminds us to look for wildness and its boons not just in exalted places, but in humble ones like mud flats. It reminds us that the flow of energy and matter—the flow represented by the Rivers of Eden—is more intricate and more ubiquitous than we can ever imagine.

Being mesmerized by flow, though, can blind us to the need for division. The evolution of species, on which Gaia's evolution depends, depends in turn on barriers between populations. So does the stability of particular ecosystems, on which Gaia's stability is built. If we think of the planet as one happy family, we may forget the dangers posed by invasive exotics, from mammals to viruses.

Then, too, Gaian science, no less than standard ecology, acknowledges that some places on earth—some high, some humble—have a greater role than others in keeping the planet running. People need to keep their distance from such places, or at least mess with them as little as possible. But if people think the Mountain is everywhere, they may have no qualms about building condos on top of real mountains, or filling in real wetlands.

Nor does science, Gaian or otherwise, support the image of Gaia as wholly benign. At most, Gaia ensures that conditions on earth will be tolerable for life in some form—not that that they will be optimal for all species, or for any particular species. Though Gaia has always bounced back after being hit by meteors, she has not been able to prevent major changes in climate or major waves of extinction, to say nothing of minor, apparently unprovoked cataclysms such as earthquakes, volcanic eruptions, tidal waves, and floods. Some of these may even serve her "purposes."

NOT AN ORGANISM, NOT A MUTUALISM

What about the central metaphor, the one that delights mystics and dismays geophysicists—the metaphor of the earth as organism? If we someday decide that it has some measure of truth to it, the reason will not be that the earth has come to seem more integrated, but that organisms have come to seem less so. Here is where the Gaia hypothesis meshes with ideas about symbiosis, immunological Darwinism, neural Darwinism, and the like. The cell, the brain, the immune system all are looking less and less like wristwatches and more like the flea markets where wristwatches and a thousand other things struggle to be "selected."

As I said, an early complaint against Gaia was that the planet as a whole is not subject to natural selection. This comes to seem less crippling when we see that natural selection operates not only on the level of the organism, but on those of the organ, the cell, and the molecule; and that a tendency toward self-organized complexity seems to exist even in nonliving matter, on every level from the subatomic to the cosmic.

Even so, an organism is far more of a functional whole than Gaia can possibly be. No single genetic program orchestrates the growth and interaction of Gaia's parts. Yet the comparison is not entirely to Gaia's disadvantage. A human body is selected for resiliency, yet can be destroyed by a small thing—a virus, or an olive pit, or a moment's inattention to the texture of a brick patio. Gaia, which is not selected at all, has survived far greater shocks—in fact, every shock that has come along over the course of three billion years. The human body is very good at some things, but living forever is not one of them. Your body cannot replace its organs with organs of new design. Your cells cannot change their genetic program (though they can shuffle the genes somewhat and play around with their application). You cannot jettison large parts of your body and grow wholly new and different parts. If Gaia were as orderly and imbued with purpose as a human body, she would probably have died long ago.

If Gaia is not really an organism, is it perhaps a symbiosis of organisms?

From space, the earth's land surface can easily look like lichen; and lichen can easily seem a near-perfect image of the biosphere. Clinging to dead rock, it is a symbiosis of species—alga and fungus—so seamless that it appears to be an organism itself. I suppose it was by a kind of clairvoyance that Thoreau, who never saw his planet from a vantage higher than Katahdin, had a special fondness for lichen.

As I said earlier, lichens are stupendously tough and self-reliant—in effect, ecosystems in a suitcase. They are often nominated as likely colonizers of other planets. Thinking of lichen, one can easily feel that every symbiosis is a kind of pygmy Gaia, and Gaia a titanic symbiosis. The feeling is bolstered by knowing that Mother Gaia's own mother, Lynn Margulis, is also a leading thinker about symbiosis. It was Margulis's theory of the symbiotic origin of the eukaryotic cell that Lewis Thomas had in mind when, in a best-selling book, he likened the biosphere to a cell.

Yet I do not agree with those who say that Gaia is simply "symbiosis viewed from space"—not if they mean mutualism. For the odd fact of the matter, as we saw when we considered "axis powers," is that mutualism can sometimes destabilize ecosystems, while predation tends to stabilize them. In other words, niceness may rock the boat; nastiness keeps it on an even keel.

Much of what Gaia does, or seems to do, is a function of life's opportunism. Nature, like capitalism, abhors a vacuum. If there is free energy to be had, something will grab it. If it is in a form that no extant thing can grab, something will evolve that can grab it. While Gaia may give the impression of a Mind at Work, this apparent mind is just the sum of interactions between self-interested organisms: a real Invisible Hand.

To anyone who has lived through an economic recession, it should come as no surprise that this invisible hand sometimes makes a hash of things. There are zigzags of boom and bust in nature as in business. Obvious as they are, these analogies are often ignored by popularizers of Gaia. The worship of Gaia's teats is only slightly more reasonable than the worship of Adam Smith's hand. Neither deity is equitable, and neither is highly reliable.

We come back to the question of chaos. For Lovelock, Gaia and chaos would seem to have the same relation in reality as they have in myth: one of opposition. The chaotic patterns that some theoretical biologists, such as Robert May, find on the screen when they model ecosystems are, for Lovelock, pure fictions. Their error lies in holding the "environment" constant. His own Daisyworld model, which takes into account the fact that organisms change their environment, produces far smoother and stabler patterns.

In fact, when you look closely at some recent large-scale computer simulations and actual experiments that have been hailed by chaos theorists as pointing to chaos in real ecosystems, you find that they really support Lovelock's view. For instance, an experiment with monocultured plots of perennial grass found chaotic dynamics brought about by the time-delayed effect of litter. Litter from one generation of grass choked off the growth of later generations, producing in one case a six-thousandfold population crash. In a real ecosystem, though, litter would be controlled by grazing, fire, and a livelier insect-and-microbial life. Similarly, a computer model of Dungeness crab populations found long periods of stability punctuated by sudden crashes, and regular cycles alternating with chaos. Even after tens of thousands of generations, the dynamics of the system never settled down. "Researchers modeled a world in which one might expect simplicity if ever it were to be found," a science reporter wrote; "the environment never changes." With this, she inadvertently put her finger on the model's flaw. As Lovelock would be quick to point out, an unchanging environment is precisely where simplicity and stability *cannot* be found. In Daisyworld—and the real world—life's effect on its environment is what keeps the system tuned.

The fact remains, though, that violent population crashes do occur in nature. They are not always (as Lovelock seems to think) a sign of ill-health or human interference. Chaos—in the lay sense if not the mathematical— does play a role in the workings of Gaia.

To Lovelock, the zigzags of the chaos theorists and their "strange attractors" look like "a demon or a dragon." He seems to fear this dragon, as if it threatened to scorch his lovely damsel, his goddess.

He ought not to fear it, I think. Despite their apparent mythic opposition, the dragon and the damsel, the monster and the goddess are one. Eve is the serpent. Melusina has scaly gams. At the very least, chaos has helped keep Gaia alive.

Lovelock admits that the chaos of the Hadean era, with its boiling lava and constant monsoons, lives on in Gaia. He admits, following C. S. Hollings, that local pockets of chaos help keep larger ecosystems stable by forcing life to probe its limits, to evolve new forms, to find new niches. But he seems unwilling to admit how close large ecosystems—and Gaia herself— come to the brink of chaos.

To understand this, it helps to look briefly at the new science of complexity, which has followed Gaia and chaos into the popular spotlight— though it has not yet caught the light in the strong way that makes science into myth. The captains of this new science study complex dynamic systems, of which chaotic systems are in their view a mildly interesting subset. Though they cannot agree on what "complex" means, and though their infatuation with computer models has led some skeptics to task them with practicing "fact-free science," they have come up with some striking speculations. One of these is the notion that complex adaptive systems—organisms, economies, maybe biospheres—reach their apex of fitness when they teeter at the edge of chaos.

While the only proof for this notion so far lies in the virtual world and not the real one, it does make a good deal of sense. One of the things that make a chaotic system chaotic is the butterfly effect, whereby a tiny flutter can be so hugely amplified that it changes the whole course of the system. But just such amplification is what makes an adaptive system exquisitely sensitive to small signals—signals that, at times, can spell the difference between life and death. The point at which the amplifier is turned up to its maximum useful power, before the feedback spirals into pandemonium; the point at which the gazelle is poised when its heart races, but does not go into convulsions, at the faintest whiff of civet—that is the edge of chaos.

If a system teeters at the edge of chaos, it may fall in. This seems to have happened to Gaia a number of times. It may be happening right now—that is, in this geological instant, a period of two million years during which the earth has oscillated violently between fever and chills. Yet throughout this period life has persisted, and has even produced, in *Homo sapiens* one of its most novel and versatile forms.

NATURE'S FOLLY

Motherliness and placidity are traits to which the popular mind is drawn, and they have earned Gaia pride of place in the New Age pantheon. By denying these traits, I do not mean to play devil's advocate. I wish only to suggest other grounds for Gaia's canonization.

We have heard enough about nature's wisdom, finely tuned and infallible. Let us now praise nature's folly, which is the secret of her wisdom. Let us praise the squabbles, the consortia and cabals, the internecine warfare, the wasted effort. Let us praise energy and information thrown to the winds: the quadrillions of sperm cells and pollen grains, of eggs and larvae and tadpoles hurled into the maw of life and ground up like so much fodder.

Nature's spendthrift inventiveness, her antic diversity—who needs 350,000 species of beetles?—give her a margin of error that human artists, engineers, and politicians can only envy. But "margin" is the wrong word: we are talking about pages of error, reams of error, continental shelves of error. In fact, error is what makes the whole thing go. Error is not just permitted by diversity; it is what permits diversity. If the mechanism of genetic transmission had been flawless from the outset, you and I would be archae-bacteria. The reign of the genes, Freeman Dyson has said, is like that of the Hapsburgs: *Despotismus gemildert mit Schlamperei*, "despotism tempered by sloppiness."

Perhaps this view of Gaia, no less than the ones I am disputing, tries to salvage the harmony and stability of nature by kicking it upstairs to a higher and more abstract level. It is a common mental tactic, this retreat to higher ground. It is what the Victorians did when the seas of Darwin's Galápagos lapped at their boots. It is how Hegel snatched divine justice from the jaws of history. Viewed from a high enough vantage point, the nastiest gut-spewing battle resolves itself into a pleasant harmony of sound and color. Let the field be wide enough and the day long enough and you will find it hard to believe in death, or change, or evil in any form.

> If the red slayer think he slays,
>   Or if the slain think he is slain,
> They know not well the subtle ways
>   I keep, and pass, and turn again.

A Hindu *je-m'en-foutisme*, an Emersonian optimism is not what I have in mind. The slayer slays, all right. The slain is sure as hell slain. Death, change, chaos, and evil are real and were real long before people appeared on the scene. Who is to say that they are justified by the fact (if it is a fact) that they make possible the persistence of life on earth? All that means (it might be said) is that death, change, chaos, and evil make possible the persistence of death, change, chaos, and evil. Speaking for myself, though, I am glad that the show has kept running long enough for me to see it.

COMPLEXITY

If it is not by happenstance that Gaia, chaos, and complexity have followed one another into the popular mind, then perhaps it is not by happenstance, either, that they have trod on one another's heels in the scientific realm. After all, science and myth are not born in cleanly isolated wards. Scientists, too, have hearts, which sometimes go pitter-pat at the same things as other peoples'.

As different as Gaia and chaos are, they are in some ways a matched pair. Both are reactions to the dominant science of the past two or three centuries: a science that is mechanistic, reductionist, deterministic, and linear. Gaia reacts by striving for a higher unity, chaos by leading us down curving halls into an infinite maze where the same turn may yield a billion different futures. One sees nature as all-powerful mother, the other as twisted, inscrutable serpent. But both assert that there is more to nature than meets the eye, no matter how powerful the lens we look through. Both deny the dream of absolute knowledge and absolute power.

What about complexity? Though its programs and motives are aptly complex, many of its boosters seem bent on reinflating the balloon that Gaia and chaos have burst. With the aid of high-powered computers, they hope to find the simple rules from which complex systems emerge. They hope to show the universal behavior of such systems. At their most ambitious, they hope to predict the particular behavior of particular systems. In short, they mean to extend the classic program of Western science—reductionism and determinism—into the nonlinear realm, which has so far resisted it. True, some fellow travelers of the Santa Fe Institute, the temple of complexity studies, abjure this program (the Nobel-laureate physicist Philip Anderson being one example) and some seem ambivalent. But the institute's presiding genius, the Nobel-crowned particle physicist and polymath Murray Gell-Mann, promotes it wholeheartedly.

By and large, complexity theorists are perfectly willing to believe in Gaia. To them, she is just a very big complex adaptive system. The objection that Gaia cannot have evolved by neo-Darwinian natural selection does not faze them in the least, since they do not believe that the human body can have evolved that way, either. In the view of theorists such as Stuart Kauffman, the survival of the fittest is a factor in evolution, but so are the patterns into which units such as proteins or nucleic acids spontaneously organize themselves—what Kauffman calls "order for free."

But if Gaia is just another complex system, then she is a system whose behavior we should be able to predict and control once we have the computational elbow-grease: in short, not a she, but an it. And here I must part company with the complexity jocks. If a system hovers at the edge of chaos, it also hovers at the edge of our understanding, or at least of our ability to predict and control.

WE SHOULD not be too eager to chisel either the myth or the theory of Gaia into a final, monolithic form. First of all, the only myths and theories that achieve such finality are dead myths and dead theories. But there is an even more important reason not to be too hasty.

The Gaia hypothesis is most useful as a hypothesis. If it fancied itself a finished structure, it could be as dangerous as the worst dogmas of the technocrats. If we knew the biosphere to be an exactly tuned machine, we would soon convince ourselves that we knew exactly how it worked, and therefore that we could safely tinker with it. If we knew that it could be trusted to safeguard our interests, we would feel free to breed, build, consume, and pollute to our hearts' content. And if we knew that it could not be so trusted, we would feel free to junk it.

The great value of the Gaia hypothesis is to remind us that we know none of these things—that we know, in fact, precious little about how the system works, or how well, or for whose weal. All we know is that it works.

A predecessor of Margulis's in Amherst, Massachusetts, once exclaimed:

> How much can come
> And much can go
> And yet abide the world!

My preference would be to change the exclamation to a question mark, and leave it hanging.

PART FOUR

EARTH JAZZ

22

MANAGERS AND
FETISHERS

O H Y E S, I've learned from my mistakes and I'm sure I could
repeat them exactly." For the student of ecological history, the
words of Sir Arthur Stehe-Greebling, proprietor of The Frog
and Peach, strike very close to home.

So far this book has been about the waves of human-led change and the
myths by which humans have made sense of them. We have let our gaze
wander from the beginnings of life on earth to the beginnings of human
life; from the first scratchings of the hoe to the harvest of great civilizations;
from the sheepy dreams of late antiquity to the cloud-draped dream of
Gaia. We have seen the vision of Eden recede ever further into the past,
even as it is projected into the future. In what is left of the book, I will try
to give some sense of how the lessons we have drawn might be applied to
the problems of the present day.

Two schools now dominate—somewhat lopsidedly—the debate about
humankind's dealings with nature. Each party claims to be defending
nature; each claims to be nature's only true knight. Naturally, they tilt at
each other.

Neither school has a registrar, and though it is easy enough to find good
examples of either type, they are nevertheless ideal types. In the real world
they mingle and metamorphose, and most people (especially those who
have not given much thought to these matters) have in their mental crops
undigested gobbets of both doctrines. Yet presenting the Managers and
Fetishers—for so I will call them—as ideal types, incorporating more or
less consistent (if not always coherent) bodies of theory and practice, may
help us clarify our own thinking.

On one side of the dichotomy, then, is the group I call the Planet Fetishers, or Fetishers for short. The name acknowledges their tendency to think
of nature as a perfect, harmonious whole—perfect and harmonious, that
is, when humans keep out. By their lights, humans have no right to a larger
role in nature than raccoons. Instead of trying to remodel our earthly

home, we ought to fade into the woodwork like the other residents. In rigorous practice, this would mean reverting to the hunting and gathering that fed us for the first few million years of our hominid career, before agriculture let us breed like mice in a seed bag. Even in half-baked practice, it would mean reverting to a far simpler way of life than our present one. In any case, it would mean returning most of the earth to wilderness.

The Fetishers' ideas can be traced back at least as far as the Romantic poets; they begin to be explicit in the writings of John Muir. Though sizable chunks of them can be found in movements as diverse as ecofeminism, ecopsychology, bioregionalism, Green politics, New Ageism, and even old-style, blue-blood preservationism, they find their most complete modern expression in a major strand of the movement known as Deep Ecology.

LET'S CALL THE HOLOCENE OFF

Deep Ecologists call themselves "deep" in contrast to the "shallow" mainstream ecologists who speak of stewardship, resources, conservation, and the like. Of the movement's American manifesto, *Deep Ecology*, published in 1985, even a convert concedes that it "shows its West Coast origins at times." The product of ivory towers hidden in the redwoods, it is one of the few books I know that manage to be flaky and moldy at the same time, like a fine old Stilton. For all its self-proclaimed frankness, it is evasive about the things that count. Squirreled away in a dark crotch of the book, in a line from an interview with the Norwegian philosopher Arne Naess, is a target for world population. It is one hundred million. There is no discussion of how that target might be reached, but the fact that the movement's totemic philosopher is a Nazi, Martin Heidegger, gives one pause. (Although their next-favorite philosopher is Spinoza, the authors quote with unblinking approval Schopenhauer's judgment that Spinoza's view of animals as unfeeling objects is "thoroughly Jewish"—when of course it is thoroughly Cartesian. The Bible and Talmud are adamant against cruelty to animals.)

When it comes to politics, Deep Ecology is deep the way an ostrich's head is deep. The same can be said of Earth First!, in effect the movement's direct-action branch. Inspired by Edward Abbey's novel *The Monkey Wrench Gang*, Earth First! specialized, until recently, in acts of sabotage (disabling bulldozers, cutting power lines) in defense of wilderness. Although members of the group have shown great courage in acts of nonviolent protest, they have also used some nasty tactics, such as the spiking of old-growth trees, which posed the risk of injury to loggers or sawmill workers. (Admittedly, there is no evidence that such injury occured: the notorious maiming in 1987, which prompted a huge backlash against radical environmentalists, involved a tree no thicker than a telephone pole, and the sheriff's prime suspect was a middle-aged conservative Republican who owned land next to the logging site.) And they have had some nasty ideas, such as the view once expressed in the group's journal that AIDS is a good thing because it reduces the human population without harming the birds and bees.

Not all people who call themselves Deep Ecologists endorse these views or these actions. The movement is far from monolithic, and many useful ideas can be found within it. Yet a certain willful blindness seems endemic. Earlier in this book we saw how useful idylls can be for those who would hide from the facts about their relation to nature and other people. The dream of a return to Eden—of a life in utter harmony with nature—is useful in this way. Already there are signs that green may replace red as the cockade color of tenured radicals at the end of the twentieth century. Radical environmentalism is like radical leftism in being too extreme to have real consequences in the lives of academics. It is better, though, in that it allows one to drop even the pretense of caring about people whose tenure on the planet is less secure than one's own.

Though a few of their ideas have been endorsed by no less a figure than the present vice president of the United States, the Fetishers will not become a majority any time soon; meanwhile, they tug usefully on the sleeve of the majority. The danger is that their doctrine may confuse more than it corrects. If you try to impose a paleolithic worldview on postindustrial people, you are asking for trouble. To the extent that it takes hold, it will only deepen the cleavage between thought and action that hobbles so many of us. Even hunter-gatherers would be hard put to meet the demands of a doctrine so self-effacing. How can the rest of us hope to measure up?

HANDS ON OR HANDS OFF?

What about the other side of the antithesis? At the far end are the crazed futurists, the sort of people who would make cows legless milk-dispensers, conveniently stackable, and replace sheep with tube-fed lamb-chop cultures hundreds of yards long. This extreme is no longer worth talking about. (Unfortunately, we will have to talk about it anyway when we get to biotechnology, where the crazed futurists have taken refuge.) Also at the far end are such recent movements as "wise use" and "takings," which use the language of freedom and property rights to disguise sheer greed and stupidity. For these people, science is just window dressing. Virtually the only scientists on their side are the lab-coated mannikins they pay to stand behind the glass.

For most purposes, though, the range of thoughtful discourse has shifted in an "ecological" direction: even most technocrats now admit that the planet is only so big and can take only so much abuse. Somewhat chastened, those who once tried to dominate nature now settle for managing it.

Planet Management has become the dominant worldview among scientists and policy-makers. It is the silicon version of the "Judeo-Christian" ethic of stewardship, which sees the earth as a garden that we are to dress, keep, and humanize. The finicky eye of God has been replaced with remote-sensing satellites that tell us, without moralizing, just where we stand. A good idea of how it all works may be had by leafing through a special issue of *Scientific American* entitled "Managing Planet Earth." Between the gaily colored advertisements in which major polluters express their commitment to a cleaner planet can be found advice on the management of water, air, population, energy, industry, economic development, agriculture, biodiversity, and climate. The charts and diagrams alone are enough to give a sense of the fearsome analytic guns these scholars pack. If you asked them how many angels can dance on the head of a pin, they would promptly devise a computer model taking into account metal tensility, mean terpsichoric kinesis, cherub/seraph ratio, and a dozen other factors the Schoolmen ignored. Much of their advice is excellent. If our leaders heeded it, the planet might stand a chance of getting through the next century in one piece. And yet there is something in the whole enterprise that rubs me the wrong way.

It has been observed that scientists, like artists, tend to fall in love with their models. The joke reminds us how differently the word "model" is used in the two realms. The artist's model is a piece of nature on which he bases his own creation. The scientist's model *is* his own creation. If one must fall in love with a model, surely the artist has the right idea.

In speaking of gardens, we noted their role as miniature worlds in which everything is under control, helping the gardener or gardenee avoid the harsh realities outside. Satellite images and computer models can play the same role, but with a difference: they make the Manager think he grasps and controls the world itself. Like Chaplin's Great Dictator spinning the globe on his finger, he gets microcosm and macrocosm mixed up. One of the many problems this causes is that, in his confidence that he has got the big picture, he thinks he knows what is best for everyone and everything. Often, it seems that the interests of the planet, the human species, and the educated elites of the industrial democracies just happen to coincide.

I would feel a lot more comfortable about Planet Management if it were hands-off management: giving as free a rein as practicable not to humans (the time for that is long past) but to nature. Many of the Managers do think in these terms, even if their jargon sometimes clouds the fact. Restoring stricken ecosystems, preserving genetic diversity (especially when this is done *in situ* rather than *in vitro*), controlling human population—all these are ways of keeping wildness alive and letting it work.

But the Managers believe that, in a world with anything like our present population and technology, human intervention will be needed not only to get wilderness going, but to keep it going. The Fetishers believe this, too— at least, the sane ones do—but conclude that we must therefore reduce both our numbers and our technological level. In other words, rather than keep wilderness in an oxygen tent, we must back off and give it room to breathe. And while they may take it too far, I think they have a point.

What worries me about the Managers is that they like managing: they have trouble keeping their hands off. Their attitude toward nature is roughly that of aeronautic engineers who have been given a chance to examine a captured enemy aircraft. They want to learn how it works and how to fly it.

A parable: A wise man of the town of Chelm moved into a new house. In it he found something he had never seen before—a thermostat. Fascinated by its ability to regulate the temperature in his house, he took it apart. As he was a wise man and mechanically inclined, it did not take him long to figure out how it worked. "I can do that," he said to himself. Whereupon he threw out all the parts except the thermometer and the wires that ran to the furnace, and stationed himself at the little hole in the wall. Whenever the thermometer needle fell below 72 degrees, he closed the circuit and the furnace went on. When the needle rose above 72 degrees, he opened the circuit and the furnace went off.

In *Gaia: A New Look at Life on Earth*, James Lovelock speculates on what would happen "if, at some intolerable population density, man had encroached upon Gaia's functional power to such an extent that he disabled her. He would wake up one day to find that he had the permanent life-long job of planetary maintenance engineer. . . . Then at last we should be riding that strange contraption, the 'spaceship Earth,' and whatever tamed and domesticated biosphere remained would indeed be our 'life support system.' " Lovelock guesses that, at the present rate of per-capita energy use, a human population of ten billion would still be living in "a Gaian world." At some unguessable point beyond that, we would be on our own.

This prospect should scare the Managers to death. Instead, it gets their juices flowing.

MANAGING TO MANAGE LESS

At times the dispute seems to be semantic, which is not to say it is trivial. The Planet Fetishers would admit that wilderness must be protected by rules—imposed or internalized—against certain kinds of human meddling. To the Planet Managers, such rules are themselves a form of meddling. At first glance this seems an example of what Bertrand Russell would call a "category mistake" (as in the "All Cretans are liars" paradox). But there is more to it than that. Some of what the Managers would call managing nature is actually managing humans to keep them from managing nature. In other words, it is hands-off management: keeping not only your own hands off, but other people's hands as well. As far as possible, the day-to-day work is left to nature. And that is just as well, since nature's store of useful information remains (despite our burning of its libraries) far greater than our own.

One Planet Manager, Walter Truett Anderson, writes that he personally favors "a world of abundant wildlife and vast wilderness spaces and human-scale cities." It sounds very like the kind of world favored by some people close to the Deep Ecology movement. Some of the Managers are as keen on wilderness as the Fetishers; they are simply more up-front about the amount of management we will have to do if we want real wilderness, as opposed to the wastelands and weedy junkyards that humans usually leave behind. The tools of ecosystem restoration have been taken up by Managers and Fetishers with equal enthusiasm. In theory the Fetishers limit themselves to rebuilding old ecosystems, while the Managers are just as happy to build brand-new ones. But that distinction breaks down fairly quickly in practice. In fact, the whilom king of the monkey-wrenchers, Dave Foreman, has lately joined forces with two conservation biologists, Reed Noss and Michael Soulé, to promote a plan that would envelop North America in a vast system of connected (and largely restored) wilderness preserves.

For the foreseeable future, we shall not have the luxury of leaving wilderness alone. Not enough of it is left, and much of what is left is damaged. There is no escaping the need to manage nature. The best we can do is to observe the following rule: So manage nature as to minimize the need to manage nature. Before any intervention, ask yourself, Will this make it less likely that we will have to intervene in the future? If the answer is no, think long and hard before proceeding.

A Latin word for wilderness, *vastitas,* hints at a crucial fact. A small wilderness is no wilderness at all—unless it is part of a larger network of wildness. The more wilderness we preserve, the less we have to manage.

We would need to do less managing of wilderness if we managed the human landscape better. If we made better use of the space and resources already at our command, we would leave more room for wilderness, which would give it a chance to function on its own. If we paid more attention to the flow of wildness in and out of wilderness areas—being wary about letting foreign species or substances in, but allowing native species and vital substances to move freely—we would have to worry less about what happened inside. Rather than managers, we would take the humbler role of doormen.

In practice, this would mean paying attention to the flow of wildness in settled landscapes. If there is to be a healthy flow between wild and settled places in a given region, then barriers against invasive exotics must be thrown up at the border of the region itself. But this runs counter to all our habits. In general, we find it easier to manage wilderness than to manage our own affairs.

In fairness to those Planet Managers who care deeply about nature—and I think most of them do—it must be said that we rarely give them much say in our affairs, or much of a chance to manage our working landscapes. We prefer to shunt them off to wild places where they won't bother us.

The ethic of stewardship finds its proof text in Genesis 2:15, in which man and woman are placed in the Garden to "work it and protect it." As I said a while ago, these two commands are contradictory. They are also complementary. We are destined to work our way across the globe, turning Eden into something else. And we are destined—in our better moments—to protect Eden against our own work. The command to protect puts upper limits on the scope of our work and lower limits on its quality. In other words, we must not try to manage too much of the world, but what we do manage—our cities, our factories, our farms—we must manage well. Little things like good workmanship and thrift in the use of energy, matter, and space can do more for wilderness than all our satellites.

But we have got things twisted around. We try to manage the whole planet when our own backyard is a mess.

LESSONS OF THE MOUNTAIN

Looking back at the ground we have covered so far, you might be tempted to say that the Planet Fetishers are devotees of the Mountain, and up to a point you would be right. The Fetishers do believe that wilderness is the center of the world. But whereas the Canaanites thought it impossible for humans to live harmoniously in wilderness, the Fetishers think it not only possible but necessary: in other words, they want to go back to Eden.

Similarly, you might assume that the Planet Managers are champions of the Tower. The Managers do believe in the Tower—the place from which one can see and control all things on earth—even if they tend to call it the satellite. But some of them, at least, have a deeper sense of the life-giving role of wilderness than the Mesopotamians did.

Neither side, it seems to me, has fully learned the lesson of the Mountain. Wilderness is not something to live in. Ideally, it is not something to manage, either. Our relation to the Mountain is trickier and more fraught with contradiction than either side admits.

What about that modern avatar of the Mountain, Gaia? Both Managers and Fetishers invoke her name when it suits their purposes. For the Fetishers, Gaia is an all-wise goddess who does what is best for all life, or tries to do so despite human meddling. For the Managers, Gaia is a machine we can fine-tune to meet our own needs: a spaceship with cruise control.

Both miss the point. Humans cannot just relax in Gaia's embrace. Nor can any other species. Every species has to look out for its own interests— that is how Gaia works. People are part of Gaia, too, though exactly what kind of part remains unclear. There is room on the planet for civilization to evolve. But this late, odd-shaped, and as yet not obviously useful tail must not get so big or so arrogant that it wags the dog. Since we don't know how Gaia works, the less we mess with her the better. And the best way to keep from messing with her is to leave vast tracts of wilderness intact and, as far as possible, unmanaged.

BY NOW SOME READERS may have noticed bits of straw in the air. Setting up my antagonists as ideal types has, I admit, made it easy to knock them down. But a straw man is a fair target if it keeps popping up again, in one form or another, in nearly every time and place. If Matthew Arnold (to take one of many possible examples) had confined himself to attacking particular persons or texts or political parties, his *Culture and Anarchy* would now be of merely historical interest, whereas flashes of his caricatured Hebraisers and Hellenisers can be sighted on our own cultural scene. I am no Arnold, but I think Managers and Fetishers, too, have been popping up, in various garbs, for centuries and will probably keep popping up for some time to come.

Managers, for example, have sometimes beaten their slide rules into pruning hooks. A case in point is the Garden Earth vision of the late René Dubos—the microbiologist who grew up in the elegantly humanized landscape of the Ile de France and who spoke ardently of "the wooing of earth." While Planet Gardening (urged in recent years by Frederick Turner and Michael Pollan, among others) sounds greener and fuzzier than Planet Management, it is open to the same objections. First off, it is risky: without the conviction that "everything that lives is holy," we are likely to humanize things too cleverly by half. Dubos himself may have had that sense of respect—that habit of *respectare,* of looking again—but those who take the license he offered probably will not. The Garden Earth will then become a Disney Earth: a high-tech theme park of simulated environments and patented animals. What is more, it may not even work.

The other objection to a Garden Earth is that it is not risky enough. "All France, it has often been said, is a garden," wrote Henry Miller, "and if you love France, as I do, it can be a very beautiful garden. . . . But there comes a day, when you are well again and strong, when this atmosphere ceases to be nourishing. . . . You long to make friends, to create enemies, to look beyond walls and cultivated patches of earth." On a Garden Earth, Miller would have nowhere to go.

At the other end of the spectrum, both Deep Ecology and Earth First! have evolved, or at least changed, since their first typical expressions made them famous (or infamous). Particular Fetishers or Managers have views more nuanced than those I have set out; particular thinkers have skirted, by deft tacking and luffing manouevres, the philosophical and practical shoals I have charted. Yet each side, it seems to me, remains stuck in a mythic and logical vessel that is fundamentally unseaworthy, and unsuited to weather the voyage ahead.

Let us try to craft another. How do you give nature room to work and still get on with the business of being human? And if part of the business of being human is to work *with* nature, how do you do that when you don't know exactly how she works or—to paraphrase Freud—what she wants?

23

BEBOP

AWHILE AGO, I spoke of music's special ability to pose and resolve problems of discord and harmony between people and nature. That is one reason for its central role in Arcadia. Music does that for modern people, too, but not half so well as it might.

In the last couple of centuries, our perception of nature has become narrowly visual. As a result, nature has taken on in our minds the frozen perfection of calendar photos, or the streamlined motion of television specials. Living in the Tower that is modern civilization, we see nature as an object to be manipulated or a view to be admired. We *hear* nature hardly at all. At the same time, we have come to hear music as a privileged language of human emotion, with no reference (except when the words or program say so) to the world outside.

Though it may be a modern specialty, the bias toward a visual perception of nature is hardly a modern invention. It is not even a human invention. As primates, we are profoundly visual creatures. Our eyes' precise stereoscopy suits us to size up the world, our hands' precision grip to seize it. We anatomize and butcher, have visions and build.

The eye is attuned to objects, the ear to process. The eye lives in space, the ear in time. It is no wonder that music, of all arts, lends itself most readily to improvisation. While the eye perches dryly at a certain point of view, the ear swims. Since the eye individuates and the ear unites, music has long been thought the art best able to give humans a sense of oneness with each other and with the universe. The word we unfailingly use for the reconciliation of unlike things is a musical word, "harmony." Music is perhaps the most social of all arts, and one of the few that admit of group improvisation.

For Schopenhauer and Nietzsche, music was the voice of the cosmos itself. For many traditional peoples it depicts the cycles and moods of nature, and often involves a delicate dialogue with birds and crickets, water and wind. By us it is bricked up in concert halls or living rooms, or blared from speakers that turn the outdoors into a living room. Meanwhile the sounds of nature are bricked out, glazed out, muffled by air conditioners and shredded by lawn mowers.

Both our perception of nature and our action upon it might be improved if we relied a bit less on sight and a bit more on sound. At least, we might use sound as a model. We might use sound in general, and one kind of sound in particular, as a model for our collaboration with nature.

EARTH JAZZ

What form should that collaboration take? The Planet Fetishers want us to sing one faint part among millions in nature's (imagined) harmony. The Planet Managers want to compose and conduct a planet symphony of their own devising. Maybe there is a third possibility: a kind of Earth Jazz.

Its advice for humankind might go something like this: Ditch your notated score—whether ascribed to nature or yourself—and learn to improvise. Respond as flexibly to nature as nature responds to you. Accept nature's freedom as the premise of your own; accept that both are grounded in a deeper necessity. Relax your rigid beat and learn to follow nature's rhythms—in other words, to swing.

A good model for the planet might be a bebop quartet led by a saxophonist. The style of each sideman pervades the whole, since the drummer, bassist, and pianist play almost all the time. Each player, though, also takes solos, stretches of music that he makes his own. During most of these solos the leader "lays out." But during the leader's solos the other musicians keep playing. In other words, they are indispensable and he is not.

If you translate time into space, the sax player into humankind, and the three sidemen into other taxa—making the piano, say, the nonhuman animals, the bass the plants, the drum set a catchall (admittedly inadequate) for fungi, protoctists, and bacteria—you get a lesson in how humans can work with nature. For humans to thrive, even the most humanized spaces must be inoculated with other species. Wildness, like swing, must flow through all things. But for other species to thrive, they must have some spaces to themselves: spaces from which humans discreetly withdraw, excuse, or recuse themselves, "lay out."

The leader of a jazz group takes a bird's-eye view of its music. On some level, he is aware of the sounds each player is making. That awareness shapes his own playing, which in turn nudges the others' in sundry ways. But he does not try to make each note that is played fit some preset scheme.

All life plays variations on the same few chord changes. Each taxon improvises, following certain rules but obeying no predetermined destiny. Each responds to the riffing, comping, noodling, and vamping of those around it. Life makes itself up as it goes along. Withal, a certain unity emerges that no one has willed.

Now, it might be objected that jazz is only a small part of the world's music. Since it grew up on a few small patches of the earth's surface—the Mississippi River Valley, the stockyards of Chicago, the steel cliffs of Manhattan Island—how can it claim to shed light on the relations of humans and nature in other places?

Jazz is not just any music, but a mongrel of splendid pedigree. It is an urban music with deep rural roots. Its rhythms arise from the juxtaposition of chicken coops and locomotives, of bayous and steel mills, of tenements and penthouses. Most of all, it springs from the meeting of Africa and Europe—the tree where man was born, and the ax that would take it down—on the soil of a new world.

Another reason jazz fits the bill is that it is a music of exile. Born of the African diaspora, it gives voice to our nomadic urges as well as our longing for a home. The bedrock of jazz is the blues—a music of yearning, of anger, of dashed hopes that (like the twelve-bar form itself) always spring back. The blues are born of exile, oppression, and shattered love. If these are the special condition of the African-American, they are also, in a broader sense, the state of all humans outside Eden.

For modern people, maybe for all people, Earth Jazz begins with earth blues. Until you have felt in your bones what it is to be a species that cannot help changing the world—a species that in making its paradise unmakes Eden—your attempts at a joyful, playful dialogue with nature are bound to ring hollow. Until you have sat down and wept by the rivers of Babylon, you will not discover that they, too, flow from Eden. When you do make that discovery, the Lord's song will rise in your throat unbidden.

JAMMING WITH THE GODDESS

"It don't mean a thing if it ain't got that swing"—Ellington's dictum fits Earth Jazz, too. What is swing? Pedants have gone gray trying to define it. For our purposes, we can think of it as a kind of suppleness, a looseness far more exact than mere exactness. To swing, one must be aware of the rhythms behind the rhythm. In nature, this means the chaos behind apparent order, and the order behind apparent chaos. A rhythm in jazz is like the coastline of Maine: no single measurement is possible: as you look closer, new convolutions appear. A good jazz solo can never be notated: you can get down to the level of dotted hemidemisemiquavers and still know that further layers of complexity lurk just beneath. Biologists trying to describe or model a natural system often get the same feeling. And Ralph Ellison may have had the same feeling about social systems when—using a Louis Armstrong record as his jumping-off point—he said of "invisibility": "You're never quite on the beat. And you slip into the breaks and look around."

We do not swing. Our science, our lifestyles are rigid. We deal with nature now as one who carries a mug of coffee in his right hand and a book wedged between his right elbow and his side. The arm is locked rigidly against the body, there is no give, the coffee sloshes wildly with every step. No wonder we lose so much.

Let me come back to a question I raised earlier. How do you collaborate with Gaia if you don't know exactly how she works, or what she wants? You do it, I think, by playing Earth Jazz. You improvise. You are flexible and responsive. You work on a small scale, and are ready to change direction at the drop of a hat. You encourage diversity, giving each player—human or nonhuman—as much room as possible to stretch out. You trade fours with the goddess: play four bars, listen to her response, respond, listen, respond. True, sometimes her response may not be clear for centuries. But then no one said this would be easy.

At any rate, it may be easier to accept our exile from Eden—and the need to intensify that exile, in a sense, by stepping back and giving Eden more elbow-room—if we see it as the premise of a creative give-and-take. The music of Eden may have been gorgeous beyond imagining, but whatever it was it was not Earth Jazz. In that tangle of world and self, trading fours was not an option. Without difference, without distance, there is no dialogue.

MIDNIGHT AT THE OASIS

Jazzmen trade fours; shepherds in pastoral trade sixes, swapping hexameters in friendly strife. Both practices may go back to the games of real shepherds. Though the static ideal of Arcadia may be a mirage, some of its ancient habits—of playful riffing, of shifting boundaries, of discord deftly harmonized—can serve us surprisingly well. Panpipes or saxophones, the basic point is the same.

Let me start giving some examples of missing the point and of getting the point. In the Sonoran Desert of Arizona, there is a place the Papago Indians call A'al Waipia where sweetwater springs trickle into a small pond, inciting a riot of green in a world of gray. For thousands of years, A'al Waipia was the site of Indian settlements. As late as 1957, Papago irrigation ditches fed more than a dozen acres of crops and orchards.

In that year, the National Park Service moved in. As one of the few true desert oases in North America, A'al Waipia had to be protected. With the connivance of an Indian who claimed title to the land, the Park Service summarily condemned the fields and buildings. The oasis was returned to its natural state, so that its full value as a refuge of flora and fauna could be realized.

Things have not turned out quite as planned. Each year, the oasis looks less like an oasis. Each year, it loses plants and animals. When the ethnobotanist Gary Nabhan visited A'al Waipia in the early 1980s, not only had the fruit trees died; so had most of the "wild" trees. Only three cottonwoods remained, and only four willows. Summer annuals, too, were scarce. A survey Nabhan undertook with the help of ornithologists at three different times of year found a total of thirty-two bird species at A'al Waipia. At a similar oasis in the Mexican Sonora, the scientists found over sixty-five.

The Mexican oasis, known to the Papago as Ki:towak, is only thirty miles from A'al Waipia, but its aspect is very different. Resplendent with palms, cottonwoods, willows, elderberry, salt cedar, date, pomegranate, and fig; verdant in summer with squash, watermelons, beans, and other crops, and with wild greens coaxed forth by plowing and irrigation; its ditches rife with Olney's tule, the edible bulrush from which the oasis takes its name, Ki:towak offers plenty of food and shelter for teal, white-faced ibis, and dozens of other birds A'al Waipia no longer sees. For this is a cultivated oasis—cultivated in the thrifty, painstaking Papago way, of which I will say more in the next chapter—just as A'al Waipia was before the Park Service arrived.

Nabhan quotes a Papago farmer: "When people live and work in a place, and plant their seeds and water their trees, the birds go live with them. They like those places, there's plenty to eat and that's when we are friends to them."

Of course, I share the Park Service's presumption that ecologically sensitive places should be left wild or allowed to go wild whenever possible. But when indigenous people have been living in a place for thousands of years, chances are the "wild" things in that place have coevolved with them. Pulling them out of the ecological structure (if indeed we have the right to do so in the first place) may cause its collapse. A park service schooled in Earth Jazz would be flexible enough to avoid such an error, or at least to correct it once its effects become clear. Unfortunately, by the time the actual Park Service concedes this specific error, it may be too late, for there may be very few traditional Papago farmers left.

A happier story—also, as it happens, an avian story—can be found in another part of the American West: the vast Central Valley of California.

Famous now as the richest farming region on earth, it was formerly one of the richest waterfowl corridors on that same planet. As late as the early 1950s, 60 million pintails, snow geese, tundra swans, and other migratory birds wintered here. In 1993, the count was down to two million. The reason was clear: California had lost some 95 percent of its wetlands. Dammed, drained, diverted, dredged, plowed, and channeled, the birds' winter homes were no longer homey.

Among the dammers, drainers, diverters, dredgers, plowers, and channelers, growers of rice were notable. Laying bare the great Western water boondoggle in his book *Cadillac Desert,* Mark Reisner reserved special scorn for the folly of growing a monsoon crop in a desert state. In truth, in most places and at most times of year, the Central Valley is dry—not the Sonora, but a long way from Siam.

Scalded by Reisner's attack, some prominent rice-growers invited him for a parley. A partnership arose between the California Rice Industry Association, Ducks Unlimited, California Waterfowl Association, The Nature Conservancy, and Reisner. Styled the California Ricelands Habitat Partnership, it promises to transform rice farming and make a lot of birds happy in the process.

Traditionally, the half-million acres of rice in the Central Valley were soaked in several inches of water from April to mid-October. Then they were drained, allowed to dry, and burned—lest the hard, silica-rich stubble stick around until spring, sheltering rice diseases. In 1991, tired of the Indian-summer pall over California's skies, the state legislature passed a law that would phase the burning out.

Enter the partnership, with a new idea: plow or roll the fields, then flood them in winter. Water would help decompose the stubble; it would also attract waterfowl, who would further break down the stubble and leave fertilizer in its place. A few growers were already doing this; their experiments and those of scientists augured well. By the winter of 1994–95, some hundred thousand acres—a fifth of the total riceland—were under water. Having reduced their pesticide use by 99 percent, the growers could play host to waterfowl without risk of poisoning them. Not only waterfowl, but migratory shorebirds, egrets, herons, terns, sandhill cranes, and peregrine falcons stand to benefit. Early counts show that they are taking up the offer.

Of course, there is the question of where the water comes from. Will taking water for the winter rice fields further threaten the runs of salmon and other anadromous fish in the San Joaquin and Sacramento rivers and their tributaries? Or might the standing water be released in the late winter or spring, swelling the streams just in time for the spring runs?

However that turns out, the partners in this enterprise are providing a master class in Earth Jazz. Maybe the ecological ideal would be to blow up the dams, smash the channels, and return the Central Valley to its natural state; but if land can be kept in production, on the tax rolls, and entice birds back anyway, maybe that is not so bad. Maybe it is madness to grow a monsoon crop in California; but if people insist on growing it, we might as well make the best of the situation.

AVOIDING THE NANTUCKET SLEIGHRIDE

The same kind of flexibility can be applied to broader issues of "resource use." In the past couple of centuries, Planet Managers have convinced themselves that they know how to manage nature "sustainably." Scientists study a given renewable "resource"—fir, cod, what have you—and arrive at a consensus as to its growth rate, recovery rate, and so on. They determine the "maximum sustained yield"—the harvest that loggers or fishermen should be able to take every year, year after year, for all eternity—and the government sets its limits accordingly.

The only problem with this system is that it almost never works. For all the sophistication of their computer models, scientists have only the vaguest idea of what is going on in the woods and still less of what is going on under the water. The manifold factors that make fisheries, for instance, surge or ebb in cockeyed cycles—prey and predators, climate and currents, toxins and dams and a hundred things not yet identified—ensure that consensus stays slippery. Controlled, reproducible experiments are out of the question. With so much wriggle room, industry can always find scientists who will testify that a higher yield would be just fine, and politicians who will believe them. Even limits set in good faith tend to put too much faith in the compliance of fishing fleets. Worse, they put too much faith in the compliance of the fish themselves. If their population has been stable for several years, it is trusted to stay stable.

You might think that, if a maximum sustained yield started to look unsustainable, it would be revised downward. In fact, just the opposite happens. What economists call a "ratchet effect" takes hold. In good years, extra boats go out and extra processing plants get built. But in the lean years that inevitably follow, those boats and plants are not put in drydock or shut down. Instead industry looks to the government, which responds with subsidies of one kind or another. Boats are kept afloat, relentlessly fishing an ever-dwindling fishery. More often than not, the net effect—seen in recent years in herring, cod, ocean perch, salmon, lake trout, sardine, anchoveta, and many other stocks—is collapse.

A society attuned to Earth Jazz would not let itself get locked into such patterns. Its scientists, knowing their own fallibility, would recommend limits below those which their computers spat out. They would be ready to revise them downward at the first sign of decline—that is, at the first sign not masked by the froth of natural variation—and would revise them upward only by small degrees. Policy-makers, knowing a thing or two about human nature, would assume that the scientists knew even less than they said they knew. Nor would they wait for perfect consensus before taking action. They would hedge their bets and make their policies as supple and reversible as possible. Enlisting the help of business and labor, they would evade the grip of the ratchet effect by diversifying local economies and spreading risk. In such a society, communities would not be hostage to the leaping and diving of a single resource like whalemen on a Nantucket sleighride—a ride that ends in the destruction of one party or both. But the society as a whole would ride the waves of nature's changes, and humankind's.

PICK YOUR EDEN

Whether we like it or not, the world culture of the near future will be, in large part, American. Both the best and the worst in American culture will be represented: the only question is, in what proportions? By speaking of Earth Jazz, I am trying to apply the greatest product of American culture to the greatest problem the world faces. It is not an answer to the problem, of course. Even to call it a model is stretching things a bit. When you come right down to the nitty-gritty of farming, industry, and the making of cities and villages, each region needs to work out its own answers. But as each of us hoes his own row (or nonrow, as the case may be), jazz may give us inspiration.

What seems at first glance to be vagueness may prove useful. Most visions of humankind's place in nature fail, I think, because they mark out too specific a place. One visionary likes the medieval city, another the nineteenth-century farming village, a third the desert pueblo. It might make a good parlor game: "Pick Your Eden." The hunter-gatherer band; the Neolithic village; the putatively matrifocal culture of Old Europe; the "Old Planting Culture" of the South Seas; the elaborate mixed farming of East Asia; the wind-and-water technology of early-modern Europe . . . Each of these has been somebody's pick, and none has been everybody's.

While each Eden may have something to teach us, none can begin to address the wide range of problems we face. Worse, each makes the mistake of being an Eden—a world made whole, once and for all. Better, I think, to accept the exile from Eden, with all the division and instability that are its baggage. Better to take as your model not a thing, time-and-place, or state of affairs, but a process.

Rather than be hostage of my own tastes, I start with the rock-bottom fact—a fact shaped only slightly, I think, by my own tastes—that we need wilderness. We need it not only for our psychological well-being (at least, this seems likely) but for sheer biological survival. The question then is, What is our proper relation to wilderness? How can we keep it alive, when our natural tendency is to overrun or smother it? Should we live in it, supposing that were logically possible? Should we live *like* it—use it, that is, as a template for our man-made world? If we cannot live exactly like it—if the man-made world has some rules of its own—how can we continue to enjoy the boons it confers on us and on all other creatures?

In seeking answers to these questions, I have tried to skirt the problems that cling to particular Edens by choosing models that are somewhat abstract, like the Mountain and the Rivers of Eden. And now I have used a model that is not a place or a time or a kind of relation between humans and nature, but something else entirely: a kind of music. Something else entirely—it is, after all, on a different ontological plane—but not something unrelated. For it seems to me that music, of all the arts, has the most to say about the relation of humans and nature. Music describes the workings of nature (including humans) at a level deeper than any particular landscape—deeper, but not more abstract. It works on a different plane from the real world, but on its own plane it is just as specific and just as exact. That is why it can mirror some things in the real world more exactly than language, which is confined to the same plane as the real world and is constantly banging into the wrong things. Music, by contrast, shares the vast sonic plane with a fairly narrow set of natural sounds readily audible by humans. It has infinite space for maneuver, infinite shades of nuance.

Because of this, and because the vast ocean of the world's music is fed by rivulets from every landscape and every culture, music has something to say about the relation between humans and nature on every inhabited inch of the earth's surface. In this way it avoids the twin perils of vagueness on the one side, and limited applicability on the other, that afflict more literal models.

A model is not worth much unless it can serve, somewhere down the line, as a guide to action. Much of the balance of this book will offer examples of Earth Jazz as it is actually played. First, though, it may be useful to listen more closely to an idea essential to Earth Jazz: the idea of learning from nature.

LEARNING FROM NATURE

"Elephant ears," Ellington was said to have—a tribute paid him for the tribute he paid other musicians, jazz and otherwise, by listening hard to what they played. To play Earth Jazz, you must first of all listen. To stand in a right relation to Eden, you must face Eden and take note of what goes on there. In other words, you must learn from nature.

The difficulty with this task is that it is much too easy. You can find in nature almost any lesson you want.

Do you copy the industry of the ant or the indolence of the cat? The energy-squandering, rock-star lifestyle of the hummingbird, or the tight-fisted sit-tight ways of the agave? The selflessness of the worker bee, or the selfishness of the egret chick that kills its siblings for a bigger share of food? The self-reliance of the orangutan, or the freeloading of the cuckoo? The innocence of the sheep, or the guile of the orchid disguised as a female bee's derrière? The sensate delicacy of a spider's web, or the blank violence of a charging rhinoceros?

Without an Aesop to guide you, how do you choose? Is it merely an aesthetic choice, or (as in Aesop's case) a rubber-stamping of your moral predilections?

It is not quite so arbitrary as all that. What we mainly learn from nature are not ends, but means to ends that we choose for ourselves. Nature is a treasure house of means, a vast patent office full of levers and enzymes, of mechanical and chemical solutions to the various problems living things have faced over the past three billion years. Unlike most patent offices, though, this is one in which we are free to rummage through the drawers and take what we need.

In this way, we learn tricks that can help us stay ahead of the game of life. We have to be careful, though, because if we get too far ahead of the game it may come crashing down around our ears. So it is important that we learn some of the larger tricks—the metatricks, as it were—that keep the game going. The principle of flow, for example, is basic to the functioning of every organism, but also to that of the biosphere as a whole.

As we rummage through nature for means to our ends, we may find that our ends have changed. An attribute that was useful as a means may become an end in its own right. For instance, we may begin studying the principle of flow as a means to solving our solid-waste problem. But the swirling dance of its patterns in nature—the water cycle, the carbon cycle, the intricate cycles of sulfur and calcium—may come to seem so lovely that we cannot help adopting flow as an end in itself. In that case, we may pursue recycling, composting, and biodegradability with the passion of artists, even when there seems to be plenty of landfill space left. We may learn, too, that means and ends are not so easily disentangled as we had assumed.

When I talk about learning from nature, I mainly mean living nature, but of course we learn from nonliving nature, too. Living nature has the advantage of a built-in teleology: natural selection compels certain ends, to which it picks out efficacious if often outlandish means. But learning the laws of physics and chemistry allows us to make up our own means. And when we pass beyond narrow notions of means and ends into the realm of complex systems, we find striking homologies between living systems, nonliving systems, and systems that mix living and nonliving, even though the first are adaptive, the second are not, and the last may or may not be, depending on whom you ask.

All technology learns from nature. Even nuclear power might be described that way: we have not only the nonliving example of fusion in the sun and other stars, but the living example of a bacterial community that, concentrating natural uranium 235 in a streambed in Gabon some 1.8 billion years ago, built a fission reactor that ran for millions of years.

It is not as if we had been ignoring nature for millennia, then suddenly noticed this wallflower, asked her to dance, and learned all about her in the space of a waltz. We have been paying court to nature all along, in our own peculiar way. And each generation fancies itself the first to notice her, the first to understand her, the first to partner her in perfect sympathy.

But we are never perfectly in step. Throughout history people have thought they were following nature: what was essential in nature, what made nature tick. Often, though, it was precisely the stuff they discarded—the stuff that didn't seem to work—that was crucial to nature's long-term workings. What was farming, after all, but a simplified copy of wild fecundity? Humans picked out what worked—seed, soil, water, sun—and set it to work for their own purposes. It did work, sometimes. But the copy was too crude. It was a line drawing, a cartoon.

The first adumbration and crude sketch is filled in by degrees, shaded, crosshatched; new details emerge. Then a new pattern leaps out at us, and we emphasize it in bold black strokes—so bold, perhaps, that some once-familiar details are obscured. This *miserlou*-like dance—two steps forward, one step back—has its advantages. For our rude cartoons of nature are also abstractions; and if these are a source of our ecological woes, they are also a source of our scientific triumphs. Rote imitation of nature makes for lame technology. As Mumford notes, it is not for his mechanical duck that

Jacques Vaucanson is remembered, but for his loom. Each abstraction is a falsification, yet abstraction is how we really get a jump on the game. Abstraction is at once the genius of modern science and its demon.

THE LANGUAGE OF BABEL

Let us think again of the Tower, the first Tower. Though God may have succeeded in knocking that one down, others would spring up like weeds. The confounding of tongues was useless, for God forgot to remove the one language that would unite all peoples in a single project. A clay tablet from Babylon dated to early in the second millennium B.C. shows a clear understanding of what would come to be called the Pythagorean Theorem. At about the same time, the Egyptians were using a triangle of knotted rope three by four by five to establish a right angle when building temples and pyramids. The Hindu sulvasutras commended similar methods, using ropes of various lengths. In China, the theorem was stated explicitly in a section of the *Chou Pei* that may antedate Pythagoras.

Mathematics was a language that God could not confound, not without recalling and downscaling the human brain he had built. Though some part of this language might be lost, sooner or later it would be recovered. In the *Meno* an unlettered slave boy prompted by Socrates puzzles out for himself some basic facts about squares and triangles. In another hour he would no doubt have reinvented the Pythagorean Theorem. To do so he would not have had to read cuneiform, any more than an American schoolgirl needs to read Greek.

Abstraction connects on one side to mathematics, the language of the Tower. On the other side it leads to one of the great ambitions of the Tower—namely, universalism. Earlier I spoke of the way modern technology imposes itself in different parts of the world, as if a machine or a chemical that works in Ohio must work just as well as in Zaire. Since the constants and coefficients in the engineer's tables do not change as he flies over the ocean, he assumes that everything else is pretty much constant, too. Many threads from many cultures have come together to inspire this confidence, among them Greek geometry, Roman imperialism, Christian evangelism, and European capitalism. Last and not least is the American anticulture. As we have seen, it was only in America, with its myth of a direct relation to nature, that the logic of universalism could attain its end. By taking a satellite's-eye view of the planet, the Planet Manager places himself above any particular culture and above any particular part of nature. That is a very American thing to do.

So far I have not made much of an effort to distinguish between science and technology. From a modern point of view, technology is just applied science. A new technology is, in a sense, an experiment in which various scientific theories are put to the test. By scientific standards it is usually a bad experiment, since it takes place in the real world, where a slew of messy variables cannot be controlled. But that is just what makes it valuable. A scientific theory—an abstraction—may be perfectly correct as far as it goes, sailing with flying colors through one well-controlled experiment after another. But the ill-controlled experiment that is technology may show that the theory has failed to take account of other things going on in the real world. A technology may achieve for a while the narrow end for which it was intended, yet gum up nature's works in such a way that in time it fails—as when pests and pathogens develop resistance to pesticides and antibiotics. Or a technology may achieve one end, but unexpectedly frustrate several others. A classic example of that is the way certain pesticides work their way through the food chain, leading by stealthy degrees from a jolly harvest to a silent spring.

You can learn a lot about a person's approach to nature by figuring out how much and what kind of abstraction he favors. Planet Managers are confident of their ability to seize upon the main rules by which nature works and use them for the good of all. They are confident, too, that the errors of one abstraction will be corrected by the next. They are very fond of mathematical models. Many are enthusiasts of the new sciences of complexity, which show how intricate systems can be generated from a few simple rules. On computer screens all over the world they are simulating the behavior of ecosystems, the human immune system, the brain itself. At last the veil of chaos that obscured living systems is being torn away, revealing a new world of mathematically specifiable order, chaos, and near chaos. Convinced that they are finally seeing nature face to face, the Managers now feel they are in a position to manage the planet properly. In other words, they feel much as their predecessors felt a century ago.

At the far end of the spectrum from the Managers, the Fetishers want to copy nature as faithfully as they can, right down to the last leaf vein. They would interfere with nonhuman systems as little as possible, and would model all human systems on nonhuman ones (or on those of indigenous peoples, which are assumed to be modeled on nonhuman ones). Thus hunting and gathering would be the preferred mode of subsistence; if gardens proved necessary, they would copy the structure and composition of the surrounding wilderness.

You might think the Managers are hardheaded pragmatists, the Fetishers wide-eyed mystics. In truth, the wish to penetrate to the bare bones of nature, to grasp the few abstract laws that support its multifarious and gaily feathered surface, is mysticism of a very deep order. Many great mathematicians and scientists—Pascal, Newton, Faraday, Pauli, even that Descartes who gets blamed dozens of times a day for parting thought from emotion, mind from body, and man from nature—were mystics. If mystics are people whose aching for wholeness cannot be eased in gardens, myths, or suburbs but insists on a deeper quietus, then mathematicians and scientists may be the wildest mystics of all.

The Fetishers might seem down-to-earth by comparison, except that their rejection of abstraction is itself driven by abstraction of a particularly naïve kind, involving notions of oneness and harmony that are untroubled by facts on the ground. In reality, the animals and "natural" humans they admire struggle with their surroundings, and succeed in changing them, far more than the Fetishers imagine.

Nevertheless, we can learn a great deal from indigenous peoples about how to learn from nature. They are pickier, more discriminating pupils than is usually supposed. Let us turn to them now for some early lessons in Earth Jazz.

24

THE WILD GARDEN

THE AMERICAN BELIEF in a direct relation to nature, described in an earlier chapter, was premised on a vision of the people who were here already. If the Indian was Adam, the white man would be a smarter and more successful Adam.

As it happens, that vision was wrong. The Indian was no Adam. The native relation to nature was complex and deeply cultural, varying greatly from tribe to tribe. America was not a wilderness when white people arrived; it was a humanized landscape, though one humanized far more subtly than Europe had been.

William Bartram's rambles in the Southeast have been cited as proof of the natural paradise that Europeans found when they got here. But much of the "wilderness" Bartram traversed had been inhabited or hunted in for twelve thousand years. The "verdant swelling knolls" were almost certainly the product of regular burning by the Indians, a practice that thinned the forest and opened up habitat for deer, birds, berries, and other things the Indians liked. This was conscious husbandry: so much so that a Spanish explorer who spoke with the Indians got the impression the deer were domesticated. Although the strawberries Bartram trampled were native, they were not a great deal wilder than commercial Maine blueberries, which are likewise stimulated by regular burning. (They were somewhat wilder, since the Indians did not use herbicides.) Most likely the trout Bartram ate were truly wild and unmanaged, for by a quirk of history most of the Southeastern Indians did not care for fish. But the "wild oranges" in whose juice he stewed them were not wild at all but, as it were, feral— that is, escaped from cultivation. Native to South Asia, they had come to America with the conquistadors.

The land the white settlers espoused was not virgin; rather, it was (in Francis Jennings's phrase) widowed. The manner of its native husbanding was hard to see not only because it was subtle, but because musket balls and microbes had felled or expelled most of the husbanders.

The Planet Fetishers who now idolize the Indians have much the same view of them as the first white settlers had—a racist view. The Indians (the settlers held) were children of nature. The land they had lived on for millennia still counted as wilderness and so could be claimed without qualm. It was "unimproved"; the whites would improve it. (Never mind that most of the first white settlements were on clearings the Indians had farmed for centuries.) The Indians had no history, or if they did it was "natural history." Their skeletons and arrowheads belonged in the same museum as the taxidermied buffalo.

"A curious thing about the Spirit of Place," wrote D. H. Lawrence, "is the fact that no place exerts its full influence upon a newcomer until the old inhabitant is dead or absorbed." The spirit of native America, which is the living spirit of the dead-or-absorbed native American, is a mantle that many white intellectuals like to put on, not least those who write about nature. The Planet Fetishers make free with native-American myths and rituals, but only those fitting their image of indigenes as invisible in the woods as fleas in fur. The Planet Managers respond that the Indians, like all other humans and all other organisms, changed their environment as much as their technology let them. Nature was sacred to the Indians, say the Fetishers; was something they made use of, say the Managers. As usual, the truth hides somewhere between these two positions.

A Tuscarora Indian named Richard Hill has observed that, for the Indian, the whole universe was civilized; if anything, man was its least civilized citizen. "One of our chiefs said it best: 'The West wasn't wild until the white man got there.' "

The Indians of North America were not ecological virgins. Their reverence for nature encompassed themselves and their needs. In a way Blake might have admired, they re-created nature in vision and action, guided by a sense that everything that lives is holy. Though they had their sacred places, they felt no need to set aside vast tracts of land as wilderness preserves, for they felt themselves to be part of the wilderness. They had few qualms about starting fires to drive game, open up habitat, or clear land for planting; it has even been claimed that the grasslands of the Midwest were their creation. At times they surely went too far: their ancestors, as I mentioned earlier, may have hunted the continent's megafauna to extinction, and some post-Columbian Indian groups equipped with Western tools (rifles, horses, snowmobiles) have engaged in comparable excesses. But these are exceptions. The point is that the Indians, far from keeping their hands off the ecosystem, were able to use it creatively without destroying it. They may even have helped it evolve in ways helpful to themselves. They did all this at population densities much higher than we generally suppose (like physicists who cannot observe a particle's state without changing it, European observers brought war and disease that erased the natives before they could be counted). After as much as twenty-four thousand years of their living in it, the land they lived in still looked brand-new.

At least, to Western eyes it looked brand-new. The same kind of misprision took place wherever the riders of one wave of change fetched up against the flotsam of an earlier one. Only gradually have Europeans come to see gardens and fields where they first saw jungle. Only in the last few decades have we begun to grasp the remarkable cunning and tact by which these gardens make nature their partner. In this chapter we will look at some of what has been learned about these non-Western gardens, and what might yet be learned from them about our dealings with nature.

GARDENING THE AMAZON

Anthropology is peculiar among the sciences in that its most out-of-date books are generally its best. The first, stumbling practitioners were the last to get a good look at the objects of study. It is as if the solar system had begun to disintegrate during the lifetimes of Galileo and Kepler and by Einstein's time consisted of the earth, the sun, and a cloud of dust.

Unfortunately, the first anthropologists did not take "primitive" farming very seriously. Its rituals, its "magic," the social roles it involved, the strange beliefs it seemed to imply—these were intriguing. Its efficacy, though, was presumed to be negligible when compared with that of modern agriculture.

For example, the most common agriculture of the rain forest, swidden or shifting cultivation, was long regarded as crude and wasteful. Its common name, "slash-and-burn," became a synonym for shortsighted mayhem. Only in the 1950s and 1960s, with the work of Conklin and Rappaport, did ethnographers begin to see just how deft a method swidden could be. New subtleties are even now coming to light, notably in Darryl Posey's studies of the Gorotire Kayapo.

The Kayapo are a fierce and ancient people who once roamed and ruled a portion of the Amazon basin as large as France. Though their lands have been eaten away by ranches and plantations, they now live in a proposed reserve of some five million acres, which includes grassland and savanna as well as rain forest. A few generations ago, the Kayapo were seminomadic; the entire tribe would trek for six to eight months, relying only on wild foods. Though they are now settled in villages, they still go on frequent hunting-and-foraging trips—including treks of two or three weeks—and spend several months of each year living in Brazil-nut groves.

In clearing a plot of forest for a garden, the men in each family fell the largest trees standing near the center of the plot in such a way that they topple outward, bringing smaller trees down with them. The result is a circle of just under an acre, constellated like a wheel. Great tree trunks radiate from the center: toward the circumference is a tangle of branches and leafage.

While all this biomass is baking in the sun, getting ready for burning, the women find their way through the outer tangle and into the open lanes between the trunks. Here they do something that textbook slash-and-burn gardeners do not do: they plant about a quarter of their root crops *before the burn*. When the fire does come, the root systems of these yams, sweet potatoes, taro, and manioc will be ready to suck up the flush of nutrients it lets loose. Moreover, they will have a jump on the weed seeds that will also want a share of that bonanza.

A slow burn is the ideal. Moving from one pile of debris to the next, the Kayapo may take the better part of a day to burn a single plot. This way, the heat is kept low and the roots of the crops already planted are not damaged.

A few days later, when the ashes have cooled, the women plant the rest of the root crops. A week or so after that, the men gather the branches and twigs that have not been thoroughly burned, make piles, and set them alight. In the ashes the women plant beans, squash, melons, and other plants that are particularly hungry for nutrients. The staggered planting helps ensure a staggered harvest—a good thing where food storage is difficult.

Now the plot looks less like a wheel than like an archery target. Allowing for variations that take advantage of the plot's various soil types, the crops are mainly arranged in concentric rings. In the outer ring, which is richest in nutrients because it is where most of the foliage fell, papaya, bananas, cotton, urucu, tobacco, and beans thrive. The next ring is manioc, the next corn and rice. In the center are sweet potatoes and yams.

After a few seasons, this garden will no longer be planted. But it will not be "abandoned"—not in the textbook sense. True, it will be allowed to revert to forest. But it will remain useful to the Kayapo for decades. Though corn and rice disappear, sweet potatoes and yams keep bearing for four or five years, bananas and urucu for eight to twelve years, kupa for thirty or forty years. The volunteer plants that begin the process of succession include fruit trees, palms, and medicinal herbs. They also include berries that attract birds and other wildlife the Kayapo like to hunt.

Once they are planted, the gardens will pretty much go of themselves. They have few pest problems, mainly because they are so small and widely scattered—in time and space—that large concentrations of pests can't build up. They can be left alone for months, and at later stages for years, which means they are well suited to a seminomadic (or hemiseminomadic) lifestyle. They can be visited and picked from during treks.

These gardens give high yields for very little work. In the balance sheet of calories invested against calories returned, they are triumphantly in the black: far more so than most modern fields. They also improve the soil of the rain forest. In fact, the "Indian black soil" found in certain places is aptly named, since it may well be an Indian creation. This, too, contradicts the textbooks, which tell us that swidden gardens lose their fertility after a year or two.

But then a question arises. If old plots are still fertile, why don't the Kayapo keep planting them? Why walk three or four hours to new gardens when you can replant old gardens that are just fifteen minutes away? In three or four years, admittedly, the nutrients released by the burn would be used up. But then why not do a new burn in a ten-year-old garden, instead of waiting twenty years as the Kayapo do?

The reason has already been hinted at. Old fields that are returning to forest are full of plants directly useful to the Kayapo, as well as berries, fruits, and browse that attract birds and mammals. The latter factor may be the key. For all its abundance, the rain forest is not rich in meat. (That is why almost no pure hunter-gatherers live there.) The more widely the old gardens are dispersed, the greater the pool of game on which they can draw.

RESOURCE ISLANDS

The Kayapo have learned to play the forest's own game, and win. Yet the forest does not lose. Though parts of the ecosystem are changed to meet the people's needs, the changes are subtle—so subtle that Western eyes can hardly detect them. The structural principles of the forest are respected, and the ecosystem as a whole keeps its integrity.

The contrast with modern agriculture, as practiced in the ranches and plantations that are carved out of the Kayapo's ancient lands, could not be more striking. Instead of an acre, thousands of acres are cleared at once. Deprived of the forest's awning, the fragile soil is baked by the sun. Organic matter breaks down rapidly and is soon leached or washed away by the heavy rains. The rains also wash away the soil itself, or pound it until it is hard as brick. In the space of a few years, forest has turned to desert.

Subtle as their swidden gardens are, the Kayapo have even subtler ways of playing the forest's game. Along their ancient paths through the forest, often near streamside campsites worn flat and hard with centuries of use, are patches of forest preternaturally rich in food plants. They did not get that way naturally (or, rather, they did, if one grants that humans are part of nature). In these places, roots, tubers, stalks, and fruits foraged from the forest nearby have been replanted to form "resource islands." For a Kayapo it is second nature to replant an unfinished bit of food near where he shits.

Nor is it only the forest's game that the Kayapo play. In the *campo* (grassland) and *cerrado* (savanna) that are part of their range, and where for reasons of health they like to site their villages, there are islands of forest known as *apete*. These are much more common near the villages than elsewhere. At first glance, these patches of forest seem natural. Only recently have anthropologists caught on that some three-quarters of them are manmade. The Kayapo make them by building compost piles from branches and leaves, inoculating the compost with bits of ant and termite nests, planting trees they find especially useful, and then allowing "natural" afforestation to take over. After a few decades, this process can result in an *apete* as big as ten acres.

When Posey and a colleague collected 140 species of plants from an *apete* near Gorotire, they learned that 138 of them were considered useful by the Kayapo, and 84 had been deliberately planted. Besides serving as supermarkets, the islands are used as shelters in time of war and epidemic, as refuges from the midday sun, as studios for body painting, as playgrounds, and as motels for trysting lovers.

Formerly, it was believed that the only way indigenous peoples managed the savanna was by burning, to keep it open and encourage the growth of fresh grass. The Kayapo do burn the *campo*, and one reason is to get fresh grass that will draw game. The other reason, though, is not to discourage trees, but to encourage the growth and fruiting of certain fire-loving trees.

Naturally, the Kayapo's ways of playing with nature have stimulated new thought about the beginnings of farming. Most scholars have assumed that people would have to be settled in one place before the domestication of plants could get started. But the Kayapo (and other peoples lately studied) show us forms of semidomestication that mesh deftly with a seminomadic way of life. Indeed, if you could shield your eyes from the slash-and-burn farming on which they have lately come to depend more heavily, and look only at their ways of moving and manipulating "wild" plants, you might think you had found the missing link between gathering and gardening.

The irony is dense. The Amazon basin, the one place on earth where nature and culture are most fiercely at odds, is also the place where the distinction between them comes to seem a fiction, thin as mist. But many other forms of wild gardening can be found in many other parts of the world. In western Java, for example—in contrast to Amazonia one of the more densely populated rural areas in the world—a form of slash-and-burn gardening known as *kebun-talun* is practiced, which yields cash crops as well as food and other items for the family's needs. Instead of native forest, the climax of succession is an imitation of the natural forest using only the most useful trees. Nevertheless, the stages of *kebun-talun* do mimic the stages of forest succession, and in particular the spatial structure of those stages, whereby sun, water, and nutrients are used in the most efficient way.

Or consider the home garden in upland Guatemala described by the late economic botanist Edgar Anderson, "a small affair about the size of a small city lot in the United States."

> It was covered with a riotous growth so luxurious and so apparently planless that any ordinary American or European visitor, accustomed to the puritanical primness of north European gardens, would have supposed (if he even chanced to realize that it was a garden) that it must be a deserted one. Yet when I went through it carefully I could find no plants which were not useful to the owner in one way or another. There were no noxious weeds, the return per man-hour of effort was apparently high, and I came away feeling that as an experienced vegetable gardener ... I had gotten more new ideas about growing vegetables than from visiting any other garden anywhere.

He goes on to list some of the plants: annona, cherimoya, avocado, peach, quince, plum, fig, coffee, giant cactus (for fruit), rosemary, rue, poinsettia, rose, hawthorn (for jam), corn, banana (the leaves used for wrapping paper and hot tamales), string bean, rampant squash and pumpkin, perennial chayote that "draped itself over everything, garden walls, trees, mature cornstalks, making the whole garden into picturesque bower." The place was "a vegetable garden, an orchard, a medicinal garden, a dump heap, a compost heap, and a beeyard" all rolled into one. Despite a steep slope, there was no erosion to speak of, and the mingling of plants meant that pests and diseases could not easily spread from one member of a target species to another member of the same species.

Like the Kayapo gardens, home gardens such as this seem to yield, among many other fruits, speculations about the origins of farming. In Anderson's case, the speculation has to do with the possible role of dump heaps on which ancient people might have tossed uneaten seeds or bits of tuber, or where volunteer weeds might have made themselves at home. He links this to Carl Sauer's theory that farming began among sedentary fisherfolk; while Posey, considering the Kayapo, concludes that farming could have started among wandering hunter-gatherers. But never mind that. What I want to point out is the odd circumstance that the wildest and most "primitive" gardens (even Anderson uses the word), gardens that conjure up visions of the first gardens on earth, are also the most advanced gardens on earth. They have neatly solved—or should I say, sloppily solved?—many of the problems that plague modern farmers and gardeners to this day.

GARDENING THE DESERT

From the examples I have given so far, you might conclude that wild gardening is feasible only in places like the humid tropics, where nature is so voluble that people need only edit her slightly to get what they need. But forms of wild gardening have also evolved in places where nature is tight-lipped.

In the Sonoran Desert of Mexico and Arizona, rainfall averages from five to ten inches a year. There are few springs. Until recently there were few wells. Summer temperatures often approach 120 degrees F. Creosote, cactus, and saltbush cling to the threadbare soil. In 1775, the Franciscan missionary Pedro Font reported: "In all this land of Papaguería which we passed through, I did not see a single thing worthy of praise."

Given that what little rain does fall may be dumped in the space of a few hours—when, instead of soaking in, much of it sweeps into arroyos and out of sight—you might think rain-fed farming would be out of the question. Yet the Papago Indians managed, long before the coming of Font and his friends, to devise a kind of agriculture that (combined with skillful hunting and gathering) allowed them to keep body and soul together in the teeth of the desert. Although largely displaced by government rations and deep government wells, the kind of farming I am talking about survives in scraps and patches to this day.

The Papago's trick was to take one of local nature's least helpful habits— that of pouring out much of the year's rain in a few quick spasms, faster than the ground could absorb it—and turn it to advantage. They would take the runoff from thousands of acres of desert and use it to irrigate a few acres of crops. To this end, low brush dikes were spread out like the walls of a funnel to draw the floodwaters in. Shallow ditches or embankments were used to guide the water to keep it in place long enough for the earth to drink it down. The efficiency of this system was vastly multiplied by canny placement of the fields and their catchments. During floods the runoff of the mountains would collect in streams that, as the land flattened out, would spread out in sheets, and it was here the Papago put their fields. Since the soil of these "arroyo mouths" was alluvial—dumped by previous floods—it was both richer and better able to hold moisture than surrounding soils.

Clever as this setup was, it was not one in which most crops would thrive. But Papago crops, like their wild desert cousins, were inspired opportunists, ready at a moment's notice to leap through the narrow window that a flash flood opened up. A prime example is the tepary (*Phaseolus acutifolius*), a bean whose small seeds come in white, brown, beige, and rust. (The word "Papago" is thought to be a contraction of *papavi kuadam*, or "tepary eaters.") Of all cultivated beans, only the tepary is known to yield seed-packed pods at temperatures above 105 degrees F. Its roots dig down twice as deep in search of water as the roots of ordinary kidney beans. When teparies find water, they drink it up fast. Instead of shrinking away from the midday sun as, say, pinto beans plunked down in the desert will do, they stretch their leaflets out and let them bask. Instead of closing their stomata, or pores, in the heat of the day—in effect, shutting their metabolic doors and taking a siesta—they keep their stomata open and keep photosynthesizing, even though this means they will lose precious moisture.

At first this might seem odd: shouldn't a desert plant use water as slowly as possible? But the desert is not a place where a plant gets a little water every day; it is a place where an annual plant may get a lot of water all at once, and not a drop more for the rest of its life. The sensible thing, then, is to use the water when you have it, and get through your reproductive cycle as quickly as you can. Most desert annuals are ephemerals—annuals with a vengeance. The motto of annuals that I mentioned in an earlier chapter— "Eat, drink, and set seed, for tomorrow we die"—has for a desert annual an almost literal application. The ethnobotanist Gary Paul Nabhan has found that, under flash-flood field conditions, teparies outyield pinto beans a hundred to one.

Although you can grow teparies with groundwater, too—as most Sonorans do today—you have to be careful not to irrigate more than a couple of times, since they need some stress in order to fruit. Moreover, teparies grown with groundwater have less protein on average than teparies grown with runoff. One reason for this may be the rhizobia bacteria that wash in from nearby wild-bean populations, set up shop in the roots of the teparies, and convert nitrogen from the soil air into a form the plants can absorb. An even more crucial factor may be the mesquite leaves, twigs, rodent dung, and other nutrient-rich organic stuff carried by the floodwaters. Some of this sudsy, life-giving scum settles naturally on the fields; some is snagged by the low brush fences the Papago set up to guide the floodwaters, then gathered and plowed into the soil. Whereas virgin soil in the Sonora typically has an organic content of less than half a percent, an old Papago field may have an organic content of 5 percent, a figure even a Midwestern farmer would respect.

In the old days, the Papago were seminomadic, moving back and forth between the Wells and the Fields. They spent the dry (or rather drier) season in the Wells—villages in the hills, as close as possible to their country's few precious springs—where they lived by hunting and gathering. In summer they moved down to the floodwater fields, getting their drinking water from natural or man-made pools that the storms had filled. But as with other peoples we have spoken of, the distinction between farming and gathering was far from rigid. For it was not only cultivated crops that throve in floodwater fields. In these seasonal oases dozens of wild plants sprang up, among them purslane, devil's claw, ground cherry, finger gourds, leafy amaranth, and the wild melon and wild tobacco that carried the epithet "coyote"—their inferiority to cultivated forms being explained by the fact that Coyote, the trickster god, had shat upon them. Since the Papago lived on wild plants and animals more than half the year anyway, they did not reject these volunteers as weeds; they harvested them. Whenever possible, they did the same with the animals that came to eat the plants: gophers, jackrabbits, desert cottontails, blackbirds, quail, white-necked ravens. If these creatures were pests, they were also "field meat."

What the Papago did in the desert, then, was not so different from what the Kayapo did in the rain forest. But whereas the Kayapo could achieve their aim by merely shuffling the natural abundance of the forest, or by making holes in that abundance and guiding the way nature filled them in, the Papago faced a tougher task. They had to concentrate in space, and stretch out in time, the eruption of life that follows a desert rainstorm.

Even by Indian standards, the Papago were poor. They went hungry for long periods, and often had to hire themselves out as laborers (or dancers) for their cousins the Pima, who lived along the rivers. Yet their economy was one of abundance, with a constant circulation of food and other gifts. In modern times a higher standard of living has become possible for farmers in the Sonora, thanks to a rate of groundwater pumping one hundred times the rate of recharge. As the amount of fossil fuel needed to retrieve this fossil water may soon cost more than the crops are worth, scientists are now busily researching the possibilities of runoff agriculture. Unfortunately, the chance to learn something about this art from its past masters is rapidly slipping away. In 1913, the Papago farmed some ten thousand acres of floodwater fields; today the total is well under a hundred acres.

Anyone who doubts that desert floodwater farming can support an advanced civilization should consider an example from the opposite side of the world. In the southern desert of Israel and Jordan, the cities of the Nabateans flourished from the second century B.C. until the Muslim conquest in the seventh century A.D. Their early wealth came from a commanding position on the southern caravan route of the ancient spice trade. Yet even after Roman sailors broke their monopoly, sometime in the first century A.D., the Nabateans managed to go on living in more or less the style to which they had grown accustomed. The secret of their success seems to have been a system of runoff farming somewhat more elaborate than, but in principle no different from, that of the Papago.

SIGNPOSTS IN THE FOREST

To be sure, indigenous farming is not always good farming. Slash-and-burn, for example, is not always done as well as the Kayapo do it. Where population pressures are too great, fallow periods are often too short and patches too close together. In tropical Africa, the forest fallow has been largely abandoned in favor of a much shorter grass fallow. In such cases, the forest is destroyed more slowly but just as surely as it would be destroyed by Western farming. Nor are Western incursions always to blame. Misuse of swidden in Africa seems to have started in prehistory, shortly after the practice was introduced from Asia. Scientists have found that sickle-cell anemia occurs mainly in those parts of Africa where slash-and-burn farming is an ancient practice. The reason is that having a single, recessive gene for sickle-cell is a defense against malaria. Where malaria is common, the gene is favored, even though people who have the bad luck to have two of them become anemic. Malaria is common in places where bad swiddening has caused compaction of the soil, creating pools of standing water that are maternity wards for mosquitoes.

In his best-selling book written before he became vice president of the United States, Albert Gore speaks of our "dysfunctional civilization," and our compulsion to plunder the natural world because of our own inner poverty. Well, I will grant that our civilization is dysfunctional if he will grant that just about every other civilization has been, too. The label would also apply, I think, to most primitive farming societies that have existed, and even to most hunting societies—the few that have survived having survived because they were tougher and happier than most (and were tucked away in marginal habitats). Most animal populations are dysfunctional. How functional is a troop of langur monkeys in which a male will kill another male's infants in order to free the mother for his own seed? The practice does nothing to help the troop, the species, or nature at large.

"Function" is a word that should be used with great caution outside the realm of machines and individual organisms. Perhaps the societies of social insects might be said to "function," but elsewhere in the animal kingdom the behavior of populations is the sum of many conflicting forces, not the neat ticking of a wristwatch. In human societies, function is something won by trial and error, and with a good deal of "moral luck." You need the right leaders at the right time, the right storytellers and the right stories.

If "advanced" societies can be subject to moral luck—if an accident of weather or biography or the timing of a fuse can make the difference between war and peace, oppression and freedom—why should "primitive" societies be any different? If anything, their smaller size makes them even more subject to accident and the behavior of a few individuals. We should guard against the idea that traditional societies are self-regulating or automatic—that people just sit back and let tradition take care of all their problems. There is a constant need for energetic leadership, inventiveness, problem-solving. And there is always the chance that a quarrel or bad decision will leave a permanent scar or deformation.

We often hear it said of indigenous peoples that they have lived "in harmony" with nature for thousands of years. This may have been true of particular groups but never, I think, of the species as a whole. The few extant hunting-and-gathering peoples are spume from the first great tide of human expansion, cast up on rocks and isolated by later, more violent tides. Meek and beleaguered they seem to us now, yet with their fire and stone, their grasses and their game, they once swept across the globe like conquerors. (Swept slowly, of course, compared with their farmer successors, who in turn were slow compared with industrial humans.) The handful of groups that are still around are ones that adapted to marginal environments, such as deserts and rain forests, which were not terribly attractive to gardening or farming peoples. In such places, the margin of survival is slender. They do not stand much abuse.

Peoples that have lived in the same place for a long time without ruining it are not "natural." They are smart and they are lucky. And because they have lived in the same place for a long time, they have been able to fine-tune their dealings with nature. No primitive people that is still around today can really be primitive. All have thousands of years of trial and error under their belts. In many cases, they have had the same basic technology for centuries, which has allowed them to work out many of the kinks—the places where technology rubbed the wrong way against nature, or against people, or against itself. From this point of view, it is we who are primitive.

Note that I am not claiming any kind of strict reward and punishment on nature's part. The notion that civilizations fall when they abuse their environment is meant to have a Jeremian knell, yet in a way is rather comforting: it coddles our sense of justice and reassures us that nature has a fail-safe mechanism that will protect it no matter what we do. Alas, things are not so neat. Any civilization with an ounce of sense knows enough to start abusing neighboring environments when its own shows strain. In fact, it is often the societies that are least provident that are also most expansionist. Gresham's law casts its grim shadow here, too: bad husbandry drives out good. All the ecological virtues in the world will not avail you if you have an expansionist neighbor.

But let me come back to the main point: that non-Western societies have not always been models of ecological rectitude. The collapse of the Maya seven centuries before Cortés may have been caused largely by the felling of trees to fuel the fires in which they made lime stucco for their vast monuments. The volcanic highlands of central Mexico seem to have lost soil at least as rapidly to pre-Columbian farming as they would when the Spaniards brought the plow. An animistic sense of oneness with nature did not prevent the Maoris from deforesting much of New Zealand and clubbing into extinction its flightless moas. The much-admired nature religion of the North American Indians did not stop them from overhunting the buffalo as soon as they got horses and rifles. Nor did believing that Buddha-mind was in all things prevent Chinese monks from shaving the mountains to feed their funeral pyres. Asian medicine, with its shamanistic faith in animal powers, has brought the Siberian tiger, the Bengal tiger, the Asian bear, the black rhinoceros, and a host of other noble beasts to the brink of oblivion. Meanwhile, the same culture that gave us Zen ink drawings is erasing with quick strokes many of the world's last forests, importing almost four times as much timber as any other country.

Suppose we grant, then, that non-Western peoples are not always perfect. Surely we have much to learn from those who, like the Kayapo, do certain things supremely well. Why not take the wild garden as our model?

We may do so, but with a grain of salt. In its purest forms, the wild garden is well suited to people whose wants are modest, whose tools are simple, and who are thinly scattered across the face of the Mountain. Most of us do not fit that description. The wild garden takes us closer to Eden than we can wisely go.

Our way of learning from nature has to be more abstract; the systems we model on natural systems have to be more concentrated, less mixed up with the extant natural systems themselves. Since we can't trust ourselves to be as smart as the Kayapo—or, rather, as much smarter than the Kayapo as our greater numbers and power would require—we had better mess with wilderness as little as possible. Jumping back to our basic metaphor, we might put it this way: our kind of Earth Jazz has to have a harder edge. If the Kayapo play a kind of New Orleans shuffle, the parts seamlessly bobbing and weaving, what we have to play is more like hard bop. The ideal would be to use our solo space—farms, gardens, factories—as boldly and economically as the Sonny Rollins of the mid-1950s used his.

Even so, we have much to learn from the Kayapo and their peers. Their paths are not our paths; but in the green depths of their paths are signs that may point our way.

25

THE TREE OF LIFE

I F T H E K A Y A P O way of learning from nature is not exactly the way
for us, then what way is?

Let us stick, for the nonce, to the business of getting food from the
earth. Until someone figures out how to eat virtual pancakes, this will
remain the business most vital to human life. What is more, the alliance of
humans with various plants and animals—an alliance formed in the course
of this business—remains the instrument by which we have most drastically rearranged the face of the planet. (Our alliance with machines and
with dead organic matter is catching up, but it has not caught up yet.)

When modern agriculture learns from nature, the lessons it draws tend
to be as abstract and as universal as possible. More than most modern technologies, though, it is sharply constrained in this matter by its raw materials. Farming cannot just learn from nature; it has to take actual, living
chunks of it and actually use them. Crops are not inventions. They are
living things with ancient and very particular evolutionary histories that
reach back long before their marriage with humankind. However cozy
that marriage, the fact remains that they have a past. As a consequence of
that past, they like some soils and dislike others, resist some pests and cave
in to others. Though a plant or an animal can be abstracted from its natural ecosystem, many of its habits remain stubbornly intact. Breeding and
crossbreeding can alter them, but haltingly, in awkward, zigzag ways.

All this is changing. In the age of genetic engineering, a trait can be
abstracted from one organism and inserted into another. The two organisms need not be related; they need not even belong to the same kingdom.
A gene from a firefly can be (and has been) plugged into a tobacco plant.
Though agriculture still learns from nature and uses pieces of it, the key
pieces are much smaller and (in theory) more neatly manipulable.

Many agronomists have long wished for this sort of abstraction. Of
course, you have to be careful what you wish for, because you might just

get it. As we shall see, the danger here lies not so much in the inevitable mistakes of biotechnology—runaway viruses and suchlike fodder for Hollywood—as in its equally inevitable successes.

Universalism has entered a new phase. While dead matter has long obeyed our algebra and our calculus with some degree of alacrity, living things have been recalcitrant. But now that we have learned, like King Solomon, the hidden language of life itself, surely they will have no choice but to fall in line. The kind of technology that emerges at this point calls for special scrutiny, since it promises to twist the links between humans and nature into pretzels of unprecedented complexity. To do justice to the whole field of biotechnology, from plastic production by microbes to gene therapy, would take a whole bookshelf. In this chapter, I want only to look at some of the things biotechnology is doing or planning to do down on the farm. Is its way of learning from nature—high level of abstraction, itsy-bitsy scale—really the best way? Or might Earth Jazz suggest something better?

A FAT MAN'S SOUP SPOON

Before we can judge biotechnology on its merits, we have to address those critics who would reject it out of hand.

While most Planet Managers are happy to enlist biotechnology in the effort to save the planet—to clean up oil spills, to use farmland more efficiently, to check global warming or mute its effects—the Planet Fetishers think this a pact with the devil. For them, tampering with the genetic code is an abomination, and a deeply seductive one: once it gets rolling, there will be no stopping it. Drought-resistant crops may seem innocent enough, but stackable cows won't be far behind.

But if there is no way to draw the line once genetic engineering gets started, then neither is there any way to draw the line *before* it gets started. Conventional breeding has performed enormities very nearly as disgusting, when you think about them, as anything the genetic engineers or their publicists have dreamed up. Imagine plucking a peaceful ruminant out of the woodland and breeding its breasts to such painful fullness that they will suckle not only a calf, but whole villages. Think of changing a wolfish predator into a chihuahua. Think of sheep with tails so fat they have to be supported by small wagons. Think of seizing on a genetic defect in some doves, a dysfunction of the inner ear, and breeding races of tumbling pigeons just for sport. Think of pygmy horses, or the hairless, hypoallergenic "hot water bottle" cat.

As disgusting as these things are, unaided nature has done things as bad and worse. For a good frisson, I would rather have a zoologist at my camp-fire than Stephen King. Among thousands of parasites, pathogens, and predators my own favorite is the candiru of the Amazon, a small fish that follows the trail of waterborne excreta into the interior of a bigger fish or the penis of a human foolish enough to pee in the water. Its spiny scales make it impossible to dislodge except by amputation. A close runner-up is the African eye worm, which the victim can actually see as it burrows across his retina.

But I am being overly generous in giving "unaided" nature credit for all these inventions. Many of them coevolved with man, as did many of the sweeter things in nature.

Bill McKibben wonders what his feelings would be on encountering in the woods a hare that had been genetically engineered to weather global warming. "Why would we have any more reverence or affection for such a rabbit than we would for a Coke bottle?" By that logic, why should we have any more reverence or affection for cats or dogs or sheep, all of which are partly man-made? Annie Dillard has a good heartless passage about steers: "They're a human product like rayon. They're like a field of shoes. They have cast-iron shanks and tongues like foam insoles. You can't see through to their brains as you can with other animals; they have beef fat behind their eyes, beef stew."

And yet there is a remnant of wildness even in them. There is a health-ier remnant in cats, which is one reason why some of us worship them. Dil-lard again: "I used to have a cat, an old fighting tom, who would jump through the open window by my bed in the middle of the night and land on my chest. I'd half-awaken. He'd stick his skull under my nose and purr, stinking of urine and blood. Some nights he kneaded my bare chest with his front paws, powerfully, arching his back, as if sharpening his claws, or pummeling a mother for milk. And some mornings I'd wake in daylight to find my body covered with paw prints in blood; I looked as though I'd been painted with roses."

Just as genetic engineering can be seen as an outgrowth of selective breeding, so selective breeding can be seen as a form of coevolution, a purely "natural" phenomenon. Ants farm fungi and raise aphids, milking them for honeydew. As we saw earlier, the alliance between humans and various plants and animals is just one of many such alliances in the annals of life on earth.

For that matter, the exchange of genes between unrelated species is nothing new, either. Conjugation—not the grammatical exercise, but the bacterial version of sex—slangily pairs members of unlike species, passing genes between them through a tube. Plasmids, which are small bundles of DNA separate from the chromosomes, flow readily through these tubes; since they often carry the genes for resistance to antibiotics, resistance can spread quickly from one species to another. Bacteria can even absorb bits of DNA from their surroundings: perhaps the genetic remnants of a dead or dying creature, briefly preserved in clay.

Viruses—wriggling at the border between life and nonlife, they are no more than packets of DNA or RNA wrapped in protein—have even looser morals, swiping whole chunks of genetic matter from one host and passing them on to the next. They ferry genes from one species of bacteria to another. They snip genes from various animals and wedge them into human chromosomes.

Viruses have been splicing genes for billions of years, but in a fairly haphazard way. Far more systematic is *Agrobacterium tumefaciens*, the soil microbe that can cause crown galls in most vascular plants. This parasite uses genetic engineering to transform plant cells into model hosts, whether they want to be or not. Sneaking in through a lesion in the plant's outer membrane and multiplying in the spaces between the plant cells, *A. tumefaciens* sends out plasmids which splice part of their DNA into the chromosomes of nearby cells. Impelled by these new genetic marching orders, the cells go into a frenzy of wild growth and division, becoming, in effect, the vegetable equivalent of cancer cells. Wild and formless as this growth seems, it serves the purposes of the parasite admirably. Not only does the tumor shelter the bacterium from any defense the plant might try to mount, but the new DNA further programs the tumor cells to produce "opines," oddball amino acids which the plant cannot use but the bacterium thrives on.

Once programmed, the plant cells stay programmed. Even if the whole population of bacteria is wiped out by antibiotics, the tumor keeps growing (and pumping out opines)—the only known case of a disease that long outlasts its germ. What is more, the new genes are passed on to the infected plant's offspring in classic Mendelian fashion.

Imagine that a burglar has broken into your home, then hypnotized you into treating him as an honored guest: so honored, in fact, that you spend your household grocery money buying cartloads of a food that he dotes on and you can't abide. Imagine, further, that all your descendants will be genetically compelled to treat his descendants in the same fashion. If you can imagine this, you may be able to empathize with a plant that has crown-gall disease.

Apparently the first true genetic engineer, *A. tumefaciens* is still the best. When biotechnologists want to introduce a new gene into a plant as quickly and neatly as possible, they call on this microbe to do the job. They insert the desired gene into the appropriate plasmid, then sic the altered bacterium on the target plant. The plasmid goes right to work, unaware (so to speak) that instead of genes for tumor formation or opine production it is splicing in genes for frost resistance, or pesticide tolerance, or something else wholly foreign to its interests. In the same way, viruses are used to splice genes into the chromosomes of mammals. For that matter, the cloning vectors and restriction enzymes that are the everyday paste and scissors of biotechnology are themselves adopted or adapted from bacteria and viruses.

In short, genetic engineering cannot simply be rejected as an abomination, sight unseen. Far from being something new and obscene under the sun, it is an instance of learning from nature and using nature. It is indeed an awesome technology, but to single it out for proscription makes as much sense as taking away a fat man's soup spoon. We are going to have to be more careful with all our technologies, right down to the simplest and oldest. Fire is now ending more nature, reckoned in hectares or in species, than all our high technologies put together.

There are many reasons to be especially careful with biotechnology. They have less to do, though, with some demonic quiddity of the science than with the institutional forces that drive it. They have a lot to do, too, with the scale on which this science attempts to learn from and work with nature: namely, the molecular scale.

CUT AND PASTE

Naturally, the one problem that has gotten almost all the media attention is the risk of creating monsters—unhealthy and unhappy monsters like the stackable cow, or monsters that may run amok. This is a risk (indeed, a reality) with conventional breeding as well. But the quick-fix mind-set of the scientists and the get-rich-quick mind-set of the corporations involved in biotechnology make the risk far greater.

Yet the real danger, as I said, lies not in biotechnology's mistakes but in its successes. Being far better equipped to deal with individual genes than with the relationships between genes—or between the organism and its environment—genetic engineers tend to take the path of least resistance, which happens also to be the path of most profit. The approach is modular. Not only interactions between the plant and its environment, but even interactions between genes get left on the cutting-room floor.

In the real world, most traits are the joint handiwork of several genes, including regulator genes (which tell the others when and whether to kick in; and suppress vestigial genes so that, for example, hens don't grow teeth) and episodal genes (which kick in at a strategic juncture to produce, say, a jolt of a certain growth hormone). After a splice has been sewn in, "an untold amount of breeding work remains before the genetic background is shaken down enough to accommodate the newly introduced gene and its regulators." And this is simply to make the gene fit the plant. How the altered plant will fit its environment is a horse of a different color.

No doubt biotechnology will solve some problems. Given the cut-and-paste approach and who is paying for the scissors, they are likely to be the wrong problems, solved in a Pyrrhic or at best Procrustean way. If a crop is threatened by weeds, the easiest thing to splice in is not the ability to repel them allelopathically, but resistance to an herbicide that kills the weeds; that way, farmers can spray to their hearts' content. It is also the profitable thing to splice in when, as often happens, the company funding the research also makes the herbicide.

A number of companies are also working on insect and disease resistance in various crops, which in theory should reduce the use of chemicals. But here again the modular approach may be troublesome. The easiest kind of resistance to splice in is not the kind controlled by several genes— the kind most often found in nature—which tends to provide partial protection against several strains of a pest or pathogen for a very long time. Instead, the easiest kind to splice in is the kind controlled by a single gene, which tends to provide complete protection against a single strain for a very short time: i.e., the time it takes the pest to evolve a resistance to the resistance. A Ping-Pong game between pest and biotechnologist is the likely result, exactly like the Ping-Pong game between pest and chemist that has been going on for years. And the farmer will pay for the balls.

ADAPTED TO A TWIG

Even if a particular wonder crop is inherently good, its proliferation is likely to be harmful, as it bulldozes under the genetic diversity needed to deal with changing conditions and mutating pests. The success of T-cytoplasm corn, a product of conventional breeding, was disastrous enough; the success of some future gene-spliced corn may be greater and so even more disastrous.

Every home gardener knows that one tree or shrub can be strangled by a pest while another—of the same species and variety, bought from the same nursery at the same time, and so close to the first that their leaves mingle—gets off scot-free. The obvious reason is that the second plant has a slight genetic edge in fending off this particular pest. An obverse and less obvious reason may be that this particular population of the pest has evolved genetically to fit that first, unfortunate plant. If black pineleaf scale insects are moved from one ponderosa pine to another, they die off. The same thing can happen if they are merely moved to a new branch of the same tree. In fact, the genetic variation between scale colonies on adjoining twigs can be so great that they are like mammalian populations ensconced in distant valleys, with mountain ranges flung between them. The difference in scale—both timewise and spacewise—between pest and host is so gigantic that for the pest a tree is a continent, oceans away from the next; the lifetime of the tree is a geological epoch. In the course of that epoch thousands of scale generations arise, each better tuned than the last (on average) to the "continent" or "valley" in which it lives.

What does all this have to do with the price of biotechnology? The key lies in the arms race between plant and pest. While the lumbering pine is genetically stuck with its broadsword and shield, the busy generations of scale insects can invent crossbows, harquebuses, and bazookas. The saving grace, from the arboreal point of view, is that various ponderosa pines' defenses are so various that most of these weapons work against one tree but not its neighbor. When, however, we fill our forests, orchards, and plantations with trees and shrubs that are genetically identical or nearly so, we are asking for a massacre.

There is a wider lesson here, too. Why does it pay for pests to specialize? Because the natural genetic variation between plants of the same species yields a wide variation in defenses. Our common field crops are too short-lived to make such specialization worth the trouble, but they, too, have (or had) wide variations in armamentarium from plant to plant. The more we replace that variety with standard-issue wonder crops, the more vulnerable our fields become to pests and diseases that are perfectly happy to spread from plant to plant—and to destroy them all. Once again, our penchant for the abstract and the universal—in this case, for the one ideal variety of a given crop, combining the best features of all the others—gets us into trouble. In a world as restless as this one, the abstract and the universal are sitting ducks.

At first glance, you might think genetic engineering would increase genetic diversity. But the tendency of twentieth-century plant breeding, bankrolled or lobbied for by agribusiness, has been to reduce genetic diversity, and there is no reason to think the tendency of biotechnology will be otherwise. Moreover, it is important to keep in mind that scientists have not been able to create any genes worth mentioning; so far, they have only juggled the ones nature provided. Once the genes are gone, the jugglers' hands will be empty. There will be nothing to do then but send in the clones.

Like all corporations, the corporations that bankroll biotechnology research care about the bottom line. They care about selling patented forms of life at a premium price. They care about selling chemicals. They care about bigger crop yields: purchased, in the grand manner of the Green Revolution, at great cost to fuel, soil, water, economic equity, and genetic diversity. The cost will be greatest in the Third World, where peasant farmers will continue to be driven off the land by capital-intensive plantations growing a single crop, often for export.

CHUNKS OF WHAT WORKS

In the 1970s, as the Green Revolution ran out of steam, a sort of vacuum became apparent in basic agricultural research. As capitalism abhors a vacuum, the molecular biologists have rushed or been pushed to fill it. Wes Jackson, a maverick geneticist who directs the Land Institute in Salina, Kansas, would like to see it filled instead with plant ecologists, evolutionary biologists, and population biologists, whose work until now has been "accumulating on the shelf more or less for its own sake." A test for telling the two groups apart is the question of whether "a crop plant should be regarded more as the property of the human or as a relative of wild things." Molecular biologists take the first view, which gives them license to manipulate the plant at will. Ecologists, evolutionists, and population biologists, by contrast, know that most of any crop's evolution took place before it became a crop. They admit that humans learn faster than nature, but insist that "nature is hard to beat because she has been accumulating information longer."

If you can't beat nature, you might as well join her. Her club is not exclusive. "The ecosystem level of biological organization is complex, much more complex than the DNA level of any species, but it is not necessarily more complicated for the human," Jackson writes. "If researchers and farmers take advantage of the natural integrities that have evolved over millions of years, they . . . will simply be dealing with huge chunks or blocks of what works."

In other words, the best scale on which to learn from and work with nature, at least where farming is concerned, is not the scale of the gene but that of the ecosystem. This is not an idle suggestion on Jackson's part, but something to which he has devoted the last twenty-odd years of his life, and to which he is prepared to devote the next fifty.

It may take that long. In essence, what Jackson and his associates at the Land Institute are trying to do is to cross a wheat field with a prairie.

Although wheat field and prairie are both, botanically, mainly grass, they are otherwise as unlike as any two things under the sun. The typical wheat field is plowed several times a year; by definition, the native prairie has never been plowed at all. While the wheat field wastes topsoil, the prairie creates it. Heat, frost, and drought often catch the wheat field flatfooted; the prairie, which has seen all these things before, takes them in stride. Left to its own devices, the wheat field is next to defenseless against pests and pathogens that, on the prairie, are naturally kept in check. Like the wheat field, the prairie is a sprawling factory for turning sunlight into fiber, starch, fat, and protein; but while the prairie relies on today's sunlight, the wheat field lives on mummified sunlight in the form of fertilizers and pesticides derived from oil and gas. The prairie lives on income, the wheat field on capital. What is more, the prairie saves.

From the point of view of the ecologist, the prairie is doing everything right. From the point of view of the hungry ecologist it has a serious flaw. What Jackson has in mind is a prairie that can feed the world.

Jackson likes to tell the story of the old Sioux Indian who watches a pioneer plowing up the prairie sod, stoops to examine the furrow, straightens up, and says, "Wrong side up." Traditionally, the joke is on the Indian.

A century after the Indian's remark, a third of our farmland topsoil, which took millennia to make, is gone. Soil erosion is only one of the sins of modern agriculture, which I have catalogued in Part One. But Jackson's quarrel is not just with modern farming, it is with all farming as we know it. He takes a long view, stretching back to the moment almost half a billion years ago when a bony planet began to clothe itself in soil. That process, which allowed intricate forms of life to wriggle up out of the oceans and plant themselves on land, began to reverse itself only ten thousand years ago with the arrival of a form of life called the farmer. It is reversing now at a roller-coaster, sick-in-the-pit-of-the-stomach rate, as if a film we had been watching all day should be run backward in seconds.

Jackson thinks the only real solution is to turn agriculture on its head. The field of waving grain he sees in his mind's eye would be plowed maybe once in five years. It would be a mixture, or polyculture, of three or four crops nourishing to humans or livestock. These crops would be bred for the purpose. Like many existing crops, they would yield at least eighteen hundred pounds of edible seed per acre. Unlike existing crops but like most prairie plants, they would be perennials. Their roots, undisturbed for years at a stretch, would hold the soil in place, and would become adept at finding moisture and sustenance. No more chemicals would be applied than nature applies to the prairie. Agronomists would call this field a "herbaceous seed-bearing perennial polyculture." With Jackson's permission, the rest of us would call it a "domestic prairie." Presumably the old Sioux would call it "right side up."

THE DOMESTIC PRAIRIE

There are good reasons to think it cannot be done. A plant has only so much energy to spend. Unlike such fictive entities as nations, it can't run a deficit; when it has earned what it can from the sun, there is no place to borrow. (In nature, that is: with man as broker, a plant can get a loan from long-dead plants in the form of fertilizer.) In view of this hard fact, there are two main schools of economic thought among plants. One advises throwing most of one's energy into seeds; the other gives precedence to roots. As an ecologist would put it, the first school recommends an r strategy, the second a K strategy. (The terms come from the basic ecological equation in which r is the intrinsic rate of a population's increase, K the highest stable population its territory can sustain. In other words, the r strategist trusts in sheer numbers of offspring, hoping that some will survive; the K strategist invests heavily in a few offspring, a number appropriate to the resources available, in the hope that all or most of them will survive.) Plants adopting the seed strategy won't live long—most will be annuals—but they may, with luck, have a lot of progeny. Plants with strong root systems will generally be perennials, able to survive drought, wind, frost, even fire; often they will reproduce, slowly but steadily, by budding from creeping rhizomes (underground stems) or roots. This is good policy on the prairie, where conditions can be extreme and competition tough.

On a level evolutionary playing field, untampered with by humans, seed strategists don't do well under prairie conditions. But seed strategists are opportunists, always on the lookout for disturbed soil and open patches. Such were the weeds that dogged the heels of early humans, some of which were domesticated as grains. By paying humans, in seeds, to husband their seeds and plant them each year in soil disturbed by the plow—in effect, giving humans a commission—they have managed to take over from the root strategists many of the prairie zones of the planet, including the Great Plains. Unluckily, the soil disturbance they thrive on has led to loss—at times slow and steady, at times cataclysmic—of the soil they and we depend

on. The great virtues of the prairie are the virtues of its roots. Breed the
prairie grasses for seed, and you lose those good roots. What is gained at
one end is lost at the other, and your high-yielding perennial turns out to
be an annual.

Things might work out that way. But biology is not a zero-sum game.
Sometimes an organism or collocation of organisms does manage, by dint
of greater efficiency or a nimbler juggling of resources and needs, to beat
the system. In her doctoral work at Cornell, Jackson's daughter Laura studied a newly discovered mutant of eastern gama grass, a tall prairie bunch-
grass that is a cousin of corn. She learned that while the mutant produces
four times as many seeds as its normal kin, it does not do so at the expense
of its root system or of any other essential function. This finding, Jackson
thinks, blows a gaping hole in old complacencies about the tradeoff
between r and K selection. At the Land Institute, attention is being given
not only to Laura's hopeful monster, but to other prairie grasses that are
relatively seed-rich; to grains that are relatively root-rich; to hybrids of
plants from both categories; to plants from the legume and composite
families, which fill out a natural prairie; and to polycultures combining
these promising plants in various ways. In short, Jackson is taking to
heart the advice of Henry Thoreau: "Would it not be well to consult with
Nature in the outset? for she is the most extensive and experienced planter
of us all."

HANGING TOUGH

There are realms of technology where things work once and for all. The
paper clip has not needed major redesign in nearly a century. The desktop
computer has been redesigned daily, but not by necessity: old models still
worked, it was only that new models worked better. Obsolescence is
planned, not a mandate of heaven.

But nothing works once and for all where living systems are concerned.
A pesticide or antibiotic loses its magic when the organisms at which it is
aimed develop an immunity to it, as sooner or later they must. Potatoes and
potato beetles, tomatoes and tomato hornworms, peaches and the crown-
gall bacterium are locked in an arms race of chemical measure and counter-
measure that can never end, except with the extinction (or retreat from the
field) of one party or the other. All natural systems must contend with
changes in climate, whether on small time scales or great ones.

From this, two consequences follow.

First, even clever imitation of nature is not enough in the long run. We
must keep trying to find out exactly how natural systems (and their man-

made adaptations) work, so that we can deal with new exigencies as they arise.

Second, we can never learn enough about how nature works to handle *all* new exigencies as they arise. Because the natural enemies of our adopted and adapted natural systems are always hatching new strategies—among them, methods of gene splicing that make our own tools look like stone hatchets—they will always be one step ahead of us. No matter how much knowledge we amass, we are always playing catch-up. If we try to manage each detail of the struggle, we are like a boxing coach who tries to prescribe his charge's every punch. Better to let him rely on his own instincts. Better to keep our living systems lithe and wily enough to block and counter whatever is thrown at them. The more wildness we leave in a system—the more diversity, the more genetic suppleness, the more ancient memories of fights lost and won in the ring of evolution—the better its chances of handling new challenges on its own. To use the language of Earth Jazz: in a world where constant improvisation is called for, it is foolish to try and write out everybody's part.

You can if you wish invoke nature's wisdom, nature's harmony, nature's benevolence. I would rather invoke nature's toughness, the toughness of the parts and of the whole: the toughness of the parts that has been earned through the struggle of the parts, the biting and gouging and poisoning and fraud, and the toughness of the whole that seems to be the net result of that fracas. Since our enemies are tough, it is good to have allies who are tough. If we coddle them with too much petrochemical food and weaken them like lapdogs with overbreeding—even if that breeding is for useful traits like yield of seed or milk—sooner or later nature will catch up with them, and with us.

The natural or cultivated toughness of our allies will not make them universally proof against all enemies and all mishap. Nor should it—for, as I said before, getting too far ahead of the game we play with nature may bring the game crashing down around our ears.

Take the "biological" control of crop pests. *Bacillus thuringiensis,* or Bt, a bacterium whose various strains prey selectively on beetles, flies, butterflies, and moths, is an almost ideal natural insecticide. It has few known ill effects on fish, birds, or mammals and it does not linger in the environment. Bt spores contain a crystal which, when activated by the host's digestive enzymes, paralyzes the gut and makes feeding impossible. The spores then germinate and proliferate, causing acute blood poisoning and releasing yet another toxin for good measure. To an assault so insidious and multifarious, it is hard to imagine how resistance could ever emerge; yet old-fashioned use (and overuse) of Bt spores has led to resistance in some

pests. Newfangled uses by biotechnologists—engineering plants to produce Bt toxins themselves, or engineering the spores to last longer in the
field, or isolating the toxins for direct application—will doubtless lead to
much wider resistance. When you spray Bt spores, you are learning from
nature in a very particular way. When you abstract the toxin and plug it into
a plant, you are learning from nature in a more abstract way. The effects are
likely to be very different. If you use the bacterial spores—assuming you
don't spray too often, and make some effort to vary the strains you use—a
fair number of your pests will survive. But that very fact makes it far less
likely that the pest will become resistant to the toxins. If, as seems likely,
resistance to Bt is a recessive trait, the surviving nonresistant pests will soon
spread their genes through the population.

Suppose, on the other hand, you plant crops that have been programmed to produce the poison. It will be produced constantly, in all the
plant's tissues; it will be active constantly, not just when activated by the
pest's stomach juices. Fewer nonresistant bugs will survive, and there will
be fewer (if any) respites in which their genes can wash back into the population. In a few years, not only will these wonder crops be useless; *Bacillus
thuringiensis* itself—a friend of organic farmers for decades, a friend of
plants for uncounted centuries—will be useless, too.

The inanity of the industry's approach is well (if in part unwittingly)
conveyed by a reporter for *The New York Times,* who writes of "the limitations inherent in biological control": "A microbe might work in one part of
the country but not another, or in one season but not another. . . . [B]iologi-
cal agents . . . do not bring about the dramatic kill that chemical pesticides
can; rather, they hold the pests within tolerable bounds. . . . Biological
agents have a big economic drawback as well: unlike chemical pesticides,
they do not control a broad spectrum of pests. A given agent works against
one pest on one crop and may require only one application. This creates a
narrow 'niche market' that many large manufacturers find unprofitable. . . .
Much of the resurgent research is aimed at overcoming these limitations."

From an ecological point of view, most of these "limitations" are
advantages. Should they be overcome, biological controls will at last be
as abstract and as universal as chemical controls, and as pernicious—
provoking resistance in target species, harming harmless or helpful non-
target species, and generally upsetting the balance of forces in the soil, air,
and water.

A better way of dealing with these limitations would be to try and
understand why they exist in nature: that is, in the wild organisms from
which our biological agents are recruited. In nature, pathogens and parasites often learn (by natural selection) not to be too virulent. They learn to

pull their punches. Otherwise, they may kill off their host population, or leave only hosts who have become wholly resistant and so are not at all hospitable. By the same token, predators rarely kill off a whole population of their major prey. When they come close to doing that, their own population crashes, giving the prey a chance to recover.

A "dramatic kill," then, is the last thing we should be aiming for. It may spur the development of resistance. It may leave the introduced predators with nothing to prey on, so they must either die off or fly away; then, if a new colony of the pest shows up, the farmer will have to buy a new batch of the predator. That may be good news for the supplier, but hardly for the farmer or the ecology of his field. By contrast, if pests are simply kept "within tolerable bounds," the predator can stick around and keep doing its job.

Understanding all this, we might not only accept the limitations of biological agents, but choose to impose further limitations on ourselves. For example, one way to keep Bt from outliving its usefulness too quickly would be to leave whole fields or parts of fields unsprayed. The unsprayed fields would serve as "refuges" for susceptible insects, from which the genes for susceptibility could spread back into the general population. As a bonus, they would serve as refuges for other predators of the pest in question.

Rather than seek a silver bullet for each problem, most scientists in the field (if not most in the laboratory) now commend a mixed bag of tricks: crop rotation, resistant crops, and the alternating use of various biological and chemical controls, each applied sparingly and at strategic moments in the life cycles of crop and pest. "Integrated pest management" is a vast improvement over the old chemical-pesticide approach—an approach not unlike that of the Crusader Simon de Monfort, who, on taking the Albigensian stronghold of Carcassone, ordered his men to "kill everyone: God will recognize his own." It falls far short, though, of the ideal proposed by people like Jackson: an agriculture in which not only chemical fertilizers and pesticides but even imported biological controls are beside the point, because the balance and vigor of a natural ecosystem has been attained. Though the full realization of that ideal may await the development of new crops and complexes of crops, we should not forget how largely it has already been realized in various forms of traditional farming and gardening.

WHAT WE LEARN from nature are mainly means to ends we choose for ourselves. Learning on a small scale, as molecular biology does, we are often blinkered to all but the narrowest means to narrow ends. But when we study nature on the larger scales—the ecosystem, the biosphere—some of our clenched desiderata come to seem less desirable. We feel suddenly like a child who has been pulling greedily at a lovely colored thread, only to find that his garment is unraveling. As we shall see, farming is not the only field of technology in which this realization may come upon us—or in which we may slowly be learning to reweave the fabric of our life.

26

THE TREE
OF KNOWLEDGE

WHY SHOULD THERE not be machines in the Garden? Leisure has always been a hallmark of paradise, and in ancient times the first squeakings of automation, such as water power, seemed to herald a new Golden Age. "Cease from grinding, ye women who toil at the mill," cried Antipater of Thessalonica, a poet of the first century B.C. "Sleep late even if the crowing cocks announce the dawn. For Demeter has ordered the Nymphs to perform the work of your hands, and they, leaping down on the top of the wheel, turn its axle which, with its revolving spokes, turns the heavy concave Nisyrian millstones. We taste again the joys of primitive life, learning to feast on the products of Demeter without labor."

Things worked out otherwise. For the masses who toiled at the mills or in the mines, the state of nature that the Industrial Revolution restored was not Rousseau's, but Hobbes's. Although things have improved in the industrial world since then, surveys now find leisure time, mysteriously, slipping away almost as quickly as meaningful work.

Whereas in some myths of the Golden Age, such as Plato's, people's wants are supplied "without toil," the Bible makes a point of Adam's charge to work and protect the garden. Earlier I gave this passage a metaphorical reading, but now I want to take it at face value. It may reflect a memory of Mesolithic gardening. It may echo the Mesopotamian notion that man was created as gardener to the gods (an inscription of Sargon the Great: "My service as gardener was pleasing to Ishtar, and I became king"). Or it may reflect the insight that leisure is meaningless without meaningful work. I assume that Adam's work was not busywork, or a hobby, or occupational therapy prescribed by the celestial therapist, but work that put food on the table.

The place of technology in the good life is not to eliminate work, or to proliferate layers of symbols and synthetics between the body and the earth. It is to restore the healthy balance of work and leisure that hunters and gatherers and early gardeners knew in good years, and prevent the fatigue and famine they knew in lean. Used judiciously, technology can help us follow the counsel of Lanza del Vasto: "Find the shortest, simplest way between the earth, the hands and the mouth."

EARLIER I RAILED against American males, complaining that they are able to relate to nature only through the mediation of a machine. (As soon as I typed the last sentence, my computer punished me by crashing.) This failing is neither exclusively male nor exclusively American. In fact, it may soon be universal. But anyone who wants to understand the state and prospects of technology in the world today must think hard about a number of American males, not all of them dead and not all of them white.

First, a step back. In the eighteenth and nineteenth centuries, the garden of Europe looked about to be crushed by industrialism. The Romantic reaction to this was expressed best, if belatedly, by Yeats:

> Locke sank into a swoon;
> The Garden died;
> God took the spinning-jenny
> Out of his side.

The problem of "the Machine in the Garden" has been picked out by the critic Leo Marx as the central American problem. But precisely because America was never a garden with a small "g" as Europe was—because it was seen as an Eden where nature could be met face to face—America's Romantics did not set their teeth so firmly against machines as did their European brothers. The seminal and somewhat promiscuous figure in America's philosophy of nature is Emerson, who inspired Thoreau and Muir on the one hand and legions of engineers on the other. "No thing shall be impossible to us," he wrote, "which we shall take upon ourselves to do"—echoing the engineers of Babel but hinting that, unlike them, we would have God on our side. Muir, who carried a rust-brown volume of *The Prose Works of Ralph Waldo Emerson* to the precipices of the High Sierra, absorbed none of its faith in progress. Thoreau was nearly as impermeable. "Instead of engineering for all America," Emerson said, "he was captain of a huckleberry party." But Thoreau, though never the booster Emerson was, was far from blind to the beauty of good technology. Going to work in his father's pencil factory, he built machines that rescued the business from an invasion of German imports. He could sit happily at the foot of a telegraph pole, savoring its spectral music made by the wind in the wires. Jefferson, of course, was an inventor (so was Muir as a boy, before a file slipped in his hand and nearly put out his eye), and the harsh view of industry expressed in his *Notes on the State of Virginia* was softened considerably in his later writings. Being Americans, the founders of our conservation ethic could hardly help having a soft spot for technology, any more than they could help reacting against it. This love-hate is at the heart of America's single greatest cultural product, the one I am offering as a model for our future. For a chorus or two, let me come back to jazz and ask what it has to teach us about technology.

THE GARDEN IN THE MACHINE

The war that freed the slaves freed Prometheus also. By the turn of the century, the piston's whump was known in every city: even in New Orleans, an island of Old World *bon temps* that was nevertheless a major port and host to a steam-powered navy. From its beginnings in that city, jazz has played with the rhythms of a mechanized society, tickling the works from within. Much American art in this century has tried to find a place for the machine in the Garden that America is supposed to be. Jazz has taken a different tack. It is the Garden, the wilderness, in the machine—at once a celebration of the machine age and a subversion of it.

It is no accident (or a very happy one) that jazz grew up alongside the phonograph. Jazz was the first music to depend on records for its popularity, its permanence, to a degree even its forms (the classic blues, for instance, bore the stamp of the ten-inch disc). With the help of a machine, an oral tradition of the lower Mississippi conquered the world. And it developed far more explosively and far more brilliantly than any oral tradition had a right to do.

The historian Thomas Hughes has likened the century of American scientific invention that began about 1870 to the efflorescences of Periclean Athens and Elizabethan England. In the heart of that era was an explosion of artistic invention that was no less astounding. America's new arts—jazz and film—were not wholly unrelated in spirit to the machines that propagated them. In each realm new forms of energy were born, new organizations of matter and mind; new rigidities and new freedoms. Edison, Chaplin, and Louis Armstrong were plumes of the same volcano.

Ushered into the machine age more abruptly than other Americans, African-Americans have perhaps felt more keenly both its constraints and its freedoms. They have perhaps understood it better, abetted by half-buried memories. Picasso, burning to express the monstrous energies of his age, turned to the carved masks of Africa. Armstrong had only to close his eyes.

Consider the saxophone, which after a shaky start became the dominant instrument in jazz, nudging aside the trumpet and unseating the clarinet. The clarinet was the first rational wind instrument, the first clear channel for the breath of the Enlightenment. No wonder Mozart loved it. Its passion is unruffled, smooth, but with hidden depths. It is the voice of a false modernity. Reborn as the saxophone, it became the voice of true modernity: cool as a sine wave, barbaric as a shaman's yowl. Like the phonograph, radio, or TV, it speaks in all voices. It can be an oboe, a clarinet, a violin, a cello. It can warble like a wood thrush or low like a bull moose; it can roar like a camp-meeting preacher, moan like a whore, or simper like a girl at first communion.

Think of syncopation. In European music it was an accentual device, a spice. In jazz it is meat and potatoes, just as it is in the ancient music of Africa. In the white mind encountering these rhythms for the first time, two images were conjured up. One was the image of throbbing drums and writhing savages ("primitive," African rhythms were called, when in fact their evolution made the rhythms of most European music look like tree shrews). The other was the image of churning pistons and clashing gears. Syncopation is the mother tongue of machines (that is, of machines not made to make music). And two machines can be far more independent rhythmically than two human beings. In the thick of mechanism, jazz finds a liberation.

Here, as elsewhere, jazz deals with the machine age far more neatly than European music has done, avoiding both the blind boosterism of the Futurists and the blinkered rusticity of a Delius or d'Indy. By the machine age, I do not just mean machines. As Hughes emphasizes, America's new machines would have been inert without other kinds of mechanisms to deploy them: power grids, assembly lines, means of social control. Indeed, Lewis Mumford noted long ago that the mechanizing of matter is as a rule preceded by the mechanizing of men. Riffs and breaks, ensembles and solos, syncopation and swing: all these are ways in which jazz plays with the problem of how to be an individual in a mechanized society.

In music, jazz did for the American landscape what George Inness's *Lackawanna Valley* did for it in art: reconciled it with the machine. But jazz's way of doing the job had more wit and more staying power. By locating the Garden in the machine rather than the other way around, jazz could make even the inner city an Eden. By viewing the landscape from the train (or bus or car), rather than the other way around, jazz could make the sleepiest hollow participate in the energies of the new century.

RISING FROM THE ASHES

Once, taking the elevated train to Canarsie, I saw stretched out below me a plant for processing scrap metal. The chutes and hoppers and crushers were set out in a touchingly linear way, so that the process could be followed from start to finish like a diagram. From high on the el it looked like a child's set of toys. Abruptly, the dump-truck-loving boy in me was jumping up with a yearning I hadn't felt in years.

A sight like that, I reflected, is not often seen by me or by other well-brought-up boys and girls. In the view of mature people, this is ugly stuff which ought to be kept out of sight, and which only people unfortunate enough to be taking the train to Canarsie should be forced to see.

What a shame. Here was a transparent view of a technology we all depend on. In this case, the technology happened to be a good one: a kind of recycling. Not only was this an ancient human activity, taking here its awesome and ungainly Industrial Age form. As well, it was a case of human participation in one of nature's deepest mysteries: the mystery of death and rebirth, or—to use a term we used earlier—of flow.

At the other end of the country, the mystery is at last being celebrated. Boys and girls are being taken to see, and participate in, the work of recycling on a scale and in a setting very nearly sublime. With an aptness that is almost embarrassing, this is happening in a city called Phoenix.

The facility I am speaking of seems to me a fine example of Earth Jazz, and especially of how to find the garden in the machine.

It all began inauspiciously enough with plans to build a solid-waste transfer station next to a 147-acre filled-up landfill on the southwestern edge of town. Hopes for the city's growing recycling program, though, inspired plans to include a recycling center that in time would handle 30 percent of the incoming trash. A penny-in-the-slot architectural firm was hired to spit out its standard design, which it duly did: a big box. With funds from the city's One Percent for Art program, two artists were hired to put a piece of art somewhere in the box's vicinity.

This was where City Hall made its fertile mistake. The artists in question, Linnea Glatt of Dallas and Michael Singer of Wilmington, Vermont, did not know their place. They failed to see the point of putting a piece of art next to a facility that no one would want to visit. Singer, a renowned environmental artist who also teaches architecture at MIT, gave the architects' plan to his students for review. They gave it back measled with red-pencil marks. To cite only one particularly fulsome bungle, the offices were placed downwind of the trash.

Armed with this critique, Singer and Glatt convinced the public-works authorities to let them earn their 1 percent by taking charge of the whole project. They worked with a team of architects and engineers to come up with a new design that would "overcome the 'not in my backyard' syndrome"—that would celebrate the infrastructure instead of hiding it. The resulting facility does just that, and (no trivial matter from the civic point of view) does it under budget.

The main building hangs from a huge, open steel truss that can be seen from the highway and functions as a kind of billboard, as if to say, "Nothing to hide—come and see." This device allows great flexibility in the use of the vast interior space, the size of two football fields, which can be partitioned and adjusted to meet changing needs. Truss, girders, and catwalks recall Frank Lloyd Wright's idea of the "building as machine," and the stepped, gray-green cement masonry is a conscious homage to his Arizona Biltmore Hotel, a Mayan fantasia on the other end of town. Everywhere, desert vines scale the ramparts. If this is a vast machine, or a monument to machinery, it is one that nature intends to reclaim: that is, to recycle.

On one side are desert and mountains, on the other side the skyline of downtown Phoenix. Behind this even-handedness is a gesture of deft irreverence: the structure's axis is skewed forty-five degrees from the rigid north-south, east-west grid on which the rest of the city is built. Traffic is carefully routed. Visitors and employees take the high road over the flank of the landfill, where they can savor the panoramic view. Dump trucks take the low road on the east side of the building. People hauling their own trash or recyclables drive down the middle, under a catwalk that leads from the parking lot to the visiting area. From this visiting area there are many views into the awesome interior, in which the work of transfer and recycling takes place, and which tinted glass bathes (as one critic said) "in a Gothic light." Appropriately, the best view is from a small outdoor amphitheater, a kind of chapel in which the visitor may seek atonement for the sins of gluttony and avarice, or think on the mystery of resurrection.

At the facility in Phoenix, schoolchildren drawn by the love of trucks will learn from nature the principle of flow. Here they may learn something about the limits of technology, maybe even about the limits of recycling. Can recycling redeem the original sin of greed, of despoiling Eden? Or will it prove to be an indulgence, in both the modern and the medieval sense? A building is mute and can give no answers. But it can ask the right questions. The place Singer and Glatt have made is a meditation on a set of interlocking themes: The good lands, badlands, and no-man's-lands between wilderness and civilization. The sublime in nature and the sublime in technology. The vastness of society and the task of the individual. The grid and the landscape. The Garden in the desert, the desert in the machine; the machine in the Garden, the Garden in the machine.

INDUSTRIAL ECOLOGY

Why all the fuss about machines in gardens? After all, the garden is already a place whose ecology has been transformed by human allies. Adding another set of allies whose sinews are steel, whose feed is coal or petroleum, just pushes things along a little further.

As with our other allies, we pretend that we control our machines utterly and that they can have no effect on the world that we do not intend. But like goats that turn forests to desert, they have a big swollen bag full of effects that we neither intend nor expect. They, too, can turn ecosystems upside down and send species packing; can allow one group of people to dominate another; can become parasites on humankind at large.

Of course, we like to think that in the garden (whether an actual garden or an idyllic landscape) we have not harmed nature but merely brightened her up a bit. We fancy we have achieved a balance—a balance which the machine, like a goat run loose in the garden, threatens to upset. Yet from nature's point of view the difference between the garden and its despoiler is at most one of degree.

Degree is important, though. The amounts of nitrogen and sulfur flowing through the global industrial system now equal or surpass the amounts naturally cycling through the living and nonliving parts of the biosphere. Industrial flows of lead, cadmium, zinc, arsenic, mercury, nickel, and vanadium are as much as eighteen times natural flows.

The global effects of these newish allies are all the more unsettling when we remember that they are only now spreading across the whole globe. Whereas the first great wave of human allies (mainly quadrupeds and perennial grasses) took millions of years to cover the earth, and the second wave (mainly quadrupeds and annual grasses) took thousands, this third wave seems poised to consolidate its holdings after just a couple of hundred.

When we find our allies having effects we never intended, our first impulse is to squeeze out of them whatever wildness is left. Wildness is the wild card: pull it out and the game will hold fewer surprises. This trick may work in the short run, but in the long run the surprises get bigger and nastier. A better strategy is just the reverse: to encourage the wildness in our allies, so that they behave more like their wild brethren, both singly and in aggregate. More exactly, they should behave less like the weedy, opportunistic, spendthrift things that human allies—and humans—tend to be, and more like the creatures they and we have been displacing. Our artificial systems should behave more like mature ecosystems.

When agriculture first takes hold in a place, it acts like an ecosystem in an early stage of succession. Energy and nutrients—stored in the soil, freed by fire, streaming unshaded from the sun—are used recklessly in the race for quick growth and reproduction, with little effort spent on durability, efficiency, or fine-tuned relations between species. Later, when local soil runs thin and virgin soil becomes harder to find, the ecosystem is forced to grow up. Various plants and animals evolve more intricate relations among them. By means of manuring, composting, and the like, a measure of the stability and efficiency of a mature ecosystem is achieved. Though a blast of cheap calories from fossil fuels has derailed this process during the past century, it may at last get back on track with the work of people like Jackson, who is mimicking the mature ecosystem of the prairie.

Powered by that same burst of cheap calories, our mechanical allies have been even weedier than their actually weedy counterparts. Drawing on the resources of the whole planet and using the whole planet as its dump, our industrial system has remained in a state so immature, ecologically speaking, that ecology has no word to do it justice. (Resorting to ontogeny, we might call it fetal: relying entirely on the maternal biosphere to feed it and process its wastes. But the industrial system is a very big fetus, and its wastes are very strange wastes.) As it spreads across the planet (in the process known as development), the system will have to grow up in a hurry, or else be dragged down by the ecological collapse it brings about.

Luckily, the system *is* starting to grow up. Inventors, industrial designers, and even managers are starting to see the advantages of making the industrial system work more like a mature ecosystem—or like the biosphere as a whole, which may be thought of as the oldest and most mature ecosystem of all. Chronicled and encouraged by such thinkers as Robert Frosch, Hardin Tibbs, Paul Hawken, and William McDonough, the movement has been called "biorealism" by some, "industrial ecology" by others.

The basic idea is simple. A mature ecosystem—and to a still greater degree, the biosphere as a whole—is mostly cyclical, using the same atoms over and over again. One creature's mire is another's meat; after many transformations, it may be the first creature's meat, too. Though energy cannot be cycled in quite that way, the biosphere does get a lot of bang for the buck; no matter how avidly one organism has popped the chemical bonds of an energy-rich compound, chances are some other organism will wring energy from the leavings. Not only does the biosphere live comfortably on its solar income, but it manages to put something aside for a rainy day: organic matter that geological chance has kept all energy from being wrung out of, and that humans call fossil fuel.

In dark contrast, the industrial system is linear: it takes in resources and puts out waste. The fact that some of the output is used by consumers before it becomes waste is irrelevant; from nature's point of view, there is no difference between plastic scraps dumped by a manufacturer and a plastic inflatable dinosaur that gets dumped a few years later. As for energy, the industrial system lives almost wholly on capital—the very capital that the biosphere has stored. Its use of energy would strike any self-respecting plant, animal, or microbe as criminally wasteful, with vast amounts lost in heat, noise, and other indiscretions.

Even our well-meant recycling programs—our Talmudic distinguishing of Number One plastic containers from Number Two, of bond from newsprint—affect only a small fraction of industrial inputs and outputs. The myth of resurrection celebrated in the Phoenix facility remains, for the most part, a myth. Though a few narrow paths double back, the industrial system remains a one-way street.

Industrial ecology aims to change all this by consciously modeling the industrial system on the biosphere. Not only would its functioning be vastly more efficient—and economical—but its relations with nature would be vastly friendlier. No longer a noxious exotic, the machine would become a mature and responsible member of nature's community.

Instead of one-dimensional recycling—paper pulped to make more (and lower-grade) paper, glass melted down to make more glass—there would be multidimensional recycling, just as there is in nature: a serried webwork in which almost any "waste" can take any number of turns and find any number of buyers. Asking an industry to eat its own waste is fine, when it works; much of the time, though, it makes more sense to ask that an industry's wastes be of such a nature, and be made available in such a form, that other industries will seek them out.

If you wanted to bet on where a modern industrial ecosystem would first evolve, the smart money would be on a place like Denmark—a small, densely settled, long-industrialized country with few natural resources and few places to stow its trash. As the immature, invasive wave of machines moves outward, it is here, near the crowded center, that maturity should first arise. And in fact it has arisen in the small city of Kalundborg, eighty miles west of Copenhagen.

An oil refinery called Statoil funnels by-products of its work to various neighbors: sulfur to a sulfuric-acid plant; heat to local greenhouses; gypsum to Gyproc, a wallboard plant; and gas, cooling water, and wastewater to Asnaes, a coal-fired electric utility. Asnaes in its turn sends waste steam to Statoil and to Novo Nordisk, a pharmaceutical plant; waste heat to greenhouses, aquaculture lagoons, and local homes; fly ash to cement-makers and road-maintenance crews, and gypsum to Gyproc. Finally, Novo Nordisk uses its sludge to make fertilizer.

To make sense of all this, you have to use the kind of flow chart, dense with boxes and arrows, that ecologists use to describe a mature ecosystem. When a group of grade-school students did so, the local captains of industry were taken aback, rather like Molière's bourgeois gentleman on learning that he has been speaking prose all his life. Kalundborg's "ecosystem" was not the product of a master plan from the Danish Environment Ministry. It had grown up piece by piece, arrow by arrow, to suit the self-interest of individual parties—which, come to think of it, is how a real ecosystem grows up.

Kalundborg's industrial ecosystem comes remarkably close to being a closed, self-contained system, but there is no reason why the arrows should not cross city limits, national borders, and even oceans. No natural ecosystem is closed—each exchanges water, carbon, nitrogen, oxygen, and other essentials with the biosphere as a whole—and no artificial ecosystem needs to be, either. Of course, exchanges between local industries cut down shipping costs and energy use, encourage self-reliance, and give citizens (like the schoolgirls of Kalundborg) a clearer sense of what is going where and why. But long-distance exchanges can be valuable, too, and sometimes nature can do the shipping: for instance, a net producer of carbon dioxide might be balanced by a net absorber somewhere else, such as the Italian agrochemicals group Ferruzi.

Since from nature's viewpoint there is no difference between a by-product and a product, products, too, must be drawn into the web. Unfortunately, the trend toward quickness and cheapness of assembly—for instance, bonding instead of using screws—means not only that products have to be junked rather than repaired, but that when they are junked it is impossible to salvage their parts for reuse or recycling. With the concept of "design for disassembly," this trend may finally be reversed. In Germany, new cars must now be put together in such a way as to be easy to take apart.

COPYING CHARLOTTE'S WEB

In the industrial ecosystem, the individual "organism" is the factory. On this scale, too, technology has much to learn from nature: first of all, about metabolism. Whereas most biological processes take place at ambient temperatures and pressures (or, in the case of warm-blooded animals, at temperatures a few degrees or tens of degrees higher), most industrial processes won't happen unless the ingredients are baked, squeezed, or both, by factors of hundreds or thousands. (In this respect we are still alchemists, except that now our ovens are stoked with vast amounts of fossil fuel.) In their use of materials, too, biological processes tend to be far more efficient than industrial ones, with a great deal of recycling going on within the organism.

On more or less the same scale as metabolism, but tending toward noun rather than verb, are materials, and here, too, engineers have been taking a closer look at nature. In so doing, they must feel something like hash-slingers who have just happened on Julia Child's spice rack. What they find is dazzling, and daunting: an inventory of some two million organic compounds, as against the mere two thousand minerals of which the inorganic world is made. What is more, the way these compounds are combined, concatenated, and deployed, even in the lowliest of creatures, makes the best efforts of human chemists and engineers look like so much scrabbling with Tinkertoys. From twenty amino acids, the cellular workshop constructs thousands of proteins, each one a cunningly folded, precisely functional work of origami. Compared with these, the best synthetic polymers are ragtag affairs, with chains of different lengths mixed up and properties correspondingly blunt and vague. Proteins are further combined and assembled into composites and other subtle structures: thus a single class of protein, known as keratin, is shaped into stuffs as unlike as horn, hair, feathers, and silk.

Silk is one natural material that leaves materials scientists agape. In both silkworms and spiders, glandular ducts squeeze out a viscous stream of keratin, its amino acids strung out in squiggly, tangly helices. A sort of extrusion nozzle applies shearing forces, tugging two ways at once and folding some of the helices into pleated sheets, which stack up to form a liquid crystal. The quick-drying product is a composite of strength and elasticity—the former due to the rigid pleated sheets, the latter to the tangly stuff in which they are embedded. To prevent cracks, a coating of lipid and protein is applied, complete with bactericides and fungicides to keep off microbes looking for a free lunch—which may explain why spiderwebs are used in folk medicine to dress wounds.

But this is just the beginning. Different spiders use different silks to solve the different engineering problems posed by their different webs (mesh, funnel, filmy dome, ladder, trapdoor, and so on) as well as by other uses (egg sacs, lassos, wrapping a box lunch). The garden-cross spider—*Araneus diadematus,* better known as the literate heroine of *Charlotte's Web*—uses seven abdominal glands to make seven silks from seven different amino-acid recipes. To span a vast expanse of space with a flat, inexpensive web—a web with enough strength to keep massive, fast-moving bugs from busting through, yet with enough give to keep them from bouncing off as from a trampoline—is an engineering challenge of the first order, to which Charlotte rises with terrific grace. The spokes of the web are relatively rigid. They are connected, though, by a "capture spiral" that is strung with tiny, sticky glycoprotein doughnuts which, besides helping to trap prey, attract moisture from the air, which seeps into the fibers of the spiral and makes them stretchy. They don't sag, though: the droplets of water that cling to the spiral act as miniature windlasses, their surface tension reeling in the cable and keeping it fairly taut—but letting it pay out when an insect hits. The droplets also add to the drag, or wind resistance, of the whole web, which helps it absorb the impact.

Not by happenstance, one of the teams that teased out these facts included a structural engineer from Ove Arup & Partners, an international engineering firm which—along with architects such as Renzo Piano—has pioneered the use of organic forms and ultrastrong, lightweight materials. But the secrets of the web are only starting to be unraveled, and engineers are a long way from fully applying them. The chemists who invented nylon in the 1930s tried to use the chemical bonds they found in silk, but their copy of nature was rough and full of snags. In general, those who want to copy the properties of natural materials can take one of two tacks: they can try to synthesize a similar structure, or they can hijack the biological process by which the stuff is made. In the case of silk, U.S. Army researchers—always on the lookout for a tough fiber that can hang bridges or stop bullets—are trying the hijack route: they have found the genes responsible for at least one type of spider silk and are inducing microbes to express them. Meanwhile, a San Diego outfit is trying to synthesize polymers more or less like spider silk by fitting various amino acids into a more or less silklike molecular frame. They have also designed generic frames that copy alpha keratin, which is found in horn, nails, and hair; elastin, which is found in skin and other stretchy organs; and collagen, the fibrous part of skin and connective tissue. ("No one understands protein folding," a spokesman for the company said, "but you don't have to be able to completely understand something to make use of it.")

Often it is hard to part noun from verb, substance from process. Part of what draws scientists to silk is that it is made at room (or garden) temperature, under modest pressure, using water as a solvent. By contrast, Kevlar is made at high pressure, at 1400 degrees F., with concentrated sulfuric acid.

The possibilities (and, in some cases, realities) of biomimesis are stunning, ranging from microscopic batteries that work on the same principles as photosynthesis, to an adhesive that sets underwater, borrowed from the blue mussel. Perhaps the most remarkable examples come from the field of biological computing, which came of age with a paper published in 1994 by the mathematician Leonard Adleman. An instance of the "directed Hamiltonian path problem"—given a set of points joined by one-way streets, is there a path along those streets that starts at one specified point, ends at another specified point, and passes through every other vertex exactly once?—was solved by a swarm of nucleotides in a test tube. In effect, Adleman tagged the points with DNA letters and let the DNA segments, following their own logic of base pairing, generate every possible path. He then used standard techniques of molecular biology to chuck out the DNA chains that failed to satisfy the problem's conditions. Though the particular instance Adleman chose was trivial, it was clear that the same techniques could be used to solve rapidly and cheaply instances of the same problem (and other problems) that would take a supercomputer months of super-expensive time. By Adleman's reckoning, a somewhat larger molecular computer—that is, one adding several more drops to the few drops of solution in his test tube—could easily perform a trillion operations per second, outstripping the fastest supercomputers by a factor of a thousand.

LIVING MACHINES

Technology is learning from nature on several scales. The molecule, the cell, the organism, the ecosystem, and finally the biosphere are all taking their turn at the lectern. Regrettably, it is the smallest scale that gets the biggest audience. Just this is what we saw happening in the quarter-section of technology that deals with the raising of food, and it is regrettable here for much the same reason it was regrettable there. The smaller the scale on which we learn, and the narrower the lessons we learn, the more readily they can be used to further our narrow, predetermined ends. A cable spun of imitation spider silk can be used to haul timber out of an old-growth forest, putting various species of real spiders out of business; a protein computer can be used to calculate the dimensions of the parking lot for a new mall. But when we learn from nature on the larger scales, our predeterminations come into question. Ends get mixed up with means; our own good gets mixed up with those of other creatures. For that matter, the linear sequence of ends and means comes to seem fairly flat-earth.

The work of John Todd and his associates at Ocean Arks International on Cape Cod is notable for mixing up ends and means in a fruitful way. Todd's enterprise is to learn the lessons of nature on all scales, from the molecular to the planetary, and apply them on the middle scale most of us live in.

The play of scales is captured in a play on words. Todd's "living machines" are machines that are more or less alive, being made up largely of living creatures. But they are also machines for living—that is, for helping humans to live in some kind of equilibrium with the living creatures outside the machines.

Why make a machine from living things? One reason has already been mentioned: living things are vats of enzymes, which can make chemical reactions happen thousands of times faster and more efficiently than they would happen in the nonliving realm. Another reason is the vast amount of free energy coursing through living things, matched in the nonliving earthly realm only by flowing lava. A third is the diversity of life: not just the diversity of species, which H. T. Odum called "an immense bin of parts available to the ecological engineer," but the diversity of organic compounds those creatures make and are made of. The exchange of matter and information, too, is far more efficient in the living than in the nonliving world, thanks in large part to the dazzlingly complex surfaces living things have coiled up inside them—surfaces which make their insides vast extensions of their outsides. Plants, for instance, are ten thousand times better at filtration than the best man-made systems.

Adding all these advantages together—or, rather, multiplying them—Todd estimates that living machines might ultimately work more than ten million times better than the nonliving ones we have been so proud of. To do so, they would have to be extremely well designed. But one of the nice things about living machines is that they can design themselves. Not only are the major parts—the organisms—products of natural selection, but so, to an extent, are the selection and assembly of those parts. A living machine can be "seeded" with various species that seem likely to work well together under particular conditions. As time goes by, some species will take, others will fall by the wayside. With a little help, the machine will organize itself much as a real ecosystem does. It will inch toward the edge of chaos, where (if the complexity jocks are right) adaptation reaches its peak. There it will perch, revising itself—for this design job is not over and done with, like most human design jobs, when the blueprints' odor has faded and the invoice is paid. Like a real ecosystem, a living machine is exposed from infancy to changes, both regular and irregular, in operating conditions and inputs; like a real ecosystem, it is constantly adjusting to the boomlets and bustlets in the numbers of its various living parts. Whereas most machines are knocked for a loop by sudden change, the living machine has a skein of feedback loops that help it adapt.

Fanciful as these notions sound, they have been realized in part by several of the living machines Todd has built to purify water: either to treat sewage or to restore the waters of a polluted stream or pond. To date, the longest-lived of these living machines has been the one in Providence, Rhode Island, which ran nonstop from 1989 to 1995. Diverting a small stream of the sewage headed for the city's large, conventional treatment plant, Todd ran it through a series of clear, upright, cylindrical, aerated tanks in a large greenhouse. At the middle and end of the series, special tanks of soil were regularly flooded and drained, mimicking a tidal marsh. Todd seeded the tanks repeatedly, and at various seasons, with chunks of life from over a dozen natural ecosystems, most of them in southern New England: streams, brackish ponds, a vernal pond, the intertidal zones of a Cape Cod salt marsh. By design, he did not know what he was doing. Snails, fish, and higher plants were put in purposely, as were commercially cultured bacteria and microbes from activated sludges and at the neighboring treatment plant; but swatches of floating scum, cross-sections of muck and ooze might contain any number of unspecified species.

As time passed, a series of miniature ecosystems took shape. Upstream, hearty pond snails and water hyacinths tucked with gusto into the rich, raw sewage; downstream, more abstemious parties sipped daintily on the clearer effluent and made it clearer still. "Sandwiches" of bacteria, algae, and protoctists, more various than in a celebrity deli, clung to the clear walls of the tanks, photosynthesizing like mad. Of two dozen snail species introduced, two families flourished, dining on sludge and thinning the dense algal mats. Water hyacinth, pennywort, duckweek, parrotfeathers, and water velvets floated on the surface of the tanks; watercress and water starwort rooted in screens, as did a fair number of the hundred or so land plants tested. Bald cypress, ornamental cedars, and willows spread their roots to the floors of the tanks, extending the living filter. In the marsh tanks, native willows and cattails succumbed to molds and spider mites, but alien bulrushes, papyrus, and elephant ears throve. Golden shiners and other fish bred among the roots and fed on decaying duckweed and other debris.

By all standard measures—biological oxygen demand, chemical oxygen

demand, total suspended solids, alkalinity, ammonia nitrogen, phosphorus, fecal coliform bacteria, and heavy metals—the thousands of gallons
that flowed from the last tank would have done any conventional treatment
plant proud. In fact, they were drinkable. And whereas the clean water
pouring from conventional plants leaves behind a great deal of sludge—
often fairly toxic, and with some of its toxicity due to the chemicals used in
treatment—which has to be dumped or burned at high cost, much of the
sludge in the living machine was converted into ornamental plants or bait
fish that people would actually pay for. True, the living machine handled a
relatively small volume of sewage, but, then, it was a very small facility. In
practice, living machines could be made bigger: better, they could be kept
small and scattered about a town or city: run in midget greenhouses along
the sidewalk, or even built into people's homes. For while a conventional
treatment plant looks and smells like a treatment plant, a living machine
looks and smells, for most of its length, like a botanical conservatory, a lush
greenhouse, the sort of place where a French café table, or even a hot tub,
would not look amiss.

Though biological treatment of wastewater has seeped into the mainstream, even the most ecologically minded designers tend to use only a few
species of microbes or plants. This they do not from conviction, or even
laziness, but simply because "simpler systems can be more readily
described to environmental regulators." Todd, for his part, insists that no
system can mimic the efficiency and resiliency of a living system unless it
mimics the diversity of a living system. All five kingdoms—bacteria, protoctists, fungi, plants, and animals—must be represented by large and
diverse delegations. Snails, for instance, are mentioned in ecological
wastewater-treatment texts only as nuisances if they are mentioned at all.
Yet Todd has found them invaluable in reducing sludge, cleaning tanks, and
producing enzymes that attack recalcitrant compounds. They act as sen-
tries, too: when a sudden jolt of toxicity flows into the system, the snails
climb out of the water and cling to floating plants, warning the operator to
recycle some downstream water into the upstream tanks. By contrast,
chemical tests of toxicity take days, during which time the shock can put a
conventional treatment plant badly off its game.

Todd's insistence on diversity, and on seeding his tanks almost randomly
and letting them evolve, means that his results are, strictly speaking, ir-
reproducible. Strictly speaking, he is not doing science at all—not in
the classic, Western, universalizing manner. To which a believer might
answer: so much the worse for science. For Todd, the living machine is a
"mesocosm"—a medium-sized scale model of an ecosystem, or of a set of

ecosystems, or better yet of the biosphere as a whole. "As above, so below": the Hermetic motto is basic for Todd, and it is hard to know whether it is a means to good functioning or an end in itself. Presumably it is both, as are most of the things Todd wants to learn from nature: diversity, resilience, self-organization, and so on. Anyway, it may be that, when a designer keeps his eye on the large-scale functioning of the biosphere, he is less likely to create something that might impair that functioning.

Having treated of the machine in the garden, then of the garden in the machine, we come now to a third possibility: the machine *as* garden. Like most pleasure gardens from antiquity to the present, the living machine is a microcosm (or mesocosm, if you like). Like them it is idealized and made agreeable to human needs. Maybe it is no accident, either, that the first living machines should concern themselves with water and its reclamation. After all, every microcosm must have at its core some simulacrum of Eden, the source of living waters.

THE LOVE OF THE NONLIVING

Hold the last few pages up to the light and you may think you glimpse through them a future truly Edenic. A future in which machine and garden mingle so intimately that cog and tendril can hardly be disentangled. In which technology marries so well with nature—is so sensitive, so gentle— that nature no longer needs a room of her own. In which the image of machines as living allies of humankind is no longer mere image but plain fact, and wolf, lamb, and laptop lie down together.

Some people do cherish such a vision, but I am not one of them. Hard, rebarbative elements both in technology itself and in the humans who make and use it will, I think, prevent this future from becoming present.

First of all, technology that is too soft, too elegant, too light, too organic may not catch on. Part of the beauty of technology is its bigness, roughness, and ugliness—its inhumanity. Although I have never worked in a factory, I learned as a gardener in the New York City Parks Department that power tools and vehicles, which at first I regarded with horror, can give great satisfaction. What makes them satisfying is exactly what makes them horrifying: noise, speed, brute power. Even the weed-eater, a deafening machine that leaves the back numb and the fingers tingling for days (and that is inferior in most ways to the hand scythe), gives a certain pleasure with its Uzi-like destructiveness.

My language makes it clear that what was awakened in me was partly the instinct for violence, which, however deeply rooted in our evolution, is not

an instinct we want to encourage. Of course, if the instinct proves in-
tractable, it is better exercised on gardening machinery than on the machinery of war. But I think something deeper is going on here.

"Soft" technology prides itself on learning from nature. The lessons it
draws have to do with harmony, elegance, and efficiency—sterling qualities
all, and qualities we need more of. But nature is also brutal, violent, and (in
its own way) wasteful. And we love that side of nature, too, as we demonstrate by naming our machines Taurus and Toro and Bronco and Hurricane. When a voice out of the whirlwind asks us, "Canst thou send forth
lightnings, that they may go, and say unto thee, 'Here we are?' "—we may
be cowed at first, but our next impulse is to give it a try. If nature is allowed
to be sublime, how can humans settle for making their handiwork pretty?

Whirlwinds and hurricanes should remind us that nature is more than
just life. Most of nature is inorganic. The human love for other forms of
life, what E. O. Wilson calls "biophilia," is not surprising, but it would be
surprising if we did not also love many nonliving things. A geologist has
observed that, if all the biomass of the world were run through a blender
and spread over the surface of the nonliving earth, it would provide a layer
thick enough for a modest peanut-butter sandwich. Living in an environment so largely nonliving, we would be very unhappy if we did not love
water and wind and stone and metal. The pinnacle of Japanese garden
design, in the Zen monastery of Ryoanji, is composed of rocks and raked
pebbles. We may think of Eden as a very green place, but Ezekiel's "Eden,
the garden of God" blazes with jewels; and even the Pentateuch cannot help
mentioning that downstream from Eden, in Havilah, there are gold, bdel-
lium, and lapis lazuli. Our search for wholeness does not stop at the tree
line. Physical scientists, as I said, may be the most thoroughgoing mystics
of all, insisting as they do on seeing the universe whole and on its own
terms. More generally, playing with the inorganic, the purely mechanical, is
part of human nature. A boyish crush on big, sometimes violent machines
may at times betray a love of death, but more often, I think, it shows a love
of the nonliving: an abiophilia.

Earlier in this chapter, I said that the best way to keep our allies—both
living and nonliving—from overwhelming the rest of nature is not to
squelch their wildness, but to nurture it. By "wildness" I meant the kind of
behaviors found in the citizens of wild, mature ecosystems, as opposed to
the weedy, opportunistic ways of pioneers and invasives, which is what
humans and their allies tend to be. But in another sense, of course, those
opportunistic ways are part of the nature, and thus the wildness, of humans
and their allies—a wildness that will never be rooted out. As allies and par-

tial creations of humans (less partial than sheep or corn, but still not made from scratch), machines partake of this wildness. They will always be a shade too brash, an inch or two too forward, a smidgen heedless of the living context around them.

But, then, this may be a sign, as well, of their deepest wildness: the wildness of nonliving matter. The earliest life on earth must have retained some of the stolid self-sufficiency of inert matter; however thin its infant membranes, it must have been, in a sense, thick-skinned. For billions of years, pioneers and invasives have kept this quality. They are always colonizing a virgin planet. They are aliens, space cadets, carpetbaggers. And humans are the champion colonizers, the great disregarders and upsetters of existing conditions—throwbacks, you might say, to the cyanobacteria that changed earth's atmosphere two billion years ago. No wonder we feel a kinship with the nonliving. No wonder we have domesticated nonliving matter, making it join in our conquest of living matter all over the planet. We think of ourselves as the summit of creation, the stamp-and-kick production number of life's musical comedy, the topmost step on the stairway from inert matter to conscious mind. When push comes to shove, though, we may have more in common with the bottom steps than with anything in between. And our particular kind of consciousness, which gives us access to the charm of quarks and the antics of quanta, also makes us as blind and deaf to the etiquette of ecosystems as if we were meteors falling from space.

The hard insensitivity that makes machines our kin can also make them seem alien and horrid to us when (as often happens) they come rumbling like juggernauts through our social systems or our man-made ecosystems—our gardens, so to speak. If you want to know how humans look to the rest of nature, think of how a machine looks to you at such a moment: when the town road crew's tractor has lopped off the daylilies at the edge of your lawn, or when automation has cost you your job, or when you are trying to convince a computer at a city revenue office that you no longer live in a place you left six years before. The blinkered, unblinking pursuit of a predetermined end seems to us "mechanical." Even when they seem most securely our allies—our domesticates or domestics, tame and housebroken—a glint of madness in our machines' aspect warns us that they may at any moment run amok.

The feeling that technology is "out of control" is common and no doubt has been common since the first stone hand-ax gashed its inventor's thumb. Technology will always have the potential to escape our control, just as a strong and high-spirited horse has the potential to throw its rider or carry him off against his will. Technology has its own wildness, which is always going to cause some trouble. But if we ever managed to wring out that wildness, what we would be left with would be pretty limp technology. And it would be technology that would not evolve.

For all these reasons, the vision of a "soft technology" or "industrial ecology" in perfect harmony with nature seems to me a shade rose-tinted. If such a technology really were achieved, it would soon stagnate. Though some may think a steady-state technology a good idea, they forget that nature is not steady-state, and that in this world you have to run pretty hard just to stay where you are. The harmony they seek will prove illusory, just as previous harmonies have been. Confident that our technology was benign, we would blithely extend its reach into the bowels of wild nature. We would not bother to keep the Mountain off limits, for the whole planet would be our garden.

Even Todd comes dangerously close to this way of thinking. As with the classic pleasure garden, there is some confusion about whether the living machine is a model of the world or a part of it: or, to put it another way, about where and what the walls of the garden (or garden-machine) might be. Here, though, the practical effects of this confusion could prove far greater. Though he seems to give native species first dibs in his living machines, Todd does not scruple to import parts. In an experimental device for breaking down paper-mill wastes, "fish from four continents" were conjoined—a scaly reminder of the Rivers of Eden. This would be risky enough if Todd's Hermetic vessels were hermetically sealed; but when he suggests linking living machines directly to nearby ecosystems, it is time to sound the alarm. "The domestic living machine and wild systems," he writes, "will coevolve to the mutual benefit of both. The mineral, nutrient and microbial diversity of the living machine, having been drawn from many regions and communities, provide beneficial feedback for the wild system that, in turn, could influence the living machine in its assigned tasks." Such glib talk about coevolution glosses over the mayhem that invasive exotics are doing to biodiversity all over the world. While the Amazonian water hyacinth, a stalwart of Todd's sewage tanks, is unlikely to play in cold Cape Cod the sort of havoc it has played in Africa, what will happen when those tanks are installed in warmer climates?

The urge to garden the planet is, of course, a form of Planet Management; it is also a form of the myth of Eden regained. What is neglected is the essential tension that has been with us since Eden: the tension between working and keeping, between changing the earth and shielding it from our changes. Like earlier waves of humans and their allies, the waves of machines surge across the globe, changing Eden (or what in retrospect looks like Eden) into something else. Exactly what that something else will be can never be fully known up front. So we have to try and keep great islands of wild nature high and dry, above the flood, as free as possible from our meddling.

STRIKING A BALANCE

But if the goal of perfect harmony with nature is not reachable, it is still worth reaching for, as long as human nature and machine nature are not strained too badly in the process. We should indeed prod our industries to grow up and act like mature ecosystems. We should indeed try to guess or assess the effects of our technologies and change them accordingly. The question is, how can we keep some kind of rein on technology without breaking its spirit?

Just as we should be careful where we let our quadrupeds graze, so we should be careful where we let our machines roam. They may be wholly useful in one place, baneful in another. Their effects on communities, both human and nonhuman; the energy they transform, the matter they ingest, and the (often new) forms of matter they expel: all these must be constantly watched with the eye of a shepherd who cares about the health of the land. Like our domestic plants and animals, our machines sometimes need to be tended, sometimes let loose, sometimes restrained, sometimes altered, and sometimes culled.

Not much of this can be done on the basis of abstract principles. Our philosophy must evolve as we experiment and see what works—not just ecologically but economically, socially, politically, emotionally. True, we have had a thousand centuries to get a general sense of what works; but each advance in technology throws all our stored-up wisdom into confusion. To the pastoral trope just employed, let me add an extra melisma: just as we saw in Arcadia, the only way to keep nature and culture in balance (the only way to keep Arcadia Arcadia) is by constant equilibration. No abstract principle can tell a shepherd just how much grazing a pasture can take, or just when or where to move his flock. Not in nature, not in culture, and least of all where they meet is there any once-and-for-all.

Where does that leave the likes of me? Here I sit making fine distinctions, drawing lines in the sand. The next tide of invention will erase them, or make them seem like fossils. Even the tide is too tidy an image—we live in a sandstorm. Hunched over, I can etch the sand in the lee of my body, but the instant I stand up it goes blank. Some equilibration.

At its most bathetic, the relationship between the humanist and the scientist seems to be this: The humanist draws a line in the sand and says, "If you would keep your humanity, forbear to cross this line!" The scientist crosses it. The humanist scampers after him, overtakes him, draws another line in the sand, and says, "Well, I suppose that's all right . . . but if you would keep your humanity, forbear to cross *this* line!"

What makes this pas de deux so bathetic is its deuxness. Not just the humanist and the scientist ought to be involved, but the whole corps de ballet. We need more democratic control over technology, in the broadest sense: our whole relationship with nature. We need time and opportunity—not just for experts, but for ordinary people—to stop and think.

In this respect, some nations are more thoughtful than others. When a new technology is presented to the Danish government for approval, the Board of Technology invites a number of ordinary people, of various backgrounds, to serve as a sort of jury. In 1992, for example, such a jury tackled the question of genetic engineering in animal breeding. The jurors sat through two background briefings, then went on to hear and cross-examine witnesses for and against: scientists, experts on the social effects of technology, and representatives of interest groups. After due deliberation, they gave their verdict to a national press conference. Among other things, they found against using genetic manipulation to make new kinds of pets, but in favor of using it to help find a cure for human cancer. Though their decision was not binding, it had enough moral weight to sway parliamentary votes on the matter.

Not only are these forums open to the public and press, but their judgments are further publicized by means of leaflets, videos, and local debates, of which more than six hundred have been held on biotechnology alone. It should come as no surprise that studies have found Danes better informed on such issues than the citizens of other nations—or that other nations, including the United Kingdom, the Netherlands, and the European Union as a body, have taken up the Danish example. Of course, this trend is being bravely bucked in the United States, where the closest thing (and not very close) to a populist watchdog of technology, the Congressional Office of Technology Assessment, was put to sleep in 1995 by a Republican Congress.

It is not enough, though, for society to choose among fully grown technologies; it must also have a hand in growing them, even in sowing their seeds. Science is already guided by social needs; they just happen to be the needs of military planners and the stockholders of large corporations. In the Netherlands, this bias has been redressed somewhat by an institution known as the science shop. Since the late 1970s, Dutch universities have set up some fifty such shops, to which community groups, public-interest groups, workers' groups, and local governments bring problems that bear on the lives of ordinary people. In all, the shops address several thousand problems a year, covering a broad range of the social and natural sciences. The cost to the universities is minimal, since students and faculty are doing what they would be doing anyway: pursuing research, writing papers, supervising research, and correcting papers. Emboldened by the Dutch example, Austria, Germany, and other European countries have set up science shops of their own. In the U.S., a similar effort has been catalyzed by the Loka Institute, a pint-sized think tank in Amherst, Massachusetts.

A HABIT OF MINDFULNESS

Wes Jackson, whose edible prairie we spoke of earlier, has proposed a moratorium in the deployment of new biotechnologies. If you can call a time-out in a basketball game, he asks, why not in this game on whose outcome so much hangs? A more sweeping idea has been put forward by the philosopher Arthur Waskow. Let every seventh year, he says, be set aside as a sabbatical in which no new technology is deployed, no new houses are built, no raw land is developed. Let people stop and think about what they have already done and what they might do in the future.

Such an idea is unthinkable in modern society, which is a good sign that it deserves serious thought. In fact, some of us already set aside every seventh day in roughly this way. For many Jews, Christians, and Muslims, the Sabbath is a time when we do no work, and reduce our dependence on technology. On that day we meddle with nature as little as we can and enjoy nature as much as we can. The Sabbath, the sabbatical year, and the jubilee year are puddles and ponds of wildness in time, fed by the Rivers of Eden. In the rhythm of Earth Jazz, they are times to lift the reed from your lips and just listen.

Unfortunately, the secular weekend is often just the reverse: a time to shop for and play with the toys you worked all week to afford. But even secular society has kept alive the tradition of using weekends and vacations as a chance to live more simply, to get closer to the Mountain and achieve some brief detachment from the Tower. Hiking, camping, and weekend or summer homes are not without cost to nature, but the cost to nature will be less if we respect the simplicity that closeness to nature demands, and really try to do without modern technology—if we backpack instead of taking a forty-foot trailer, or build a rude cabin instead of a plush second home.

Anyway, the cost to nature is amply repaid if people end up thinking harder about what they are costing nature the rest of the year. When television tells you what you need and politicians tell you what the economy needs, it is good to have first-hand experience of needing less. In this respect, a vivid alternation between Tower and near Mountain may be more useful—as well as more satisfying—than the usually futile search for a middle landscape that is perfect all year round.

But to become mindful, to detach oneself from the rush of technology, may not always require that one venture into the wilderness, or even into the countryside. The Sabbath, or something like it, can allow one to step back from the Tower without actually leaving it: to be in the modern world but not of it. Instead of moving toward wildness in space, one allows wildness to well up in a particular volume of time. To the receptive soul, a walk in the park can be as restorative as a trek in the wilderness. Even the city streets wear a new face when you carry no money or credit cards and travel only on foot.

This last example refers to the traditional Jewish Sabbath, when Jews are commanded to do no work. Though the Bible does not specify what it means by work, the rabbis interpreted the word to cover all the activities involved in the building of the tabernacle in the desert, which later became the Temple on Mount Zion. So while the need to build the Tower is acknowledged as a part of human nature, from the very start there is a counterweight: a reminder not to take our Towers too seriously. God is a wild creature, a creature of the Mountain, who cannot be cooped up in human constructions, no matter how grand. And humans, so long as they retain some trace of God's image, will not be fully confinable either.

A habit of mindfulness, I said, may be more important than either rigid habits or rigid beliefs. A habit of setting aside a time or place for mindfulness can be part of this. Such a habit may help to overcome the spoiled-child greediness, broken by spasms of rejection, that now marks our dealings with technology. The habit should be applied, though, not only to technology in the narrow sense, but to economics, land use, lifestyle—the whole breathless march of civilization.

But maybe "mindfulness" is too solemn a note on which to end a chapter that started so jazzily. Instead, let me invoke Enki, the Sumerian god whose randy adventures with his wife, the Earth Mother, and her daughters and granddaughters were retailed in Part Two. Enki was the best-loved of Mesopotamian gods, for two reasons: he was the god of water and the god of technology. Technology was very, very good to the Mesopotamians.

In the story of Enki and Ninhursag, I argued, it was the risks of technology that were metaphorically explored. Pushing a good thing too far— that good thing being irrigation, or the dousing of female earth with male fluid—Sumer was rewarded with saline soils. Though Enki is bailed out at the end, the story leaves a bitter taste in the mouth: the taste of roadweed, caper, cassia, and other plants that grow on ruined ground.

On the whole, though, the Sumerians were very happy with Enki and the technologies that he gave them, or that they gave themselves in his name. In Sumer, humankind was on a roll: brainstorming, improvising, one-upping and amazing itself. Not only large-scale irrigation but wheeled vehicles, animal-drawn plows, sailboats, the arch, the vault, the dome, metalworking of all kinds, surveying and mapping, mathematics, writing—courtesy of the last, the list goes on and on. If Enki's antics and inventions sometimes got him (and humans) in a tight corner, he was generally clever enough to wiggle his way out, or un-stuck-up enough to beat a retreat. For like many "culture heroes"—givers of technology—in ancient lore, he was also a lovable rogue, a madcap improviser, a joker and jokee: in short, what folklorists call a trickster.

In the mainstream of Western tradition, the prime culture hero is a very different figure. He is a titan who steals fire from the gods, gives it to humankind, and is punished by being chained to a crag where an eagle tears at his ever-regenerating liver. Though Prometheus starts out in Hesiod as a trickster, by the time of Aeschylus he is a tragic hero: fiery, willful, single-minded, undaunted—in a word, Promethean. This is the self-image of Western science and technology when it triumphs, and even when it fails—for it always fails nobly, bloody but unbowed.

Chopped liver notwithstanding, it is easy to see how seductive such a self-image might be. It is easy to see, too, how it might lead one astray. When nature "fights back"—when leveed rivers flood twice as fiercely, when pesticides appear in human breast milk, when sealed office buildings make people sick—the scientist sees only the petty jealousy of the gods in the face of progress. His spine stiffens. He marches onward, changing his tactics but never changing course.

A trickster, on the other hand, is quick to see when his trick has failed and quick to try something else. He does not take himself so seriously. If he can pull a fast one on nature and get away with it, fine. If he must humble himself before nature to get out of a scrape, he is down on his knees in a flash.

Fortunately, the image of technologist as trickster has not vanished from the Western mind. In America, Enki is reborn in the inventor, the handyman, the *bricoleur* who improvises wonders from tin cans and baling wire. Eyeing this figure, you can see yet another reason why the jazz age was also the age of invention.

But every age is an age of invention to some degree. Reckoned by person-hours, the rate of invention seems to have been as high in the Old Stone Age as it is today. Invention is a bedrock trait of human nature and one of its chief glories. Take him for all in all, the jazzy, tricky inventor is not a bad model for technology in our day. But for the inventor to do the work of Earth Jazz, he must be aware that the balance of power between humans and the rest of nature has shifted somewhat, and must shift his tricks accordingly. Pulling a fast one on nature is still fine—all technology does that—but the standards of "getting away with it" must be more stringent. (As I said, there are times when you get so far ahead of the game that the game collapses on top of you.) The inventor must listen more carefully to nature's response. If nonhuman nature cries out, chances are good that sooner or later people, too, will feel the pain.

27

THE URBAN ANIMAL

F OR SEVERAL CHAPTERS now I have gloried in the conceit of collaborating with nature; but I have taken pains to point out the dangers, too. Like any model, Earth Jazz can mislead. Carried away by mastery of our instrument, or by the mystic communion we feel as the riffs fly, swift as thought, from one player to the next, we may dominate the music to an obnoxious—and, in the long run, perhaps catastrophic—degree without any inkling that we do so. That is why it is so crucial that we stop playing now and then and just listen: that we give the other players room to stretch out, as free as possible from our influence.

The sixteenth-century kabbalist Isaac Luria of Safed, known as the Holy Lion, taught that, in order to create the world, God had to draw himself inward—to take a step back, as it were. Since God was everywhere, he had to retract himself in order to leave a space where other things could exist. This self-retraction Luria called *tsimtsum*.

The time has come for a human *tsimtsum*. The present lord of creation, humankind, must take a step back and give the rest of nature room to breathe. Man must tighten his belt so that the rest of creation (by which I mean both the object and the activity) can go on.

In saying this, I am calling neither for a return to primitive living, nor for a "decoupling" of humans from nature. Even as we make more room for wilderness, we must collaborate with the rest of nature in new and creative ways—some of which, I hope, have been noted in the foregoing chapters as instances of Earth Jazz.

In the Lurianic system, *tsimtsum* is only the first stage in the process of creation. Though the universe takes on a life of its own, it is formed and nourished continuously by the emanations of godhead. To speak of a human *tsimtsum* is not to deny our creative partnership with nature, but to admit that true creation begins with self-limitation: with the acknowledgment that there is something in the world other than the self, and that this something has its own creativity. Only with such an other is partnership possible—not with a mere reification of our needs, desires, and fears.

What is nature's creativity? A poet could give a thousand answers, but I will give just one. Nature's creativity lies in the ecological and evolutionary processes that give the planet its thousands and millions of faces, as well as the single, remarkably consistent face that some call Gaia. Since we are dependent on those processes for our very survival, *tsimtsum* has an urgency far stronger than the courtesies of "I" and "Thou."

The concept of "limits to growth" is one that unites Fetishers but divides Managers. One prominent Manager, the ecologist Barry Commoner, argues that the notion of carrying capacity is meaningless when applied to a planet run by clever humans. "In an abstract sense, there is a global 'limit to growth,' but this is determined not by the present availability of resources, but by a distant limit to the availability of solar energy. . . . That distant limit is irrelevant to current policy. . . . The question is whether we can produce bountiful harvests, productive machinery, rapid transportation, and decent human dwellings sufficient to support the world population without despoiling the environment." The answer, of course, is yes—provided we take Commoner's advice.

His view of population growth is not quite so cavalier as that; he takes the line that economic development will at some point stabilize population more or less automatically (the "demographic transition": see below). Still, I am troubled by his denial of limits, by his vision of a planet wholly humanized: a vast organic garden festooned with appropriate machines. For an ecologist, Commoner is oddly silent here about such matters as species extinctions and the loss of wilderness. With the best intentions and the best technologies, large numbers of humans weigh heavily on the ecosystems they invade. Farming, even the organic farming that Commoner favors, takes up space. If there is one thing ecologists have learned in recent years, it is that wilderness cannot be kept in boxes; and the smaller the boxes, the more frantically they must be managed to retain even a semblance of life. Today, many wilderness areas are being managed very frantically indeed. Wildlife biologists now speak of wilderness areas as "megazoos" in which endangered species are tagged, tracked, and provided with dating services.

People involved in such projects, and in the many other projects Managers get involved in, may think they are collaborating with nature. But management is not collaboration. Few humans can do creative work with the boss looking over their shoulder, correcting each dubious move. In this regard, human and nonhuman nature are very much alike. Before we can hope to collaborate with nature, we must give her some elbow room.

PROSPECTS FOR *TSIMTSUM*

Would a human *tsimtsum* mean rolling back the waves of human-led change? If so, the notion might seem as facetious as King Canute's when he sat on the shore and commanded the tide to turn back. But the truth is that the tide of human advance has ebbed many times, in many places. War, famine, and plague have been responsible most of the time, but not always. For the kind of rolling back we need, a decrease in human population is not essential, though it would certainly help. While the sheer number of people on earth matters, where they are and what they are doing matters just as much.

Oddly enough, the waves of human-led change may in effect be rolled back or annulled by new waves of human-led change. Industrialization can often bring about a "demographic transition" that lowers the rate of population growth to something very near zero. As wealth and health increase, people switch from an *r* strategy (having as many kids as they can) to a K strategy (having a few kids in whom they invest heavily). (In ecologists' patois, *r* and K are similarly used to describe the difference between annual and perennial grasses, or between frogs and rhinoceri.) Though the switch is far from automatic—hinging on the specifics of economic security, the position of women, and other social factors, as well as the availability of contraception—it has been thrown in a number of newly industrial nations.

Industrialization's older sibling, urbanization, can concentrate people in a smaller area and so, in theory, leave more room for wilderness. More productive farming can squeeze more food from less land: in India, high-yield grains are reckoned to have spared for nature's own use more than a hundred million acres that would otherwise have been plowed. In all of these ways, the expansion of humans and annual grasses may be reversed.

In practice, it is not easy to know whether a real *tsimtsum* is taking place. Though people of industrialized nations may have fewer offspring, each one uses much more of nature—both as larder and as dump—than preindustrial people do. Squeezing people into cities does no good unless their needs for energy, sanitation, and so on are met sustainably, a condition few Third World cities are set up to meet. As we have seen, the effective base of the Tower generally extends far beyond the city limits. And modern agriculture does not cut nature any slack if it depends on overuse of chemicals and abuse of soil, and so is unsustainable in the long run. Nor does nature gain much if the retraction of civilization leaves only a wasteland. Sometimes, it is true, nature heals itself, spontaneously reclaiming its lost realm; more often, perhaps, humans must help it along by the practice of "restoration ecology."

In the last few chapters of this book, we will look at some of the forms a genuine *tsimtsum* might take: in particular, the spatial relations between the Mountain, the Tower, and the places in between. For nature's sake and our own, we must give nonhuman nature space in which to go about its business. On the simplest level, this means preserving wilderness. But it also means arranging our own, human space in intelligent ways.

ARE CITIES UNNATURAL?

Humankind's career has been a checkered one. We started out scavenging like hyenas, then learned to hunt in packs like wild dogs. Much later we took to standing in place like cows, placidly mouthing grasses. And shortly after that we began to act like bees and ants, swarming all over each other in tenements and factories and mines, enacting cramped and furious dances of labor and caste.

The quickness of the leap from cow to bee—a matter of a few thousand years, as against the millions it took to reach cowdom—is notable. The city is the fruit of farming in more ways than one. Farming does not simply make cities possible, it makes them necessary: not only to use the surplus of wealth in a suitably unequal fashion, but to absorb the savage energy that farming pens up. In fact, the jump to cities is so quick and so strong that it has almost the look of a reaction. And it looks less like the swinging back of a pendulum than like the explosion of a spring compressed too far.

Are cities unnatural? They have been around for six or seven thousand years. Farms and villages have been around for ten or eleven thousand years; viewed through the million-year spectacles of the anthropologist, that is not much longer. It could be argued that cities are more natural than farms and villages, because they bring us closer to our hunter-gatherer past. That the chaos of city life is more congenial to our primitive needs than the strict, property-bound order of village life. That grocery aisles are our foraging ranges, skyscrapers our mahogany trees, leveraged buyouts our hunt.

In Part One I talked about the notion that farming and village life have distorted human nature, running it through a set of funhouse mirrors in which people get fatter, greedier, squatter, stolider, more earthbound and hidebound. While all these tendencies are part of human nature to start with (as are some of the nicer, more nurturing sides that farming brings out) it is true that village life often exaggerates them. It is often assumed that city life stretches them even further, adding concavity to the mirror in proportion to the city's size and density. In many ways, though, city life reverses the trend.

Lewis Mumford saw the city as offspring of a marriage between the two previous "stages" of human living: the hunting-and-gathering band of the Paleolithic and the farming village of the Neolithic. By this he meant that it weds the female principle of home and nurture to the male principle of predation and control. But there are other ways in which the Paleolithic makes a comeback in cities. People are less tied to particular patches of ground (or pavement). Wealth comes and goes quickly, and is less easily defined by large, visible things. Different tribes, bands, and families meet on equal footing (or lack of footing): who is the native and who the outsider is not always clear.

A crowning glory of the settled life, the city is oddly tied to the wandering life. (As we shall see, the same paradox plucks at the pleasure garden.) In the Bible, the first city is founded by a wanderer, Cain. Some great cities have been born of the conquest of farming peoples by marauding horsemen, who needed a headquarters from which to control scattered villages. Others have sprung up at the crossroads of trade routes, getting their start as glorified bazaars and parking lots for caravanserais. Cities excuse, even encourage, coming and going. What is more, in a great city one can experience the joys, dangers, and cultural jars of travel without ever leaving the city limits. (As I write, my glance falls on the headline of a newspaper calendar section: "Night Nomads: The Everchanging Face of Clubs and Crowds.")

URBAN BY NATURE

Man, Aristotle said, is by nature a *politikon zōon*—a phrase often rendered "political animal," which in modern lingo means someone keen to elbow and scheme his way to the top. A better counterpart of the Greek might be "urban animal," a creature whose nature it is to live in cities.

Aristotle knew of peoples—the Scythians, the Sarmatians—who did not live in cities. He knew that even the Greeks had once been scattered thinly in forests and grasslands. But humans in that state were not yet fully human. They had not realized their nature.

The dietary laws of modern thought forbid such an argument. For us, human nature is something shaped by the past, not something to be attained in the future. If humans did not evolve in cities, it cannot be our nature to live in cities.

Yet even strict Darwinians speak of "preadaptation," whereby a trait that has been selected because it is useful for *x* turns out, down the road, to be useful for *y* as well—turns out, perhaps, to be the germ of a new and undreamed-of alphabet. So the primate hand, which evolved to grasp the branches of trees, turned out to be good for grasping sticks and stones.

With some modification, it turns out to be admirably adapted for bashing the keyboard of a computer.

Humans are preadapted for city life in various ways:

We are social animals, and smart ones; not only our smarts but our very self-consciousness may have evolved to let us play the social game more cannily. The chance to play that game with more people, and by more complex and variable rules, gets our juices flowing.

We are creatures of an edge habitat, and we like to be able to move quickly from one kind of space to another.

We are predators. We enjoy getting our livelihood by taking it from other animals—even if they happen to be people—instead of making it grow from the ground.

We are foragers. We like to be able to find food when we want it instead of having to store it from one harvest to the next.

The foregoing list may remind you of a list in Part One that ticked off some of the deep human needs that farming meets. You may now conclude that you should have skipped both lists, since they obviously cancel each other out. But that is just the point. Neither farming nor cities existed during the millions of years in which most of hominid evolution took place. Yet neither is "unnatural." Which is "more natural" is a question for rhetoricians, not biologists. Which makes us happier is a question for each of us.

Both the city and the country answer needs that are deep in human nature. So it is not surprising that we bounce from one to the other, either seasonally or at odd intervals, or that we try to still the bouncing by finding a point of repose somewhere in between.

Once the city is born, there is no going back, just as there is no going back on language, or laughter. The city can no more be tucked back into the village from which it burst than our swollen frontal lobes can be stuffed back into the hindbrain. It may get swallowed up in the maw of a giant mall, but it will never be digested.

While Americans and some Europeans are busy moving out of cities, the rest of the world is busy moving in. The wave of urbanization and industrialization that began somewhere in the vicinity of Manchester is just now reaching the outer reaches of Africa, South America, and Asia. So while some of us see the city as a lost cause, a matter of interest only to historians and police chiefs, it is a matter of consuming interest for most of the world's population. Megacities such as Mexico City, Calcutta, and São Paulo, with their sprawling shantytowns, snarled traffic, child prostitution, open sewers, and air that is darkness visible, give us an unwelcome glimpse of what the general human condition may be in the next century.

As the city pushes its way into the countryside, it exerts not only an outward pressure but a kind of suction as well. In modern Africa and South America as in ancient Latium, subsistence farmers displaced by great plantations, and replaced by slaves or machines, have been pulled into the tenements and shantytowns of the swelling cities. In England at the dawn of the Industrial Age, enclosure drove herds of peasants into the waiting folds of the industrialists. Often the depeopling of the countryside has been deliberate policy, meant to ensure a steady flow of cheap labor to mills and factories. It was so in America in the 1940s, when business leaders (such as those who formed the Committee for Economic Development) pushed through government policies that pushed farmers off the land.

But country folk are not always pushed off their land. Sometimes they are pulled by the attractions of the city—attractions that are not purely economic. Young people, especially, are drawn to cities because they offer one of the glittering lures of Eden: freedom.

A young man can have few more paradisiacal experiences than that of living for the first time in a big city, sharing a small apartment with a woman he loves. A squalid studio becomes a tropical island; one lives in the nude, bathed in sweat and distant salsa music, foraging at midnight in the refrigerator.

In the city a couple can feel beleaguered, yet sheltered and alone—sheltered by the thin walls of an apartment from the beleaguering noise and tension and violence of the city—all of these being still perceived, but perceived as "out there." In the country it is nature itself that is the shelter. A couple feels alone only if it *is* alone. Noise and tension and violence must not be perceived at all. The exception (a big one) is bad weather: cold and wind and rain. A house in the country shelters one from inclement nature just as an apartment in the city shelters one from inclement culture.

Today, the radical privacy that Adam and Eve enjoyed is most swiftly recaptured in a big city, where the faces rushing by are as alien as animals, waiting to be named. A journalist visiting the Serengeti was reminded of "the cave paintings of Cro-Magnon man and his powerful dream sense of being surrounded by streams of running game." With its thundering herds of buses, taxis, and humans, the city awakens us to that dream.

The city is rarely likened to a savanna, but often to a jungle. Usually what is meant is that it is ruthless as a jungle, red in tooth and nail. But it also has a jungle's energy and density of life, both day and night. For those with a modicum of money, the city is magically abundant; in all seasons, at all hours, its eggroll trees give fruit. Once I found myself unable to leave a certain room of the Museum of Modern Art, because Rousseau's *The Dream* was in it. The scene was familiar, and not simply from a dream; but when had I walked in bright moonlight through a jungle brimming with life? Never. Only in Manhattan had I seen such alert, feral faces, such serpentine motions, and half-naked women, and a black musician playing, all in the magic circle of a streetlamp, as in a moonlit clearing.

There are feelings our forebears got in wilderness which we get only in great cities. The feeling of being lost in something huge, something other, something that is not of your own making and that does not give a hang whether you live or die—and yet not so lost that you have no hope of finding your way. Of being alone and invisible, far from the eye of family and friends. Of being up against hardship and danger, but having the skills to surmount them. The feeling that anything can happen—that at any moment a magus may appear and transform you into a prince or a fleabitten cur. That the elevator into which you now step may take you to heaven or hell.

To say this last thing is to say that the city is what wilderness was—a world-pole. In times past, young men braved the wilderness to prove their manhood. The wild world-pole was the place of passage. Today the other world-pole, the big city, is where young men and women go to prove themselves, to make their fortune and win their mates.

WHAT KIND OF WORLD-POLE?

So it is "natural" that the city should play a central role in human life. But what exactly should that role be? And how can the city best perform it?

It is easy to chide the Mesopotamians for making the city the world-pole, yet it was a necessary step. They would not have been able to do the remarkable things they did if they had not been able to abstract human culture from nature to some degree. The same can be said of the Western civilization they helped shape. To view human culture as having its own logic, its own map, its own font and its own channels is both useful and true— up to a point. If a child does not assert its independence from its mother, it does not grow up. True independence is an illusion, but a useful one.

The human imagination needs a place where it can play with blocks— with hard, man-made objects. A place where technology is given free rein to make life as convenient, fast-paced, baffling, luxurious, tough, or decadent as it likes. A place where culture can obey the call of its own nature—casting filaments of symbolism, spinning webs of meaning from the rooftops. A place sealed off to some degree from the messiness of nature, so that humans can wallow in their own mess. A playroom into which Mother Nature may peer disapprovingly, but which she may enter only as a guest.

The city is to culture what wilderness is to nature: the epicenter of diversity, the hotbed of evolution, the font from which energy and information flow to the ends of the earth. A great city is a cultural rain forest, endlessly diverse, every niche stuffed to the gills. The eager, scrounging species that amuse tropical ecologists have their urban cousins among the con men, the hawkers, the microscopically small businessmen who offer services no one guessed were needed: making weights to hold down papers at newsstands, painting trade names on windows too high to read. Commerce does not care for vacuums any more than nature does.

The ecologists themselves have kin in the city. *Flâneurs* like Baudelaire, collectors of arcana like Walter Benjamin, lepidopterists of eccentricity like A. J. Liebling and Joseph Mitchell, mappers of food chains like Balzac, bitter gourmets of folly like Karl Kraus, chroniclers of mute, lost objects like the comic-strip artist Ben Katchor—such naturalists of culture are drawn to Paris and New York as surely as ornithologists are drawn to Ecuador. It is not that interesting tidbits of human culture cannot be gleaned in the suburbs or in rural villages; it is just that the city gives you more bang for the block.

The city is the place where culture gets to work out its complexes, to take its thoughts to their logical ends, unchecked (almost) by the "yes, but" of nature. It is the laboratory of a culture's future. It is also a cabinet of its past. Any city worth its salt is cluttered with junk; what is sad is when the junk is sloughed off too quickly, before it has had a chance to earn its patina or to work out its relation to unrelated things. Then it joins the junkheap, the city's *Doppelgänger*, where only God and garbage men get to savor its odd propinquities. The sloughing off will always seem too quick for the cabinet and too slow for the laboratory (though even a laboratory may find that old parts come in handy at odd moments).

Socialism, Engels said, would abolish "the contrast between town and country, which has been brought to its extreme point by present-day capitalist society." According to Raymond Williams, this contrast is the culmination of the division of labor—to which I would add that it may be just as necessary. To see the naturalness of the city, you must accept the need for division: not only between human and nonhuman realms, but between various human realms. You must see that Eden is not for us—but that there are all kinds of paradises and near paradises we can make, some of them as different from Eden in outward form as one place can be from another.

CITY LIMITS

The Tower should be seen as a deputy world-pole, independent in some ways though finally dependent upon the Mountain. This view is not to be confused with the more common Western (and originally Mesopotamian) one in which the city is the one and only world-pole, surrounded (at some distance) by wilderness. Quite apart from the values it implies, the mere geography of this view may be leading us astray in an age when there is precious little wilderness left. If we assume that there is a sea of wilderness around us, what is to stop us from enlarging the island on which we stand? Why should not the concentric rings of city and suburb expand at will?

In the third millennium B.C., the time of the historical Gilgamesh, cities began to have walls. Like the membranes of cells or the skin of the human body, city walls had a double function. They not only kept alien things (marauding nomads, wild animals, the armies of other cities) outside; they also helped concentrate and organize what was inside. Although ancient cities often spilled over their walls (the spillover forming the first suburbs), they had to retain most vital functions inside, so that the whole population could retreat behind the walls in time of siege.

When Hellenistic and Roman engineers began building cities according to plan, the walls were the matrix. The layout of streets, fountains, baths, temples, gymnasia, theaters, markets, and apartment blocks was shaped by that first square stamp. Aqueducts, sewage systems, streets, and public buildings were all planned to meet the needs of that number of people which would comfortably fit inside the walls. Once the population exceeded that number, the Romans would build a second city a few miles away rather than strain the infrastructure of the first. (They were not so scrupulous about older cities, least of all Rome itself, which grossly outgrew its facilities.)

Even the medieval cities of Europe, many of which were largely unplanned, were lent form, density, and liveliness by their walls. True, the wall could be moved outward to make room for more people: Strasbourg's, for instance, was rebuilt four times between 1200 and 1450. But this was enough of a nuisance that a good deal of infilling and piggybacking would take place first. When, in the age of the cannon, walls became elaborate fortifications—often covering more ground than the city itself—moving them was pretty much out of the question. Thus, while Strasbourg's population trebled between 1580 and 1870, its acreage stayed the same. Some would argue that in many cities the squeezing went too far, destroying open space and cooping many people up in lightless, airless tenements. On the whole, though, the walls forced cities to go in the right direction: upward. By the time the great bulwarks, made risible by modern warfare, were razed—leaving great curved boulevards in their place—most cities of continental Europe were well trained in the habits and methods of upward growth. Though they might still spread and throw out suburbs, a dense and walkable center was firmly established.

Not so in England or in America. In all of the Americas north of Mexico, there is only one walled city (Quebec), which helps explain why there are so few cities of character.

Bounds beyond the walls can be useful, too. The Greeks had ridges and ranges that acted as natural barriers between one polis and another, as natural limits to cropland, and as natural wilderness preserves. Although these limits were transgressed by logging and grazing, they did have a strong inhibitory effect. Large parts of Europe and America are flat and lack that advantage. There are few natural limits on urban sprawl—or on political agglomeration: and indeed, a case could be made that the proper relation between town and country was lost when city-states were swallowed up by empires and nation-states. For then the wild places between towns (and their outlying fields) were no longer no-man's-land, where shepherds of rival towns might cross paths and trade gibes; instead they were parts of the king's domain, or "the people's," that had not yet been settled or made profitable.

As long as human settlement has a dense and distinct center, it is not too dangerous. Nature knows where it stands. But when settlement becomes a thin gray wash over the landscape, nature does not know where to turn.

The city, I said, is a laboratory or playground where culture can play without too much interference from nature. But this does not mean nature can be ignored. Both laboratories and playgrounds tend to have walls around them, to keep dangers from getting in and mischief from getting out. So the first thing to recognize is that this province of culture—this "free city," like medieval Hamburg or Dubrovnik—must be geographically limited. Outside its walls, nature reigns.

Nature reigns inside, too, but less conspicuously. The laws of physics and chemistry apply, as do (still less conspicuously) the laws of ecology—for instance, that everything must go somewhere, and waste products do not just vanish into the thick air of garbage trucks. So the bounds that must be respected are not just geographical. A city that does not respect them can bring disaster upon itself and upon vast regions of the earth under its sway. But a city that does respect them can find ample room within them for the play of pure culture.

Outside the city, nature must be attended to more deeply and more constantly. There the play of pure culture would be a slap in nature's face. There the point is to play *with* nature—to jam, so to speak. It may be a sign of America's folly that we now do much of our playing with pure culture in suburban strip malls and industrial parks and alongside highways—out in the wide-open spaces, where nature's voice ought to be heard but instead is gagged with asphalt.

STALKING THE WILD METROPOLIS

We generally think that a city is a city: if nature is going to get in, a special place (such as a park) will have to be reserved. This premise is rejected by proponents of the "ecocity"—a city that would not be a city in the usual sense, for it would welcome nature in its every nook and cranny. It would be a marriage of Mountain and Tower in which the Mountain wears the pants.

The roofs of buildings would sprout gardens that would provide most of the city's food. Vines, shrubs, and trees would caress the organic contours of the walls. Vast tracts of urban land, as well as nooks and crannies, would revert to wilderness. In the drawings of one ecocity theorist, the imagined metropolis looks like a Mayan ruin reclaimed by the rain forest.

Let us grant that all the technical hitches can be fixed: that Styrofoam pellets in the soil will make roof loads bearable, that building materials will be found that can cope with dampness and probing tendrils and roots. Will these organic forms so appealing to back-to-the-landers appeal to people who like cities? Many city-dwellers may be city-dwellers because they want to escape the messiness of nature and dwell in a man-made order, or at least a man-made mess. The influx of insects and other flying, crawling, and swarming things so relished by the planners may strike them as a plague. The riot of green and gold may seem so much tinder for hay fever. "Wildlife," to them, means deer bearing Lyme disease, raccoons bearing rabies, owls eating cats, and bears eating babies. I am not just talking about neurasthenics in cork-lined rooms, fur-warmed women who "just adore a penthouse view," nerds who had their shorts run up the flagpole at summer camp, or exquisites like Oscar Wilde who define nature as a place where birds fly around uncooked. I am talking about ordinary city people.

Some of the most loved cityscapes in the world—the skyline of Manhattan, the markets of Canal Street, the medieval warrens of Prague's Hradcany, the walled Old City of Jerusalem, or for that matter a winding back alley, with its collage of chimney pots, clotheslines, brick and cement and fading paint, in almost any city you can name—have no visible nonhuman life in them at all, except for cats and pigeons and maybe some carp kept alive by a fishmonger's tender mercies.

As I noted when speaking of technology, most of nature is inorganic. Biomass accounts for one part in ten billion of earth's mass, and (as far as we know) virtually nothing in the universe. We would be deeply unhappy if we had no love for stone, sand, clear water, or metal. In the nonliving environment we have created, in cities, we would surely die.

But we don't die. Many of us find our biophiliac needs met by an occasional saunter in the park or jaunt in the country, and by the company of cats, dogs, and other human beings. Anyone who has sat around a kitchen table with friends or relatives, talking deep into the night, has probably felt that a deep human need—perhaps the deepest of all human needs—was quietly being met. The same need undoubtedly kept our oldest, shaggiest relatives up and gabbing around the campfire. Beyond their circle of light, grass whispered and animals moaned; beyond ours are brick and glass and the squeal of taxis. They got up in the morning to hunt mammoths, we get up to hunt contracts. But the basic human ritual stays the same.

Wildness must penetrate civilization, but it must not overrun it. Civilization, too, has a right to its own center and its own province. It even has a right to its own forms of wildness. The reckless play of culture, of technology, of trade, of social forms, of human types in collision: these are things that ecological puritanism would confine in the public stocks.

Even in the biological realm, it is very hard to say what wildness in a city is or ought to be. "If we want to find sustainable landscapes in cities," the Canadian landscape architect Michael Hough writes, "it is not to the lawns, fountains, and flower displays of pedigree parks and formal avenues that we must turn, but to the fortuitous landscapes off the beaten track, in the forgotten and waste places of abandoned industrial areas, in the nooks and crannies behind the corner gas stations, or in the down-at-heel residential areas where the maintenance man doesn't venture." Such landscapes, he says, "have a greater sense of place." If so, then the place they have a sense of is not New York, Toronto, or Berlin, but The City of the Temperate Zone. The flora of that "place"—Norway maple, black locust, quaking aspen, the redoubtable tree-of-heaven—crops up in junkyards and vacant lots in cities of similar climate all over the world. Though it has little in common with the "wild" or "natural" or "native" flora of the larger place any particular city sits in—or of the city's own acreage before it was settled—it is arguably the "wild" and "natural" and "native" flora of the city. Should we welcome it as the fair-and-square victor in the Darwinian tournament, urban division? Or should we condemn it as the product of environmental folly—as the vanguard of an alliance that will soon upset ecosystems in the hinterland, too—and suppress it, hoping to restore the preurban plant communities? There are no easy answers.

As I said, the city must learn to live within certain ecological bounds. It must not strain the carrying capacity of the land or lands to which it is economically and ecologically tied. It must not take more from the earth than it can put back, and what it does put back must be digestible. It must not impinge too much on the wild places on whose ecological mercies it depends. Of the technologies and design forms that can help it live within these bounds, some may be based on the close imitation of nature, but others may abstract from nature in outrageous ways. No rules can tell you in advance which will work better or which will make for a more satisfying urban life.

When the ecocity movement insists that cities live within certain ecological bounds, it is absolutely right. When it demands that they do so by means of "organic design," "natural process," or the wholesale importation and slavish imitation of wild nature, it goes too far. The deep nature of cities is denied. The needs of people who like cities are denied, too—and they are the people who need to be won over.

CITY WATER

The ecological problems that cities bring upon themselves (and upon other places) are legion. At various points in history, however, some cities have solved these problems or have avoided creating them in the first place.

In usurping the role of the Mountain, one of the things the Tower takes upon itself is the control of water. Often, it seems the very source of water—as the British phrase "the Company's own water" suggests. But the city's liquid hubris causes all kinds of trouble, from good old-fashioned filthy water to the sinking and cracking of pavements and buildings.

Boston offers a good look both at the problems and at one precocious effort to solve them. Founded on a readily defensible, bottlenecked peninsula in what is now Boston Bay, the city grew without much regard for its surroundings. As the Good Book promised, every valley was exalted, every mountain and hill made low. Most wetlands and backwaters were filled in. From being a peninsula in a bay, Boston became a nearly continuous expanse of settled ground mildly inconvenienced by rivers and inlets.

In the 1820s, the back part of the bay became the Back Bay—a neighborhood that, at the time, had little cachet but considerable bouquet. Shrunken, fouled, and cut off from the tides that had formerly flushed them, the remaining tidal flats stank to high heaven. Nor did the foul water know its place: shorn of its wetlands, the Charles River was now prone to sudden flooding.

In the 1880s, Frederick Law Olmsted took on the problem. One purpose of his Emerald Necklace for Boston—a garland of parks and greenways—was to restore some of the region's natural hydrology so as to limit floods and keep the water pure. Part of this enterprise was his plan for the Fenway: a long, kidney-shaped valley scooped out of the tidal flats. Within the valley, a meandering channel was dredged for the Muddy River, and the sewage that was helping to muddy it was diverted by underground pipes directly to the Charles. Normally the Fens' water would cover thirty acres, but when the river was in spate another twenty acres could safely fill up. With its cunningly (because naturally) shaped banks, the Fenway could double its liquid contents without raising the water level more than a few feet. The banks were planted with estuarine species to which it was a matter of indifference whether the water was fresh or salt, tickling their roots or over their heads. At the junction with the Charles, a tidal gate managed the flow of water to prevent flooding and keep the basin flushed.

A pioneering piece of ecological engineering, the Fenway would also be a park. As such, it had to be sold to Bostonians bred on the tidy Public Gardens. Olmsted knew which buttons to push. A salt marsh, he wrote, "would be novel, certainly, in labored urban grounds, and there may be a momentary question of its dignity and appropriateness . . . but [it] is a direct development of the original conditions of the locality in adaptation to the needs of a dense community. So regarded, it will be found to be, in the artistic sense of the word, natural, and possibly to suggest a modest poetic sentiment more grateful to townweary minds than an elaborate and elegant gardenlike work would have yielded."

The Fenway was built, and proved a triumph. The Back Bay soon became the blue-blooded neighborhood we now know, famous for its high-purposed matrons with fastidious, patrician noses. In 1910, though, Olmsted's masterpiece was trashed by the building of the Charles River Dam several miles upstream. Thirsty for water, unvisited by the tides, used as a dump for fill from subway excavations, the Fenway sickened and—for all intents and purposes—died.

Olmsted was ahead of his time, of course, but it may be that his time has finally come. In 1965, the master dam-builders of our age, the Army Corps of Engineers, advised against building a new, $100-million dam across the Charles. Instead, flooding from urban runoff should be curbed by buying or otherwise protecting eighty-five hundred acres of wetlands and other natural storage areas, at one-tenth the cost. In 1977, the corps began to implement its new plan—which, like "radical" efforts in Denver, Chicago, Portland, Toronto, and other cities, merely confirmed Olmsted's wisdom.

In chapter 29, I will come back to the larger role of greenbelts, green-ways, and other wild corridors, including those that follow the course of rivers and streams. Such fingers of wildness not only serve the city's needs, but tickle its pretensions. They remind the Tower that, however lofty it gets, it can never float free of its dependence on the Rivers of Eden, and so on the Mountain.

Beneath the asphalt of the densest city, streams still flow. Studying a low-income housing project in Minneapolis, Catherine Brown and William Morrish, who together head the Design Center for American Urban Land-scape at the University of Minnesota, found that the cracked walls and other signs of decay were not wholly the tenants' fault; a long-paved-over but still-urgent stream made the ground unstable. Brown and Morrish proposed, among other schemes, a plan in which the stream would be dis-interred, brought back to life and light, and made the heart of a vest-pocket wilderness. Something like this has already happened in Berkeley, where a long-forgotten stream called Strawberry Creek, freed of its concrete coffin, has become the nub of a neighborhood park.

Whatever its practical value, a place like that is vital as a symbol. How can city folk be reminded of their place in nature and still keep the freedom of their urbanity? Strawberry Creek is one answer. Another is offered by Rome's Trevi Fountain.

In 1723, Giovanni Poleni presented to the Royal Society of London a paper that, building on the much earlier work of Bernard Palissy, set out for the first time a modern view of the water cycle. Seven years later, the poet and philosopher Nicola Salvi marmorealized that view in his design for a renovation of the Trevi. In 1762—years after his death, and at the end of a tempestuous public-works process that lasted over two centuries, stretched across the reigns of fourteen popes, and drew on the talents of Alberti and Bernini as well as Salvi—his vision was realized in the fountain we now know.

At the center stands Oceanus, his draperies aswirl. At his beck, two winged horses led by tritons erupt from the stony earth, one acting the part of a prancing stream, the other of a galloping torrent. Actual water seethes about them, sliding into a pool that is big enough by fountain standards to signify the ocean. From the pool, jets of water jump skyward, enacting—in symbol and slightly in fact—the evaporation that closes the cycle. An urn wreathed in ivy thrums the praises of stored water, and carved foliage whispers of water's role in sustaining all life. As Salvi explained: "The Sea is, so to speak, the perpetual source which has the power to diffuse the various parts of itself, symbolized by the Tritons and the sea Nymphs, who go forth to give necessary sustenance to living matter for the productivity and conservation of new forms of life, and this we can see. But after this function has been served, these parts return in a perpetual cycle to take on new spirit and a new strength from the whole, that is to say from the sea itself."

Yet the Trevi is not simply an exercise in iconography. Off to the side are spouts you can drink from, and Romans do. Not long ago Rome had more than thirteen hundred working fountains at which people drank, bathed, did their laundry, and filled jugs for household use. Enchant as they might a Shelley or a Respighi, the fountains of Rome are infrastructure.

Americans have tended to frown on such splashy mingling of business and pleasure. Hawthorne called the fountains of Italy "monstruous devices in marble, all and one of them superfluous and impertinent." He preferred "my own town-pump in old Salem." But the Trevi is pertinent in the deepest sense. Infrastructure—like parks, buildings, and other features of urban design—can make clear the lifelines that join the city to suburb, countryside, and wilderness.

A small relief on the upper right of the Trevi's façade shows a young woman pointing to the ground while soldiers look on with interest. It is the legend of Trivia, the virgin sprite who, on a sweltering day in 19 A.D., led a band of legionnaires to the source of a secret spring in the hills east of Rome. Spirited to Rome by aqueduct, this Aqua Virgo is the very stuff that plashes in the Trevi. Trivial in the grand scheme of the planet's water cycle, it is crucial to the Roman's sense of place: his sense, you might say, of the local play of Mountain and Tower.

Is the Trevi so different from other waterworks I have spoken of— spoken of slightingly, more often than not? Not *so* different. They, too, play with the imagery of the cosmic waters. But where the main intent of their builders seems to be to take upon themselves the role of the sky god, here pure science (in the form of a theory of the water cycle) seems to breed a certain humility. True, the façade of the palazzo from which the Trevi bursts is blazoned with the names of popes; but the palazzo itself seems to dissolve into rocky surf. However glorious our Towers—Salvi seems to say—they come from the Mountain, and unto the Mountain they shall return. Water gives form; water takes form away.

THE MOUNTAIN'S BEST FRIEND

Wilderness holds the germ of the city, of every city that has been and every city that will ever be. Not only the biological stuff of the city, but the very traits that define its cityhood—the division of labor, the honeycombed thickness, the tangled exchange of goods and information—have their origin in wildness. And without wilderness to do the global housekeeping, no future city will long survive.

The city, for its part, contains in its stones and flesh and frenzy the memory not only of the wilderness from which it was once hewn, and of the wildernesses that are still chipped away to make it grow, but of the many earlier and huger wildernesses that helped push evolution in this peculiar direction. The city may even be said to hold the germ of future wildernesses. For without big, vital, and well-ordered cities, the tide of human population must soon overwhelm whatever wilderness is left.

DESPITE ARCADIAN CANT, the city is not a place you must escape from if you want to live a fully human life. Cities are natural. Even their unnaturalness is natural, for it springs from our nature and (if kept within bounds) can meet our quirky needs without doing nature too much harm. And it turns out that the Tower, which figured early in our story as a bastard pretender and enemy of the Mountain, can be its best friend and staunchest defender. Ideally, it concentrates both the warm bodies of humans and their steamy cultural energies in a small, bounded, insulated place, so that wild nature need not take the heat.

28

RECLAIMING ARCADIA

I F THERE IS one thing many Managers and many Fetishers seem to agree on, it is this: the middle landscape must go.

A return to primitive living is what many Planet Fetishers propose. They ask not simply that we take a step back, but that we make ourselves vanish—that we melt into the foliage, as certain American Indians were able to do. If they had their way, the swollen sphere of civilization—the range of the Tower—would be not just restrained but excised like a tumor. In that case, of course, the middle landscape would vanish, too.

In response to this vision, some Planet Managers have pointed out that, at present population levels, a return to primitive living—or even to some kind of soft-tech homesteading—would tax nature still more than what we do now. They have proposed their own form of *tsimtsum*, which they call "decoupling." This would involve concentrating human population in cities, and using the most advanced technologies to minimize the impact of their needs and wants on nature. The rest of the world could then revert to wilderness, which people could visit once in a blue moon.

Any symmetry between these two visions breaks down when we consider their value. The first is simply an egregiously goofy form of the dream of Eden and need not detain us. The second, though, bears some thinking about.

The decouplers have got hold of some home truths. They see that the Mountain and the Tower must have their own bailiwicks, and that the dreamily slatternly urge to mix them is behind many of our problems. They see that cities can be good for nature and good for us. They see that efficiency in the management of our own backyards will leave much more room for real, unmanaged wilderness. They see that the only alternative to civilization is better civilization. But while decoupling makes more sense than the idea of returning to Eden, it, too, may be pushed too far.

DANGERS OF DECOUPLING

One problem with radical decoupling is the excluded middle: middle landscape, middle lifeways. Not everyone wants to live at the same distance
from nature. A finely graded spectrum of distance—from the harsh green
of pure wilderness to the steel gray of the central city—has been part of the
mindscape and landscape of every civilization. In a form somewhat shifted,
but perhaps more finely grained, it has been part of more primitive cultures, too. Without it, some of the color is leached out of life.

To use the language we used earlier in the book, we might say the decouplers see the need for the organs of wildness and civilization—for the
Mountain and the Tower—but not for the network of nerves and arteries
by which they interpenetrate.

An even greater danger of decoupling is that we may lose for good—or
fail to relearn—the art of living with nature. Decoupling may present itself
as an about-face in our relations with nature, but in a way it simply speeds
up the march we have been on for the past century or two: a march away
from any real contact with nature. The few people who still have such
contact—loggers, hunters, small farmers, indigenes—are stragglers who
must be mopped up. Then we can settle down to purely aesthetic enjoy-
ment of nature, entrusting the extraction of food, energy, and so forth to a
handful of technocrats.

Like the extinction of species, the loss of our rural culture beneath the
treads of agribusiness is both tragic in itself and tragic in its effects. One
effect is the ecological damage done by agribusiness; another is the hemor-
rhaging of ancient know-how and local intimacy with the lay of the land,
which will be hard to replace if and when we want or need to go back to
small farming. Wendell Berry has eloquently described both those effects.
Another, subtler effect becomes apparent when one hears our intellectuals
talking about man, nature, and the future of the planet. "I am a child of the
suburbs," one writes, "and even though I live on the edge of the wild I have
only a tenuous understanding of the natural world. I can drive past hundreds of miles of fields without ever being able to figure out what's growing in them, unless it's corn." Although I now live among (dwindling)
farms, and have written about agriculture, I guess I am only slightly better
at this. I have done some gardening, but have never had to depend on it
to keep me alive. How on earth are the likes of us going to act as poet-
legislators for the planet, or as counselors in the marriage of man and
nature? For all our tramping around in the woods, we are ecological
celibates.

The contempt that many farmers, ranchers, loggers, and hunters feel for armchair environmentalists—even for active backpackers and birdwatchers—is not based solely in their wallets. In their heart of hearts, they feel that they know and even love nature better than the environmentalists do. In a sense, they are right. The backpacker loves nature chastely; the farmer wants to get a child upon her.

Just this, Wendell Berry would say, is the problem with modern nature-love, the love of Sierra Clubbers who look but don't touch. For we must touch nature somewhere, even if that somewhere is conveniently out of sight. And the more out of sight that somewhere is—the more removed from the gleaming nature of pinup calendars and television specials—the deeper the wound is likely to be.

THE CAREFUL SWAIN

Even if we could, then, we should not choose to argue the middle landscape out of existence. There are many ways of living on the earth, and many places to live. No one point on the spectrum is the right place for everyone. Wherever you choose to live, though, you have to accept the responsibilities that come with the territory—a small price to pay for the privilege of living on a planet as hospitable as this one.

What are the responsibilities of people in the middle landscape? To find them out, it helps to look first for irresponsibilities. Latent in the middle landscape—despite its humble airs—is a tendency to use resources with a lavish hand. Where settlement sprawls, both land and energy are squandered. Time, too, is squandered in schlepping, chauffeuring, and commuting, which means that a premium is placed on convenience (or imagined convenience) in eating and other functions. This in turn means that the tiniest item is swaddled in reams of packaging.

An oddity of our modern Arcadia is that it is crammed to the brim with commodities. Arcadia is supposed to be a place where just enough technology is used and no more. In suburbia, though, the appetite for gadgets is more avid than in the heart of the city itself—boredom, no doubt, being the main reason. Even those who leave the city in search of "the simple life" often end up using more resources (especially fossil fuel) and making more waste than they did before. The process goes something like this: In the first stage, the middle landscape looks perfect. In the second stage, you become aware of the tradeoffs. In the third stage—by degrees—you reject the tradeoffs, and help yourself to everything you want. Instead of a balance, you want the best of both worlds: and not just the best, but everything that is good. A horse, a snowmobile, a set of English garden tools, a pickup truck, a random-play compact-disc changer—your middle kingdom grows and grows. Rather then give up the stimulations of the city or the tonic of deep country, you drive vast distances to get both.

The first responsibility of the modern swain, then, is to resist the suction of glut: to be more swainlike (or perhaps swinelike, pigs being the least wasteful of higher animals). The second is to channel your consumption in useful ways. If part of the charm of the middle landscape is that it is, in part, a working landscape, then it is in your interest (as it is your responsibility) to keep it working. This means buying local produce and other farm and forest products. It means supporting land trusts, public purchase of development rights, and other methods of keeping farmers from getting squeezed to death between high land prices and low food prices. It might mean joining a land trust in which you and others build your homes on small parcels nested within an endangered farm, leaving the rest to be farmed in perpetuity. Or it might mean becoming partners with a farmer, in effect, by taking part in what is known as community-supported agriculture (CSA).

Seeded in Switzerland, Germany, or Japan (depending on which account you read) and transplanted to North America in the mid-1980s, the CSA movement now involves over 560 farms in the United States and Canada and countless more (countless partly for reasons of definition) worldwide. For each farm, a group of families, ranging from 30 to 130 or more, shares the expenses, guaranteeing the farmer or farmers a basic income. They also share the harvest, which they may pick themselves, pick up, or have delivered. Though the harvest will vary from year to year and from crop to crop, it nearly always supplies each family with an abundance of fresh, organically grown fruits and vegetables, and in some cases meat, milk, honey, or other good things. No only is this arrangement good for the farmer, who won't get rich but won't go under, either; it is also good for the community. Sliding scales may give poorer families a break, and overflow crops may go to soup kitchens. Occasional farmwork and festive gatherings (a first-plowing potluck, a midsummer bonfire, a fall harvest festival) bring people together, and the whole arrangement brings people closer to the land and its pungent, balky workings than a misted supermarket case could ever do.

To say that people come to the middle landscape to escape responsibility is only half true. Many come to find responsibility of a certain kind: responsibility for one small piece of the earth's surface. They like to live in a place that responds visibly to their efforts—their mowing, clipping, painting, tinkering. Isn't that what ownership of a free-standing single-family house is all about?

Commendable as far as it goes, this notion of responsibility does not go far enough. As Henry Thoreau demanded, "What's the use of a house if you haven't got a tolerable planet to put it on?"

Besides nurturing the health, both natural and cultural, of the middle landscape, the dweller in that landscape has a responsibility to nurture the health of both ends: wilderness on one side, the city on the other.

Human life is a current that runs between two oppositely charged poles, the Mountain and the Tower. Let either terminal corrode and the current runs down. Reduce both poles to the same flat valence and there is no power at all. Those who think they can live in the middle of the circuit and enjoy human life in its fullness forget, perhaps, where the power comes from.

UNSPRAWLING THE SUBURB

In the age of the steam railroad, a suburb was a small town surrounded by countryside. In the age of the automobile, a different kind of suburb has filled in the spaces between the old ones and oozed past them into the hinterland. It has also destroyed many of the old suburbs, or made them over in its image, by sucking the life from their centers.

The automobile suburb scatters its houses thinly along curving roads, many of them culs-de-sac. From each market-segmented "pod" of a development, "collector" roads lead to an "arterial," a multilane highway with synchronized stoplights. The arrangement looks good on paper, or did thirty years ago. Within the isolated pods, households would be brotherly as peas; a sense of community would ripen. Children would be safe from through traffic. Homes would not be troubled by the clangor of work and commerce. Nature (as we learned from Olmsted, who had learned it from the English landscape gardeners) would rejoice in the curving roads and great swards of grass. Traffic (as we learned from the highway engineers) would move with maximal efficiency, just as water moves through a hierarchy of mains and pipes. Best of all, developers would not have to waste a lot of money (or valuable acreage) on asphalt.

In practice, things have not worked out so well. With large lots and houses set well back, people keep to themselves. Since each pod (and often a whole development) is pegged to a particular market segment, families feel bound to move out as soon as their incomes no longer fit. Walking anywhere outside the pod is circuitous, tedious, and often hazardous. What ought to be a half-mile walk to a friend's house becomes a three-mile drive. Children are prisoners. Parents are paddy-wagon drivers. Clogged with everyone who has to go anywhere, the arterial grows sclerotic. Commutes once absurdly long are now obscenely long. Since only the arterial gets enough traffic to support stores, it is soon lined with them; and since this prime real estate is beyond the reach of local merchants, they are soon driven out by chains and franchises. The strip is born.

As for nature, whether she rejoices in this state of affairs is anyone's guess, but clearly the room she has to rejoice in gets smaller every day. Partly to give her more room, but mainly to give people a decent place to live, a school of architects and planners known variously as traditionalists, neotraditionalists, and New Urbanists is turning back to the model of the small town. In a small-town-style suburb, an old-fashioned grid gives people various ways of going from point A to point B. Car traffic is dispersed. On residential streets, it can be slowed down by means of jogged corners, planters, or plain narrowness. Foot and bike traffic can take as direct or as rambling a route as it likes. With houses close together, close to the road, and often wearing the broad grin of a front porch, almost any route is friendly to the eye. Some are friendly to the ears and mouth as well, offering chances to stop and chat. There are plenty of points B within walking distance: schools, parks, playgrounds, libraries, churches, stores, restaurants, coffee shops, bars, small factories, and other businesses. Many of these condense about the train or bus station, forming a dense town center. Here, too, are apartments for the elderly and the nonwealthy, who may also find accommodation in "granny apartments" or furnished rooms rented out by homeowners.

So far, most New Urbanist communities have been new: built from scratch, from raw countryside. While they may reduce the need for other, more damaging developments, they do not really provide the *tsimtsum* we need. For that, what is called for is "infill"—growth directed not into the countryside, but into thin spots (of which there are many) in the suburban fabric. This can mean reviving old town centers, or creating new ones in the maw of centerless sprawl, or adding houses in the gaps between existing houses. It is not an easy sell—least of all in residential districts. Still, changes in zoning and tax laws (on various levels) could encourage the building of front porches, in-law apartments, solar greenhouses, and other additions that might help fill some of the yawning voids in suburban space-time, raising population densities and making streets friendlier. The possibilities are greatest in declining suburbs, which in the United States account for one-third of the total.

BUNCHES, NOT PODS

For the foreseeable future, of course, development will continue to flow outward into the countryside. Some of this may be absorbed by "new towns" or the revival of old towns, but many people will be wanting to live "in the country," amid forests and fields. Is there any way to keep them from wrecking the very Arcadia they seek?

One way, pioneered in the 1980s by Randall Arendt, Robert Yaro, Harry Dodson, and others connected with the Center for Rural Massachusetts, is cluster development. Instead of letting houses line up like parked cars along rural roads, or fan out in space-wasting, farm-and-forest-killing subdivisions, the idea is to cluster a half-dozen here, a half-dozen there on small lots, tucked among trees or modestly shielded from the road by some landscape feature or other. As far as possible, wildland and farmland are kept wild or at work. Some of the saved land neighboring the houses may be owned in common or in trust for the families to recreate in. Instead of peas in rows of "pods," metaphorically speaking, you have bunches of grapes—which turn out to leave a lot more room for leafage.

A famous series of drawings by Dodson's firm shows a typical western New England rural landscape before development, after standard development, and after cluster development. The difference is striking—between standard and cluster, that is; the difference between before-development and after-cluster-development is slight, which is precisely the point. Where cluster development has been tried, the view on the ground confirms Dodson's vision.

Right now, the biggest obstacle to cluster development is not greedy developers or pigheaded home-buyers, but well-meaning zoning boards. Many rural townships and counties try to stay rural, or at least rural-looking, by mandating big house lots—a half-acre, an acre, even two acres. The result is an upscale kind of sprawl that looks rural only to those inured to its ubiquity, and which gobbles up farmland at a furious rate. Cluster development requires cluster zoning: allowing, or even encouraging, the building of homes on much smaller lots when they are clustered and properly sited.

Cluster development fits snugly into land-trust arrangements. A group of friends may buy a tract of forest and agree to keep it forest, bunching their homes in one section. A county land-trust may broker a deal in which a beleaguered farm, instead of being broken up into subdivisions, gets clusters of houses tucked into some of its less productive parts, while the rest goes on making milk or meat or corn. The latter sort of arrangement can even be combined with community-supported agriculture, so that the people living on the farm form the core group of shareholders. Something along those lines has been done in Grayslake, a rural township forty miles north of Chicago (and a few miles from the world's largest outlet mall).

The basic layout of cluster and open-space development lends itself as well to cohousing. In this somewhat more communal form of community, already common in Europe (especially Denmark) and now making converts in the States, families own their own homes but share some meals, child care, and other activities in a common social hall. Such a hamlet might seem custom-made to serve as the mainstay of a community-supported farm, and in Ithaca, New York, a cohousing group is playing just that role.

SUMMER PLACES, REVISITED

Though the suburban trend may be nearing its peak in America, it will be swelling elsewhere in the world for some time to come. As long as this is so, we can only do our best to strengthen what is good and weaken what is bad in the suburban way of life. It is a tall order, though, to save the suburbs from the deepest pitfall of Arcadia—the idea that you can find a single perfect place that mixes the best of nature and culture, which is tricky enough for one person and baldly impossible for masses of people.

Despite our best efforts, the overall effect of blending Mountain and Tower in this way is to denature both. My own growing conviction is that we do better to accept the split between city and country. If we need both—and many of us do—then the sensible thing is to move back and forth between them.

So far I have talked about reclaiming the side of Arcadia that stays put, or tries to—the side misguidedly enshrined in the modern suburb. But what about the other side: the seminomadic side, enshrined in all kinds of restless and hopeful moving about, but most of all in summer and week-end places?

In ecologically correct circles, summer and weekend places are a guilty secret. For all their sweet breezes, they are in a bad odor—elitist, greedy, ecologically unsound. Worst of all, they show a lack of stick-to-itiveness, of commitment to a single patch of ground.

In most modern ecological visions, there is not much room for moving about. It is frowned upon with special harshness by people we might call neo-agrarians, such as Wendell Berry, and by people who call themselves bioregionalists, such as Kirkpatrick Sale. The neo-agrarians take as their models some of the more skillful and forbearing peasants and small farmers of Europe and European America, while the bioregionalists like to talk about the native Americans. Both schools hold that any place is best cared for by people who have lived there a long time and plan to stay—who do not flit about, seasonally or otherwise.

The command to put down roots—to "stay where you are," as Gary Snyder says; to "stay home and be decent," as Wes Jackson advises—can easily get a grip on consciences of a Puritan bent. It singles out one side of human nature, the nesting and hunkering-down side, and calls it natural.

But the other side is just as "natural." Let us take a moment to explore it.

NOMADIC NOSTALGIA

All the world, the Talmud says, is watered with the dregs of Eden. On the simplest level, this reflects the image of Eden as a source of waters, from which great rivers rush to the ends of the earth. More mystically, it foreshadows the kabbalistic idea that everything in the world, down to the lowliest lump of coal and ugliest deed, is charged with the scattered sparks of a godhead which, in the first instant of creation, exploded like a supernova with the force of its own holiness. But more than anything else, it rubs salt in the wound of our exile.

Few wounds, and few truths, are as widely shared by humans as the exile from Eden. But when you look at it closely, this exile turns out to be an exile of a very peculiar sort. If Eden is viewed as a lost state of wildness—as the state of the Old Stone Age, or even of apehood—then it is a state of nomadism: if not in the strict sense, which involves herding, then in the loose sense of being often on the move. Even when Eden is read as a cipher for childhood, the same point can be made. Though most modern infants do not see the world as a Bedouin infant, jogging to a camel's trot at his mother's necklaced breast, sees it—as "a swaying nipple and a shower of gold"—even so, they, too, get carried around a lot. Without leaving home, a toddler travels distances that must seem immense.

Greek and Latin pastoral is based on the simple, rugged, and frequently rejuvenated life of the seminomadic herdsman. The Golden Age of Hesiod and others, when people slept on the ground and ate its spontaneous fruits, sounds a lot like the life of itinerant hunters and gatherers. So there is a paradox. We are exiled from a time or place—Eden, Arcadia, the Golden Age—which itself may represent a state of wandering.

Nostalgia for that state of wandering pops up in the oddest places. It pops up, for instance, in pleasure gardens, most of which—unless they are impossibly small—are made for walking. The exceptions prove the rule. In a Persian *chahar bagh*, or paradise garden, one did not stroll, but sat in shady meditation. Yet such gardens were traditionally part of far larger parks in which walking, riding, and hunting were assiduously practiced. The Iranians are supposed to have been pastoral nomads from the South Russian steppe. As agricultural society closed in on them, the rulers could indulge their nomadic nostalgia only behind the walls of their paradises.

The Mogul rulers of India were Mongol nomads who, by a twist of history, had adopted Islam and the Persian garden style that went with it. In India, they migrated seasonally from garden to garden, from paradise to paradise. The Emperor Jahangir, for example, would travel each year from Delhi to Kashmir, accompanied by thirty-five thousand cavalrymen, ten thousand foot soldiers, and heavy and light artillery.

Or consider the Emperor Hadrian. His life was, as Gibbon says, "almost a perpetual journey." He had "marched on foot, and bareheaded, over the snows of Caledonia, and the sultry plains of the Upper Egypt; nor was there a province of the empire which, in the course of his reign, was not honoured with the presence of the monarch." When he was not on the march, he was busy building his villa at Tivoli: a theme park featuring replicas on various scales of famous places he had seen on his travels—not just buildings, but whole topographical features, such as the Vale of Tempe in Greece. Did his villa represent a respite from wandering, or an apotheosis of wandering? To me it looks more like the latter. In any case, when we look at the great gardens of the Renaissance and later, we find them inspired more by the pastoral ideal than by any other. And pastoral, as we have seen, is full of nomadic echoes.

Can it be that gardening, the art of a sedentary people, reaches its peak in the recovery of nomadism? Ornamental gardens are the hallmark of a settled and stable society. "Cultivating one's garden" is the very image of a settled-down, domestic life, a life lived happily ever after. Yet it is also a means of escape from the impedimenta of civilization, most of which grew out of sedentism.

Even God walked in his garden in the cool of the day. Did he remember fondly his wander-years, before he was tied to a particular universe? Before he had a family?

Civilized peoples are not the only ones who feel exiled from the wandering life. Several times in every century, a prophet would arise among the Tupi-Guarani, a people of the South American rain forest, and promise to lead them to the Land Without Evil. In that land (he would tell them) there is no hunger, no sickness, no death, and no work. Hoes and spears guide themselves. Crops spring up like weeds. All laws of clan and village are suspended and the days are spent in sacred dancing and feasting. If the prophet was any good at all, they would follow him—sometimes for ten years at a stretch.

CAN HUMANS GROW ROOTS?

The restlessness of modern people is nothing new. Wanderlust is one of the oldest and deepest lusts around. As we have seen, even civilizations notable for their *Sitzfleisch* are pricked in tender places by pangs of nostalgia for the wandering life. We move between the Mountain and the Tower, between many Mountains and many Towers. Humankind does not belong in one or the other, or at one measured point in between. Nor is any people or person consigned to a certain spot or a certain way of life. If you find a place and a lifeway you like, and are able to keep it the way you like it, more power to you. But you must not set that stability up as the human norm.

In its sternest form, bioregionalism denies not only our human nature but the bare fact that we are animals. To insist that we "put down roots," that we be "connected to the soil," is to transplant *Homo sapiens* from kingdom *Animalia* to kingdom *Plantae*. Animals—as every schoolchild used to know—are motile. Our kingdom is of this earth but not in it: or at least, not in any specified square foot of it. What may be the most wrenching image in all art of our animalhood—of "unaccommodated man"—is spoken by a man cast out, a wanderer: Lear.

Animals are motile for the same reason they have walls (that is, membranes): it lets them maintain the conditions they like and avoid the conditions they don't. Instead of just walling out heat or cold or noxious salts or obnoxious predators, they can run away from them. If you haven't the knack of photosynthesis, the best way to maintain your negentropy—the basic state of life—is to go find something to eat. For that matter, even plants follow the sun, water, and soil conditions they like by means of tropism, of runners and other forms of long-distance vegetative propagation, and (on behalf of the next generation) of the dispersal of seed.

Migration is often a pursuit of sameness. Storks, whales, and nomads do not seek change. Nature changes, so they move. The change may be seasonal, or it may follow a longer curve. Farmers move, too, when climate changes or when the waves of change (partly their own doing) make the soil too poor or too dear. In the same way, whoever seeks Arcadia must stay a few steps ahead of his fellow seekers. Ideally, Arcadia stays the same and the Arcadian stays put; but the distant whisper of nomadism reassures him that if change does come—as, deep down, he knows it must—he can escape it.

For most species, staying put is feasible only when the climate agrees to stay put, too. When climate changes, species adapt or migrate—or perish. As it happens, the past ten thousand years have been an interval of climatic stability unmatched in the planet's recent history. Anthropologists have long known that sedentism and its partners—farming, cities, civilization—could have arisen only during an interglacial period, but they have long wondered why this particular interglacial period got the nod. The answer may be that only this one was stable enough. Before this one, people were too busy chasing the fickle rain, grass, and herds of game to settle down for more than a few months at a time.

We have no way of knowing how long our strange luck will hold out. But reason suggests that, if we try to base our civilization on the idea that people ought to stay put, we may be in for a rude awakening. A steady-state earth is as much a myth as nature's harmony is. To keep the earth as it is (or was) forever would take human intervention on a scale to boggle the mind even of the busiest technocrat.

TSIMTSUM IN THE CATSKILLS

Moving about is nothing to feel guilty about. But the idea of moving back and forth between city and country faces a hundred other objections. Even when practiced only by the well-heeled, it grinds down the social and economic pile of the countryside. Practiced by the masses, it would surely be a disaster. In fact, it would negate the ecological usefulness of cities. If each city-dweller must have his own chunk of countryside as well, what good are all those high-rise apartment buildings? Of course, not every city-dweller can afford his own chunk of countryside, but that is not a fact to congratulate ourselves on; the Kantian test (What if everyone did that?) is one to which, in this case, we must grit our teeth and submit.

Remember the dilemmas of the summer Arcadian recounted earlier. Seeking authentic but comfortable country, he brings the city with him. Seeking the unspoiled, he becomes a germ of spoilage.

Perhaps, though, we are painting with too broad a brush. What richly deserves blackening is a particular type of summer home: the single-family home, equipped with all modern conveniences, on a big piece of private property. Obviously, it is a type modeled on the suburb.

There are other traditions. Start with the most obvious: the villa tradition that runs from ancient Egypt and Mesopotamia to modern Europe. By way of the English country house, this is the tradition from which the suburban-style summer home mainly derives. But this older tradition has one great advantage: a single house can serve an extended family, as well as miscellaneous guests and retainers. Of course, if that house is a great house sitting on a great estate, there is not much *tsimtsum* to be gained. But if the house and grounds are modest—as they can be, since rarely are all members of the extended family in residence at the same time—we begin to see something that may point us in the right direction.

Even in America this tradition persists, and not only among the rich. A place to which family members return (however briefly) year after year, generation after generation, is one of the most precious things a family can have, whether it is a mansion on the ocean or a cabin in the woods. When families are scattered across the continent by the winds of livelihood or longing, such a place is a rock of stability. Where such homes are common, the countryside itself is anchored to some degree against the waves of change.

Most of these arrangements—like shared rentals, or time-shares, or rentals for part of a season—are considered stopgaps for those who can't yet afford a nuclear-family summer home of their own. But in fact they are closer to the sort of thing nature can afford. Not only that, but in their sense of festive community they come closer to the true spirit of Arcadia.

This brings us to another crucial point. The main upscale stream of the villa tradition—that of the *villa pseudourbana*—takes pride in bringing all the delights of the city into the heart of the country. Even the alternative school of the *villa rustica* has often had a rather lavish notion of rusticity; but at least it is on the right track. Different points on the gamut from Tower to Mountain call for different ways of life. The closer you get to the Mountain, the more you have to rein in your wants, your impacts, and your level of technology.

While the suburban-style summer home boasts all modern conveniences, others preserve a simpler way of life. Hunting and fishing camps, cabins, bungalows—their primitivity is not only appropriate to their place, but part of what makes them special. Every child knows this, even if grown-ups tend to forget. Though the "simple life" may be far more complex in some matters (such as those surrounding ingestion and excretion) than city life, it does tend to make fewer demands on nature. Moving back and forth between the two ways of life, you are prompted to think about both as you would never do in a full-time middle landscape. By setting aside a time and place in which you do without some of the objects you thought you needed, you may come to wonder whether you needed them in the first place.

Even if exercised only in a limited time and place, this material self-restraint—purposely denying oneself certain things and technologies, as if stepping backward in time—is as much a form of *tsimtsum* as the spatial self-restraint I was just talking about. Both forms are represented in that now decaying staple of the American vernacular landscape, the bungalow colony. Rampant, for example, in the Catskills both before and during the age of the great Borscht Belt hotels, this was a cluster of tiny houses (rented or owned), with shared open space and a shared "casino"—a battered common room with card tables and maybe a Ping-Pong set. An even better case in point, common in the Catskills around the turn of the century, was the *kochalayn,* a large farmhouse, perhaps with outlying cabins, in which several families shared a kitchen and dining room. And then there were the working farms that took in summer boarders, and sometimes accepted help with the chores in partial payment of the bill. The most famous of all Catskill hotels, Grossinger's, started out in just that way.

AT PLAY IN THE WORKING LANDSCAPE

A certain ambiguity in this section may not have escaped the reader's notice. Am I talking about moving back and forth between city and wilderness, or between city and country—that is, the rural middle landscape, the classic stomping ground of the summer Arcadian?

Well, I am talking about both. Which of them one chooses as a foil to the city is largely a matter of taste. As a matter of ecology, I would rather see people spending most of their leisure time in the rural landscape, making only brief forays into real wilderness. That way, wilderness is protected. But what happens to the countryside?

As I have argued in response to the decouplers, we need a middle landscape. The kind of middle landscape we really need, though, is not the suburban kind, but the kind closer to the Mountain: the true rural kind, in which people wrest a living from nature. Summer people like such places—or their images of them—and gaily invade them. In short order, the working landscape becomes a playing landscape. In the midst of his glee, the summer Arcadian is deeply uneasy: Do I really belong here? At best he feels like an outsider, at worst like an occupying power.

Though relations between city and country have never been without tension, in some times and places they have been closer and more friendly than they are here and now. In many cities of Renaissance Italy, each neighborhood would "adopt" a certain outlying village; even today, a flow of relatives back and forth, for both work and play, shows that such relationships persist. As late as the fourteenth century, English city folk of all classes were bound by law to help with the harvest in nearby fields. According to Mumford, "The summer exodus of the East Londoners to the hopyards of Kent is perhaps the last survival of that medieval custom." In Mongolia, many city-dwellers still return each summer to their herds, their tents, and their ancestral steppe.

When the urbanization of America was still in its early stages, it was common for people who had made the migration to spend their holidays with relatives who had not. Children would commonly be sent to spend their summers "back on the farm" with their aunt and uncle or their grandparents. Both young and old visitors often helped with labor-intensive jobs like haying or the harvest.

Such customs have been revived, in a way, by European and American outfits that fix middle-class city people up for working vacations on small, usually organic farms. Though the "yuppie migrant worker" may cut a laughable figure, the arrangement is a good one. The farmers get free labor; the city people get room, board, and a taste of the rigor, beauty, and tenuousness of the way of life that makes their food.

If community-supported agriculture can help to redeem the suburbs, it may play a like role in the summer and weekend world. A city family might buy a "share" in a farm, entitling them to a share of the produce. According to preference, they might have the produce shipped to them, or they might make periodic day trips to pick it themselves; they might take long working vacations on the farm, or they might be allowed to build a cottage somewhere on the grounds. Simply buying a building lot on a farm, perhaps as part of a land-trust arrangement that keeps the farm viable, is another option.

Nor do city people always need a farmer to intervene between themselves and the soil. In Austria today, many large cities are ringed by farmland where burghers (many of modest means) tend small allotments. On some plots, the weekend gardeners have built huts or cabins where they can spend the night. Similar arrangements are common elsewhere in Europe.

Many farms have long donated bruised produce to food banks; now some have revived the biblical practice of gleaning, allowing the poor to pick over the fields after harvest. More to modern taste, perhaps, is the arrangement in Pittsburgh, where patrons of the city's main food bank can work off their chore hours on one of four nearby farms (one of which is leased by the food bank). Six hundred volunteers work on the farms each

season, and a number have been hired as paid farmhands. Since the volunteers are, in effect, paid in kind, nutritionists at the food bank help them
figure out what to do with odd-looking vegetables.

In the Middle Ages, cultural differences between country and city folk
were far greater than they are today, yet peasants came to the city regularly
to sell, buy, and take part in festivals and fairs. Nowadays many country
people (like many suburbanites) think of the city as an alien, perilous, over-
sexed and overpriced realm where, if you don't get rolled, you'll surely get
fleeced. (If medieval peasants had similar feelings, necessity brought them
to the city anyway.) The new city greenmarkets where urbanites buy their
arugula from the man or woman who grew it, rather than from a lab-
coated cashier, are a only a first step—though a crucial one—toward a
reimagining of the great medieval fairs and markets. Much needs to be
done to make the city a reasonably friendly, accessible center of trade and
culture for its hinterland.

Of course, the world does not divide neatly into city people and country
people. You may be a city person at one point in your life, a country person
at another. You may grow up in the country, move to the city to make your
fortune and find a mate, then move back to the country (or the suburbs) to
raise your kids. In doing so, you are no more unnatural than a salmon who
swims upstream to spawn.

LOVE OF PLACE

Some would say that all of this moving about keeps people from forming a
deep attachment to any one place. And this raises a broader question:
Should an environmental ethic be founded on love and knowledge of a
particular place, or of the planet as a whole?

Of a particular place, say the bioregionalists and the neo-agrarians. Of
the planet as a whole, say others, among them such unlikely bedfellows
as the Planet Managers and the New Agers.

Perhaps the question can be answered by default. It is probably too late
in the day for people to start putting down roots that will deepen with each
generation. Hunter-gatherers and nomads, it is true, did most of their
migrating within one bioregion. But their wanderlust did not have television and movies to excite it or cars and planes to gratify it. The day is fast
coming when people will choose their landscapes, just as they now choose
careers rather than inherit them from their fathers. For some, the landscape
of choice will be the landscape of childhood. For others, finding the landscape of their dreams will be the work of a lifetime. "There is undoubtedly
a deep affinity," writes John Cowper Powys, "probably both psychic and

chemical, between every individual human being and some particular type of landscape. It is well to find out as soon as possible what kind this is; and then to get as much of it as you can."

It is true that people who move around a lot do not always feel deeply responsible for the places they pass through. Often, though, a place finds its staunchest champions and acutest students in people who come from—or end up—someplace else. Often, the most rapturous writing about a place is done by newcomers to it and exiles from it. Of course, exiles from one place are newcomers to another and often write feelingly of both.

Among the Australian Aborigines, a folktale is told of an Aranda people who, being forced to leave their ancestral land, looked back from a mountain pass to admire its beauty. Had they ever admired it before?

No people on earth has sprung from the soil of its place. Even the Aborigines are not aboriginal, but came to Australia within the last sixty thousand years. No culture is pure. All are products of history, of migration, of hybridization.

Against the false cosmopolitanism that would paint the world with one broad brush, the best defense is not provincialism, it is true cosmopolitanism. Only by understanding and respecting other cultures can we hope to make a world in which our own culture stays vital. By the same token, love of a particular place is not, by itself, enough to keep the planet in shape. No doubt it would be a good thing if regions were more self-sufficient, if they did more stringent ecological accounting, and so minimized their ill effects on other regions. But in any practicable future the planet and its various peoples will stay enmeshed in a net of cause and effect. We will have to find a way to feel responsible not only for the places we pass through, but for the places we never come near that are nonetheless touched by our actions.

Nature itself may have found a way to throw that responsibility in our face. The flocks of refugees that flutter against our shores are often nothing more or less than our own political, economic, and ecological fecklessness coming home to roost. But perhaps there is some instrument other than this least welcome form of nomadism—something that can kick in sooner, before so much damage has been done.

Will an abstract love of Gaia do the trick? Probably about as well as the abstract love of humanity serves to prevent war and exploitation. A better bet, for my money, would be a love of particular *places*—not one, but several—and at least a vague sense of how they figure among the many different places needed to make a functioning world. And this in turn is possible only for those who travel, migrate, move back and forth between city and country, or otherwise get around.

We carry around so much guilty baggage about moving or staying put. If only we could accept that all of us, natives and nomads alike, are sojourners—that we have the land on loan, and must hand it on in good condition. Whether to our own children, or someone else's, or back to the wolves and snails should make no matter.

29

TWO NETWORKS

ARCADIA DRAWS US because it is the place where nature and culture most richly interpenetrate. We draw Arcadia, and paint Arcadia, and prosodize Arcadia for the same reason. The way a pasture laps the shore of a wood; the way a hedgerow skirts a field of grain; the way a village clings to the flanks of a stream—the particular charm of a middle landscape lies in its particular geometry.

A part of the aesthetic pleasure we take in these shapes may come from imprints of ancestral landscapes that shaped us. Another part may be more intellectual—the hint of an answer to a problem we are not aware of posing, but on which our fate partly depends. Roughly, that problem is this: What is the best spatial relation between wildness and civilization? In what topologies and in what shapes must they interpenetrate, if each is to be true to itself and neither is to denature the other?

Strange to say, it is only lately that this problem has come to the attention of Western science. Even the arts and letters have dealt with it only glancingly. We tend to think of the relation between nature and culture, when we think of it at all, either quantitatively or qualitatively; we rarely think of it geometrically. Not the least important fact, however, about the shape of things to come is, literally, their shape.

Although I have sometimes spoken of the waves of human-led change as if they were concentric circles, they are of course nothing of the kind. They move in splashes, trickles, narrow channels, spurts. They follow tracks, highways, rivers, ocean currents, power lines. Being mindful of this network of civilization—and of its inexact inverse, the network of wildness—can help us check the brutal effects of the waves of change.

It may even help us roll them back. In the Lurianic system, *tsimtsum* is not just a retraction; it is also a sharpening of structure. As the godhead pulls itself inward to make room for the universe, its diffuse infinitude crystallizes in a structure: the *sephirot*, which are imagined as the head, trunk, limbs, and genitals of a macrocosmic man. Similarly, in cells, organisms, cities, empires, and other living systems, a phase of expansion is often followed by a phase of retraction or retrenchment in which structure is clarified. For its own good if not the planet's, the empire of humans and their allies should be desperately eager to enter such a phase.

As the network of civilization is refined, the network of wildness must be extended. Wildness must flourish on every scale, from the continental to the vest-pocket and backyard, and on every inch of the continuum from the peak of the Mountain to the bowels of the Tower. Flowing into the places civilization has claimed, it may or may not reclaim them, but at least it will keep them alive.

A HEDGE AROUND THE TREE OF LIFE

Start with the Mountain. The last few waves of people and their allies have spread so far and surged so high that most of the world's wildernesses have become islands in a sea of human activity. Like real islands, they are losing species. Big as they seem, most are not big enough to give full range to their top predators, or to permit seasonal or long-term migration, or to provide many species the genetic hedge they need against evolutionary faltering and ecological mischance.

Unlike real islands, these terrestrial islands do have a good deal of intercourse with their surroundings. But intercourse that would be helpful, such as the inflow of native organisms from nearby wild places, is often blocked, and the intercourse that does take place is often harmful. Just outside the boundaries that politicians draw to mark wilderness preserves, mining and logging injure the larger watershed; straying bears and wolves are shot by ranchers or stopped by their fences; streams are dammed and water is drawn off for homes, farming, or industry; invasive exotics are given free rein. Some of these things go on even within designated wilderness areas: for instance, one-third of such areas in North America are grazed by non-native domestic animals.

"Make a hedge around the Torah," advised the rabbis of the Talmud. They took their own advice, setting up a barrier of law that would keep Jews several steps away from a slothful or thoughtless or hotheaded transgression. For them the Torah was "a tree of life," so precious that it must be "hedged about with roses." Surely that other tree of life—the one that stands in the midst of Eden—deserves no less. Indeed, the Midrash says the Tree of Knowledge formed a hedge around the Tree of Life; is this not a fitting role for our knowledge and our technology?

The granddaddy of all wilderness areas in the lower forty-eight is also the first to hedge itself around—to "buffer" itself, as ecologists say. Covering a total area of ninety-four hundred square miles, or about six million acres, New York's Adirondack Park holds nine-tenths of all the wilderness east of the Mississippi and north of the Mason-Dixon Line. With a little pushing and twisting, Yellowstone, Yosemite, Olympic, and Grand Canyon national parks could all be jammed within its borders. Only 42 percent of the land within those borders, though, is publicly owned and thus subject to the state constitution's pledge, made in 1894, that it be "forever kept as wild forest lands." The rest was subject to a crazy-quilt of local and state regulation until the mid-1970s, when the park's chief planner, George D. Davis, drew up a regional master plan. Aware that a landscape is as various in its sensitivities as a human body—gnarled and callused here, tender there—Davis had the park's anatomy mapped. On the basis of that map, a zoning map was drawn that determined where, in what forms, and in what quantities farming, logging, and building would be allowed.

Though sorely tested by the second-home boom of the past decades—with as many as a thousand houses going up a year—the plan has been hardy enough to spawn offspring all over the world. In 1991, Davis and the firm he now heads, Ecologically Sustainable Development, teamed up with thirty Russian and American scientists to design a regional plan for what may be the biggest region for which a regional plan has ever been designed: the 150-million-acre Baikal watershed, which sprawls across parts of Mongolia and three provinces of Russia.

The better we manage our own activities and those of our allies in the neighborhood of wild places (and everywhere else, for that matter), the less we have to manage the wild places themselves. And that means we can let them *be* wild.

The debate about endangered species bears scrutiny in this light, too. The worldwide emphasis on endangered species has been enormously useful in tugging at heart strings and purse strings. And because a species is (generally) a distinct entity that can be scientifically and legally defined, and is made up of individuals that can be counted, it is a useful sort of thing to write laws about. Both these merits have their flip sides, though. Cute or charismatic species may do well, but the fuss over a snail darter leaves the public bemused, and even a passably attractive species such as the northern spotted owl can be tarred and feathered by those who think their livelihoods threatened by its protection. Then, too, what lends itself to legislation often lends itself to litigation.

But the biggest problem with trying to protect endangered species is that, in the nature of the case, it generally happens too late. The U.S. Fish and Wildlife Service has a backlog of thirty-six hundred candidates for endangered status. Many have died in the waiting room. By focusing on ecosystems, we can protect vast numbers of plants, animals, fungi, and microbes—many of which we have not yet even identified—before their state gets too desperate. With a bit of foresight, we can do it before development pressures have made land acquisition prohibitive and planning a minefield.

BUILDING BRIDGES

Maybe the best thing you can do for islands of wildness is to build bridges between them. In the terms I have used before, these are the arteries of wildness that flow from the heart of wildness to the uttermost limbs; these are the rivers that flow from Eden to the ends of the earth. Known to ecologists—who have lately taken a great interest in them—as landscape linkages or corridors, they are stretches of land, and sometimes water, that have been preserved in or restored to a state of wildness—at least relative to the land that surrounds them—and that connect one patch of wilderness (or other habitat) to another.

What exactly are their functions? One is to give wide-ranging animals, such as large mammals, the wide range they need to forage—to hunt shifting game, to follow shifting grass—and the choice of habitats they need for breeding. When that ranging follows the rhythm of the seasons, we call it seasonal migration. Then it may span a much larger area—in the case of many birds, the length of several continents. Birds, or course, can fly from wild patch to patch, but even they will use a corridor, such as a mountain range or river valley, whenever they can, since this makes it much easier to stop for a nap or a snack.

Another function of a landscape linkage is gene flow. A population of a given species on a terrestrial island of wildness, like a population on a real island, is likely over time to become inbred, losing the genetic variability that would keep it tough and resilient. But if cousins from off-island can find their way on, they can provide the needed new blood. A like process takes place one notch up, at the ecosystem level. As species in a wild patch weaken and die out, whether because of inbreeding or random disasters, the ecosystem as a whole loses diversity—something that island-biogeography theorists, in a strange semantic flip, call "relaxation." But a bridge can allow members of the lost species (or, failing that, species in the same line of work as the lost species) to recolonize the island. As missing players are replaced, the team is brought back to midseason form.

Another kind of migration takes place in response to long-term changes in climate, available nutrients, or other vital statistics. Faced with global warming, for example, animals may move northward or uphill, perhaps over the course of many generations; plants may do the same, by dispersal of seed, over the course of still more generations. But species stranded on an island—real or terrestrial—are trapped. If, on the other hand, there are bridges of habitat they can inch along, generation by generation, they may just outrun the fiercening sun that pursues them. Similarly, a species may have to decamp in response to a catastrophe: a volcanic eruption, a hurricane, even the periodic, healthy purgation of fire. If the exits are locked, you end up with the ecological equivalent of the Triangle Shirtwaist Fire. A linkage not only lets members of the species escape, but also lets the species recolonize the patch when the successional time is ripe.

HEDGEROWS

Just as islands of wildness occur on many scales, so do the bridges between them. In 1993, for instance, the provincial government of British Columbia set aside the watershed of the Tatshenshini and Asek rivers as a national park. A mere 2.5 million acres itself, the park joins wild lands in Alaska and the Yukon territory to form a reserve of twenty-one million acres.

As we move farther from the Mountain, the patches of wildness get smaller and so do the links between them. In the rural landscape, linkages of various kinds have long been a tool of game management. When most of the American Midwest was falling to the plow in the early part of this century, a practical interest in game birds and game squirrels (of all things) gave a reprieve to some stream banks and strips of woodland. The fencerows and shelterbelts of osage orange, cottonwood, ash, and cedar that went up after the Dust Bowl years on the Great Plains served not only to stop wind, hold water, and check soil erosion, but also to shelter and give passage to wildlife.

The ancestors of these linkages, like the ancestors of the people who made (or did not unmake) them, can be found in the Old World. Hedgerows may be almost as old as farming, at least in once-forested places like Europe; the compound of hedgerows and fields now covers about a tenth of the earth's land surface. Barriers to domestic animals, hedgerows are thoroughfares for many wild ones. In England, the rambling medieval hedgerows, with their motley assemblage of trees and shrubs, were homes and highways for a wider range of animals than the straight hawthorn hedgerows of later enclosures. With time, the two types have somewhat run together, as both attract birds and the birds disperse seeds. English hedgerows embrace as many as six hundred species of vascular plants and offer breeding sites to four out of five English woodland wildlife species. Two out of three lowland terrestrial bird and mammal species patronize them. The French *bocages* of oak, chestnut, wild cherry, roses, and other plants are similarly hospitable.

Wrens, butterflies, snails, and shrubs are believed to colonize hedgerows from adjoining woods. The flow goes the other way, too: while populations of chipmunks and white-footed mice in small, scattered woodlots often suffer local extinction, woodlots with hedgerows attached are soon recolonized. Of course, these species are perfectly able to move across open fields, but they do so at peril of their lives.

For all their service to wildness, hedgerows do not have to be particularly wild. The vast majority start out planted, as lines drawn across fields rather than lines left when forests were erased. Their service to wildness is ultimately a service to humankind, yet they help us in more direct ways as well. Apart from keeping the cows out of the corn, they arrest soil erosion by wind and water; they retain soil moisture; they slow the runoff of minerals and other nutrients, keeping farmland fertile and streams clear; they baffle winds that would parch or chill crops. Hedgerows give shade to people as well as animals; they provide both meat and the firewood to cook it over. A study of a French parish in which nearly every household heated with wood found that nearly all the wood came from hedgerows. Coppicing—cutting trunks in a way that makes for resprouting—can give a reliable supply of wood that is just the right size for the stove. Pollarding—lopping off each branch's new growth—can furnish food for livestock. Chestnuts provide mast for beasts and flour for people, and other nuts, fruits, and berries from hedgerows are similarly versatile. Though too-greedy use or too-willful management of hedgerows by humans can impede the flow of wildness through them, a long history shows humans and nature sharing them with great amicability.

Hedgerows and shelterbelts are not the only kinds of linkages needed in the middle landscape, of course. For reasons that are fairly obvious but will be detailed a bit later, much wider swathes of ground and water are needed, too. Yet it is interesting that, as we move from the middle landscape toward the Tower, the minimum width of an effective linkage may actually need to go up. By and large, plowed or grazed fields are a friendlier matrix for wild things, and so for paths of wildness, than are asphalt, concrete, or monocropped lawns. If we want wildness to penetrate the very marrow of civilization—and we do—we are going to have to allow it some fairly big entrees.

GREENWAYS

As a very general rule, the network of wildness and the network of civilization are inverses, like the goblet and kissing couple in a figure-ground picture, or the bats and birds in an Escher drawing. Roads for people are barriers for other animals. A German study found that forest beetles and mice almost never crossed two-lane roads, and shied away even from an unpaved forest road that was closed to public traffic. Small forest mammals in southern Canada stayed clear of roads when the road clearance was more than twenty meters. Even desert mammals unfazed by open spaces are often fazed by highways. And we all know what happens, often enough, when an animal fails to be fazed.

In wilder places, though, what is even more damaging than roadkills or the slicing up of habitat is the human activity that roads bring, which often leads to the dicing, mincing, and pulverizing of habitat. The role of roads in bringing settlers, loggers, ranchers, miners, and drillers into the Amazon rain forest is well known.

Nevertheless, the networks of wildness and civilization can sometimes, in some places, coincide. From the days of the *voyageurs* to the mid-nineteenth century, the map of interior America (both North and South) was largely the map of its rivers and their banks. Rivers were the main arteries of Western civilization, the routes by which European flora and fauna, led by people, invaded the hinterland. Yet they were also major arteries of homegrown wildness. As the hinterland has filled in, their role in the flow of wildness has grown even more vital. They remain major arteries of civilization, too: their waters are less used for transport, but rails and highways line their banks.

More often than not, city greenways hug the sides of rivers and streams: these being, if they have not been channeled or forced underground, the last routes open to wildness in an urban setting. In an earlier chapter I spoke of Olmsted's Emerald Necklace, which not only provides flood control for the Boston area but also entices wildlife—birds above all—into the city. Tucson, Toronto, Washington, D.C., and other cities have shown similar foresight.

The networks of Mountain and Tower can coincide in drier places, too. A swathe of tallgrass prairie that runs through some of the densest development in the upper Midwest is known as the Amtrak Preserve: a passenger-railroad right-of-way managed by The Nature Conservancy, which saves Amtrak the trouble of spraying herbicides once a year. The rights-of-way of railroads that radiate from London allow wildlife to commute from the countryside. In sandy soil, they harbor rabbit warrens and fox dens. High-voltage power lines move the very juice of civilization, yet their shrubby cuts through suburbs and even through forests are cheerily used by songbirds, deer, and many other creatures.

HIGHWAY OR HOME? AND FOR WHOM?

Some biologists worry that landscape linkages might be used as readily by invasive nonnatives (including pathogens) as by the natives they are meant for. Generally, the more a linkage approximates wildness, the less likely it is to beckon the exotic traveler. A cognate concern is that linkages may link too much—that they may mix up populations or subspecies that were formerly distinct; that they may bring together species from different regions that were better off apart; that they may reduce allopatric speciation, the forming of new species that sometimes happens when populations of a species are separated. When you think what an unbroken tract of near wilderness this continent was when white people arrived, such worries seem misplaced. As for allopatric speciation, our present splintering of the landscape seems to have created far fewer species than it has destroyed.

We come back to the old problem of balancing the need for flow against the need for division. In the hierarchy of natural systems—cell, organism, ecosystem, bioregion, continent, biosphere—the units at each level naturally do more exchanging of matter and information internally (that is, between the lower-level units they contain) than with other units at the same level. The broad tendency of human meddling over the last ten thousand years has been to cut off flow within bioregions and to increase flow between bioregions. Whatever we may be thinking, this is how we act: we flow globally and choke locally. Both ways, we reduce biodiversity. The task that faces us now is to backtrack a bit and bring back, as best we can, the balance of flow and division, locally and globally, that had obtained for the previous two hundred million years of earth's history: that is, since the breakup of Pangaea. That balance of flow and division—of rivers and walls, so to speak—produced a reasonable balance of speciation and extinction, giving us the pied and dappled, wildly melodious world we now take for granted.

Abstract principles alone, though, are not enough to guide the making of landscape linkages. Much depends on which species you hope to help. A planned network of linkages may be insufficient for some creatures (such as top predators), unnecessary for others. Some species may not use a linkage that seems custom-made for them, or may use it only sparingly.

Though it has deep roots in folklore, the scientific study of landscape linkages is just beginning. We are just starting to get a spatial picture of how wildness and civilization might rub shoulders without too much friction, and the picture will keep changing as society, technology, and the state of nature change. One rule of thumb, though, seems hard as nail: the bigger our wild places, and the wider and wilder the linkages between them, the less management we'll have to do. If we now have no choice but to act with specific threatened species in mind—gauging their rates of movement, mortality rates, germination rates, and so on, and planning linkages accordingly—this is a job we should turn over to nature the first chance we get.

GREENBELTS AS GIRDLES

A good network of wildness, then, should have in it the seeds of self-expansion. Linkages should bring wildness into the heart of the Tower; they should also help to restrain or turn back the Tower's own relentless self-expansion. The kind of greenway known as greenbelt is exactly that: a green belt that holds back the gluttony of the city, for its own good as well as nature's.

Perhaps the first greenbelt bylaw can be found in Numbers 35, in which the Children of Israel are told to give the Levites cities with "open land" about them "from the wall of the city and outward a thousand cubits round about." The open land would be pasture "for their cattle, and for their substance, and for all their beasts." God was not singling the Levites out in this respect; he was just making sure their cities would have what most ancient cities had. One reason for a belt of open land was defense: in medieval times, too, there were often zones outside the walls *non aedificandi*—not to be built on, lest they give cover to attackers. Another reason was convenience: husbandmen in the city could be close to their pastures, burghers could have a pleasant walk or ride outside the walls. In prosperous cities, suburbs sprang up anyway: in 1288, Friar Bonvesin da la Riva wrote of Milan, "Outside of the wall of the moat there are so many suburban houses that they alone would be enough to constitute a city." Under population pressure, walls could jump outward and greenbelts would have to retreat proportionally. In that case the amenity of open land might be retained, but the growth of the city was not restrained.

By the late nineteenth century, the amenity was being lost, too. Cities such as London rumbled outward, chewing up the home counties at such a pace that only a very determined walker from the inner city could hope to take a stroll in the countryside. In 1898, alarmed by this juggernaut and hoping to reclaim the balance between nature and culture that small cities of the past had known, Ebenezer Howard published the book that would later be known as *Garden Cities of To-Morrow*. Fleshed out with social theories and economic projections, Howard's basic ideas were two: First, that a small, collectively and flexibly planned city, partly industrial but girdled with an ample, inviolate belt of farmland, would be, for most modern people, the perfect place to live. Second, that when cities start to swell beyond their ideal size, they should spin off satellite cities to which they would be linked by rail lines.

Just as the first idea had its origins in the ancient world, so did the second. Growing Greek cities routinely gave birth to daughter cities, though mostly at a colonial distance. Roman planners, too, would found new cities when old ones got too fat (a regimen never imposed on the city of Rome itself). The idea did not exactly die in the Middle Ages, which was after all a time of busy city-founding, but it did lose much of its rigor. Leonardo da Vinci in his notebooks conceived a plan for relieving the overcrowding of Milan by building ten cities of five thousand houses and thirty thousand inhabitants each, with horse and foot traffic kept apart, and with gardens watered by municipal pipes.

Unaware of Leonardo's, Howard managed to get his own ideas off the page and on the ground. Rallying private investment, he founded the Garden Cities of Letchworth and Welwyn—both great successes, neither widely copied. London's swelling went on apace. Not until 1944, facing a rush of discharged soldiers, did Parliament pass the Town and Country Planning Act, which formally enshrined both of Howard's core ideas. Like a rising loaf thrust into the oven, London's spatial expansion has stopped just where it was in 1944. The suburbs suddenly drop off. Beyond them is countryside—a greenbelt of five to ten miles in width, with a much wider aureole beyond it that is almost as sacred. Beyond these, in a ring stretching from twenty to forty miles from central London, are the London New Towns—ten of them, not counting the old towns that have been allowed to expand.

The plan has been a great success—in some ways, too great a success. Industry has thrived in the New Town ring, drawing population not only from London but from other parts of the country. The shift toward a service economy keeps much of the job growth in London, and the rail system is so good that a commute of forty or fifty miles is considered a trifle.

Between these factors and the postwar baby boom—which, oddly enough, the planners had not planned on—the New Town ring (to say nothing of the ring beyond it) has gotten far more crowded than the planners planned. Though the New Towns themselves have—as Howard hoped—their own greenbelts about them, many of the older towns (Reading, High Wycombe, Luton, Bishop's Stortford, and so on) are now ringed with their own miniature suburban sprawl.

All in all, however, the success is still a success. Suburban sprawl, if it is not wholly prevented, is at least interrupted. And if there is not much land anywhere near London that a Yank would call wild, the ancient countryside still harbors a great many wild things. In wilder America, a greenbelt can give city folks access to correspondingly greater wildness—as in Portland, Oregon, whose citizens live cheek by jowl with elk, bald eagles, and chinook salmon.

A SPATIAL SABBATH

Probably the most sweeping vision of *tsimtsum* yet proposed in any detail is the North American Wilderness Recovery Project, known to its friends as the Wildlands Project. Under this plan—noted earlier as a joint effort of Managers and Fetishers—at least half the land surface of the lower forty-eight would be devoted to core reserves and the inner parts of corridors between them. Within these areas all roads, dams, power lines, and other prying fingers would be dismantled. Outside these areas, like a bulky parka keeping off the chill of civilization, would be buffer zones of limited human activity. But the parka image is not quite right, for civilization would no longer be the outside and wilderness the beleaguered inside; instead, civilization would be confined to pockets within the matrix of wilderness. In a way, the buffer zones would be more like the lining of the stomach, keeping acids and toxins away from the rest of the body.

The Wildlands Project speaks deeply to my wilder self. It may even flow logically from some of the premises I have set out. Given the political realities of the moment, though, I want to propose something slightly more modest.

In an earlier part of the book I said that, as the Rivers of Eden distribute wildness in space, so the septenary cycles of jubilee, sabbatical, and Sabbath distribute wildness in time. They make ponds and puddles of wildness on every scale, from which the rules and uses of human economy and technology are excluded, or in which they are mirrored softly, lazily, and upside down.

The analogy, too, can be turned on its head. If the Sabbath is a wilderness in time, then wilderness is a Sabbath in space. Why not use the interlocked cycles of Sabbath, sabbatical, and jubilee as a model for the way wildness ought to be distributed in space?

Take any geographical unit—country, state, province, county, town, borough, precinct, block, backyard. Let each unit devote one-seventh of its land to wilderness, or something as close to wilderness as circumstances permit. If the wilderness is there already, let it be preserved; if not, let it be created.

The result, you might think, would be that one-seventh of every country would be wilderness. You would be deceived, for two reasons.

First, what I have in mind is not a simple fraction, but a sum of fractions. If a county did not already have a seventh of its land in state or federal wilderness, it would add some reserves of its own to fill out its quota; and so on down the line.

Second, on the continental scale the portion of wilderness could and should be much greater than a seventh, at least in some cases. Thus a continent such as North America, South America, or Eurasia (to say nothing of Antarctica) would ideally preserve vast tracts that either resist settlement and exploitation, or are thought to be vital to the workings of the biosphere. At the risk of stretching the analogy past the breaking point, such tracts might be compared to the Edenic or the messianic era, "the time that is all Sabbath." As these stand beyond the geometry of time, so the great wildernesses of the world stand beyond the normal geometry of human space. They are the Mountain itself.

A cynic might ask: Why all this *tsimmes* (Yiddish for a stew, metaphorically a fuss, not to be confused with *tsimtsum*) about the number seven? Should a question of great ecological weight rest on the slender reed of a Chaldean superstition? Is this notion any less arbitrary than, say, the recent insistence of a Republican Congress that the federal budget be balanced in exactly seven years—an insistence based, as the Speaker himself admitted, on a hunch?

Well, it is *somewhat* less arbitrary. Ecologists estimate that, at a bare minimum, 5 to 10 percent of an ecosystem must be preserved if it is to stay at all healthy. Make it a seventh and you have a margin of error. (While trying to set aside one-seventh of every "original" ecosystem might raise more problems than it solves, it is plain that, in choosing which parts of a state or county or town to set aside, we should make sure that key ecosystems are represented—just as we should take into account all the questions of spatial relation, connectivity, and so on that were touched on above.) Besides, it is hardly arbitrary—or it is arbitrary in a useful way—to join a culture's sense of space to its sense of time, and to ground both in the bedrock of ancient symbols. Even those self-styled conservatives who despise conservation might warm to the idea of a network of wildness, were it presented to them in these terms. Many of them claim to know what the temporal Sabbath is all about; should not a spatial Sabbath be right up their alley? If we can set aside sevenths of our time for holiness—that is, for purposes higher than human aggrandizement—why not sevenths of our space?

TSIMTSUM IN YOUR OWN BACKYARD

On so large a scale, bringing about a healthy spatial relation between Mountain and Tower, between the network of wildness and the network of civilization, takes political action. Without shrinking from this, some of us may be forgiven if we wonder what we can do with our own two hands, in our own backyards. How can we set aside a seventh—or, if we choose, more than a seventh—of our own turf for nature's use?

In her marvelous book *Noah's Garden: Restoring the Ecology of Our Own Backyards*, Sarah Stein gives the beginning of an answer. Slowly and with many missteps, Stein has transformed her own six acres in New York's Westchester County from the museum of invasive exotics typical of suburbia to something an Indian might have recognized. By bringing back native trees, grasses, and wildflowers, she has brought back as well such native Americans as the bluebird, the meadowlark, the great blue heron, and numberless butterflies.

Not everyone with a backyard has so many acres to work with, or the time and resources (including a son who is a plant molecular biologist) to work with them so well. But if all the hours and dollars now spent on suppressing native plants (above all by lawn-mowing) were spent instead on encouraging them, a fairly spectacular *tsimtsum* could be achieved in a fairly short time.

Many people, told that their grounds must be 100 percent native and natural, rebel—and rightly. They like their tulips and tea roses, their tonsured hedges and lawns. But one can have such things and still be ecologically correct. The key is to keep them in their place.

Ancient Roman estates could have tousled meadows and groves as well as formal topiary gardens. Renaissance villas could have sylvan *boschi* as well as formal parterres. In thinking that his naturalistic lawns had to run all the way to the door—that every thread of knot garden had to be ripped up, every boxwood fancy lopped down—Capability Brown was, at best, confused. And some advocates of the wild garden may be said to share his confusion.

Every lived-on property, whether an estate of ten thousand acres or a brownstone lot with a backyard a few yards square, has its progression from inside to outside, from hearth to wilderness. Metaphorically it has that progression; it ought to have it in practice, too. Close to the house, artifice is in order: the well-tended garden, the stretch of lawn for children to play on. Farther out (and, generally, farther from the road) wildness should be invited to make itself at home.

In all this, one rule must be respected. Exotic plants are fine, but only if they are a lot of trouble. The kinds of exotics that garden stores and catalogues are most eager to push, and many people are most eager to buy—hardy, spreading perennials, eagerly self-seeding annuals, "low-maintenance" plants in general—are exactly the kinds that out-Herod Herod and must be avoided, because they may be invasive. They may escape into the wild.

By and large, this means that exotics belong in the high-maintenance, relatively formal part of your garden. If a nonnative plant needs cosseting to stay alive in your climate, it is not likely to survive in the wild. If, on the other hand, a nonnative plant can be "naturalized" in a naturalistic setting, it does not belong there: it is invasive enough to be shunned.

Is it only for reasons of mythic convenience that I put the formal garden near the house, the wild stuff farther off? Not at all. Other sorts of convenience also come into play. There is your own convenience in having things that need tending close at hand, as well as a relatively controlled environment for sitting, reading, playing, eating. There is the convenience of wild things, many of which are shy of houses and people. And there is the convenience of wild things in a larger sense: the convenience of connection, of flow. The back of your property may border a woods or a meadow; more likely, it borders the back of someone else's property, which—for reasons of mythic or personal convenience—is likely to be the wildest part of that property. In many very ordinary suburbs, the result is a series of woody or shrubby corridors that stretch most of the length of a neighborhood and link up with larger, often public wild patches.

Mostly, this happens by accident. If people did it mindfully—making an effort to restore native ecosystems and maximize the flow of wildness—the result would be that much better. Stein suggests a way of doing this with little or no palaver or prearrangement. Let us suppose you live in a suburb of squarish lots in a region like the Eastern United States, which naturally wants to be forest. You preserve or restore a margin of woods and thicket along the back and sides of your property, with as large a patch as possible in one corner. If everyone in the neighborhood does this, the effect is a network of corridors with large nodes at many of the intersections—a suburban version of hedgerows and woodlots. A tiling of the suburban plane, in which each lot is a tile. The more tiles in place, the more complete, beautiful, and useful the pattern.

30

HOT AND COOL

No one has told my cat about global warming. Confronting December in the country, she regularly grows a winter coat, something she never did in Manhattan. Her thighs are shaggy as a musk ox's. The flea comb no longer plumbs her depths. When she grooms herself she sounds like a rooting pig. The wildness in her, sprung from its dungeon, had already spoken up in a hundred ways that were fearful to watch and sometimes unpleasant to clean up. The new coat is the most striking thing, though, because it is so physical.

The rest of the household is worried about global warming. We knew all along that there was work to do on the earth's behalf (and our own) and that much of it was urgent. Still, it seemed manageable. We would chip away a little at this problem, a little at that, and as more and more of us applied our chisels the shape of a sane planet would emerge. But now the solid material we had been working with seems to melt under our hands. Our problems multiply, ramify, run in rivulets of unpredictability.

Nuclear power rises like a phoenix from its contaminated ashes. Hydro-electric dams rear their massy heads. Nature preserves that looked worri-somely small now look hopelessly small. Forests weakened by acid rain stand helpless before the gathering heat; wetlands besieged by humans on land are now to be attacked by the rising sea. If there was hope that agri-culture might finally shake its addiction to chemicals, it now looks as if new pressures (drought, floods, the northward advance of southern pests) may cause it to backslide. Environmental cooperation among nations, which was just hatching, seems likely to be pecked to death in a flurry of new inequities and recriminations.

Briefly, the facts are these. Over the past century, humankind has added large quantities of carbon dioxide, methane, halocarbons, nitrous oxide, and other "greenhouse gases" to the earth's atmosphere. These gases let the sun's energy in but do not let much heat energy out. All other things being equal, temperatures at the earth's surface should rise as a result. All other things are not equal: a skein of feedback loops involving cloud cover, ocean currents, even the activity of soil bacteria complicates things almost beyond imagining. Some of the loops are negative and will make things better, some are positive and will make things worse, and some have not yet declared their valence. Add to this the cooling effect of sulfates and sundry man-made dust, the lag in air temperature caused by the ocean's absorption both of carbon dioxide and of heat, and the fact that we are due for an Ice Age in a few centuries, and you have some idea of the mess of entrails our haruspices must read. Computer models requiring billions of calculations predict that, unless rigorous action is taken, global mean surface temperature will rise 1.8 to 6.3 degrees F. between 1990 and 2100, with wild variations (in rainfall and hurricanes as well as temperature) from place to place. Although this may not sound like much, the total global warming since the height of the last glaciation, eighteen thousand years ago, has been about ten degrees F.

Supposing the models are right, other models try to guess the effects: forest death, forest fires, extinctions of flora and fauna, the spread of tropical diseases, drought in the world's breadbaskets, mutant typhoons thawing of the polar icecaps, climbing sea levels, Venetian cities. Even the possible benefits of global warming (fewer winter illnesses, stimulation of crop growth by carbon dioxide, longer growing seasons in Russia and Canada) are likely to be tainted by social disruption, for we are talking about a rate of climate change far greater than anything mankind has seen. And there will be aggravating factors: farmers will have to deal not only with heat and drought but also with increased ultraviolet radiation (caused by ozone loss in the stratosphere), soil erosion, acid rain, and air pollution (such as ozone in the troposphere).

The models are only models, but they have grown markedly in sophistication over the past few years; more to the point, events have begun to bear them out. A number of studies indicate that temperatures have already risen an average of one degree F. over the past century. Many of the predicted effects of global warming—the storms, the crumbling ice shelves, the migrating mosquitoes—are already manifest. In 1992, the 2,500 scientists who advise the world's climate negotiators were collectively perched on the fence; in 1995 they came down in a body with the resounding (for scientists) pronouncement that "the balance of evidence suggests . . . a discernable human influence on global climate."

Oblivious, my cat grows a winter coat. The prospect of global warming might not have affected her plans in any case, since New England winters are still plenty cold. It has affected my plans, ironically, by making me think harder about how to thicken the fur of my house. If I insulate better I can burn less wood and oil, and belch less carbon dioxide. Keeping warm this winter—and cool next summer—I can do my bit to keep future summers cool for everyone else.

That I am copying my cat does not bother me at all, in fact it soothes me. With all this talk about the end of nature, it is good to be reminded that there is still a quantity of nature in each living thing and that we can learn from it. Is it true, as some think, that one smudged human thumbprint, one sour breath, is enough to denature nature? No truer than that one touch of nature is enough to naturalize civilization. The wildness in our genes and in the genes of our cats and cabbages affects civilization far more than our dams and dark emissions affect nature. We modify nature, but nature makes us from scratch.

MANAGERS AND FETISHERS, AGAIN

Climate is likely to be one of the main fields on which, in the next decade and the next century, Planet Managers and Planet Fetishers—and other parties—will fight it out. Most of the books so far published on the subject are by Managers, which is hardly surprising; not only do they outnumber the Fetishers to start with, but it is just the sort of problem that gets their workstations humming. A taste of their approach is offered by a piece of junk mail I got a few years ago from the Union of Concerned Scientists which, in hectic letters on the outside of the envelope, asked: "IS EARTH'S CLIMATE OUT OF CONTROL?"

Had I stopped beating my wife? As far as I knew, earth's climate had never been under control. Certainly it had never been under mine. It had, perhaps, been under the control of the earth itself, if one accepted some version of the Gaia hypothesis; but that was not what the question on the envelope seemed to imply. It seemed to imply that, if we were not at present controlling the climate, we should be.

In practice, most Managers' goals turn out to be more modest. Admitting that global warming is not certain and that its extent and results are wildly uncertain, one prominent Manager, the climatologist Stephen H. Schneider, puts his recommendations into two broad and overlapping classes. The first consists of "anticipatory adaptations"—things we can do now that will help us adapt if and when global warming becomes a problem. The second category is made up of things that we would be wise to do anyway, for independent reasons, even if global warming fizzles out.

Fortifying our cities to resist higher sea levels is an example of the first; improving our energy efficiency, of the second. Making our agriculture more diverse and adaptable falls into both classes.

On the Fetisher side, the main contribution to the debate has been Bill McKibben's book *The End of Nature*. A lucid and deeply moving account of the causes and likely effects of global warming, the book also responds to the problem, both practically and philosophically, in ways shaped by Deep Ecology. McKibben does not think much of technical fixes. By concentrating on the kind proposed by the crazed futurists, such as covering the oceans with Styrofoam pellets to reduce their absorption of sunlight, he makes all technical fixes sound repugnant. By arguing that, no matter what we do, it is too late to stop global warming altogether, he makes both technical fixes and government action sound hopeless. Instead, he rests his fainting hopes on a sort of religious revival, a millennial fit of repentance such as the West has not seen in centuries.

He asks that we return to a "humbler" way of life, a "path of more resistance." What he has in mind is not hunting and gathering, exactly, but a Gandhian or Thoreauvian life of homesteading and abstemiousness. To begin, we can turn the heat and air conditioning down, mothball our cars, take long walks instead of airborne vacations, and stop having kids.

So the battle lines are drawn. What follows will plot a path of moderation between the two poles that have been set up—between the teeth of wildness and the suction of total control. It will also apply the principles (and antiprinciples) of Earth Jazz to a problem that desperately needs them.

THE GREENHOUSE PRINCIPLE, OR THE NEED FOR WALLS

There is nothing unnatural about the greenhouse effect itself. It has been in place for eons. Without it, Earth would be nearly as chill and lifeless as Mars. As I mentioned in the chapter on Gaia, three and a half billion years ago the concentration of carbon dioxide in the atmosphere was hundreds of times greater than at present; that is how Earth was kept warm enough for life to emerge in the sun's salad days, when that star was perhaps one-quarter less bright than it is now. The planet has been shedding atmospheric carbon dioxide steadily, like a man peeling off layers of clothing as the morning warms up. The problem now is that humanity, like an overprotective mother, is adding layers. The expected (and partly confirmed) result, global warming, may be unnatural or artificial, but the greenhouse effect itself is not.

In fact, the greenhouse effect displays some of nature's most ingrained habits. Erwin Schrödinger, seeking to define life to a physicist's satisfaction, found its signature in the ability to resist or reverse the flow of entropy. Among other things, this can mean maintaining an energy gradient between inside and outside. The relevant inside can be the cell, the body, the dwelling, even (if the Gaia theorists are right) the planet. This is one aspect of the need for walls, which I have spoken of several times as counterweight to the need for flow.

The art of maintaining such a gradient is in part the art of insulation; more broadly, it is the art of energy efficiency. The greenhouse effect embodies both. Although no one has told my cat about the greenhouse effect, it is clear that she grasps the basic principle. Mammals are expert in the art of insulation, being distinguished by hair or blubber as well as by milk. Carnivorous mammals are doubly expert when it comes to energy efficiency (nap, pounce, nap). But every life form masters this art in some way, for none can afford to do otherwise. Only man, with his knack of fire, has had the luxury of getting sloppy about energy efficiency. And only modern man, sinecured by fossil fuel, has been able to slough off the discipline altogther.

Except in my capacity as mammal, I am not an expert on energy efficiency. But some people who are think that world carbon-dioxide emissions could be reduced enough to arrest global warming simply by the use of existing energy-conservation technologies, such as superinsulation, efficient engines, industrial heat recovery, materials recycling, and low-emissivity window coatings. As the last example makes clear, it is a matter of fighting the greenhouse effect with the greenhouse effect. A study by Amory and L. Hunter Lovins of the Rocky Mountain Institute and two German scientists argues that a planet industrialized to western Germany's level, with a population of eight billion and a gross product grown fivefold, could use one-third less energy than it uses now. The secret of this prestidigitation lies in the fact that efficient technologies would pay for themselves (some sooner, some later) in lower fuel bills. Far from requiring brutal government intervention or a gutted standard of living, the stabilization of global warming could be achieved by helping the free market get its hands on the right devices, which it is now prevented from doing by hidden energy subsidies and other political baffles.

Though other experts are less confident of this, most seem to agree that the cheapest, quickest, and cleanest weapon against global warming, and therefore the one that should be deployed first, is energy efficiency. I like that idea, and not because it promises to perpetuate the Western way of life we all know and sometimes love. After all, some of the forms that energy efficiency can take are not mere technical fixes, but involve changes in our way of life, such as growing more food locally (in solar-heated greenhouses, to give an apposite example) instead of shipping and trucking it outlandish distances. No: the reason I like the idea of energy efficiency is that it is an instance of learning from nature. And learning from nature—listening to nature—is a prime technique of Earth Jazz.

THE PRINCIPLE OF FLOW

Another thing we can learn from nature is the principle of flow. Our technologies must take account of the fact that their products come from somewhere and have to go somewhere. In particular, fossil fuels must give way to renewable energy. The new solar-thermal plants, in which sunlight heats oil that turns water into steam that drives turbines, are already as cost-efficient as late-generation nuclear plants (which, admittedly, is not saying much). In some places, wind power is now as cheap as electricity from coal. Photovoltaic, geothermal, tidal, hydrogen, and small-scale hydro technologies are also coming down in price (despite the malign neglect of the Reagan years) and do not release any greenhouse gases. (Nuclear power does not, either, but it does produce radioactive waste that has to go somewhere and as yet has no safe place to go.) If hidden subsidies to fossil fuel were removed—and if the true costs of global warming were reflected in a carbon tax, as in Denmark—renewable energy would soon stand revealed as a remarkable bargain.

Since vehicles account for one-fifth of the world's carbon-dioxide emissions, the replacement of gasoline with ethanol and other fuels made from recent biomass—corn, sugar cane, algae, agricultural wastes—could make a big dent in global warming. (To be sure, support for public transportation, in both the developed and the developing world, could make a much bigger dent.) As long as the biomass is constantly replaced, the principle of flow is honored and there is no net addition of carbon dioxide to the air. Even something as atavistic as wood, if it comes from a properly managed woodlot, is preferable to fossil fuel. Wood, too, releases carbon when it is burned, but the carbon has been captured from the atmosphere recently and can quickly be captured again by new saplings. The carbon in much of our oil and coal was captured by the vast rain forests of the Mesozoic, whose like we will not see again unless global warming gets far worse than we expect. Locked up for millions of years, it floods the carbon cycle—as if the dead should arise from their graves and enter the job market. Likewise, vast amounts of carbon locked up in tropical rain forests are now being released by burning, not for energy but simply to clear land. So we might add a corollary to the principle of flow, and call it the Pandora Principle:

what nature has locked up should be unlocked only with the greatest caution.

We come back to the Rivers of Eden. Though most rivers flow one way, they could not keep flowing unless evaporation and rain closed the circle. All rivers have banks, which should tip us off that flow in nature is not random. In the very idea of a river, the need for flow is balanced by the need for walls. When it comes to elements and compounds more specialized than water, the need for walls is even greater. Cycles must be carefully channeled. Some natural substances must be left where nature has put them, and some man-made ones should not be made in the first place.

Unlike the Lovinses and most other Managers, I think some cutting back of consumption (among the well-off) will be called for. The exact mix of cutting back and technical fix is partly a matter of public policy and partly of personal taste. Some may think superinsulation unnatural (though they would have trouble convincing an otter or a musk ox of this) and prefer to shiver a little. If the hairshirt fits, I say, wear it. But in general what is needful is not to live poorer but to live smarter. For our species to try to insulate itself from the scrabbling and squashing that is nature would be the most unnatural thing of all. We cannot walk on air, but we can look where we are treading.

THE PRINCIPLE OF DIVERSITY

How is it that my cat, raised in Manhattan walkups, knows how to grow a winter coat? Although she is a domestic animal, the wildness has not been utterly bred out of her. She still has some "ecological genes" in her, as a breeder would say, that let her adapt to changes in the environment. Her species as a whole, moreover, has a great deal of genetic variation in it, making it infinitely more adaptable than any individual cat could be. On a still higher level, the diversity of species within an ecosystem allows it to adapt to disturbances that might otherwise be fatal.

Here, then, is another lesson we can learn from nature. If the buildup of greenhouse gases that has already happened means that we are "committed" (consider the exemplary career of this word: insanity, politics, romance, and doomsday in just four decades!) to several degrees of global warming, then we shall have to adapt. And the key to adaptation, especially in the face of the unknown, is diversity. We must do everything we can to stanch the loss of diversity in flora and fauna, domestic as well as wild. The diversity of human ways of life must also be preserved, in fact multiplied, above all where food and energy are concerned. Diversity, like energy efficiency, is one of nature's fingerprints, as visible on the blown-glass surfaces of human culture as on a meniscus of seawater squirming under the microscope. And as beautiful, as Gerard Manley Hopkins knew:

> Fresh-firecoal chestnut-falls; finches' wings;
> Landscape plotted and pieced—fold, fallow, and plough;
> And áll trádes, their gear and tackle and trim.

Diversity may prove useful not only in adapting to global warming, but also in stopping it. We do not know how the levels of carbon dioxide and other greenhouse gases in the atmosphere, and thereby the earth's temperature, are regulated (assuming that they are regulated, which some scientists dispute). But it does seem clear that loss of diversity in a given ecosystem usually brings about instability. Each species we lose (and we are losing dozens each day) is a piece knocked out of the structure of our house. We do not know which ones are holding up the roof; in a sense, all of them are. A plantation of cloned trees may lock up as much carbon dioxide as a native rain forest, but its role in the larger mechanism that regulates carbon dioxide and other greenhouse gases is likely to be very different.

THE PRINCIPLE OF SELF-REGULATION, OR LETTING NATURE WORK

This brings us to a third thing we can learn from nature: to sit back sometimes and let nature do the work. Learning from nature need not always mean imitating nature. It can also mean leaving big chunks of nature in place to do what they have evolved to do.

We tend to think of nature as a warehouse of raw materials awaiting our shaping hand and animating breath. In the light of insights provided by the Gaia theorists, among others, it now appears to be more like a greenhouse—an oddly commodious shelter in a cold and gasping universe. You need not buy the Gaia hypothesis lock, stock, and barrel to see that many things in nature, from cells to ecosystems, tend to maintain stability in the face of disturbance. Any adjustment of, or adjustment to, climate change will need the help of the animals, plants, fungi, and microbes of the rain forests, wetlands, oceans, and continental shelves. In providing that help, they will move in mysterious ways. Our attempts at management will only get in the way.

Most important, then, is that we get out of the way: in other words, that we perform a *tsimtsum*. This means not only taking up less space, but reducing our impact in other ways. Efficiency, recycling, thrift, good workmanship: all these things can help. As I said, the better we manage our own affairs, the less we need to manage wilderness. The better we do our own job, the freer we leave nature to do hers.

LIKE A VIRGIN

Twenty years ago, the hot topic in popular climate books was not global warming; it was the coming Ice Age. The geological clock said we had run through most of a typical interglacial, and for a while the thermometer seemed to agree. Paradoxically, it may turn out that both predictions were right—that we shall have both global warming and a glaciation in rapid succession. In *The Ages of Gaia*, Lovelock writes, "Humans may have chosen a very inconvenient moment to add carbon dioxide to the air. I believe that the carbon dioxide regulation system is nearing the end of its capacity.... The perturbation of a system that is close to instability can lead to oscillations, chaotic change, or failure." A lurch in one direction may trigger a wilder lurch in the opposite direction.

In the face of such uncertainty, are all our well-meant actions so much stumbling in the dark? Not at all. At the heart of Earth Jazz are improvisation, flexibility, responsiveness. We need to respond to the threat of global warming on the basis of the best information we have now. But we need to respond in a flexible way, so that when our information changes we can modify our response. If we use this opportunity to limber up our industry, tune up our transportation system, and diversify our energy sources, we will be better able to deal with global warming, oil embargoes, economic turmoil, and even a new Ice Age. (If we decide we need to let more carbon dioxide into the atmosphere, we can do so at a moment's notice. Taking it out is not so easy.)

The burden of Earth Jazz—a burden at first heavy, then amazingly light—is to accept that nature is in flux and culture is, too. We cannot spray a fixative on nature so that it holds the shape we are comfortable with. Nor can we force ourselves to match nature's stasis, which is only a mirage anyway. The earth's climate will change drastically, at some time or other, with or without our help. We can fight the change, adapt to it, or do both. If the change is partly our doing, we can change what we do. But we must always be ready to adapt, for sudden change is not a one-time emergency; it is a fact of life.

DEEP DOWN THINGS

Although stabilizing global warming may be the toughest environmental problem we have ever faced, the answers are at hand. We do not even need

to draw a pentagram on the floor and summon up the nuclear and genetic
engineers. The answers are not easy, but they are not horrible or unnatural
or Procrustean. Some of them might even be pleasant.

This is not to say that I join the Planet Managers in swelling with adrena-
line at the prospect of adventure ahead. If we see global warming as yet
another challenge, yet another chance to show our stuff, we are bound to
make yet another hash of things. The optimism of the Managers has been
shaken by global warming, and that may be a good thing. We need to be
taken down a peg. Whether we need to be ground under the heel of a brutal and now somewhat deranged nature (or ex-nature) is another matter.

Recoiling from control, which they think impious, the Fetishers want to
renounce influence as well, which cannot be done. I think they are right to
recoil from control. The total control promised by some—all the earth a
formal garden, or a tight and tidy spaceship—is impious; worse, it is
tedious. It is an ancient dream, but to my mind a bad one. A full-time job
as "planetary-maintenance engineer" is not something I would highlight in
the help-wanted pages. If the Fetishers want to cuddle up in a childhood
memory of Eden, the Managers want to grow up, but with a vengeance: to
become not only their own fathers, but the fathers of all life.

I would like to believe that nature will always elude total control—that
the dream of absolute power is not only impious and tedious, but fatuous.
At the same time, I would like to believe that, just as we can never wholly
subjugate nature, so nature can wholly abandon us. Hopkins again—

> Generations have trod, have trod, have trod;
>   And all is seared with trade; bleared, smeared with toil;
>   And wears man's smudge and shares man's smell: the soil
> Is bare now, nor can foot feel, being shod.
> And for all this, nature is never spent;
>   There lives the dearest freshness deep down things . . .
> Because the Holy Ghost over the bent
>   World broods with warm breast and with ah! bright wings.

Holy Ghost or no Holy Ghost, I am encouraged in this belief by trivial
miracles like my cat's winter coat. If, as the climate models suggest, she and
her progeny have to trade it in eventually for something lighter, I like to
think that the services of Genentech will not be required—that wildness, if
we give it some room, will take care of itself.

31

THE FOOTHILLS
OF EDEN

I N WRITING THIS book, I have been trying to shake myself awake
from the dream of Eden. Or maybe I have been fighting the urge to go
back to sleep.

Yet the dream of Eden remains alive, and instructive. The dream of living "in harmony" with nature, as animals do, is dangerous only if we forget that it is a dream. It is dangerous if we pretend that we are just animals; or that animals live in anything that one can with a straight face call harmony with other animals, or with plants, fungi, or microbes. But if we strive to figure out what this thing we want to call harmony really is—if, in short, we study ecology—we can try to reproduce it, as far as humanly possible, in the humanized landscape.

The blade that whirls at the gates of Eden is not our enemy. It protects Eden and it protects us. If we try to get around it, we end up trampling either Eden or our own humanity.

Leaving Eden was a condition (even a definition) of being human. But that does not mean that the idea of Eden no longer matters. The most ancient idea of Eden—of a wild place at the center of the world from which all blessings flow—matters more than ever in a world where wild things wear radio collars and blessings flow through pipes and cables.

We cannot go back. What we can do is stand in a right relation to Eden, so that it stays alive and well and we can enjoy its bounty.

WE HAVE COVERED a lot of ground, more or less on foot. Perhaps it would be useful to retrace our route by air, pointing out the major landmarks and drawing the major lessons.

We began with harmonious dreams and harsh facts, both arising from the same dynamic. As the edge of a new wave moves across the earth, it first "improves" and then ruins each place it passes through. To make sense of these changes, we make up stories and assign names: Eden, Arcadia, the Golden Age. We may not use these names every day, but the stories are always buzzing in the idle corners of our minds.

We humans tend to think of ourselves (pridefully or guiltily, depending on our mood and the spirit of the age) as unique in our remaking of the globe. But we have not remade it single-handedly: we have had the help of annual grasses, perennial grasses, quadrupeds, and a host of other organisms, living and dead. What is more, our various alliances are merely the latest of many examples, over the course of earth's history, of conglomerates of species that conquered the earth. Seen from this point of view, our own victories seem a shade less grand and a few degrees less shameful.

That is not to say they are harmless. By allying ourselves with annual grasses—which thrive in disturbed soil, and put their energy into seeds rather than roots—we have thrown into reverse a process of soil formation that had been going on, in fits and starts, for nearly half a billion years. By taking allies and hitchhikers with us on our travels, we have reversed the breakup of Pangaea, and are well along toward making the whole world a single ecosystem, a single market, and a single culture. At first glance, the result seems to be an explosion of diversity; as the dust clears, we begin to see how much we have lost, and are losing every day. By allying ourselves with long-dead mollusks, cycads, ferns, and other species squeezed by the ages into energy-rich carbon compounds—and so joining the distinguished company of mushrooms and other saprophages—we have gained the power to disturb the soil and displace living species at a still faster clip. Meanwhile, vast amounts of carbon that had been buried for ages are abruptly let loose in the atmosphere, with results we have only begun to guess.

Our remaking of the globe, then, is neither so remarkable nor so reprehensible a feat as we tend to think. The conquest of fire, the Neolithic Revolution, and the Industrial Revolution are generally cheered—or damned—as triumphs of culture over nature. In fact, they are examples of the kinds of strategies that living things have always employed in the fight for energy, nutrients, space, and offspring. But to call them natural is not to stamp them with a seal of approval. The rhetorical use of this word is almost always bogus, for both nature and culture can be found on both sides of any debate. Though invoking them is sometimes helpful, the last appeal must be to experience. In this case, experience tells us that our empire has expanded too fast and too far for our own good, to say nothing of nature's. For the most practical of reasons, it may have to be restrained, restructured, and even rolled back.

THE MOUNTAIN AND THE TOWER

To make sense of the waves of change, people tell stories. Among the first recorded efforts are some of the myths of the Near East, which have had a lasting influence on the mindscapes and landscapes of the West.

The Mountain and the Tower are the poles between which human culture has shuttled for the past six thousand years. One can learn a great deal about a culture, or a piece of a culture, by asking which claim it accepts. Does it follow the Canaanites and consider that the center of the world is wilderness? Or does it side with the Mesopotamians in judging that the center of the world is the city?

Both sides are right. Both city and wilderness are sources of life, sources of weal, centers from which human waves have moved outward. The question is whether one looks to the proximate source or the ultimate source. Looking to the proximate source is useful in many ways and may leaven the growth of civilization; but in the long run it is dangerous. Convinced that our well-being springs from our own cleverness in reshaping the world around us, we are tempted to reshape more and more of it, extending the reach of the Tower into every corner of the world. And that is biting off more than we can chew.

For the myth of the Mountain is rooted in ecological fact. Man-made landscapes survive only at the sufferance of the wildness around them, or the wildness that remains in them. The flow of energy, water, nutrients, and genetic information; the maintenance of temperature and the mix of atmospheric gases within narrow limits; the fertility of the soil—all these are achieved by wild nature in ways we do not fully understand. Since we do not know how the job is done, we cannot do it ourselves. Even if we could, we would end up spending most of our waking hours working for something that we used to get for free.

In other words, humans and their allies are able to conquer the world, but they are not able to run it all by themselves. If the waves of human advance go too far or run too deep, they may finally bring about their own undoing.

Whatever else Eden may be, it is first of all an avatar of the Mountain. And the main lesson of the story of Eden is that we cannot live there. Although it is the source of human life, it is not a place for humans to live. A place for gods and animals, but not for us. As soon as we become fully human, we begin to destroy Eden and thereby expel ourselves. Only by keeping our distance from some of the wilderness that remains can we keep from fouling the wellspring of our own life. The fiery sword (whatever it is: our awe of wilderness, our fear of its dangers, our dismay in the face of its grueling beauty) is the best friend we have.

If the first lesson of Genesis is that we cannot live in Eden, there may be some comfort in the second lesson: we can yet enjoy Eden's benison, if only we let it flow. That is a big if, however. The four Rivers of Eden are ensigns of the flow of wildness. Dam that flow, and the man-made world must dry up and blow away; and so, at last, must wilderness itself. Even the biggest wilderness preserves are not big enough to stay healthy if migration, gene flow, and the circulation of energy and nutrients are blocked beyond their borders by highways, dams, development, ranchers' fences, the dredging of wetlands, and the poisoning of waterways. Wilderness is the heart of the world, but a heart is not much good without arteries and capillaries that touch every cell of civilization with wildness. There must be Mountains on every scale, from the Amazon rain forest and the Arctic tundra to the vest-pocket park in the inner city and the thicket in your backyard: swathes and patches and pockets of wildness, representing not only literal mountains but ecosystems of all kinds.

The obverse of this network is the network of civilization, which likewise has its own centers, its own branches and subbranches on every scale. But there is, or ought to be, an asymmetry. While every part of civilization must be touched by wildness, the deepest parts of wilderness must remain (as far as possible) untouched by civilization. Even the flow of wildness must be limited, if we include under that heading the invasions of nonnative species that humans tend to encourage.

This, then, is the view of one looking up at the Mountain, a view colored by awe, gratitude, and regret. As for the view from the Tower, it is not, as one might think, wholly triumphal. Only in history books does a whole civilization have a single, unmuddied worldview. The first civilization's triumphs had a dark underside that spawned a swarm of nagging doubts. As the same is true of our own civilization, these myths touch us more deeply than we might expect.

Nowhere is this clearer than in the story of the fabled King Gilgamesh and his expedition to cut down the great trees of the Cedar Mountain and kill their guardian, the monster Huwawa. When Gilgamesh refuses Huwawa's plea for mercy and pledge of fealty, he is in effect refusing the ecosystem services that wild nature is only too glad to perform, as long as it is kept alive. As the annals of deforestation from that day to this confirm, the attempt to take Eden by force of arms—to fight the fiery sword with fire and ax—can only lead to ruin.

Another story—bawdy, slapstick, and somewhat gruesome—is told of the technological god Enki and the earth goddess Ninhursag. Outwardly about onanism, incest, and disease, it is really a parable of the way Sumer's greatest triumph, irrigation, led to its greatest disaster, salinization—the deep soil spoiled by a crust of salt. Confident that they had mastered the secret of fertility, the Mesopotamians thought they could manage without the earth's help. The same mistake has been made many times since by adepts of "scientific" agriculture.

The Tower's mistakes, as well as its triumphs, have shaped the Western world and are quickly reshaping the rest of the world as well. For that we can thank the momentum of technology. But we can also thank Christianity, which not only chose the most Mesopotamian of the Hebrew Bible's many ideas about humans and nature, but added some Hellenistic and Roman ideas that would prove equally fateful. Chief among these was universalism: the belief that one culture, one religion, one urban plan, one landscape grid, one system of roads and aqueducts could be imposed on every land and people. For it is one thing to think, as the Hebrews did, that a deep unity lies beneath the diversity of nature, and quite another to impose a unity of one's own devising. The highways of Rome paved over the Rivers of Eden. They are doing so to this day, as the twin steamrollers of capitalism and Christianity turn the world into a single market, a single culture, and a single ecosystem—a new, improved, impoverished Pangaea.

IDYLLS

Though the exile from Eden is the bedrock fact about humankind, it doesn't feel like bedrock: it feels like a chasm. To close this fissure in the world and in ourselves, all kinds of effort have been expended: mythic, literary, architectural, horticultural, even scientific.

The simplest and maybe commonest way people try to resolve the tension between Mountain and Tower is by seeking a kind of arithmetical mean between them. The idea is that standing in the right relation to Eden is just a matter of standing at the right distance. In other words, if we cannot live in Eden, maybe we can live in Arcadia. Unfortunately, Arcadia is not a place but a phase—a phase that must pass, if indeed it has not passed already. The closer one looks, the briefer it seems to be, until at last it vanishes altogether. By the time a place has become civilized enough to make the poet (the urban observer) feel at ease, it has lost its charm.

Could we live in Arcadia, if only nature and culture had the good grace to stand still? Not even then. For the real problem with the Arcadian dreamer is that he tries to give responsibility the slip—to solve the puzzle of his relation to nature by finding the place where it is already solved. He seeks the perfect place, the fulcrum on which nature and culture are balanced. But standing in the right relation to wilderness is not a matter of standing at the right distance. At almost any distance, one can stand in a right relation or a wrong one.

This error would not interest us much if it were confined to poetry, or to the past. But it is common today, in real life. In suburbia—Arcadia in its major modern form—the error is all the more deadly because everyone is making it at once. In order to have a middle landscape, you must also have both ends. But under the dispensation of suburbia, the lifeblood of the city is sapped by strip malls and the flight of the middle class, while farmland and wilderness are gobbled up by subdivisions.

Trying to strike a balance between nature and culture—the basic idea of Arcadia—is not a mistake. It is a necessity—a part of the search for wholeness in the world and in ourselves. Where the mistake lies is in trying to strike the balance once and for all, when it really needs continual equilibration. A little scraping at the buried roots of pastoral, not least in the actual, geographical Arcadia, shows that the ancients knew this (at least, some of them knew it some of the time). The role of music is revealing: in music, disparate things are harmonized, but not once and for all. Without new discord to deal with, concord loses its meaning.

Finding a midpoint between Mountain and Tower is one way of resolving the discord we feel in the world and in ourselves. Another way is to make a microcosm of the world, and a macrocosm of the self, that is whole, perfectly centered, and sheltered from the winds of change. From ancient times to the present, people have made pleasure gardens for that purpose, among others. In the great prototype, the Persian paradise garden, the four Rivers of Eden become timid, lisping channels, giving lip service to the wildness but behaving in the most civilized manner imaginable. This sort of garden joins Mountain and Tower not by finding a midpoint between them, but by being an image of one and an instance of the other. The Romans took a more direct approach: in their villas and gardens they wanted the best of Mountain and Tower, no compromise necessary. And they wanted it—material girls and boys that they were—not on the symbolic plane, but in dumb, literal fact. Their Renaissance descendants were at least as ambitious but subtler, making world and self whole in a way at once sensual and magical, with Mountain and Tower metamorphosing one into the other as if conjured by Ovid. As for the English landscape garden, here wholeness was sought in the form of a literal Arcadia, complete with sheep.

After the Fall, though—maybe before the Fall, too—wholeness is achieved chiefly by walling out what does not fit. In most Western languages, the word for "garden" comes from a word for "enclosure." From the ideal world and ideal self that the garden is, whatever is troubling—change, wildness, the loss of wildness, the blatant exploitation of nature or other humans—is cast out. Like anything cast out, it becomes a demon, skulking in the shadows and waiting its chance.

In parks and gardens, the rulers of Persia could escape the hot and gameless tabletop they had made of their plateau. Medieval nobles could flee the overcrowding and famine that followed the rapid settlement of Northern Europe, once its soil was broken by the heavy plow. Florentines could find refuge from plague, proletarian unrest, dispossessed peasantry, and the tragedy of Arcadia they were financing in far-off Spain. The Sun King could hide from his failure to thaw the Little Ice Age, which again and again plunged his teeming nation into famine—famine that his rational meddling (the baroque structure of taxes, tariffs, and controls, the mercantilist stress on manufactures over farming) only made worse. And the English landlord, looking over the landscape that stretched out lazily like the common pasture of comradely shepherds, could overlook (ha-ha!) the enclosure of common land that made it all possible.

On the simplest level, the garden offers physical and psychical escape: out of sight, out of mind. But to pretend that one was simply in nature would not work. One would be haunted by the feeling that one must stand in *some* kind of cultural relation to nature, and ghosts of one's actual relation would crouch in wait behind each shrub. So nature in the garden is not only walled, but framed and even screened by the fiction of another culture. Almost always it is a culture of the past: a mythic past or a mythicized past, of course, for a real past would probably not be much better, taken all in all, than the present. As Americans now model their grounds on eighteenth-century England, so the English aped antique Italy, the Italians ancient Rome, the Romans classical Greece.

Love of nature is always love of another culture. Those who want to go back to Eden think they love nature itself. They think they can deal directly with nature, bypassing the cultural middleman. All they are doing is disguising the cultural basis of their fantasies. This means they are not thinking clearly about the kind of culture their fantasies would require—or the effect that culture would have on nature.

In America the dream of Eden was tested: the dream of a direct relation to nature, unmediated by culture past or present. The true Yankee hated the Indians, but did not have much use for Europe, either. His relation to nature might be mystical or, more likely, practical, but in either case direct:

just take it and use it, culture be damned. The upshot was a gobbling up of nature such as Europe had never seen.

Like the garden, like Arcadia, like America, the myth of Gaia tries to restore a lost unity. Gaia is the Mountain whose slope is the whole globe; or, to put it the other way around, the Mountain is the peak of the greater mountain that is Gaia. In each case it is the power of wildness that is acknowledged: the order hidden in disorder on which all life depends. But the myth of Gaia glues back together the unity smashed in our expulsion from Eden. For if the Mountain is everywhere, then perhaps we are not expelled after all.

Like all too-neat unities, the myth of Gaia has its cast-out shadow in the myth of chaos, which likewise has a science attached. But a third fledgling science—that of complexity—suggests that Gaia and chaos are not really the bitter enemies they seem. Like many complex adaptive systems, the biosphere may be poised at the edge of chaos, and once in a while may fall in. At the very least, dark veins of chaos give Gaia some of her strength. And some of her strength will, I think, remain hidden in darkness, beyond the reach of the modelers' computer screens. In truth, the Gaia hypothesis is most useful as a hypothesis: a suggestion that there are more things in heaven and earth—and more links between those things—than are dreamt of in our philosophy. It is a suggestion worth keeping in mind as one moves from observing the human role in nature—real and imagined—to deciding what one's own role should be.

EARTH JAZZ

Of the two schools of thought that now dominate the debate about humankind's place in nature, neither has learned the lesson of the Mountain. The Planet Fetishers know that wilderness is the center of the world, but make the mistake of thinking that we can live in it, in some kind of primal harmony: in short, that we can go back to Eden. The Planet Managers, though they have grasped the vital role of wilderness more deeply than the Mesopotamians did, make the old mistake of thinking they can control it. They still put their faith in the Tower, the place from which man sees and controls all things on earth. Only now they call it the satellite.

Right now, wilderness is so worn down and battered that there is no escaping the need to manage it, to some degree. The best we can do is to follow this rule: So manage nature as to minimize the need to manage nature. What this means, first of all, is stepping back and giving wilderness a chance to breathe, to dust itself off, to get back on its feet.

To create the world, God had to draw himself inward. The time has come for such a *tsimtsum* to be performed by the present lord of creation, humankind. The great insight of the kabbalists is that true creation begins with self-restraint—with the granting of an independent reality to what is not the self.

Even as we give wilderness space in which its own deep creativity can emerge, we must collaborate with the rest of nature in new ways. *Tsimtsum* is only the first step in creation.

Pastoral poets have shown us the role of music in working out (often unconsciously) the problem of discord and harmony between humans and nature. Carrying this metaphor into a modern context, I said that the Fetishers want us to sing one faint part among millions in nature's (imagined) harmony, while the Managers want to compose and conduct, measure by measure, a planet symphony of their own. But there is a third possibility: one in which we accept nature's unpredictability, and learn to improvise. In which we respond flexibly to nature, as nature responds to us. In which we relax our rigid beat and learn to follow nature's rhythms—in short, to swing. In which we no longer see a death struggle between nature's freedom and our own, but celebrate nature's freedom as the premise of our own. In which we accept the role of leader, but allow the other players a full measure of creativity. In which—not least—we learn to take our mouth from the reed once in a while and just listen. This is what I call Earth Jazz.

To play Earth Jazz, you must first of all listen. To stand in a right relation to Eden, you must face Eden, look at Eden, learn from Eden.

Strictly speaking, we cannot learn ends from nature, only means to ends we choose for ourselves. We can learn what works. We can learn some tricks that work for a species that wants to stay ahead of the game; and if we step back for a moment, we can also learn what helps keep the game going. At that point, we may decide that the ends we started with were too narrow.

The idea of learning from nature is hardly new; humans and hominids have been doing it for millions of years. Copying nature by rote can be useful, but it is only by abstracting general ideas from nature that we get a real jump on the game. Abstraction has been the genius of modern science, and its demon. In the realm of living nature, each abstraction is a falsification. We think we have got the basic rules of plant growth figured out, so we build hugely simplified ecosystems that promise to meet all our needs. For a while, they do. Then it turns out that something was missing.

Most scientists, and most Managers, would say we just have to keep whittling away at our abstractions, getting a notch closer to truth (and to what works) with each shaving that falls. The Fetishers think we should copy nature word for word—or simply insert ourselves in the extant text, displacing a leaf here and a twig there. Between these extremes are wiles and stratagems innumerable.

Some of the subtlest of these can be found among indigenous peoples—but only if you know what to look for. Only lately have we begun to see how deftly complex is the "primitive" farming and gardening practiced by native peoples all over the world. By and large, the Wild Garden mimics the structure of the surrounding ecosystem. It lets crops keep some of their wildness. It preserves the fecund chaos of a natural ecosystem instead of imposing a sterile order. Though the Wild Garden takes us closer to Eden than we, with our numbers and power, should dare to go, its tangled depths are rich in tantalizing hints.

So far, most of those hints have been ignored by modern agriculture, which desperately seeks abstraction—even though its stock in trade consists of specific living things, from specific places, with specific evolutionary pasts. Biotechnology offers the neatest way yet around these niggling specifics. The great danger from biotechnology lies not in its inevitable mistakes, but in its just as inevitable successes. Thanks to the cut-and-paste approach inherent in genetic engineering and the fact of who owns the scissors, these successes—pesticide-tolerant crops, pesticide-laced crops, genetically uniform supercrops—impose so crudely on the subtleties of nature (not to mention those of culture) that they may leave in shreds the fabric on which our life depends.

Rather than copy nature on the molecular scale, it might be wiser to do so on the scale of the ecosystem, deploying "big chunks of what works." That is the tack taken by Wes Jackson, pioneer of the edible prairie, as well as by other stubborn, underfunded researchers in other parts of the world. On this larger scale, means and ends tend to get mixed up, which is a good thing: in nature, they *are* mixed up.

People like Jackson are trying to make our living allies behave more like members of mature ecosystems. As we have seen, other people are attempting the same thing with our mechanical allies. Going far beyond ordinary recycling, they are trying to weave different industries into a tight food web in which one's waste product is another's raw material. On smaller scales, too, both processes and materials are being copied or adapted from nature.

Looking at these trend lines, some thinkers see at their convergence a world in which technology is so elegant, so gentle, so organic that it poses no threat at all to wild nature. Like other dreams of harmony, this one may end up harming what it wants to cuddle with. Even if we *could* create a perfectly gentle technology, it might not satisfy our deeper needs. What we love in nature is not just her elegance and gentleness, but her violence and waste: the negligence of the leopard who shrugs off the leftovers of his kill, the clouds of sperm that never reach their goal. Nor is it only living nature that we love. Stone, sand, fire, and water speak to us, too. A sound vision of technology must take into account our love of inorganic things, whose blind, dumb, violent wildness—perhaps the deepest wildness of all—is oddly akin to our own.

Machines will always be a little reckless, changeable, prone to wander off and make trouble. Like goatherds, we have to keep an eye on them—"we" meaning all of us, not just scientists or bureaucrats. Like Arcadians, we have to learn the art of striking a balance, of constantly judging and adjusting our labile technologies. There are many means of doing this: technology juries, science shops, community-based and participatory research, Sabbaths and sabbaticals, even a simple habit of mindfulness. Instead of the proud Prometheus, science might take as its model Enki, the trickster and mud-flat jazzman who, when he gets into a scrape with nature, is not too proud to bow and scrape his way out.

If this is a sort of self-retraction—a drawing in of the puffed-out chest— it reminds us that *tsimtsum* is the premise on which Earth Jazz rests; and that *tsimtsum*, too, must work itself out in everyday life. If anything, the consequences of *tsimtsum* touch us even more closely, for they touch us where we live. They bear on the whole range of earthly spaces, from the Mountain to the Tower.

Though the Tower is guilty of many sins against nature, to deny its place at the center of civilized life—to soak it in wildness and pound it into "organic form"—would be a misguided penance. Culture needs a space where it can play freely, without too much meddling by nature—but within the limits nature sets. Such a space has been furnished by various cities, past and present, that have solved many of their ecological problems without sacrificing their essential cityhood.

But city life is not for everyone. Without the whole gamut of tones between Mountain and Tower, the music of human life on earth would be an empty octave, bare of interest. Without the middle landscape in its various forms, most people would lose any sort of practical contact with nature. We need Arcadia—but not on the old, carefree Arcadian terms. With the freedom to live in the place you choose comes the responsibility to live in a way appropriate to that place. In the case of Arcadia, this means cutting consumption, fostering wildness, and keeping the working landscape in working order. With the privilege of living between Mountain and Tower comes the duty to promote the health of both.

Tsimtsum does not mean evacuating the middle landscape, but it does mean keeping the human footprint as dainty as possible. Neotraditional planning, infill, cluster development, and community-supported farming are some ways of doing this which, as it happens, also make human life more pleasant.

For many people, though, trying to find the perfect blend of city and country will make less sense than moving back and forth between them. Adapting old traditions to modern needs—living simply, sharing digs, supporting farming in various ways—the summer Arcadian may feel less like an invader and more like a friend of the working landscape. Nor is it only seasonal movement that humans have a weakness for: nostalgia for the nomadic life turns up in all sorts of places, from the depths of the Amazon rain forest to the nooks of the pleasure garden. To be workable, a model for human life on earth must leave room for humans to move around. Somehow, it must find a way to make us feel responsible for the places we leave and the places we arrive in, as well as for distant places touched by our actions.

If Arcadia is the place where nature and culture ideally mingle, then reclaiming Arcadia means that we have to get the mingling right: not just the proportions but the patterns, shapes, spaces. Not just in the middle landscape but in every landscape, it matters a great deal how the network of wildness and the network of civilization are disposed in space.

Most of our wildernesses are too small, too fragmented, or too isolated from one another to keep their health and their wildness. We must clear the blocked vessels of migration and gene flow: vessels that come in all sizes, from continent-spanning watersheds to hedgerows. But even as we free up the flow within bioregions, we must try much harder (which should not be too hard, since at present we hardly try at all) to block the global flow of our allies and hangers-on. Nature needs walls as much as it needs flow. Without walls, there is entropy—of which the New Pangaea may give us a global example.

Of all the gigs Earth Jazz has to play, none is trickier than the problem of climate change. Fetishers and Managers agree on the need to do something, but they disagree bitterly about what needs to be done. Fetishers want to banish human influence from the surface of the earth, Managers to extend it (mindfully) into the skies. In reality, a few basic principles gleaned from nature can help us respond sanely to global warming in ways that will be useful even in the unlikely event that the warming turns out to be a flash in the pan. Major climate change of some kind will happen sooner or later; the stability of the last ten thousand years has been a fluke in earth history. If we have given her enough elbow room, nature should be agile enough to respond; and if we have learned her lessons, it is just possible that we may be agile enough, too.

THE VIEW FROM HEAVEN

No wonder the Planet Managers are fond of satellite imagery. They are sons of the sky god. They know that seeing is controlling. The international space station is the ultimate Tower—or, rather, it is the first Tower, finished after a delay of five thousand years and major cost overruns. When Russian, American, Japanese, and German astronauts chat merrily as they take in the view, the project of Babel will be realized at last. Language barriers will seem as wraithy as political borders, burned away in the sphere of blue flame.

And this is the joke history has played on God: He worried that if man reached the heavens and looked down, he would assume the mantle of the sky god. He would move the pieces of creation as if on a chessboard. He would play with thunderbolts, taking potshots at anything that moved without his leave. He would see, and he would control.

Man did all these things, in fact, from lesser perches, platforms commanding views of cities and states. But when, at last, he reached the heavens and saw the view he had ached to see—the whole planet abject as a dog at his feet—he felt a strange stirring within him that could not be attributed to zero gravity.

For thousands of years the Great Mother had murmured throaty murmurs about the oneness of nature. For centuries man had slapped her and told her to shut up. Seeing is believing, he said. The evidence of his eyes proved that nature was many different things, and himself the most different of all.

Now he saw, and he believed.

IN CICERO'S FAMOUS "Dream of Scipio," the dreamer is plucked up by his great ancestor Scipio Africanus—like Scrooge kidnapped by the Ghost of Christmas Past—and flown to a perch in the stars from which the earth looks laughably small. The lesson is that he must not take delight in earthly things, but must devote himself to public service and the hope of a heavenly reward. The lesson we draw from the same vision is different: that we should take delight in the earth, and for its sake should devote ourselves to public service in a new and broader sense.

Africanus is the Father of Fathers, the great general who humbled the ululating goddesses of North Africa more utterly than Aeneas humbled Dido. By lifting his descendant high into the heavens, he shows him that the Mother of Mothers is less than she seems. It is the gesture of all fathers who lift or toss their infants high in the air, and thereby demonstrate that the mother is not all in all.

But we who have been tossed into the heavens not just in dreams but in real life, have been moved in unexpected ways. We feel a rush of love for the earth's smallness and fragility, as when one first notices that one's mother is aging. If it is plain that the earth is not all in all, it is even plainer that it is all in all for us, our brave house on the whooshing prairie of space. And not just our house, but our body: a truth we had to leave our body to see.

NOTES

THESE NOTES CONTAIN seven kinds of information: (1) general sources; (2) specific references; (3) suggestions for further reading, not always distinct from (1); (4) scholarly quibbles, qualifications, equivocations, and pygocryptions; (5) tangents and excursus; (6) current developments that, in a few years, may be so much water under the bridge; (7) tidbits too tangential to put in but too juicy to leave out.

When some new subject matter heaves into view, I give a list of sources I have found generally useful, sometimes specifying those which I think a reader new to the subject—as I often was—will profit from. Thereafter, specific references are given only for quotations, and for facts or statistics that might raise a flag in the reader's mind. Wherever possible, I have referred quotations to their original source, whether I stumbled upon them myself or found them in the secondary literature. I have tried to verify each quotation myself, in the original language if possible. Mention of secondary authors at the beginning of a section will acknowledge my debt to them.

Books and articles are referred to by the name of the author, or, if a single author has more than one work cited, by title, sometimes in a shortened form. For full information the reader may refer to the bibliography, which includes nearly all works mentioned. (The exceptions are classical, rabbinic, and other standard works, which are listed only if I have quoted a published translation.) If a note gives only the name of the author, but the bibliography lists several works under that name, the reader should assume that all are relevant.

For certain classic or classical works I cite the chapter, section, or line numbers in common scholarly use. The same goes for chapter and verse in the Bible.

INTRODUCTION

xvii **chunks of dream you remember on waking:** Freud, *The Interpretation of Dreams*, chap. 7.

  xx **Depending on context, "nature" may mean:** John Passmore writes (p. 5n.), "I wish I could wholly avoid the word 'nature.' But if it is one of the most ambiguous it is also one of the most indispensable words in the English language."

PROLOGUE: PERSONS FROM PORLOCK

 xxi **Coleridge says:** The poem seems to have foreseen its own scattering amid the business papers of Porlock—for it is not only the larger part of the poem that has been lost, but the poem or song beneath the poem. If only the poet could recapture the song he once heard sung by an Abyssinian girl, all who heard him would say that

> . . . he on honey-dew hath fed
> And drunk the milk of Paradise.

Dreamtime of the Aborigines: See Strehlow; also the striking fantasia on these themes and others by Chatwin, *The Songlines*.

other indigenous peoples are sometimes able: See Eliade, "Nostalgia for Paradise."

near universality of Golden Age and Fall stories: Eliade ("The Myth of the Noble Savage") notes that "the 'good savage' of the travellers and theorists of the sixteenth to the eighteenth centuries already knew the Myth of the Good Savage; he was their own ancestor who had really lived a paradisiac life, enjoying every beatitude and every freedom without having to make the slightest effort. . . . The savages, for their own part, were also aware of having lost a primitive paradise." See also Eliade, "Nostalgia for Paradise"; J. Z. Smith; Ries. On Golden Age and paradise stories in Western civilization, see Lovejoy and Boas, who assemble a wide range of Greek and Latin texts; Manuel and Manuel.

xxii **In the King James text:** Genesis 3:17.

Milton's image: *Paradise Lost* 12. 645–49:

> Some natural tears they dropp'd, but wiped them soon;
> The World was all before them, where to choose
> Their place of rest, and Providence their guide:
> They hand in hand with wand'ring steps and slow,
> Through Eden took their solitary way.

xxiii **"live in an environment infinitely superior . . .":** Strehlow, p. 35.

1. THE MARRIAGE OF GRASS AND MAN

3   THE FALL OF MAN: I use this term interchangeably with "the expulsion from Eden." No endorsement of the Christian concept of original sin is implied.

bacteria at deep-sea vents: Microbes residing in the gills of clams and the trunks of giant tube worms use sulfur compounds, welling up from the earth's crust, as energy sources for themselves, their hosts, and (indirectly) most of the other highly specialized creatures that live in these warm oases on the ocean floor.

"Hot and crisp, mine burned my mouth . . .": E. M. Thomas, *Reindeer Moon*, p. 336.

The trick is to find the most efficient path: A paraphrase of Lanza del Vasto: "Find the shortest, simplest way between the earth, the hands and the mouth" cited in Wendell Berry, *The Unsettling of America*, p. 96.

4   basic moves in the game of life: On the chemical genius of bacteria, see Margulis, *Early Life;* Sonea and Panisset; and, on a more popular level, Margulis and Sagan, *Microcosmos.*

trophic: From the Greek for "food."

transition from apehood: On human and hominid evolution, see Leakey; Stringer and McKie.

aided by fire: While the deliberate setting of fires to improve pasture for game is well documented for hunting peoples from Argentina to Manchuria, the question of just when the practice began, and how great a role it has played in creating and maintaining grasslands and open woodlands, remains controversial. See Sauer, *Land and Life;* "Man's Dominance"; O. Stewart. For more on fire as an ally of humankind, see note below, p. 448.

5   Ice Age extended the African savanna: See W. C. Brice; H. E. Wright. For a vivid view of the North American scene, see MacLeish; for an equally vivid (if specifically fictitious) view of the Eurasian scene, see E. M. Thomas, *Reindeer Moon.* Thomas, who began her career as an anthropologist in Africa, bases many of her guesses about daily life in the Pleistocene on the ecological likeness between the savanna and the Ice Age steppe; see her postscript.

The clearing of the forests of Europe: See G. Clark.

6   "Agriculture" from the point of view of the plant: For other attempts to take this point of view, or simply to see agriculture as a case of coevolution, see Rindos; D. Harris and G. Hillman. For more traditional views, see Struever; Reed; and the very readable Heiser.

A classic (and highly engaging) exposition of the dump-heap theory of the origin of agriculture can be found in E. Anderson, chap. 9.

herding the quadrupeds instead of chasing them: Note that the word "drive" is used both of the herding of cattle over long distances and of the leading of game, by fire or other means, to the slaughter.

7   Natufians of the Levant: Taking advantage of a rich ecological mosaic in the hills of the Near East at the end of the Ice Age, the Natufians were among the first

known peoples who could afford to stay put. They did so by collecting, milling, and storing wild foods: grains (harvested with stone sickles) as well as acorns, almonds, and pistachios, together with meat and fish. Their cornucopia remained more or less full from about 12,500 to about 10,500 years ago, at which point a drier climate seems to have forced the gradual transition to farming. See D. Henry.

8 **Weeds are pioneer species:** On weeds, see E. Anderson, who writes (p. 15), "the history of weeds is the history of man." On pioneer species generally, see almost any ecology text: for example, Odum, *Fundamentals*, chap. 9.

2. AXIS POWERS

9 **Cenozoic:** The present age is called the Cenozoic, or "common era of animals," by the same logic that replaces A.D. with C.E., or "Common Era." For a brief history of Cenozoic life, see Cloud, chap. 16.

Ophrys **orchid:** Depending on species, the flower may mimic a female bee, wasp, or fly. The orchids bloom early in the spring, when male bees, wasps, and flies are impatiently waiting for their female cospeciants to appear. When an insect makes love to an orchid, a pollen sac may cling to his body and so be transported to the next orchid he accosts. See Raven et al., p. 603.

mutualism: On this and related terms (neutralism, competition, amensalism, parasitism, predation, commensalism, protocooperation), see Odum, chap. 7, sec. 16. On the tricky relation of mutualism to coevolution, see Nitecki, especially the essay by Schemske. See also J. Thompson; Boucher.

Some authors (e.g. D. H. Lewis; Janzen) prefer to define a mutualism as a relationship between individuals; but this raises a number of problems, such as the use of the rubbery "potential fitness" as a yardstick.

insect taxa do not seem to have expanded much: Recent studies of the fossil record (Labandeira and Sepkoski) suggest that the great radiation of modern insects started 245 million years ago—early in the Mesozoic—and was not speeded up by the advent of angiosperms in the Cretaceous. Insects, then, are the senior partner in the present alliance; for the most part, flowering plants have adapted to them, not the other way around.

10 **vast fermentation vats on legs:** Farlow.

oxygen, which is toxic: See text, p. 55.

symbiosis: For general surveys, see Ahmadjian and Paracer; D. C. Smith and Douglas.

11 **Giant tube worms:** Sulfur power may be peculiar, but it is nothing to sneeze at. Though lacking eyes, mouths, or apparent means of locomotion, giant tube worms are perhaps the fastest-growing invertebrates on earth. Studying a region of the Pacific floor that had been sterilized by lava in 1991, scientists who dove almost a mile and a half in the submersible *Alvin* found that tube worms had formed dense forests, with worms nearly five feet long waving gently in the liquid breeze, in the astonishing span of two or three years (Lutz et al.).

The symbiosis between giant tube worms and sulfur-metabolizing bacteria seems to have started in the Carboniferous period. Although nearly all of the symbioses described in this chapter remain current, exactly when each began is a matter of educated guesswork. Since the inhabitant, or smaller party, tends to be small and soft, direct fossil evidence is rare: a few insects and nematodes in amber, traces of mycorrhizal root systems. For the rest, we depend on indirect evidence gleaned from the fossils of the exhabitant, or larger party—its morphology, its geographical distribution—and what we know about the symbiosis as it now stands. See Bermudes and Back.

Deep-sea cold seeps: In the late Cretaceous period. See Beauchamp et al.

Corals and giant clams: Corals seem to have done this in the late Triassic, giant clams in the late Cretaceous. See R. Cowen; Fitt and Trench; Trench; Yonge.

hindgut of a modern termite is a zoo: See Margulis, *Early Life*, chap. 4; Margulis, "Spirochetes"; and the popular treatment by Lewis Thomas, pp. 26–30. At least three other species of symbiotic microbes live in or on *M. paradoxa*.

lichen: See Ahmadjian; Kappen and Lange. Some lichens are unions of fungi and cyanobacteria.

Vascular plants first colonized dry land: Pirozynski and Malloch; Atsatt. Long after the two partners had (unlike lichens) become a single organism, some plants enlisted distinct fungi to extend and ramify their root systems. With the aid of these fungi, called mycorrhizae, such plants had and still have an advantage in colonizing sand dunes, glaciers, and other harsh environments. See Malloch et al.

12 **Every cell of every "higher" organism is an alliance:** General assent to this once-heretical statement is owed to the brilliant, dogged work of Lynn Margulis. See her *Symbiosis in Cell Evolution*, as well as the popular account in Margulis and Sagan, *Microcosmos*. Margulis herself gives full credit to early formulations by K. S. Mereschkovsky (before 1910) and Ivan E. Wallin (in 1927).

All other creatures are eukaryotes: Recently, some biologists have been inclined to supplement the five kingdoms of living things—bacteria, protoctists (the new name for eukaryotic microbes), fungi, plants, and animals—with a sixth, the Archaea, otherwise known as archaebacteria. (The five kingdoms are mapped in the mind-opening field book by Margulis and Schwartz; for the sixth, see Bult et al.; Olsen and Woese.) Viruses, which are just packets of freelance genetic material and (though capable of reproducing inside host cells and causing a great deal of mischief) have no metabolism, are not generally defined as living.

a very tough bacterium: Something like the extant but very ancient *Thermoplasma*, which lives in hot springs and other unlikely places and is a member of the archaebacteria.

a new species was born: Of the sort we would now call a mastigote.

a whiplike oar: Called an undulipodium. Not to be confused with the flagellum of certain bacteria, which has a spiral motion more like a propeller.

voracious oxygen-breathing bacteria: Similar to the modern *Daptobacter*, *Paracoccus*, or *Bdellovibrio*.

13 **small photosynthetic bacteria:** Similar to the modern *Prochloron*.

keeping them alive inside itself, busily photosynthesizing: The cell membrane of the single-celled organism would, of course, be permeable to light.

supplanting the older photosynthetic bacteria: Mainly the cyanobacteria, formerly known as blue-green algae. Bacteria remained dominant in waters close to shore. On fossil evidence that might be used to infer the conquests of new turf by these various mongrel microbes, see Knoll.

the birth of a new, chimerical species was actually observed: Jeon.

14 ANTS AS FARMERS AND RANCHERS: Much of the following account is drawn from the magnum opus of Hölldobler and Wilson, *The Ants*. Although that magisterial work is very readable, those intimidated by its bulk might try the same authors' *Journey to the Ants*.

It should be noted, though, that Hölldobler and Wilson themselves might not accept my parallel between human-led alliances and those in which humans play no part. They write (*Journey*, p. 206): "We are the first species to become a geophysical force, altering and demolishing ecosystems and perturbing the global climate itself. Life would never die through the actions of ants or of any other wild creatures, no matter how dominant they became. Humanity, in contrast, is destroying a large part of the biomass and diversity of life . . ." Although this view is common enough, it is surprising to hear it repeated by two such knowledgeable biologists. Many species, notably of microbes, have been geophysical forces; and the cyanobacteria very nearly put paid to life itself.

Leafcutter ants are the dominant ants of the New World tropics: They also range as far north as the Pine Barrens of New Jersey and as far south as the chilly deserts of central Argentina. On leafcutters, see Hölldobler and Wilson, *The Ants*, pp. 596–607; M. Martin; Chapela et al.; G. Hinkle et al.

fungus gardens: Varying with the species of ant in a way that suggests long coevolution, the fungi involved have recently been identified as homobasidiomycetes of the order Agaricales.

wide variety of plants: Just how wide the variety is depends on the species. Some are specialists in grasses, some prefer dicots, and some have a weakness for human crops.

breakfast, lunch, and supper: Equipped with this excellent food supply, a single nest—which can be the size of a large basement—can produce many millions of workers and many thousands of queens. Each queen on her nuptial flight carries a wad of fungus in a pouch in her throat. When she sets up housekeeping in her new nest, the first thing she does is spit out the fungus and commence fertilizing it with drops of fecal liquid. Within a month, her eggs, larvae, and pupae are cradled in a queen-sized bed of fungus. As the workers mature, they begin to feed themselves and the larvae with the "heads of kohlrabi" (also known as staphylae). Shortly thereafter they start cutting leaves. Once these farmhands are on the job, the queen can devote the rest of her life to laying eggs. (The description here is of *Atta sexdens*.)

15 **lose some of the digestive enzymes:** Precisely because they produce so few digestive enzymes, they are able to take in, from the "balloons," enzymes produced by the fungus, pass them through their bodies unharmed, and return them to the fungus by means of fecal droplets. So the droplets not only fertilize, but serve to recycle enzymes as well.

raising a crop: If leafcutter ants are the main reason farming is hard in the American tropics, they can claim precedence as their excuse: they have been farming fungi for something like 50 million years. Whereas species of the genus *Atta* are stubborn monoculturists, apparently cultivating the same strain of fungus for 25 million years, other leafcutter species are always on the lookout for new fungi (and perhaps new flavors). The catholic leafcutters seem to be the more ancient type. See Chapela et al.; Hinkle et al.

a florist's arrangement: While only a few species of ants have green thumbs of such exquisiteness, a great number perform horticultural services for the plants in which they nest—fertilizing them, carrying and planting their seeds, protecting them from predators, even weeding out or pruning back their competitors. As they are often rewarded not only with room but with board—in the form of special food packets and nectar secretions—they, too, can be thought of as farmers.

dairy farmers: Most members of the three most "advanced" ant families, the Myrmicinae, Dolichoderinae, and Formicinae, are involved to some degree in the raising of aphids and their cousins—scale insects, mealybugs, treehoppers, and leafhoppers—which are known collectively as homopterans.

mana: Mana was probably produced, as *man* is, by a scale insect that feeds on tamarisks.

16   **30 million years ago:** Ants and aphids have been found cheek by jowl in a block of Baltic amber thought to date from the early Oligocene. Having had so long to practice, ants have brought ranching to a remarkable pitch of excellence. Queens of some species take aphids in their jaws on their nuptial flights, so that the new colony will not lack its herd. Often the ants build rude barns for their livestock. Some species keep the aphid eggs in their nests through the winter, then carry the newly hatched nymphs to nearby plants in the spring. While the aphids graze, the ant workers protect them from predators and parasites, just as shepherds protect their sheep from wolves. And domesticated aphids need protection: like sheep, many have lost the ability to flee—in their case, to jump.

The aphid is not ungrateful. Instead of defecating at random as wild aphids do, it defecates on command. When its rear end is tickled by an ant's antenna, it gently extrudes a droplet of honeydew, which is held in place while the ant laps it up. An unmilked aphid of a domesticated breed can be as sore beset as an unmilked cow. Having lost the wild aphid's ability to eject honeydew some distance from itself, it ends up befouling itself and its friends, sometimes with fatal results.

The most remarkable herding ants of all have only recently come to light in the Malaysian rain forest. They appear to be the only true nomads in nonhuman nature. Their nests are made of their own bodies, which cling together like Chinese acrobats to form a chain-mail shelter for the ant brood and the herd of mealybugs. The mealybugs hatch in the nests and are ferried to feeding sites on the tender young shoots of various flowering plants. If the plant is shaken or otherwise disturbed, each aphid climbs aboard an ant's mandibles and rides to safety, lightly caressing with its antennae the head of its savior. When the herds have made full use of one pasture, the ants carry them to a greener one; and when nearby pastures have been exhausted, the acrobats unscramble and the nest and its contents move to a new site. The mealybugs do not survive in the wild, untended by ants. Nor do the ants last very long when their herds have been rustled by entomologists.

bees, wasps, beetles, and butterflies: Most of these take the honeydew from the homopterans without doing anything for them in return.

lycaenid-butterfly larvae: It has been estimated that 100 *Polyommatus coridon* larvae can produce enough sugars to cover the energy needs of 1,400 to 2,800 workers of the ant species *Tetramorium caespitum.*

Ants have also been found tending hemipteran, plataspid, and coreid bugs.

mutualism can sometimes destabilize them: In 1972, Robert M. May—a physicist invading the ecosystem of the ecologists, which he gratefully noted (*Stability*, p. vi) had "not yet reached (or exceeded) its natural carrying capacity"—caused a stir by using mathematical models to produce counterintuitive results: for example, that species diversity does not in itself make ecosystems more stable, but rather less so; and the mutualistic interactions between two species tend to be unstable. Whereas in case of a predation, negative feedback tends to damp the cycles of population boom and bust, mutualism involves positive feedback. As a result, simple models yield "silly solutions in which both populations undergo unbounded exponential growth, in an orgy of mutual benefaction" (*Theoretical Ecology*, p. 95). May could correct this by allowing for diminishing returns in at least one of the benefits, but the system remained more vulnerable to disturbance than it would have been without the mutualism.

Squaring this result with the facts—the widespread presence of mutualism in nature—has kept ecologists busy for the past quarter-century. Some have made their models more stable by including a third species, a competitor or predator of one of the mutualists (Heithaus et al.); some by adding a third and fourth (Ringel et al.); some by allowing the mutualist populations to migrate or spread out (Hutson et al.). This last points in the direction I want to go; but I take the long road of evolution and biogeography rather than the shorter one of population ecology. What I am claiming is that the runaway growth May found has, at times, been something close to a reality: to wit, when a new mutualism was just hitting its stride. Rather than strip their local environments bare, the populations in question have expanded geographically, and in so doing have often destabilized the ecosystems they invaded, displacing other species as well as non-mutualistic populations of their own species. Eventually they have settled down, becoming pillars of a more-or-less stable community—the sort of community that ecologists observe and try to model. And their mutualism, bound up with other mutualisms and other interactions in the refashioned ecosystem, may then become a source of stability.

Germany, Italy, and Japan: One might counter with the claim that NATO has done much to stabilize the world. But NATO is a defensive alliance, and no alliance in nature is purely defensive. Nature does not have neatly marked and recognized borders. Every species is constantly struggling for *Lebensraum*, or at least trying to increase its numbers at the expense of other species.

Invasions by single species are common enough: Though most common where man is involved—which throws into question whether they really *are* invasions by single species. I will have more to say about this in chapter 4.

17 **cyanobacteria poisoned almost all their contemporaries:** Schopf. For a popular account see Margulis and Sagan, chap. 6.

what a lone species does is just freeloading: Nevertheless, even mutual aid within a species can be destabilizing, if it is highly developed. Ants and modern humans are successful (at other species' expense) not only because they are agricultural, but because they are urban.

as long as they get to set seed: In theory, no organism should care about survival beyond reproductive age, except insofar as this may help its offspring or related offspring survive. But in fact many do, because the survival instinct, once set in motion, is a machine that will go of itself unless there is some imperative to turn it off.

far more of it germinates than without human help: Even if the plant retained its brittle rachis, or fruiting stalk, which would allow its seeds to be sowed by the wind.

18 **the "dump-heap" theory:** See E. Anderson, chap. 9.

the main thing Thoreau deigned to grow: "This was my curious labor all summer,—to make this portion of the earth's surface, which had yielded only cinquefoil, blackberries, johnswort, and the like, before, sweet wild fruits and pleasant flowers, produce instead this pulse" ("The Bean-Field," *Walden*).

rhizobia: See Smith and Douglas, sec. 4.2; Madigan et al.; Sprent. For speculations on how this vital symbiosis might have evolved, see Sharifi; Sprent and Raven.

mycorrhizae: See Ahmadjian and Paracer, chap. 7, sec. C; Harley and Smith.

19 **the host of creatures in the soil:** See chapter 3.

lost in the general tangle: To add a further knot, consider the following. We should not imagine that an alliance of species is formed in one place and then moves outward like a phalanx, trampling everything in its path. It is far cleverer than that. As new ecosystems are encountered, new allies are recruited and old allies dropped. Native species make themselves useful to the conquerors. In political terms, we might call them quislings; but as biological conquests tend to be irreversible anyway, such collaboration is really in everyone's best interest. By this means, the disruption of the old ecosystem is somewhat eased; some threads, at least, of the fabric are left intact.

When humans with their wheat and barley moved northward through Europe some eight thousand years ago, they found a couple of hardy weeds asserting themselves in their grainfields. The wet clay soils and cloudy skies that made a sun-burnished Levantine such as wheat so uncomfortable were sweet home to these native grasses. As it emerged that the seeds of these grasses could be eaten with profit by humans as well as by livestock, their status was upgraded from weed to crop. Oats and rye became reliable allies of humankind in Northern Europe. (In one respect rye was fatally unreliable: it sometimes harbored an enemy of humankind, ergot, the toxic fungus that caused outbreaks of St. Anthony's fire in the Middle Ages.) In the same way, the European axis of grains, cattle, and pale people that conquered America gathered to its colors maize, potatoes, pumpkins, squash, and numerous other native plants. Some of these have turned the tables by becoming the dominant plants in parts of Europe. Potatoes in Ireland, Poland, and Russia, tomatoes in Italy, maize in Hungary are like slaves from the colonies who sailed to the mother country and made a fortune.

other humans (or hominids) and their allies: For decades, anthropologists have debated whether major innovations tend to spread by cultural diffusion (from one group of people to another) or by demic diffusion (one group of people displacing another: driving them out, killing them off, or simply outbreeding them). In the two crucial cases of farming and the domestication of horses, the genetic studies of Cavalli-Sforza and his colleagues have decided the matter in favor of the latter, nastier alternative. But note that though nasty, this scenario need not be violent: computer models (see Leakey) suggest that with a competitive advantage of only 2 percent, one population might easily replace another within the span of a millennium.

While Cavalli-Sforza does not speak in terms of coevolution or alliances between species, his vision of successive waves of advance by successive human groups is very much in harmony with the vision put forth in this book.

its own parasites and pathogens: See McNeill, *Plagues and Peoples;* Crosby; Ewald.

20 **In computer simulations of life:** Of many instances, I will cite what is probably the first: the work of Barricelli at Princeton in the early 1950s.

mutate to take advantage of this new host: It is as if Tartuffe, while sponging off a Catholic family, should meet a Texas millionaire who happened to be a Baptist. In no time flat he would be speaking in tongues. (See Morse.)

There are exceptions to this pattern: See Ewald; Mitchinson. If long-housebroken pathogens can still decimate populations newly exposed to them, then clearly they have not moderated as much as they seem to have done; rather, it is the long-time hosts' evolution of resistance that has done most of the work toward an "arrangement." Still, the evolutionary trend among diseases transmitted by human contact seems to be toward moderation, as the host and pathogen have, in part, a shared agenda. In the extreme case of a dormant virus that lives in the host's gametes, the agendas of host and parasite are virtually identical (Dawkins, chap. 12).

21 **The infamous exception, syphilis:** See Wilford.

the Indians' allies were no match: Some parts of America, such as tropical rain forests, had parasites and pathogens of their own which took a heavy toll on Europeans. But such places could usually be avoided. Biological countermeasures against European invasion worked somewhat better in Africa, the primordial home of humans and human afflictions. Even today Africa is conspicuous, though hardly unique, as a place where large numbers of people moving into long-isolated environments seem to be opening a Pandora's box of germs.

3. DIRT CHEAP

22 **a population of five million:** Cavalli-Sforza et al., "Demic Expansions."
   Humans now control at least 25 percent: Ehrlich.
   An acre of good topsoil may house: Wes Jackson, *Altars*. A wonderfully vivid guide to life underground is Farb, *Living Earth*. Those seeking more recent and detailed information might turn to Paul and Clark; Wallwork; Richards. A well-illustrated brief account may be found in Raven et al., chap. 26.

23 **One teaspoon of good grassland soil:** Raven et al., chap. 26; Paul and Clark, p. 71.

24 **Aristotle called earthworms:** *Historia Animalium* 570[a] 16.
   If you could somehow unravel the fungi: Farb.
   root system of a single four-month-old rye plant: Raven et al., chap. 21. As far as I know, nobody has tried a comparable reckoning for a mesquite, whose roots have been found to reach a depth of more than fifty-three meters, or for a giant sequoia.
   by staying very small: According to two well-worn laws of zoology, Bergmann's rule and Allen's rule, animals closer to the equator tend to be smaller or else more elongated than those in chillier climes, because they want to increase surface area for cooling. By analogy, one might say that bacteria use the pygmy strategy, fungi the Bantu strategy.

25 **Surface area is needed for life:** In the daylight world, many creatures—among them humans—have taken the opposite tack: getting very big, establishing a much smaller ratio of surface area to volume (or weight), reducing metabolism to a pace that in the night world of the soil would seem like death. Even allowing for the work of bacteria in our raveled guts, our bodies have comparatively little chemical effect on the world around us. It is lucky the soil organisms have not taken that tack; if they had, we would all have been smothered years ago in leaves, corpses, and dung. Though technology extends our bodies in a number of ways, vastly stretching our effective surface area, so that we do in fact affect the world powerfully, we are still (at best) a distant second to the bacteria in this respect.
   one of the things that makes this planet hospitable to life: Hutchinson, *The Ecological Theater*.
   researchers tried to measure the surfaces: Farb, p. 9. Admittedly, this is an upper limit, reached only by very clay-rich soils. The surface area of a more ordinary soil may be only a tenth as much—but that is still plenty. (Daniel Hillel, personal communication.)

26 **They are colloidal:** On soil structure and chemistry, see the classic text by Brady.

27 **Earthworms literally eat their way:** In his ground-breaking work *The Formation of Vegetable Mould, Through the Action of Worms*, Darwin wrote (p. 313): "The plough is one of the most ancient and most valuable of man's inventions; but long before he existed the land was in fact regularly ploughed, and still continues to be thus ploughed by earth-worms. It may be doubted whether there are many other

animals which have played so important a part in the history of the world, as have these lowly organised creatures." He neglected to observe that, whereas earthworms have never, to anyone's knowledge, harmed the soil, plows undoubtedly have. Of course, such an observation would have been unusual in Darwin's time, but it would not have been unheard of: see, for instance, the work of the Vermont scientist, linguist, diplomat, and statesman George Perkins Marsh.

28 **without decay, we would all drown:** In natural ecosystems as well as managed forests and grasslands, most of the net primary production falls to the ground. Even in heavily grazed grasslands, large herbivores rarely eat more than 50 percent of the greenery.

29 **not until vascular plants colonized the land:** The first solid evidence—of creeping bryophytes—dates from the middle Ordovician; see Cloud, chap. 13. It should be noted, though, that apparent fossils of filamentous cyanobacteria, together with the carbon content of rocks compressed from ancient soil, have recently led some scientists (see Horodyski and Knauth) to conclude that the land may have been densely settled as long as 1.2 billion years ago, in which case dirt may be older than we think.

30 **in the last ten thousand years:** For a dizzying overview of the rise and fall of the soil, see Wes Jackson, *New Roots,* pp. 5–10. The first real history of soil erosion can be found in G. P. Marsh. Among the handful of studies of soil's role in history, Hyams is quirky but insightful; Carter and Dale is more mainstream but not wholly reliable; Hillel, *Out of the Earth,* is current, scientifically informed, and most valuable where this distinguished soil scientist speaks of his old stomping ground, the Near East.

by a factor of thirty: Pimentel et al.

31 **what industrial agriculture is really doing:** The classic critique of industrial agriculture (in its classic American form) is *The Unsettling of America,* by the Kentucky farmer, poet, and essayist Wendell Berry—a book whose grace, humanity, and clarity of vision bring to mind Matthew Arnold's *Culture and Anarchy.* (Most of Berry's books touch on farming in one way or another and all are worth reading; if he repeats himself, he does so out of passionate conviction.) Wes Jackson's *New Roots* is a still more sweeping critique—not of industrial agriculture alone, but of virtually *all* agriculture. (I will have more to say about Jackson in Part Four.) The book of essays edited by Jackson, Berry, and Bruce Coleman is also valuable. A more dispassionate (but far from passionless) account of American agriculture is M. Kramer.

For every bushel of corn: W. Jackson, *Altars,* p. 35.

a third of its farmland topsoil: Pimentel et al.

On nine-tenths of U.S. farmland. . . . The cost of erosion: Pimentel et al.

conservation practices reduced soil erosion: Gardner, p. 44.

32 **tilling less and spraying more:** The extreme form of conservation tillage, "no-till," entails the use of 30 to 50 percent more pesticide than conventional tillage (M. Hinkle).

Kansas farmers exposed to herbicides: Hoar et al.; Zahm et al.

In 1948, at the dawn of the chemical age: David Pimentel, personal communication.

Livestock and the feed grown for them: U.S.D.A., chap. 10, sec. 4.

salmonellae: See Davies; Holmberg et al.

each . . . drinks 325 gallons of water a day: U.S.D.A., chap. 7, sec. 2.

Ogallala Aquifer: Reisner, *Cadillac Desert*, introduction; chap. 12.

saline or sodic conditions: U.S.D.A., chap. 2, sec. 4.

more irrigated farmland is lost: Gardner.

552 million hectares had been damaged: Oldeman et al.

33 **cost of soil erosion worldwide:** Pimentel et al.

fire, an aerobic creature: In the muster of our allies in the last chapter, we might have added one that is not, strictly speaking, a species: fire. Most creatures on earth get the energy they need by respiration, which is a slow and precisely controlled form of oxidation. Fire is fast, reckless oxidation. Earth—thanks to the hard work of living things—is eccentric among planets in having an atmosphere rich in oxygen: an "oxidizing" atmosphere, to use the chemist's term. In such an atmosphere iron rusts, wood burns, and creatures breathe.

In physiological terms, fire can be thought of as an aerobic species of peculiarly violent metabolism. In trophic terms, it can be thought of as a super-predator that seizes and uses the energy amassed by other organisms. In terms of ecological succession, fire is the species that turns the great wheel, returning a climax ecosystem to its humble beginnings.

It is not so far-fetched, then, to think of fire as one of humankind's allies, and by no means the least of them. Sometimes the energy released by fire is directly used by humans, as when we warm ourselves by a fire, saving calories we would otherwise get from food. Sometimes fire predigests our food, as cows predigest grass for us by turning it into milk and meat. But sometimes fire just wastes a lot of energy. Why would we want an ally that does that?

Much of the energy in mature ecosystems, especially forests, is locked up in things inedible to humans. Either relatively little new energy is being captured (as in temperate forests) or it is captured and turned over so fast (as in tropical rain forests) that humans barely get a swipe at it as it goes by. If you live in a country where all the wealth is held by a class to which you do not belong, you may welcome a revolution in which a great deal of wealth will be destroyed, if it gives you a prospect of getting some wealth later on. Fire is that revolution.

When it comes to nutrients, the advantages are much clearer. Nutrients, too, get locked up in mature ecosystems. Fire lets them loose. A great deal of carbon flies away, but some of it—along with most of the other nutrients—remains earth-bound in the form of ash, ready to generate new growth that humans can use.

Agriculture is such a system: Cf. Rindos, who speaks of a "Maladaptive Paradox."

similar logic guides parasites: See text, p. 20, and note.

34 **scrupulous use of condoms:** See Ewald, "The Evolution of Virulence."

The more you demand: From this point of view, mug-shooting humankind as a parasite of the whole planet is misleading, for it misses the waves that play upon the planet's surface. Humans, like other species, parasitize or exploit particular regions or ecosystems, then expand into other regions. Although we now exploit almost the whole surface of the globe (and many regions above and below the surface), it has taken us a few million years to reach this point. Of course, on the scale of the life history of the planet, that is simply the time a pathogen takes to infect a host, incubate, breed, and overrun its system. But such an image fast-forwards over the serried patterns of our expansion, and blurs the link between expanding and demanding.

Incidentally, the image of man as a parasite on the planet, though it has a modern ring, goes back at least to ancient Chinese cosmogony (Gascoigne). According to a Shang creation myth, the macrocosmic primal man, P'an Ku, was hatched from a primal egg. After growing for a while he suddenly fell apart, and the world we know is made out of his pieces. The mountains are his limbs, the rivers his blood, the wind his breath, the thunder his voice, the sun and moon his eyes; and humankind is derived from the parasites on his body.

4. THE NEW PANGAEA

35 **Great Library of Alexandria:** Popular myth places the burning in 640 A.D. and the blame on the Caliph Omar, who was supposed to have reasoned that if the books contradicted the Koran they were wrong, and if they agreed with the Koran they were unnecessary. Unfortunately for those who savor a good *mot*—and for those who would like to put the blame for all history's evils on turbaned heads—the actual burning was done by Westerners. The larger portion of the library, known as the Brucheion, was destroyed by Aurelian in 272 A.D. The smaller portion, known as the Serapeion, survived until 391 A.D., when it was torched in a riot led by Bishop Theophilus on orders of Emperor Theodosius I. (See Wellisch; Tarn, chap. 8.)

As to information content, my calculations are as follows. A bit of information decides between two possibilities. Since there are four nucleic-acid base-pair possibilities, each base pair represents two bits. An average species has one billion base pairs (Wilson, *The Diversity*, chap. 8), or two billion bits of information. (A "higher" plant or animal has something closer to ten billion base pairs.) The ancient Greek alphabet had twenty-six letters (including the archaic digamma and koppa), and a handful of other symbols (letters were also used as numerals), so each letter contains roughly five bits of information ($2^5 = 32$). (See Smyth, secs. 1–3 and 347.) One "book" of the *Iliad*—one book being a typical scroll—contains some 20,000 characters, or 100,000 bits of information. Thus the information from the genes of a single species would fill 20,000 scrolls.

I have, of course, made some simplifying assumptions, including the absurd one that all the books in the library were in Greek, and that all Greek letters occur with equal frequency.

how many species we are losing daily: Wilson (*The Diversity*, chap. 12), using conservative assumptions, comes up with a loss of seventy-four rain-forest species a day; because the rain forest has so many creatures to start with, and because their homes are being demolished so rapidly, that is where most extinctions are taking place. Almost all reputable estimates of total species loss are so far above the estimated "background" or "natural" rate of extinction that, for all practical purposes, we can assume that all current extinctions are man-made.

Informed by the joy and dismay of a lifetime in the field, Wilson's book is a fine introduction to biodiversity and its loss. More dated, more sensational, but still useful (particularly for its opening image of "rivet-popping") is Paul and Anne Ehrlich, *Extinction*. For a rigorous approach to some tough questions about extinctions over the course of earth's history, see Raup.

37 **Yuri Knorosov:** P. Morrison.

predator overhunts its prey: The violent oscillations that result were modeled in a classic series of differential equations by Lotka and Volterra in the mid-1920s. In real life, a number of factors can damp the oscillations, among them coevolution of the two species; other species thrown into the mix; and the effect of the species on their "environment." See Odum, chap. 7; Wilson and Bossert, chap. 3.

march of well-armed humans: For this "Blitzkrieg" theory, see P. Martin and R. Klein. Their position is explained and buttressed in Diamond's very readable *Rise and Fall*, chap. 18. Raup (chap. 5) sits on the fence, but cites with approval the conclusion of the great ecologist Robert MacArthur that only humans are able to bring about the extinction of a widespread species by predation alone.

to garnish women's bonnets: The first *Bulletin of the American Ornithological Union* (1884) estimated that five million birds gave their lives each year for the sake of women's hats (Matthiessen, *Wildlife*, chap. 8).

38 **so that Chopin's waltzes could be heard:** Coniff.

INVADERS: For a problem of such magnitude, biological invasion has been shamefully scanted by ecologists. After a promising start (in this as in so many matters) by Marsh in the mid-nineteenth century, the problem of biological invasion was largely ignored—except as a nuisance that had to be dealt with here or there, as it cropped up—until Elton's pioneering work in the 1950s. As far as I know, there has been no major synthesis of the field since then. A useful overview is provided, though, by Vitousek et al. And I must add, belatedly, the work of an historian, Crosby (*Ecological Imperialism*), who I now find spoke of the "reconstitution of Pangaea" long before I did.

Schieffelin resolved: Plimpton.

39 **it can follow him anywhere:** Compare the case of pigeons, which arrived in London in the thirteenth century, as church steeples were going up.

40 **forty-five hundred alien species:** Office of Technology Assessment, *Harmful*, chap. 1. Of the most recent class of invaders studied—the 205 species introduced or detected between 1980 and 1993—fully 59, or almost 29 percent, had done or were expected to do harm.

Invaders now make up one-fifth to one-third: Stuckey and Barkley.

Pangaea began its slow breakup into continents: See Cloud, chap. 14; J. Walker, chap. 13. If systematists had lived in the Mesozoic, they would have found their work physically thrilling but mentally untaxing. Dinosaurs in the part of Pangaea that is now China were much of a muchness with those in the part that is now Peru.

six realms of life: Besides Wallace, see Elton (chap. 2), who would add a seventh realm, the Eastern Pacific beyond Australia's continental arc.

after a mere twenty-two years of work: Counting from the opening of his first notebook for the systematic study of the origin of species, in July 1837, to the publication of the first edition of the book in November 1859. See Darwin, *The Autobiography,* last sec. ("My Several Publications").

41 **from first impact to full squash:** Indeed, the case could be made that in some places the pileup continues: for instance, in South Central Asia, where the Himalayas are still rising as a result.

When ecologists use their standard species-area curves: Vitousek et al., "Summary"; Westbrooks. The number of species representing a given taxon varies approximately as the cube root to the fourth root of the area of land; see Wilson and Bossert, chap. 4.

Permian extinction: E. O. Wilson, *The Diversity,* chap. 3.

42 **Isthmus of Panama was broken:** Elton, pp. 37–40.

Just how neatly some species can stow away: Elton, pp. 111, 29.

aphid that lived an obscure life: *Phylloxera vitifolii.*

fungus that lived on Asian chestnut trees: *Endothia parasitica.* The American chestnut is *Castanea dentata,* the Asian *C. crenata* and *C. mollissima.*

43 **woolly adelgid:** *Adelges tsugae* Annaud.

Asian tiger mosquito: *Aedes albopictus.* Vitousek et al., "Summary."

mitten crabs: *Eiriocheir sinensis.*

zebra mussels and Asian clams: *Dreissena polymorpha* and *Corbicula fluminea.* See Vitousek et al. "Summary"; *Biological,* app. 2. Laboratory sampling of ballast from 159 Japanese cargo ships pulling into Coos Bay, Oregon, turned up 367 different species of plants and animals, most of the latter in larval form. All the major marine habitat groups were represented. See Carlton and Geller.

European gypsy moth: *Portheria dispar.* Raver describes a still more terrifying prospect: the spread of the Asian gypsy moth, a subspecies also classified as *P. dispar,* which is not only more voracious than its European (and now American) kin, but far more mobile. While the European female is flightless, the Asian can fly up to thirty miles to lay her eggs. Since 1991, there have been sightings in Washington, Oregon, North Carolina, and New York. Russian cargo ships and an American military-transport ship returning from Germany are among the identified vectors.

Bt: See p. 331.

golden-apple snail: *Pomacea canaliculata* is host of the lungworm, *Angiostrongylus cantonensis,* which causes eosinophilic meningoencephalitis. See Vitousek et al., "Summary"; Naylor.

44 **water hyacinth:** *Eichhornia crassipes* is also a pest in Florida and along the Gulf Coast, where it is known as "million-dollar weed"—a reference to the cost of keeping it under control. See J. McKinley; E. Palmer.

cichlids: Seehausen et al. In the United States, freshwater ecosystems are the most endangered of all. Of the eighty-six species of freshwater fish listed as threatened or endangered in 1991, at least forty-four were victims (in part) of introduced fish. In at least twenty-nine cases, the introductions were made for the purpose of sport fishing. While federal and state officials who deal with dry land have learned something about the dangers of exotics, many fisheries managers remain oblivious. See Wilcove.

"Whenever we try . . .": Muir, *My First Summer in the Sierra,* chap. 6, entry for July 27.

interaction between higher-order systems: See Eldredge and Salthe.

Gaian worldview: For an attempt at fuller digestion of the Gaia hypothesis, see chapter 21.

46   a competitor for human patronage: It should not be a zero-sum game, given that millions of humans are hungry; but it often seems to be.

47   nature, culture, and economics: For a brief attempt to distinguish the appropriate responses to invasion in each of these realms, see note, pp. 554–5.

48   centers of plant and animal domestication: First mapped in the 1920s by the great Russian geneticist N. I. Vavilov, the centers of plant domestication have been remapped, and the world's crops rejuggled among them, by Harlan and by Sauer (especially in *Agricultural Origins*), among others. Though Vavilov's guiding assumption—that the place where a particular crop is most diverse is the place where it was first domesticated—has been challenged, his basic conclusions have held up fairly well.

loss of cultivated varieties: See Mooney; Eisenberg, "Save the Veggies" (not my title!), from which some of this section has been adapted. Though I deal here with crops, the loss of domesticated animal varieties is equally grave (see Raloff). At least fifteen hundred of the world's five thousand or so livestock breeds are now considered rare: that is, they have been reduced to fewer than twenty breeding males or one thousand breeding females. In the United States, Holsteins now make up 91 percent of all dairy stock—a flagrant case of putting all one's milk in one (admittedly capacious) pail.

On indigenous farming, see chapter 24 and notes.

49   Little seed companies were swallowed: On this, and on the seed business generally, see Doyle, *Altered Harvest.*

50   T-cytoplasm hybrid corn: Ibid., chap. 1; National Academy of Sciences, *Genetic Vulnerability.*

5. THE HUMAN MUSHROOM

53   every American ingests 160 pounds of fertilizer: Bayliss-Smith, p. 108.

we invest three calories of energy: Cox and Atkins, p. 618.

Tsembaga farmers: Bayliss-Smith, p. 108.

54   sudden rush of energy has destroyed biological information: Wes Jackson's statement of "a law at work in the world, a law of human ecology: high energy destroys information" (*Altars*, p. 15) is my point of departure here, but I find it needs some qualification. After all, a more or less steady, sustained flow of energy can locally create a "dissipative structure," of which hurricanes, patterns in boiling water, predator-prey population cycles, and you the reader are all examples. Life on earth can be seen as a giant dissipative structure that takes shape from the energy of the sun and reradiates that energy into space.

The concept of entropy is notoriously tricky, its link to information even trickier. Apart from having a small child in the house, the best introduction I know of to the concept of entropy is Atkins, which pointedly avoids talking about information. See also text, p. 54.

55 **sustained fission by uranium-isolating bacteria in Gabon:** See Lovelock, *The Ages,* chap. 5.

  diving for cover in the mud: Later, many anaerobes would find shelter in the guts (and other organs) of animals, especially large herbivores.

  No fuel is unequivocally "clean" or "safe": Wind power, as clean a source of energy as one could want, has brought about a great slaughter of raptors, which get caught in the blades of the turbines. According to a study released by the California Energy Commission, 567 birds of prey fell victim to the Altamont Pass wind farm in 1989 and 1990, at least 39 of them golden eagles. See Weisman.

57 **"the machine in the garden":** The title of Leo Marx's fecund book on nature, technology, and the American mind. Its notion of the "middle landscape" informs my discussion of Arcadia in Part Three.

6. LIFE ON THE EDGE

59 **if such a creature there be:** Even the celebrated "wild children" barely qualify, since most were exposed at some point to a modicum of either human or animal culture.

  Most of the time, when I speak of "culture" in this book, I am speaking of human culture. But other animals have culture, too: that is, they can pass information "by behavioral rather than genetic means," in the words of John Tyler Bonner (*The Evolution,* p. 4).

  ancestor of our word "nature": Apart from the *Oxford English Dictionary,* Liddell and Scott, and other standard reference books, insight into the history of the concept can be gleaned from Glacken, Passmore, and K. Thomas.

60 **Simha Bunim of Pzhysha:** Buber, *Tales,* vol. 2, pp. 249–50.

61 **"information gets its foot in the door . . .":** Chris Langton quoted in Lewin, p. 51. See also Waldrop, chap. 6; Kauffman.

  even animals are not happy animals: On chimpanzees, see Goodall. On the emotions of animals, see Hearne; Masson and McCarthy.

  Farming—some have argued—is deeply unnatural: The longest and best-argued brief is Paul Shepard's brilliant if wrongheaded book, *The Tender Carnivore and the Sacred Game,* which should be read (as should his other works) by anyone interested in humans and nature. A very different argument—less psychological, more ecological—can be found in Wes Jackson, *New Roots.* For much earlier expressions of the same view, see text, pp. 95–96, and notes.

  As for those who call farming a splendid triumph over our animal nature, they are too numerous to cite. For centuries, this has been the majority point of view.

62 **lucky thread by which democracy hangs:** A contingent fact, as Stephen Jay Gould has noted ("Human Equality").

63 **Bushmen will note a particular root:** "When we came to the center of an enor-
mous plain with no tree or bush to mark the place, Gai stopped and, glancing
around for a moment, pointed suddenly with his toe. After trying hard to see, we
noticed a tiny shred of a vine wound around a grass blade; no part of the vine
still touched the ground, as the vine had dried and parts of it had blown away.
Gai had known where the bi was, he told us, because he had walked by it months
ago in the last rainy season when the vine was still green, and he had remembered. He had assumed that it was still there because only his own people used
the territory around it and if one of them had taken it he would have heard.
Bushmen talk all the time about such things. He had had it in his mind to come
back and get it when the tsama melons were gone. . . ." (E. M. Thomas, *The
Harmless People*, pp. 112–13.) See also R. B. Lee; Lee and DeVore. The seamless (or
delicately seamed) continuum between foraging and farming is further probed
in chapter 24 of this book.

To paraphrase Nietzsche: He is speaking (2nd essay, sec. 19) of the "bad conscience": "*Es ist eine Krankheit, das schlechte Gewissen, das unterliegt keinem
Zweifel, aber eine Krankheit, wie die Schwangerschaft eine Krankheit ist.*"

contest of the male and the female: My speculations here are just that, of course,
and the transitions I speak of would be matters of degree. Certainly I do not
endorse the idea of a matrifocal, peaceful, ecologically benign Neolithic culture,
devoted to a unitary Goddess, that was swept away by evil Semites and Indo-
Europeans. Despite the hard work of Gimbutas, the evidence for such a "civilization of the Goddess" remains soft. By the time we get to cultures whose writing
we can read, mother goddesses and earth goddesses are integrated into the larger
pantheon, often serving the needs of the patriarchy (see Frymer-Kensky).

64 **God with the Sickle:** Gimbutas, *The Language of the Goddess*, pp. 83–4.

march of farmers across Europe: Colin Renfrew has made the case that the
instrument by which the Indo-European languages conquered Europe was not
the sword or the battle ax, but the hoe. The received view, honed in recent years
by Marija Gimbutas, is that warlike nomadic herdsmen swept across the continent in several waves, beginning about 3500 B.C., bringing their languages with
them. Renfrew points instead to the Neolithic farmers who, in the fifth or sixth
millennium B.C., brought wheat and barley to Europe from Anatolia (where Hittite, the oldest attested Indo-European language, was later spoken) and then,
breeding like mice, swarmed more or less peacefully to the shores of the Baltic
and the Atlantic, overwhelming by sheer numbers the hunters in their path.

Though the recent genetic studies of Cavalli-Sforza and others tend to support Gimbutas on the Indo-European question, they are consistent with Renfrew's view that agriculture conquered Europe not by cultural diffusion but by
"demic diffusion"—that farmers outbred and displaced hunter-gatherers,
instead of passing their trade secrets along to them. If this is generally true, the
idea that farming has some genetic basis may not be so far-fetched after all.

On the Indo-European question, see also Mallory; Zvelebil and Zvelebil; the
reviews of Renfrew by Anthony et al.

it might be easier to take bottomland: Cf. Carneiro.

vampire, witch, or whore: For a nuanced account of these recrudescences, see Duerr; for a crude account, see Paglia.

65 **earth monsters slain by heroes:** See Scully, *The Earth, the Temple, and the Gods.*
and maybe snakes: Twined with the Great Mother from Minoan and perhaps Neolithic times; see Gimbutas, *The Language of the Goddess.*
private parts of the Great Mother: Ancient Greek charms for pregnant women carried the image of a Gorgon and words to this effect: "O womb, dark and black, like a serpent you writhe, like a dragon you hiss" (Peter Brown).
to stifle our guilt about penetrating our mother: "Plowing the furrow" is a sexual metaphor at least as old as Sophocles (*Antigone* 569). Cf. Rank and Sachs: "Noteworthy in this connection is the fact that both in Greek and Latin as well as in Oriental languages, 'ploughing' is commonly used in the sense of practicing coitus . . . and according to Winckelman . . . the expressions 'garden,' 'meadow,' 'field' in Greek denoted the female genital organ in jokes, which in Solomon's Song is called vineyard. The neurotic counterpart to this symbolizing personification of the earth is found among the North American Indians whose resistance against cultivation by ploughing is explained by Ehrenreich that they are afraid to injure the skin of the earth-mother; here, the identification has succeeded too well, as one might say." For more on this "neurotic" identification, see my text, p. 328.
Other myths: For a reading of these and other farming-related myths from the somewhat different viewpoint of an eminent classicist, anthropologist, poet, and goddess-worshipper, see Graves.
certain of the rabbis claim is actually wheat: See text, p. 95, and note.
Many cultures see culture as male, nature as female: For a plethora of examples, see Duerr (whose civilized text is flanked by a boundless wilderness of endnotes). But there are many exceptions, and even a single culture may flip-flop wildly on this question.

7. THE MOUNTAIN OF THE GODS

69 **earlier human and hominid waves:** Among them, arguably, were *Homo ergaster*, *H. heidelbergensis*, and *H. sapiens*, allied with various perennial grasses and quadrupeds. See Tattersall; Stringer and McKie.

71 **That is the world-pole:** Though I have borrowed the concept from Eliade (*Cosmos and History,* pp. 12ff.), I have chosen a cruder term than his *axis mundi.*
For examples of World Mountains and other world-poles, see Eliade; Butterworth; Campbell, *The Masks of God,* vol. 1; as well as the splendidly illustrated book by Bernbaum.
In the Zoroastrian Avesta: Boyce, pp. 16–17.
When the shaman finds or creates the world-pole: On the role of the shaman, see Eliade, "Nostalgia for Paradise"; Campbell, *Historical Atlas,* vol. 1.

72 **THE CANAANITES:** A good introduction to the lands and peoples of the ancient Mediterranean, several of which play major roles in our drama, is Grant. Rang-

ing farther east, but with a more old-fashioned approach, is Moscati, *Ancient Semitic Civilizations*. Readers who like maps will like McEvedy, who also provides a pithy and interesting text. On the geography and ecology of the ancient Mediterranean and Near East, see Semple; Aharoni; Hillel, *Out of the Earth;* and the well-illustrated book by Attenborough.

Several thousand years of bad press in Hebrew and Latin sources have helped keep the Canaanites, and their cousins the Phoenicians and Carthaginians, from getting the serious study they deserve. For general treatments, see Harden; J. Gray; Anati, *Palestine Before the Hebrews;* Moscati, *The World of the Phoenicians*. Canaanite texts are translated by H. L. Ginsburg in Pritchard, *Ancient Near Eastern Texts;* short selections are translated, and discussed, by Cross and by Clifford.

Canaan: To avoid the political freight, as well as the imprecision in an ancient context, of such terms as Israel, Palestine, Syria, and Syro-Palestine, I will use the most ancient name for the region that still means something to modern ears.

Amanus: The ancient name; modern names include Alma Dag, Nur, and Hatay.

"the source of the Two Rivers . . .": Clifford, p. 49. This may be the site in the Lebanon range now called Khirbet Afqa, where the river called Nahr Ibrahim appears to rise dramatically from a great cave, or it may be some vaguely defined part of the Amanus range. The alternative rendering is Ginsburg's.

Mount Zaphon: It is usually identified with Jebel 'el-Aqra', at the mouth of the Orontes, in Syria. The Hebrew word *tsaphon*, "north," is derived from the name of the mountain, not the other way around.

73 **"He takes a thousand pots of wine . . .":** Pritchard, *ANET*, p. 136.

"Pour out peace in the depths of the earth . . .": Clifford, p. 68.

role of Zaphon as world-pole: As we noted, among the things the world-pole conjoins are the organic and inorganic. This may help explain the role of the temple ("the speech of wood and the whisper of stone") and of precious stones on the World Mountain.

74 **zone of sterility and death:** Mot's home is seen sometimes as a dry and rocky steppe, sometimes as a bog (Clifford, p. 81):

> In the midst of his city, Ooze,
> Decay, the throne on which he sits,
> Slime, the land of his heritage.

Although the mingling of dry and wet has been taken to show that Mot's home is being confused with Yamm's, there may be more to it than that. Too much water, or the wrong kind—salt rather than sweet—can kill as surely as too little.

"He plows his chest like a garden": Pritchard, *ANET*, p. 139.

"She seizes Mot . . .": As I cannot locate the source for the translations I have offered of this and the following passage, I will assume they are my own adaptations of several published translations. If I am wrong, I hope someone will correct me.

"**The skies rained oil . . .**": Clifford and Cross make this El's dream, Ginsburg someone else's.

75 **nip-and-tuck fight:** The poem says that the decisive contest takes place "in the seventh year." This may be an idiom for "after a while," or may mean that a mock battle was staged every seven years. In that case, the logic would be that a famine was bound to come at least that often (compare Pharaoh's dream of seven fat years followed by seven lean) and had to be fought back. The sabbatical year described in the Bible (Exodus 23:11, Leviticus 25: 3–4) may have been based on the same premise, except that instead of fighting back the power of death the Israelites defanged it by letting the land rest.

offspring of the first farmers: The growth of the cult of Baal in Syria around the fourteenth century B.C. has been linked (by Cross, p. 48) to the appeal of the seasonal myth recounted above, which agreed so well with the needs of settled farmers. It often assumed that earlier Canaanites were seminomadic herdsmen, but of this there is scant proof. It is true that El is called "Bull," but such gods are common among mixed farmers. That he lives in a tent is a stronger piece of evidence, but may tend to show only that his origins were not Canaanite at all.

for an ecological fact: In *Memories, Dreams, Reflections* (p. 251), Jung says that on a visit to the Pueblos of New Mexico he "stood by the river and looked up at the mountains, which rise almost another six thousand feet above the plateau. . . . Suddenly a deep voice, vibrant with suppressed emotion, spoke from behind me into my left ear: 'Do you not think that all life comes from the mountain?' An elderly Indian had come up to me, inaudible in his moccasins, and had asked me this heaven knows how far-reaching question. A glance at the river pouring down from the mountains showed me the outward image that had engendered this conclusion." For attempts at a more scientific assessment of the services rendered by wilderness, see Daily.

76 **maelstrom of gene flow:** In a series of reliefs in the temple of Amon at Karnak publicizing the triumphs of Thutmose III (reigned 1504–1450 B.C.) in Canaan, several panels display specimens of flora and fauna which the pharaoh carried back to Egypt from his third campaign. (The Egyptians were avid gardeners and the first collectors of exotic plants and animals.) The caption reads in part, "All plants that [grow], all flowers that are in God's land [which were found by] his majesty when he proceeded to Upper Retenu." (Borowski, p. 4.)

During the Pleistocene Ice Age: See D. Henry; W. Brice.

77 **plants and animals alike:** An individual plant does not have the luxury of moving around. A population of a given species of plant can, of course, migrate very slowly in response to slow (not seasonal) changes in climate. And populations of animals at opposite ends of a species' geographical range will vary in accordance with climate, much as plants would. In general, though, one can say that plants must do by variation what animals do by migration. It was the telescoping of both these processes into small geographical areas, courtesy of Humboldt and hillsides, that made settled farming possible.

anatomically modern humans: Bar-Yosef and Vandermeersch. Creatures once dignified (if that is the word) with membership in the human race, and classified as archaic *Homo sapiens* and *H. sapiens Neanderthalensis*, are now generally placed in separate species.

78   **analogy of rain to semen:** On the other hand, oceans are often thought of as female: Tiamat, Rahab, and many other examples might be cited. In European legend, it sometimes seems as if the "matrifocal" world of Old Europe has retreated underwater: a consequence, perhaps, of the vanishing of the Minoan island of Thera, which may have given rise to the story of Atlantis. Byatt recounts a Breton legend of the City of Is, an underwater world that is mirror and antithesis of this one—of Paris, which is the Indo-European, patriarchal world of technology. When Paris is destroyed for its sins, Is will take its place.

Sometimes underground waters, too, are seen as female: see Schama's dazzling, meandering book (which, for fear of undue influence, I avoided reading until mine was almost done). See also my text, chapter 11.

79   **Minoans, too, had sacred mountains:** See Scully, *The Earth.*
Among the cattle-mad Dodoth: E. M. Thomas, *Warrior Herdsmen.*

8. THE TOWER OF BABEL

80   **When farmers first wandered down:** On the unpromising natural endowments of ancient Mesopotamia, see Hillel, *Out of the Earth,* chap. 11; S. Kramer, *The Sumerians.*

On ancient Mesopotamia generally, see ibid.; Oppenheim; Crawford. Sumerian, Akkadian, and Babylonian texts are translated in Pritchard, *ANET,* as well as in books by Kramer, Jacobsen, Clifford, Ringgren, and Heidel.
grainfields, date plantations, fishponds: Kramer, *The Sumerians,* p. 109; but for caveats as to when some of the items on this list may first have been introduced, see Mauer.

81   **a series of trenches:** The best account of ancient Mesopotamian irrigation is in Hillel, *Out of the Earth,* chap. 11. See also Adams and Nissen.

82   **small-scale irrigation:** As practiced, for instance, by the Samarran culture of the Zagros foothills.
ziggurat: See Kramer, *The Sumerians,* chap. 4; Ringgren; Jacobsen, *The Treasures of Darkness.*
Nebuchadnezzar planted them for his Persian bride: Moynihan, pp. 23ff.
Every great temple claimed to stand: Half a century ago, it was widely believed (in the tight circle of scholars who care about such things) that the Mesopotamians saw the world as a mountain engirt by the oceans, with its base in the underworld and its peak in the heavens. The ziggurat was supposed to be a mock-up of this *Weltberg,* the best that could be managed on the mud flats of Sumer and Akkad. All of this now seems less clear than it once did. What remains clear is the role of the world-pole.

Eliade, by the way, makes no distinction between "cosmic mountains" that are real mountains and "cosmic mountains" that are fakes. "The names of the Baby-

Ionian temples and sacred towers themselves testify to their assimilation to the cosmic mountain," he writes, which is true enough as far as it goes.

Enuma Elish assigns to Babylon: In the fifth tablet (Pritchard, *ANET*, suppl., p. 502), the victorious Marduk tells the other gods of his plan to build a temple and a "luxurious abode" for himself and his cronies:

> When you come up from the Apsu for assembly,
> You will spend the night therein, (it is there) to receive all of you.
> When you des[cend] from heaven [for assem]bly,
> You will spend the night there[in], (it is there) to receive all of you.
> I will call [its] name ["Babylon"] (which means) "the houses of the great
>   gods" . . .

In the same way, a city might claim—what amounted to the same thing—that it was the navel of the world. This was no figure of speech, for Sumerian priests held that in the temple of Inanna in Nippur lay the scar that was left when the great god Enlil parted heaven and earth. Although the birth cord had been cut, the temple remained a kind of cable between the worlds. The precinct was called Dur-an-ki, "the bond of heaven and earth." Its center was called "flesh producer" or "the place where flesh sprouted forth"—for it was here that human beings first sprang like seedlings from the earth ("The Creation of the Pickaxe," in Clifford, p. 14).

In Egypt, another flatland, it was thought that at the center of the world stood a hillock. Straddling this hillock as it arose from the primal waters, Amun-Re began the work of creation. Though it may have been inspired by the fertile mudpies that rose each fall as the Nile's flood bubbled away, the myth took on more glorious colors. Hermopolis claimed to be built on the holy hillock. So did every other Egyptian city with an ounce of self-respect. (Clifford, pp. 25–28.)

the countryside is godforsaken: Though the gods seem to be omnipresent, or heavenly, in the sense that one can pray to them anywhere and be heard.

83 **They knew of Bedouin:** In a Sumerian poem (S. Kramer, *The Sumerians*, p. 253), a friend tries to dissuade the goddess Adnigkishar from marrying the Bedouin god Martu:

> A tent-dweller [buffeted] by wind and rain, [he knows not] prayers,
> With the weapon he [makes] the mountain his habitation,
> Contentious to excess, he turns against the lands, knows not to bend the
>   knee,
> Eats uncooked meat,
> Has no house in his lifetime,
> Is not brought to burial when he dies.

they saw humankind itself as a crop: For example, in the myths of the pickax and of Enki and E-engurra. Cf. a later creation fragment in Heidel, *The Babylonian Genesis*, pp. 68–71.

Surely the city was the source of all life: "And, by the way, who estimates the value of the crop which Nature yields in the still wilder fields unimproved by man?" To the question posed by Thoreau in *Walden* ("The Bean-Field"), a first effort at an answer was recently made: a systematic attempt to estimate the annual economic value of the earth's (more or less) natural ecosystems. Though the researchers cast their net wide—taking in such varied things as recreation value, pollination services, forest timber, the effect of wetlands on shrimp harvests, and the role of the oceans in regulating atmospheric carbon dioxide— plenty of things slipped through, among them nonrenewable fuels and minerals and the merits of such relatively unstudied ecosystems as deserts, tundra, and urban parks. Despite its conservative assumptions, the study came up with an annual value of $33 trillion. By contrast, the GNPs of all nations on earth total about $18 trillion. In short, the Mountain has challenged the Tower on its home turf, economics, and bested it nearly two to one. See Costanza et al., as well as the comments by Pimm ("The Value"); see also Daily.

84   **Canaanites were a culture of the first type, Mesopotamians of the second:** The Canaanites, too, had their artificial world-poles, the stone pillars which the Bible calls *matsevot*. So did the Old Europeans, the people who lived in southeastern Europe in the Neolithic and are supposed to have revered nature and the Great Mother. (Gimbutas says these were not world-poles, but offers no evidence.) Artificial world-poles were common in the ancient world, as they are in surviving shamanic cultures. You can't always find a World Mountain when you need one. If you can't come to the World Mountain, the World Mountain will come to you.

But there is a difference between these jerry-built world-poles and the monstrosities heaped up in Mesopotamia. The former may be seen as signs of rugged individualism, or of a god's ubiquity, or of his whimsy—his love of sudden cameo appearances, of leaping out from behind a rock or setting a bush on fire, of dropping a ladder from the sky a pillow's length from your head. ("Surely the Lord is in this place; and I knew it not": Genesis 28:16.) The latter show only the self-importance of city, priest, and king.

In the matter of the "sacred marriage," too, resemblances between Mesopotamia and Canaan are only skin-deep. Earlier I spoke of the sacred marriage between Baal and Anat, a marriage that was probably restaged each year in the temples of cities such as Ugarit. Most likely the king played the part of Baal, a priestess or temple prostitute that of Anat. Meant to ensure nature's fertility, perhaps by exciting her lust, such rites have been common in farming cultures all over the world (see Frazer). A few years ago, the new emperor of Japan was rumored to have performed a Shinto ceremony of this sort before his coronation. (In a common variant, the king or chief, or a surrogate, buys this brief surge of pleasure—and nature's bounty for his people—with his life.)

In Canaan, the human enactment might take place in the city but the real show, the divine tickling of nature's womb, took place on Baal's mountain: that is, in the realm of wild nature. In Mesopotamia, things were ordered differently. In a ritual libretto from the city of Isin (Jacobsen, *Treasures*, pp. 34–38), the king takes on the role of Dumuzi-Amaushumgalanna, an amalgamation of the shepherd god and the god of the date harvest. He weds Inanna, the sexpot goddess whose epithet is Ninegalla, "queen of the palace." (In Semitic lore, Inanna becomes Ishtar or Astarte, who in turn is confused with Anat.)

> In the palace, the house that administers the nation . . .
>   (therein) has the dark-headed people,
>   the nation in its entirety
> found a dais for Ninegalla.
>   The king, being a god, will sojourn with her on it.

> That she take care of the life of all the lands . . .
> The king goes with lifted head to the holy loins,
> goes with lifted head to the loins of Inanna. . . .

But Inanna, unlike Anat, does not have much to do with wild nature. She seems to have gotten her start as the spirit of the communal storehouse. A poem from Uruk describes the bedecking of Inanna for her wedding in terms suggesting the decking of the storehouse shelves with dates:

> A date-gatherer is to climb the date palm
>   for holy Inanna.
> May he take fresh ones to her!
> The man has decided to take them . . .
>   to the gem-revealing heap.
> On the surface of the heap he is gathering
>   lapis lazuli for Inanna.
> He is finding the "buttock beads,"
>   is putting them on her buttocks!
> He is finding the "head beads,"
>   is putting them on her head!

Whereas in Canaan it is the coupling of rain and earth that matters, in Mesopotamia it is the coupling of crop and storehouse. Attention shifts from natural cause to cultural effect. The real show takes place in the city.
skyscraper rests on thousands of acres: The ecological "footprint" of a typical U.S. city-dweller has been estimated at twelve acres (Callenbach, *Ecology*). Suburban feet are even larger.

9. THE FIERY SWORD

86   **The more we know about the Israelites:** In the first half of the twentieth century, the German school of Alt and Noth predominated: this school denied the historicity of the Bible, and spoke of a gradual infiltration of Canaan by semi-nomadic tribes, which in time formed a loose confederation. About midcentury, a counterattack was led by Albright and the more radical Kaufmann, who

propped up the patriarchal stories, and those of the conquest of Canaan, with sherds and other evidence newly unearthed. In the last couple of decades the winds have shifted yet again, with Dever, van Seters, Gottwald, and Mendenhall, among others, placing the Israelites more firmly in a Canaanite context. (Mendenhall goes furthest, making the Israelites out to be downtrodden peasants of wholly native origin.) For a clear presentation of this general approach, see Coote.

Atlases of the Bible generally take the Albright-Kaufmann line, for two reasons: it gives them more to work with, and it pleases their readers, most of whom are believers who want a scorecard while they read the Holy Writ. If allowance is made for this bias, atlases can be very useful: for example, Pritchard; Grollenberg.

When I had finished my text and was revising these notes, I came across the book published in 1996 by Hiebert, who, drawing on many of the same sources, appears to have reached some of the same conclusions as the present work about the role of nature in the Hebrew Bible. Although it is reassuring to have my lay speculations confirmed by a biblical scholar working at Harvard, it is also a blow to my vanity—which, outweighing shame for my tardiness, compels me to point out that Part Two of my book, in substantially the present from, was circulated in manuscript as early as 1992. Clearly Hiebert was as oblivious of my work as I was of his.

87 **not so much settling as resettling the uplands:** In the late Bronze Age this region had thirty settlements that we know of; in the early Iron Age it had 240 (Coote, p. 124, drawing on Finkelstein). The last period of dense settlement had been the Hyksos period (1650–1580 B.C.), when Canaanite warlords held sway not only in Palestine but in Egypt itself. In the New Kingdom period that followed, Egypt rebounded with a vengeance. Not content to let his Canaanite vassals and Nubian garrisons keep a lid on things, a pharaoh such as Thutmose III (1504–1450) would personally conduct a campaign each spring, like a bandleader taking his band on tour. As soon as the mud had dried on the roads, his army would be sweeping along the coastal plain, subduing cities, torching grainfields, pillaging granaries, cutting down fruit trees, and generally making war in the approved ancient manner. The walls of the Temple of Karnak list the spoils from a single siege, that of Megiddo, during Thutmose's first campaign. They included 340 living prisoners and 83 severed hands; 2,041 horses, 191 foals, and 6 stallions; 1,927 cows, 2,000 goats, and 20,500 sheep; some 450,000 bushels of wheat, "apart from what was cut as forage by his majesty's army"; 1 "fine chariot worked with gold belonging to the prince of [Megiddo]. . . , and 892 chariots of his wretched army"; "7 poles of meru-wood worked with silver, of the tent of that enemy." And so on. (Pritchard, *ANET*, pp. 237–38; Coote 36ff.)

Thutmose and his army returned to Palestine each spring for the next fifteen years. The habit was passed on to his successors. Meanwhile, the Hittites encroached from the north, tangling with the Egyptians and entangling the Canaanites in alliances on one side or the other. Whichever side the Canaanites joined, they came out the losers.

Egypt's sway in Palestine had two effects. On the one hand, the quality of visual art improved; on the other hand famine, poverty, and oppression were rampant. This went on until the twelfth century B.C., when the assaults of the so-called Sea Peoples—European and Anatolian pirates and marauders like Odysseus and Menelaus, known as the Philistines in Palestine, to which they gave their name—forced both Egyptians and Hittites to retrench. Now the Canaanites were able to breathe again. Although the Philistines drove them from parts of the seaboard, for the most part they and their domesticates were able to go about the business of breeding more or less unmolested. These two factors combined to push settlement into the hills.

THE HILL FARMERS: On the ecology and husbandry of ancient Israel, see Aharoni; Hopkins; Borowski; Reifenberg; Hillel, *Out of the Earth;* and the eccentric but interesting works of Hareuveni, founder of the biblical landscape garden Ne'ot Kedumim.

88 **"And the children of Joseph spoke unto Joshua ...":** Joshua 17:14–15.

bronze shares: In the early part of the so-called Iron Age, bronze was still the metal of choice, for the techniques of steeling, quenching, and tempering were as yet unknown, and without them iron is brittler than bronze.

89 **"a cemented cistern, which loses not a drop":** *Pirkei Avot* 2. 11. A century or two later, Israelites would use the same principle to make farming possible in the Negev Desert. Runoff from the rare and violent rainstorms would be channeled into sunken fields and orchards, so that the rain falling on an acre of ground might sustain a single tree. Mastery of this technique by the Nabateans of the Hellenistic period allowed them to plunk cities down in the heart of the desert, and so to inhale the wealth of the spice trade. See Hillel, *The Negev.*

that Samson should run into a lion: Judges 14:5.

bet av: Hopkins, pp. 252ff.

"be fruitful and multiply": Genesis 1:28.

90 **"Woe unto them ...":** "Woe unto them that join house to house, that lay field to field, till there be no place, that they may be placed alone in the midst of the earth!" (Isaiah 5:8.)

farm produce was among their primary wares: This became true of the Israelites during the monarchy, when the kings wanted cash crops for export.

MOUNT EDEN: On Eden as the Mountain of God, see Levenson.

91 **from Philo's time to the present:** This was the received opinion in Philo's day, though Philo himself was unconvinced (*Questions and Answers on Genesis,* cited by Manuel and Manuel, p. 43). Those modern scholars, such as Speiser and Zarins, who take the four rivers to be tributaries, and situate the Garden at the point where their joined waters debouch—that is, at the head of the Persian Gulf—have the weight of ancient tradition against them.

less to do with oil or blood than with water: See Hillel, *Rivers of Eden.*

identified by the rabbis with the Nile and the Ganges: Ginzburg, vol. 1., p. 70; Josephus *Antiquitates Judaicae* 1. 39; Rashi *ad loc.*

"a tree that spreads its canopy ...": *Bereshit Rabbah* 15. 6. (*Bereshit Rabbah* is the first section of *Midrash Rabbah* and consists of commentaries on Genesis.) Others render "to walk around its trunk" as "to walk a distance equal to the diameter of its trunk." The legends and homilies in the Midrash were strung together over the course of fifteen hundred years, from the time when the Pentateuch was compiled to the eleventh century A.D. Some of its traditions may be as old as anything in the Bible itself.

92 **"Thus saith the Lord God . . .":** Ezekiel 28:12–16.

the king of Egypt is compared to Assyria: Ezekiel 31:4–9, 15–16:

> The waters nourished it,
> The deep made it to grow. . . .
> Therefore its stature was exalted
> Above all the trees of the field. . . .
> All the fowls of heaven made
> Their nests in its boughs,
> And all the beasts of the field did bring forth their young
> Under its branches,
> And under its shadow dwelt
> All great nations. . . .
> I made it fair
> By the multitude of its branches;
> So that all the trees of Eden,
> That were in the Garden of God, envied it. . . .

Thus saith the Lord God: In the day when he went down to the nether-world I caused the deep to mourn and cover itself for him, and I restrained the rivers thereof, and the great waters were stayed; and I caused Lebanon to mourn for him, and all the trees of the field fainted for him. I made the nations to shake at the sound of his fall, when I cast him down to the nether-world with them that descend into the pit; and all the trees of Eden, the choice and best of Lebanon, all that drink water, were comforted in the nether parts of the earth.

lost Hebrew epic of Eden: See Cassuto.

"the son of morning": Isaiah 14:9–15.

93 **jewels for fruit:** The *avnei esh* (literally, "stones of fire") of Ezekiel 28:14–16 may refer to "stones of lightning" or "stars of El" mentioned in various Ugaritic and Akkadian texts; they may be the lightning that flashes from the Mountain of God; or they may simply be "fiery jewels." Jewels, especially sapphire or lapis lazuli, are found in Baal's temple on Mount Zaphon, as well as downstream of Eden in J's account (see below). Lapis even makes its way into accounts of God's appearance on his other great mountain, Sinai.

Family likenesses between Bible stories and the myths of the Near East are usually pointed out with a purpose: either to debunk the Bible or, contrarily, to show how it moralizes and demythologizes its sources. When the latter tack is taken, it is noted that in Genesis the Mountain of God is not mentioned, no doubt because pagan deities clustered on it like figurines on a wedding cake, or like flies a few hours after the reception; that the garden is made not for gods but for a man and a woman; that it is a man and a woman, not gods or angels, who sin and are cast down; that there is no such tasteless and nonnutritive frippery as trees bearing jewels; that gold and jewels are withdrawn from Eden altogether and deposited in the land of Havilah, which may or may not be nearby. (The phrase *asher sham hazahav*, "where there is gold," in Genesis 2:11, can be read to mean roughly, "That's where the gold is—not in Eden!") As for the creation of humankind, P's account is so abstract that it reads like a philosophy text (or a matchbook ad for the Famous Demiurges School: "Want to create a world and not get your fingernails dirty?"), and J's, messy as it seems to us, is antiseptic next to the blood-and-sperm-drenched mayhem of most cosmogonies.

Some of this is plausible, some of it sounds like special pleading. For our purposes, what is left out of the Bible may be as important as what is left in. Whatever official line P was toeing, whatever ironic point J was making, the fact remains that their pollarded trees stood in a vast forest of popular ideas about Eden. The canonic text may have leveled that forest for a time, but the stumps keep sprouting new shoots.

A note on this note. Source criticism of the Pentateuch teases its rough weave into four or five strands: J, the Yahwist; E, the Elohist; P, the Priestly source; D, the Deuteronomist; and (in some versions) R, the Redactor, who also did the weaving sometime in the fifth century B.C. For a fine summary of recent thinking about who these authors were, when they wrote, and why they wrote what they wrote, see Friedman. An imaginative reconstruction of one strand is Rosenberg and Bloom's *Book of J*.

what chance did humans have: According to one legend (*Pirkei de-Rabbi Elie-zer*, chap. 20, cited in L. Fine, p. 99), Adam was not just expelled from Eden but, in effect, cast into the netherworld, like Lucifer. Begging forgiveness, he stood in the waters of upper Gehenna for seven weeks, "until his body became like a species of seaweed." Of course, you might say he was trying to merge with nature—with the primordial ooze—as a penance for parting from nature.

vision of paradise as a forested peak: True, the word *midbar*, or wilderness in the sense of desert—an arid or deserted place—is sometimes used in opposition to Eden; but to conclude from this (as Nash, for example, does) that Eden and wilderness in the more general sense are opposites in Hebrew thought is a mis-understanding. See my chapter 12.

not meant for humans at all: Cf. Thoreau, "Ktaadn," *The Maine Woods:* "The tops of mountains are among the unfinished parts of the globe, whither it is a slight insult to the gods to climb and pry into their secrets, and try their effect on our humanity. Only daring and insolent men, perchance, go there. Simple races, as savages, do not climb mountains,—their tops are sacred and mysterious tracts never visited by them. Pomola is always angry with those who climb to the summit of Ktaadn."

"All the world is watered . . .": Talmud, *Taanit* 10a.

Humans cannot see God's face and live: The voltage of pure godhead would cinder us like Semele, like Nadab and Abihu. To prevent this, the *sephirot*, or divine spheres of Kabbalah, serve, one might say, as step-down transformers. For a general theory of "the holy" along these lines, see Otto.

Bushman story of beginnings: Bleek and Lloyd, pp. 54ff.

94 **room at the top, but not for us:** Later, there would be hints that humans, a lucky few, might climb the holy mountain and enjoy its blessings at full strength, instead of settling for the dregs.

> Who shall ascend the mountain of the Lord
> And who shall stand in his holy place?
> He who hath clean hands and a pure heart . . .
> He shall receive a blessing from the Lord. . . . [Psalms 24:3–5.]

> How precious is Thy lovingkindness, O God!
> And the children of men take refuge in the shadow of Thy wings.
> They are abundantly satisfied with the fatness of Thy house;
> And Thou makest them drink of the river of Thy pleasures.
> For with Thee is the fountain of life;
> In Thy light do we see light. [Psalms 36:7–9.]

The word for "Thy pleasures," *adanekha*, is the plural of "Eden." According to the rabbis, the light nurtured in the Temple was the first light of creation, by which Adam could see from one end of the world to the other.

Of course, the lucky few who fed on the plenty of the Temple were priests. (In Ezekiel 28, the gems that "cover" the privileged god or angel match those on the breastplate of the high priest.) A son of Voltaire might note dryly that in fact riches flowed from the land uphill to the Temple, not the other way around.

"fill the earth and subdue it": Genesis 1:28.

95 **the Fall:** As I noted earlier, I mean by this the expulsion from Eden; there is no reference to the concept of original sin.

"Cursed is the ground for thy sake . . .": Genesis 3:17–19.

Agriculture as we know it: Other peoples, too, have identified farming with the Fall. The Bobo of Burkina Faso believe that the primal harmony of the world was destroyed when humans began farming. Similarly, the Khasi of Assam say humans were once able to reach heaven by climbing a great-tree—until they cut it down to clear space for a garden. In a Dogon myth from Mali, heaven and earth were once adjacent, but God parted them (and made humans mortal) after being annoyed by the din of women crushing millet. (Ries.)

"Rabbi Meir said: It was wheat . . .": *Bereshit Rabbah* 15. 7.

96 **"The Martu eat bread . . .":** Adapted from S. Kramer, *The Sumerians*, p. 287.

snakes were used to protect granaries: They are still used that way. In Iowa, farmers keep bull snakes to rid their barns of rats and mice (Palmer, p. 543).

To dig the great irrigation canals: At one time it was thought that the digging and periodic dredging of large canals, and the divvying up of water from them, required a ruthless centralization of power: what Karl Wittfogel called "oriental despotism" or "hydraulic society." It now looks as if large-scale irrigation can get under way with a fairly loose power structure, such as the alliance of priests and princes that ran the towns of fourth-millennium Mesopotamia. But if the canals did not really demand centralized power, they could be its instrument once it got started. Water could be used to reward friends and punish enemies. So could surplus grain and other provender piled up in central storehouses. These could also be used to correct inequities caused by the river's caprices. Although most canals could be dredged and maintained by loose-knit cooperation among villagers, the "need" for large-scale maintenance work was a good excuse to draft peasants for corvée labor. Not only was this a good way of keeping tabs on people, but it made it easier to organize labor gangs for harvest and other work on plantations owned by the crown or the priesthood. With the population strung out along the banks of navigable canals as if along a subway line, it was easy to send troops or tax collectors to any village that got out of line. On the other hand, it was easy for rival city-states to do the same, so a village often found itself tugged at from two or three sides. Moreover, the great tracts of semidesert that hemmed in these ribbons of green were populated by herdsmen. These people, whose flocks were vital to the economy, might be loyal peasants or they might be fierce tribesmen. Or they might be loyal one week and fierce the next.

Even a loose sort of "hydraulic society" was pretty hard on the landless peasant, who would get unsteady work on gangs that dredged canals or moved northward along the river to bring in the harvest. Some inkling of what his life was like can be gleaned from the "Georgica Sumerica," a sort of farmer's almanac in which a "farmer"—that is, the owner of an estate—advises his son as follows: "While the field is drying, let your obedient (household) prepare the tools for you, make fast the yoke bar, hang up your new whips on nails, and let the hanging handles of your old whips be mended by the artisans." Lest we think that these are meant only for oxen, the farmer goes on, ". . . (After) having had it (the field) harrowed (and) raked three times and pulverized fine with a hammer, let the handle of your whip uphold you; brook no idleness. Stand over them (the field laborers) during their work. . . . They must carry on by day (and by) heaven's stars for ten (days). . . ." (Kramer, pp. 105–9, 340–42; Salonen, pp. 202–12.)

As soon as a society has a surplus of food that can be stored, the possibility arises of dividing it unevenly. Stratification rears its lopsided head. We find it among the flush hunter-gatherers of the Pacific Northwest; we find evidence of it among the Natufians. But only with large-scale agriculture does it reach its full potential.

"Cursed is the ground by thy passing over": Gesenius, the great nineteenth-century Hebraist, notes that *avur* can also mean "produce" or "yield" (cf. Joshua 5:11ff.). Thus the phrase might be read, "Cursed is the ground in (i.e., with respect to) thy produce"; or, more intriguing for our purposes, "Cursed is the ground by thy produce."

97   **"to work it and protect it":** Genesis 2:15, my trans. There are, of course, many other licit interpretations of the two Hebrew words. Among the more interesting is Cassuto's: he makes *l'ovdah* refer to "divine service" (following *Bereshit*

Rabbah 16. 5: "These denote sacrifices"). And he makes *l'shomrah* refer to the task of guarding which was formerly (in Babylonian tradition and the lost epic tradition to which Ezekiel alludes) that of the cherubim—to whom it reverts when Adam is expelled.

"So He drove out the man . . .": Genesis 3:24.

98   **"I will not let thee go, except thou bless me":** Genesis 32:26.

10. THE RIVERS OF EDEN

99   **One breeding pair of spotted owls:** Hunter; cf. Pimm et al.

     Grizzlies need nearly as much elbow-room: Mattson and Reid.

     broken into fragments: On habitat fragmentation and its effects, see MacArthur and Wilson; Soulé and Wilcox; and the popular treatments by Weiner and Quammen. Unfortunately, most writers on this subject ignore the other side of the coin: the joining together by humans of regions formerly separated. On the local scale, we make islands; on the global scale, we bridge them; and the combination is lethal.

     Warblers and other songbirds: Robinson et al.; Rich et al.

100   **there must be arteries as well:** A much more detailed consideration of the form these arteries might take, and of the scientific principles underlying them, will be found in chapter 29.

101   **It gives us a model to work from:** Among the doughtiest champions of this view (and the corollary about particular places) are Wendell Berry and Wes Jackson.

     Sabbath, sabbatical, and jubilee: See chapter 29.

102   **Kern County:** Maize.

     The wall defines a living thing: This is implicit in the physicist Erwin Schrödinger's definition of life (discussed in my text, pp. 263, 416) as the ability to resist and indeed reverse—locally, of course—the flow of entropy. It is explicit in the definition of life as "autopoiesis" or self-making: the first thing that organic molecules do, in the computer simulations of incipient life by Maturana and Varela, is join hands and form a circle, or, rather, a membrane (see Varela et al., as well as the slightly goofy popular treatment in Maturana and Varela). Many less abstract (though still speculative) theories of the origin of life also put walls at the very start of the process, whether these are protein "tennis balls" (Rebek), or the oily droplets known as coacervates (Oparin), or plain old surface-tension bubbles in water (L. Lerman). Even if life is imagined to have formed on a solid surface, such as crystalline clay (Cairns-Smith) or pyrite (Huber and Wächtershäuser), the principle is the same: to distinguish self from nonself, keeping vital molecules together long enough to engage in metabolism and replication. One might say that the wall is clung to, rather than bounced against; or that the "wall" lies between two dimensions of space and a third.

     While there are accounts of the origin of life, such as Eigen's, that put replication of nucleotides first, I do not find them convincing. Of course, evidence for any account remains slender at best. On theories of life and its origins, see also

Horgan; Schapiro; Dyson; Margulis and Sagan, *What Is Life?* Incidentally, it now seems possible that membranes are the fundamental premise not only of life but of all physical reality. In physics, the leading contender for a Theory of Everything holds that the basic components of the universe are not particles or strings but p-dimensional objects, or "p-branes"—specifically, membranes in eleven dimensions which, when rolled up, produce superstrings in ten dimensions. See Duff.

103 **degree of closure increases as one rises in the hierarchy of living systems:** The "Vernadsky Paradox" propounded by Barlow and Volk—how can Gaia be a living system if it is "appreciably closed from the standpoint of matter exchange"?—more or less dissolves when one sees that such closure is a matter of degree.

104 **no easy answers:** Hard ones will be sought in chapter 29.

105 **sloppy rhetorical habits:** In particular, we should remember that humans have their own nature and that many other animals (including those which are not human allies) have their own culture.

107 **A property of several acres can run the gamut:** This subject is further addressed in chapter 29.

108 **most primal of man-made world-poles:** It is not, though, the only smallish world-pole of a civilizing kind. The magic even a minimal world-pole can work is displayed in Stevens's "Anecdote of the Jar":

> I placed a jar in Tennessee,
> And round it was, upon a hill.
> It made the slovenly wilderness
> Surround that hill.
>
> The wilderness rose up to it,
> And sprawled around, no longer wild.
> The jar was round upon the ground
> And tall and of a port in air.
>
> It took dominion everywhere.
> The jar was gray and bare.
> It did not give of bird or bush,
> Like nothing else in Tennessee.

Plainly the jar is a Tower, though a very small and portable one.
postmodernists telling us wilderness is a myth: See, for instance, Cronon, ed.

11. STORMING THE MOUNTAIN

112 **best-seller in the Bible's world was the Epic of Gilgamesh:** Versions and scraps of versions have been unearthed as far afield as Assyria, Anatolia, and Palestine,

most copied from the Akkadian but some translated into Hittite and Hurrian. See Tigay.

Unless otherwise noted, the translation quoted is that of Speiser (in Pritchard, *ANET*), though I have also consulted Heidel's and sections of others; for the Sumerian version, it is S. Kramer's "Gilgamesh and the Land of the Living" (also in *ANET*). Among secondary sources, I have found Jacobsen, *The Treasures*; S. Kramer, *The Sumerians*; and Tigay generally useful. The ecological interest of the Gilgamesh story has not escaped modern writers: see Perlin, *A Forest Journey*; R. P. Harrison.

reigned for thirty-six thousand years: My calculations are based on the King List, translated in S. Kramer, *The Sumerians*, app. E.

"two thirds of him is god . . .": I. ii. 1–2 (tablet I, col. ii, lines 1–2).

Adam Kadmon, **the macrocosmic primal man:** See Scholem, *Major Trends*, lecture 6; *On the Kabbalah*, chap. 3.

Adam himself, who before his Fall: *Hagigah* 12a. Burt Visotzsky, who found the source for me, convinces me that this reading is probably based on a mistranslation, but I like it too much to give it the boot.

Gilgamesh may be such a shaman: See Butterworth.

113 *jus primae noctis:* The *droit du seigneur* that is the mainspring of *The Marriage of Figaro* and is hinted at in *Don Giovanni*. The citations from the Epic are I. ii. 12, 16; from the Old Babylonian version, iv. 32–34. See also Tigay, p. 182.

just as the first humans were in Mesopotamian myth: "Creation of Man by the Mother Goddess," in Pritchard, *ANET*, pp. 99–100.

"After he had had (his) fill of her charms . . .": I. iv. 22ff. Enkidu's seduction is both touching and arousing: as recently as 1963, a major English edition (Heidel) printed it in Latin.

The motif of the hairy wild man lured into human society is common in the Far East and seems to have been inspired by the orangutan. See Jacobsen, *The Treasures*, p. 214.

114 **Enkidu's state before his "fall":** See Tigay, p. 203; but cf. S. Kramer, *the Sumerians*, pp. 220ff.

women and children force us to be civilized: Another twist on the question of which sex is wilder; see text, p. 65, and note.

The harlot leads Enkidu to the city: She says (I. iv. 34–39):

> "Thou art [wi]se, Enkidu, art become like a god!
> Why with the wild creatures dost thou roam over the steppe?
> Come, let me lead thee [to] ramparted Uruk,
> To the holy temple, abode of Anu and Ishtar,
> Where lives Gilgamesh, accomplished in strength,
> And like a wild ox lords it over the folk."

Now that Enkidu has become godlike, he ought to live in the city, where the gods live. We are reminded (some of us) of the kinfolks' injunction to Jed Clampitt when he has struck oil and struck it rich: "Californey is the place you oughta be."

Although Enkidu's life on the steppe may sound like fun, it does not take long to convince him that city life will be even more fun. If sex is what civilizes him, the prospect of more sex is one of the things that lure him to the city. Contrary to pastoral lore, sexual opportunities are far more plentiful and varied (within our species) in the city than in the country.

The hooker (an apt term here) says (I. v. 6–12):

> Come then, O Enkidu, to ramparted [Uruk],
> Where people are re[splend]ent in festal attire,
> (Where) each day is made a holiday,
> Where [. . .] lads . . . ,
> And la[ss]es [. . .] of figure.
> Their ripeness [. . .] full of perfume.
> They drive the great ones from their couches!

The gaps tantalize, like swirls of diaphanous cloth. How much more must these images new to Enkidu tantalize him! "Each day a holiday" is a good working definition of paradise. But this Eden is clearly urban. Four thousand years later, the call of the unwild will hardly have changed. One thinks of Sportin' Life's pitch to Bess: "You and me will live that high life in New York . . . And through Harlem we'll go struttin', / And there'll be nuttin'/ Too good for you."

There follows a sort of proto-pastoral interlude at a shepherd's hut in which Enkidu is introduced to the pleasures of bread in its solid and liquid forms: "He drank beer—seven kegs. . . . His insides felt good, his face glowed." (Jacobsen, *The Treasures*, p. 198.) At the same time, the pastoral realm is introduced, for the first time in world literature, to its role as a halfway house between wilderness and civilization. Then Enkidu is led to the city.

Gaster's rendering: In his edition of Frazer, p. 182.

115 **as Mel Brooks has said:** *History of the World, Part One*. To be sure, modern royals are somewhat more closely observed.

"Land of the Living": Jacobsen (*The Treasures*) renders this "The Land Where the Man Lives." Could this conceivably mean "the land where man (first) lived"—something on the order of Eden?

"I peered over the wall . . .": S. Kramer, "Gilgamesh and the Land of the Living," pp. 25–29.

"I said that you are gods . . .": Translation mine.

116 **knowledge of death makes Gilgamesh want to make a name:** This is clear in the Sumerian version, implied in the Old Babylonian (III. iv. 3–25).

cedar was the wood of choice: On timber in the ancient Near East, see Perlin, *A Forest Journey;* Meiggs; Thirgood. On timber in the Gilgamesh story, see Hansman.

***Juniperus excelsa*, the Sumerian term *gish erin*:** Hansman.

117 **monument erected by Yahdun-Lim of Mari:** Pritchard, *ANET*, p. 556.

Huwawa has been identified with the chief god of Elam: Hansman.

"E[nkidu] killed [the watchman] of the forest . . .": Old Babylonian version,
V. C. reverse, 12–13; translated by A. K. Grayson in Pritchard, *ANET*, Suppl.,
p. 504. In this, as in the psalm that follows, the name that sounds like Syria is
thought to refer to Mount Hermon, in the northeast of Israel, or perhaps to the
whole Anti-Lebanon range, of which Hermon is the southernmost part.

118 **Shamash sends eight mighty winds:** In the Hittite version; see Tigay, pp. 78ff.
Huwawa offers a sustained yield of timber: V. iv, Hittite version. Fragment. 22–26.

> "Let me go, Gilgamesh; thou [wilt be] my [master],
> And I shall be thy servant. And of [the trees]
> Which I have grown, I shall [. . .]
> Strong . . . [. . .]
> Cut down and houses [. . .]."

relief from the great wall of Karnak: Pritchard, *The Ancient Near East in Pictures*, p. 100, no. 331.

119 **accomplices in the rape of their homelands:** Of course, Enkidu would be more
like the local elite (the Wabenzi, as they are called in Africa, for the Mercedes cars
they drive) than like the slave or wage-slave workforce.

120 **Gilgamesh "opened up the secret dwelling of the Anunnaki":** Oddly, the Anunnaki do not seem overly upset at having their roofs ripped off. If they are, the
urban gods must overrule them. (Besides, one gets the impression from *Enuma
Elish* that the Anunnaki spend as much time in town as they can, like businessmen on expense accounts.) The high gods are admiring; the goddess Ishtar is
aroused not to anger but to lust. Seeing the sweaty muscle-man wash his hair after
his exertions, she invites him to become her lover. But Gilgamesh snubs the love
goddess, reeling off a list of lovers who paid for her favors in baneful ways—the
most famous of these being the shepherd god Tammuz, later known as Adonis.

No wonder Gilgamesh rebuffs Ishtar—he knows her all too well. Her town is
his town, so he knows all the dirt. As Ivana Trump and Leona Helmsley can testify, no one is a goddess in her own city. Already we suspect that, for the goddess,
the move to the city has been a mistake: once her feet are clear of the soil, are
shod and tapping on tiled floors, she is easily swept up, ravished, and laughed at.
Her only consolation is that in the long run the gods will not fare much better.
They will either be smothered by the trappings of power, or fade until they are
no more than pale ideas.

121 **"Enki and Ninhursag":** See the translation and commentary by Kramer and
Maier (chap. 1, representing Kramer's latest thoughts on the poem); his paraphrase in *The Sumerians*, pp. 147–49; Jacobsen's in *The Treasures*, pp. 112ff.; and
the very different rendering by Alster, which recalls a much earlier interpretation
by Jacobsen ("Mesopotamia," p. 173). See also Kirk, who unlocked for me a
poem with which I had struggled for months and opened the door to my own
interpretation.
Ninhursag is the mother goddess: In this poem she is called by another of her
names, Ninsikilla.

that some have called the Sumerian paradise: See, for example, Kramer, *The Sumerians,* pp. 281–84; his translations of "Enki and Ninhursag," which are subtitled "A Paradise Myth"; and the more personal accounts of his "quest for paradise" in "Dilmun" and *In the World of Sumer,* pp. 190–203. Though the idea did not originate with Kramer, he does seem to have had considerable stake in it, and his scholarly weight has given it wide acceptance. See also Bibby; Rice; Potts; Khalifa; Alster.

The identification of Dilmun with paradise, and more particularly with Eden, led me down what I now think is a blind alley. It is, however, a blind alley with remarkable and revealing views, and readers who would like to take a brief version of my journey are invited to read the rest of this note.

After some throat-clearing, "Enki and Ninhursag" begins:

> The land Dilmun is pure, the land Dilmun is clean.
> the land Dilmun is clean, Dilmun is bright! . . .

> In Dilmun the crow screams not,
> the *dar* bird cries not *dar-dar,*
> the lion kills not,
> the wolf snatches not the lamb,
> the dog knows not to subdue the kid,
> the swine knows not to eat barley . . .

> the eye-sick one says not, "I am sick-eyed,"
> the one suffering from headache says not, "I am headache-sick,"
> its old woman says not, "I am an old man,"
> The old man does not say, "I am an old woman,"
> the young girl places not water for her bath in the city,
> the ferry-man says not, "It's midnight,"
> the herald circles not round himself,
> the singer sings not elulam,
> at the outside of the city no shout resounds.

In this rendering by Alster, Dilmun seems less paradise than potential: a seed that unfolds only when water is added. Yet in this respect it is like Eden, a place that may have been similarly inchoate until God "caused a mist to rise from the ground." Even if we had never heard of Dilmun, this feature of the Eden story might lead us to suspect a Mesopotamian influence. In Canaan, most farming was rain-fed. But springs and oases were valued, too, as elsewhere in the Near East; and it is to these, rather than to Mesopotamian-style irrigation, that the "mist" in Eden probably refers.

Kramer (*The Sumerians,* p. 149) points out another plausible link to Eden. The deity created to heal Enki's rib is named Nin-ti, which can mean either "the lady of the rib" or "the lady who makes live." This hoary pun might explain why

Eve, "the mother of all life," is fashioned from Adam's rib. In Hebrew the pun falls away, leaving the speechless bone.

In the Sumerian flood story, which clearly had great influence (at however many removes) on the one in the Bible, the counterpart of Noah, known as Ziusudra, is rewarded for his piety and prudence by the gift of "breath eternal like a god." He is made to live "in the land of crossing, the land of Dilmun, the place where the sun rises." (The word rendered "crossing," which Kramer takes to refer to the rising of the sun, can also mean "rule"; the word for "land" can also mean "mountain.")

In the Akkadian version of the same story nestled in the Epic of Gilgamesh, the lucky man, now called Utnapishtim, is made to live "far away, at the mouth of the rivers." It is here that Gilgamesh, wrenched by the death of Enkidu and intent on learning the secret of eternal life, seeks him. To get here Gilgamesh crosses in pitch-darkness a cosmic mountain range guarded by scorpion-men. He then confers with Siduri, the barmaid of the gods, in a garden of lapis lazuli and carnelian by the side of the sea. When pressed, Siduri tells the king to pole his way across the Waters of Death in a boat using twice-sixty poles of sixty cubits each. (Each thrust requires a new pole, lest a drop of death-dealing water touch his hand.) He does so, and finds Utnapishtim.

Since many details of the story have changed from Sumerian to Akkadian and Babylonian times, there is no good reason to think that Ziusudra and Utnapishtim have ended up in the same place. The only overlap between the phrases used to describe their respective hideaways is the idea of "crossing." But while Ziusudra lives in "the land of crossing," Siduri speaks of "the place of crossing" as if it were the narrow point in a body of water—evidently the Waters of Death—that lies in Gilgamesh's way.

Heidel (*The Gilgamesh Epic*) thinks that Utnapishtim's abode was first imagined at the mouth of the Tigris and Euphrates rivers. As that area became too familiar to be mythical, the man and his family were relocated from the place where the sun rises to the place where it sets. The mountains that Gilgamesh crosses Heidel identifies with the Lebanon and Anti-Lebanon ranges; Siduri's garden he sets on the shore of the Mediterranean. (One is tempted to place it on the slopes of Mount Carmel, the "vineyard of God," beneath which the bay reclines like a huge wine bowl.) Utnapishtim's home would then be somewhere beyond the sea. (Others put Siduri's garden on the island of Failakah, which the Greeks later took to be sacred to Artemis.)

There are problems here. Since the mouth of the Tigris and Euphrates, the Shatt al-Arab, was near the first Mesopotamian cities and regularly navigated by their seamen, it is hard to believe that it was ever mythical. Perhaps "the mouth of the rivers" had always meant a faraway place beyond the sea—much like the Isles of the Blessed that Hesiod (*Works and Days* 166ff.) locates "at the ends of the earth," "beside deep-eddying Ocean."

But if Heidel is right, then we may see here the first instance of a rule that will later become familiar—namely, that paradise follows profits. Such quasi-paradisiacal places as the Cedar Forest and the Land of Crossing seem to have ambled from east to west in step with the interests of Mesopotamian business-men. The interest of loggers shifted from the Zagros range, which had been pretty well shorn by the end of the third millennium B.C., to the mountains of Lebanon and Syria; the interest of merchants in general shifted from the Gulf to the Mediterranean. Both changes were related to the gradual slide of power and population northward along the Euphrates (Sumer, Akkad and Babylon, Nin-eveh). This in turn was linked to environmental degradation in southern Mesopotamia and points south, such as the Indus Valley.

Whatever their locations, the places to which Ziusudra and Utnapishtim have retired are functionally more like the Isles of the Blessed, or Homer's Elysian Plain—exclusive resorts for a few select friends and relatives of the gods—than like Eden or anything else we would call paradise.

What really makes Dilmun seem an odd kind of Eden, though, is that it actu-ally existed. No shimmying mirage, it was a place of hard stone and harder calculation. To be exact, it was a hub of commerce.

Imagine if we found in some Judean tel a sheaf of business papers refer-ring to "merchants of Eden" and "ships of Eden" and imports of copper ingots from Eden. In Mesopotamia, in regard to Dilmun, that is exactly what we have found.

Some of the earliest writing in the world graces what we would call business papers, except that they are made of baked clay. (If poetry, storytelling, and the life of the spirit were our only concerns, our mouths might still be serving us well. Dr. Johnson's dictum, "None but a blockhead ever wrote except for money," may or may not be true today; it was certainly true in the fourth millennium B.C.) The very earliest texts as yet found in Mesopotamia or anywhere else, tablets recovered from the temple district of Uruk and dated to about 3200 B.C., conjoin the early cuneiform sign for Dilmun with the signs for such items as axes, tex-tiles, and tax collectors. An inscription on a temple-door socket dedicated by Ur-Nanshe, king of Lagash, around 2500 B.C. records that "the ships of Dilmun brought him wood as a tribute from foreign lands." Two or three kings later, a merchant named Ur-Enki was importing as much as a hundred kilograms of copper from Dilmun in a single shipment. Among his steady customers was Luqunntu, the wife of King Lugalanda. Apart from copper and copper ore, the dates, onions, and textiles of Dilmun were in demand. Returning to Dilmun, Ur-Enki's ships carried wheat, shelled barley, oils, ointments, cedarwood, and cheese.

As Sumerian rulers gave way to Akkadians, Elamites, Amorites, and Babylo-nians, the Dilmun trade had its ups and downs but mostly remained a good racket, at least until the middle of the second millennium B.C. In Babylonian, *lu-tilmun-a,* "the men of Dilmun," came to mean "merchants"—much as "Canaan-ite" would among the ancient Jews and "Jew" among later Christians.

Although the two faces of Dilmun—paradise and mart—might seem miles apart, in Mesopotamian eyes they seem to have been twins. In a version of the story of Enki and Ninhursag from Ur, dated to 1800 B.C., there is an interpola-tion that is probably meant to follow the part of the poem in which Dilmun, hav-ing been watered by Enki, is called "the bank-quay house of the land."

The land Tukris shall transport gold from Harali,
Lapis lazuli and bright . . . to you.
The land Meluhha shall bring cornelian, desirable and precious,
Sissoo-wood from Magan, excellent mangroves,
On big ships!
The land Maharshi will (bring) precious stones, dushia-stones,
To hang on the breast.
The land Magan shall bring copper, strong, mighty,
Diorite-stones, na-buru stones, shumin-stones to you.
The land of the Sea shall bring ebony, the embellishment of (the throne)
 of kingship to you.
The land of the tents shall bring wool, and fine powder to you.
The land Elam shall transport its load of exquisite wool to you.
The sanctuary Ur, the dais of kingship, the pure city,
Shall bring to you barley, tree-oil, precious clothes, exquisite clothes on
 big ships to you.
The broad sea shall bring its abundance to you.
The city, its dwelling places shall be pleasant dwelling places,
Dilmun, its dwelling place shall be a pleasant dwelling place.
Its barley shall be very fine barley,
Its dates shall be very big dates!
Its harvest shall be threefold. [Alster]

All this is splendid but by no means mythic. What we have here is a textbook account of Dilmun's role in the commerce of the ancient world. Its only home-grown products seem to have been dates and onions (its barley, however fine, was supplemented by imports). Otherwise its wealth came from the processing and trading of raw materials brought in from points beyond. The source of its good fortune was twofold: its happy situation between Mesopotamia and those points, and Enki's gift of fresh water, which made it a perfect place for ships to put in.

Almost all scholars now agree that Dilmun was the island of Bahrain, though it may have included parts of the Arabian mainland as well. The identification was first made in 1880 by Henry Rawlinson (the British officer and prodigious amateur scholar to whom cuneiform first surrendered its secrets), on the strength of a chunk of temple found on the island and bearing a dedication to Enshag (or Inzak), the patron god of Dilmun. Other evidence, both textual and archeological, has supported Rawlinson's guess.

The oddest thing about Bahrain from an archeological point of view is something that can be seen from the air. Large parts of the island's surface look like sheets of sand-colored bubble wrap, or the surface of a pimpled-rubber Ping-Pong paddle. They are covered with as many as two hundred thousand burial mounds built of stone but partly submerged in sand.

To some eyes (for example, those of Lamberg-Karlovsky) these hives suggest a larger death industry than the native population of the island could have supported. Not that there would not have been enough native corpses to fill the mounds, but that it would have taken too much money and labor to build them. Some archeologists think, therefore, that Bahrain must have been a kind of posh necropolis in vogue among the Mesopotamian elite—something like the funeral islands near Venice—and that this had something to do with its role as a holy place, a paradise, the place where Ziusudra found his reward. Like any place of pilgrimage worth its salt—Rome, Jerusalem, Mecca, Benares, Lourdes—the island milked this business and its spillovers for all they were worth, selling religious *tchatchkes* as well as food and lodging to the dying, the dead, and their relatives and well-wishers. In time those little mounds were parlayed into a commercial empire of mountainous heft.

For all its grim appeal, this theory does not stand up to scrutiny (see Alster). Apart from ignoring the fact that no identifiably Mesopotamian grave-goods have been found in the mounds, the theory flies in the face of everything we know about Mesopotamian funerary practice and belief. Moreover, Dilmun was not a kind of heaven where the righteous went when they died. The whole point about Ziusudra was that he did not die. True, the dying might have hoped their afterlife would be improved somehow by proximity to a place where someone lived forever; but the brutally sharp divide the myth draws between Ziusudra and all others of mortal birth, including the mighty Gilgamesh, makes this seem unlikely.

What is far more likely (and supported by the evidence) is that Dilmun first grew fat on trade, then saw its luster of gold and jewels flare into an aura of holiness. The building of great temples is one way this might have happened; the Sumerian Flood story is an example of another way. If eternal life was to be a reward and not a punishment, the gods had to find Ziusudra a nice place to live. With its wealth, its cosmopolitanism, its fresh water, its healthful climate (so unlike the marshes of home), Dilmun fit the bill nicely. Just as we might like to retire to St. Croix or the south of France, so Ziusudra would spend his golden eons in Dilmun.

As the home of a friend of the gods, Dilmun may have come to seem a little holier. But even this should not be overplayed. Dilmun was called "holy" but so were many other cities.

Around the middle of the second millennium B.C., a cabal of factors—the collapse of the Indus Valley civilization, the rise of new trading partners to the west, the northward slide of power in Mesopotamia, and maybe a change in climate on the shores of the Gulf itself—conspires to depose the Gulf from its primacy in world trade. The Mediterranean usurps its place. At this point Dilmun sinks into darkness. We still see the name padding a list of some king's conquests or inflating his title, but we do not see it on billing statements or tax rolls or the other documents that really matter. Rummaging in the relevant layers, the archeologists agree: the great cities are abandoned.

We would not expect such unsinkable merchants as the "men of Dilmun" to go gentle into that darkness, and there is some evidence that they did not. Herodotus relates that the people of Tyre believed their ancestors had come from an island in the Persian Gulf called Tylos, which is the name the Greeks used for

Bahrain. At first this sounds like the sort of fabulating based on bad etymology that the Greeks, like the Hebrews, went in for. But as G. W. Bowersock has noted, Herodotus said many foolish things that have turned out to be true. Bowersock finds good reason to think this is one of them. If he is right, then the Babylonian term "men of Dilmun" and the Hebrew term "Canaanite" denote not merely the same profession, but the same ethnic group. When the trade routes migrated from the Gulf to the Mediterranean, the traders followed them. The Dilmunites, who had dominated trade in one epoch, became the Phoenicians, who dominated it in the next. The people we find doing business in the world's earliest texts went on to exploit the alphabet for similar purposes.

Dilmun was not the ancestor of Eden. It may have been *an* ancestor, but hardly the most direct in lineage or closest in features. The rope linking Dilmun to Eden is long and frayed. At its near end, it is braided with several others that appear to be stouter, one of which leads to the Canaanite—and Israelite—Mountain of God.

The Mesopotamians did not need a paradise, it seems, because they already had one: namely, Mesopotamia. It was a man-made paradise, but that is no objection. Many "natural" paradises are man-made. Most of South Florida is naturally swamp, fine habitat for waterfowl and gators but not for people.

Still, it is hard to believe the Mesopotamians could have been so blissfully contented with their lot. Were not these the first human beings to be fully estranged from Eden—the first to suffer the indignities of taxes, big government, forced labor in muddy canals, and plantation-style work in the fields?

Yes, but they were not the people who wrote down the myths. For priests, landowners, wealthy merchants, and scribes, a city such as Uruk or Lagash may indeed have been paradise. These were the first human beings to taste the delights of a leisured city life: art, music, literature, fine clothes, catering, and temple prostitutes, "their ripeness [...] full of perfume." The people who probably had the biggest share in shaping the myths, the priests, were among the most pampered of all. (The climate may have been on the warm side, but when you officiate in the nude that is not a huge problem.)

Even for the peasant, life in Sumer and Akkad had its compensations. Public granaries made him more secure against famine than his forebears had ever been. When he sweated in the fields or stood chest-deep in muck to dredge the canals, he had the satisfaction of being part of an army that was subduing nature. He had the satisfaction, too, of feeling superior to the nomads and Stone Age farmers in the surrounding hills and steppe.

The fact that Abraham, Isaac, and Jacob's sons had to go to Egypt for bread, rather than to some lowland town in Canaan, shows that famine was widespread among the settled peoples of the region as well as the seminomads. If we can rely on the social background of these stories (as opposed to the stories themselves), it would seem that the peasants of the great river valleys, with their vast granaries under public control, really did have a more secure life than those of Canaan. As the story of Joseph and the seven lean years suggests, the price of this freedom from nature was bondage to other men.

122　**"Enki and the World Order":** Translated in Kramer, *The Sumerians,* chap. 5.

123　**"During the nights...":** "Atrahasis," trans. Speiser, in Pritchard, *ANET,* pp. 104–6.
"blossoms out in mockingly beautiful floral patterns": Hillel, *Out of the Earth,*
p. 83. His account (chap. 11) of salinization in ancient Mesopotamia is both
expert and readable.

124　**ratio of wheat to barley:** Jacobsen and Adams.
salt must have caused failures of the wheat crop: First proposed by Jacobsen
and Adams in 1958, the theory has intermittently been attacked (for example, by
M. A. Powell) and defended (for example, by Artzy and Hillel) in the years since.
What is in question is not that salinization occurred, but whether Mesopotamian farmers had effective methods to arrest, reverse, or soften its effects. Most
recently, the trend has been to explain the vicissitudes of farming in various parts
of Mesopotamia, and the attendant fates of cities and empires, in terms of
climate change; see, for example, H. Weiss et al. (see also the commentary by
A. Gibbons). Though no single factor can explain the whole course of Mesopotamian history, my own (lay) hunch is that salinization played a major role.
a cautionary tale: In Kirk's reading, the myth links irrigation and sexual excess,
hinting that the pursuit of fertility, if pushed too far, can lead to infertility. But
Kirk never mentions the real-world fact that makes sense of the analogy—
namely, salinization.

12. THE HIGHWAYS OF ROME

127　**vigorous hybrid of Mountain and Tower known as Zion:** On the tangled web
of writ and legend that connects Zion, Sinai, and Eden, see Patai, *Man and Temple;* Levenson; Clifford.
"Great is the Lord, and highly to be praised...": Psalms 48:1–3.
to make Mount Zion the new cosmic mountain: "Har Tsion" might even have
sounded like a pun on "Har Tsaphon"—a deliberate tweaking of the Canaanite
peak.
　　Isaiah (33:20–24) speaks of Mount Zion in much the same way:

> Look upon Zion, the city of our solemn gatherings....
> But there the Lord will be with us in majesty,
> In a place of broad rivers and streams;
> Wherein shall go no galley with oars,
> Neither shall gallant ship pass thereby....
> Then shall the blind man take large spoil,
> The lame shall seize the plunder,
> And none who dwells there shall say, "I am sick,"
> The people that dwell therein shall be forgiven their iniquity.

Even one who begrudges Dilmun the name of paradise, as I do, cannot help
hearing here an echo of its praises: "Its sick-eyed says not 'I am sick-eyed.' " What

is most audible, though, is the sound of the great rivers of Eden. In Ezekiel, their sound becomes a roar.

" 'These waters issue toward the eastern region . . .' ": Ezekiel 47:8–12. If, in Ezekiel's vision of the river flowing from the Temple, *ets rav meod* is taken to mean "a very great tree," the allusion to Eden is clear as day—not least because there are two trees, one on each side of the torrent. If instead "a very great forest" is meant, the healing of nature becomes less of a metaphor and more of an ecological need. Although the Judean hills had been partly deforested by the Israelites who first tilled and terraced them, it is likely that forests near Jerusalem suffered far more from the sieges of Sennacherib and Nebuchadnezzar (and later of the Romans). Whether in the future tense, present, or past, the Bible never tires of describing the trashing of the land by invaders. It is often hard to tell what is metaphor and what ecological fact. The same is true of verses that tell of healing and redemption and the kingship of God.

128   **It was here that Solomon was taken to be anointed:** I Kings 1:33, 38, 45.
author known as J: See above note to p. 93.
"On that day living waters shall issue . . .": Zechariah 14:8.
he controls and doles out the primal waters: For one brief moment before the Temple arose, the Talmud says, that task devolved upon David: "At the time that David dug the foundations [for the Temple], the watery abyss [*tehoma'*] came to the surface and sought to flood the world. David recited the fifteen [songs of] ascent [Psalms 120–34] and brought them [i.e., the waters] down." (*Sukkot* 53a; see Patai, *Man and Temple*, pp. 55–57.)
They call Zion the navel of the universe: "Just as the embryo begins at the navel and proceeds onwards from there, so the Holy One (blessed be he) began to create the world from its navel and from there it spread out in different directions" (*Midrash Hashem Behokhma Yasad Arets,* Jellinek 5:63). "The world was created from Zion" (*Yoma* 54b). "Just as the belly-button is positioned in the center of a man, thus is the Land of Israel positioned in the center of the world. . . . And Jerusalem is in the center of the Land of Israel, and the Temple is in the center of Jerusalem. . . ." At the center of the Temple was the Foundation Stone, "and beginning with it the world was put on its foundation" (*Tanhuma: Kedoshim* 10). See Levenson, *Sinai and Zion,* p. 118.
To make it seem to flow, instead, from a skimpy hill: If Zion was the cosmic mountain, how did one explain those taller mountains to the north? The Israelites seem to have skirted this embarrassment by making the northern mountains pillars of the sky. They are like the poles that hold up a bridal canopy, or the posts of a four-poster bed: important for structural reasons, but the real action happens in the center.
   The success of the Hebrew poets in making Jerusalem the center of the world may be measured not only by the number of medieval and Renaissance maps that adopted this plan, but by the hogsheads of blood spilt in that neighborhood over the past few millennia.

129   **Sinai remains untamed:** It should be noted that Mount Horeb plays the role in the E and D sections of the Bible that Sinai plays in J and P. See note to p. 93.

no one knows where it is: Some archeologists claim to have found it in the modern Har Karkom, in the western Negev. See Anati, *The Mountain of God;* Levenson, *Sinai and Zion.*

a place where God lets himself be known: The word for "wilderness," *midbar,* has been related to the root *dbr,* "to speak." The Midrash comments, "Whoever does not make himself like a wilderness, open to everything, will not be able to acquire wisdom and Torah" (*Bemidbar Rabbah* 1.7).

130   **"They shall sit every man under his vine . . .":** Micah 4:4.

There were such clans: One reason the J author played up the story of the wandering patriarchs may have been to help the house of David cement its control over these restless elements, which were a critical part of its power base. See Coote.

Those scholars who see the Hebrews as nomads may be swayed by the sight of the wandering Jews of diaspora. Albright is more blatant when he paints the patriarchs as donkey caravaneers (once a merchant, always a merchant).

seminomadism was an escape hatch: On the relations between nomads, seminomads, and settled peoples, see Khazanov; Bar-Yosef and Khazanov.

In the sabbatical year, the land itself is allowed to revert: How exactly this might have jibed with the biennial fallow common in the ancient Near East is something of a puzzle; see Hopkins.

131   **Bob Marshall:** By far the least famous of these figures, Marshall is by no means the least noteworthy. A New York Jew—son of the jurist and Jewish leader Louis Marshall—he became a botanist, a devout backpacker, founder of the Wilderness Society, forestry official under Franklin Roosevelt, and architect of wilderness protection in America. A great wilderness in Alaska bears his name. See Nash, chap. 12.

Song of Songs: One might say the functions of the World Mountain have been dispersed among Sinai, Zion, and the "mountains of spices." For years I pictured these as the sort of heaps you might find in a warehouse in Cochin, like the heap on which Rushdie's Moor is conceived; they are, of course, real mountains on which fragrant plants grow wild. The gazelle on the mountains may be a fleeting reminder of the Goddess: Scully (*The Earth,* p. xiii) speaks of a stone rhyton found at Kato Zakro showing a temple in a cleft gorge "crowned with a heraldic grouping of wild goats." Art Waskow (*Godwrestling,* chap. 6) has written of Canticles as "Eden for adults"—transmuting into a higher sensuality the Garden's childlike polymorphous perversity, its confusion of species, sexes, and senses. (Queerly, synesthesia—"All the people saw the thunder"—also played a role at Sinai, the place where distinctions were made.)

gashing of sycamore figs: Borowski, p. 129. Herdsman-poets were found among the Greeks, no doubt, as they were found until recently in Macedonia. But by the time they find their way into writing they are myths or literary constructs. It is not at all clear that they had much effect on the genre named for them, the bucolic. See Halperin.

In looking at the Hebrew view of nature: A large and prestigious school of thought blames the Hebrew Bible for most of our present ecological ills. Tracing the history of this charge is a task at once fascinating and infuriating. Though hints of it can be found in ancient polemics, the idea is first made explicit by writers such as Hegel who, while not quite thinking in terms we would call ecological, contrast the mean abstractions of the Hebrews with the sensuous fullness of the pagan pantheon. Common among more or less anti-Semitic writers for the next two centuries, the idea took root in the Nazi blood-and-soil cult and among its highbrow sympathizers, for most of whom the nature-hating seminomads of ancient times merged with the "rootless cosmopolitans" of the present. Regrettably, the great Toynbee, too, played a part.

The first clearly "ecological" formulation can be found in the famous essay of Lynn White, Jr., "The Historical Roots of Our Ecological Crisis." Polemics along the same lines, some of them far less temperate than White's, have since disfigured fine works by Paul Shepard (*Man in the Landscape, Nature and Madness*), Frederick Turner (*Beyond Geography*), and the influential landscape architect Ian McHarg. (The Frederick Turner in question is the Frederick Turner who was born in 1937, writes about American Indians and jazz as well as humans-and-nature, and is not to be confused, but constantly is, with the Frederick Turner who was born in 1943 and writes epic poetry and arty essays as well as pieces on humans-and-nature. The first Frederick Turner has a middle initial, W., but uses it only in his Library of Congress listings. It is very provoking for readers in this field that two distinguished authors insist on using exactly the same name. In my bibliography, I distinguish them by means of that furtive "W.," in brackets.)

Deep Ecologists (see Part Four) have taken up the chant with a vengeance, getting their script mainly from Heidegger. New Agers have been inspired mainly by Jungians such as Joseph Campbell, whose books (whatever his personal attitudes may have been) fairly glow with anti-Jewish animus (see, for instance, the typically ignorant comments in the *Atlas*, vol. 2, part 1), and by other professional mythographers who have found slim pickings in Hebrew fields. Ecofeminists lump the Semites together with the Indo-Europeans as nature-hating, warlike seminomads who dethroned the Goddess and set sky gods up in her place. (So reluctant are these writers to acknowledge male influence that one of them cites the author of "The Historical Roots" as Lynne White.) As Daum and Plaskow have noted, the new accusation against the Jews is: You killed our Goddess.

Perhaps because overt anti-Semitism is no longer so fashionable as it once was, many of these writers make their ostensible target the "Judeo-Christian tradition"—a seesawing couple, balanced on the hyphen, one going up while the other goes down; or a sort of vaudeville team in which one plays straight man while the other takes the pratfalls.

The response from the "Judeo-Christian tradition" has been mixed. Some writers have answered with thunderbolts and anathemas. Some have tried to climb off the seesaw, leaving their whilom playmate in the lurch. Others have searched their own traditions, and hearts, for the seeds of an ecological ethic that might blossom in our own time. Among Christians, noble attempts have been made by Bill McKibben (notably in his eloquent *The Comforting Whirlwind*), by

René Dubos (who provides a useful bibliography on the question), and by the two Berrys, Wendell and Thomas. Among Jews, Arthur Waskow, Everett Gendler, and Arthur Green have explored paths opened up by such earlier thinkers as Buber and Kook, and feminists such as Judith Plaskow and Marcia Falk have probed the boundaries between immanence and outright paganism. There have also been more or less disinterested scholarly responses, such as that of Passmore, which finds the Greeks and Romans, rather than the Hebrews, mainly responsible for the Western tendency to treat nature as a set of objects to be used and abused at will.

The modern ecopolemic against Judaism does not deal with the ideas of the Mountain and the Tower, or with anything so subtle as the actual relation of the Israelites to the land. Instead it relies on a few simple ideas. The first of these—that the Hebrews were nature-hating nomads—was dealt with in chapter 9. The second idea has to do with the demographic profile of the Hebrew god: male, single, and transcendent. If the Mesopotamians scorned the Great Mother, the argument goes, then the Israelites abolished her. A single, transcendent god takes the sacredness out of nature and leaves it dead—a heap of matter to be exploited as one pleases. To anyone who has actually read more than a few pages of the Bible, this is eyewash, as I hope this section of my text ("The Still Small Voice") shows.

There is one last point in the ecological indictment of the Hebrew Bible. It is based on the six days of creation in Genesis, which seem to erect a "ladder of nature" on which humankind occupies the highest rung. One passage is singled out (Genesis 1:28): "And God blessed them; and God said unto them: 'Be fruitful, and multiply, and fill the earth, and subdue it; and have dominion over the fish of the sea, and over the fowl of the air, and over every living thing that creepeth upon the earth.' " (See Jeremy Cohen.)

In Judaism, neither the sequence of creation nor these marching orders were ever taken to mean that humankind could do as it pleased. The fact that man is the last creature to be made was read by the rabbis in diverse ways. Is the rest of nature a table set for the honored guest? Or is man meant to remember that he is an afterthought, a Johnny-come-lately, a kid brother to the worm? (Shaped from a hunk of clay left amid the sawdust on the workshop floor. From spare parts, from outtakes. From earth that was itself created, in fact excreted, by worms.) "Man was not created until the sixth day so that if his pride should govern him, it could be said to him, 'Even the tiniest flea preceded you in creation' " (*Tosefta Sanhedrin* 8:4–5). This is a dichotomy even more pronounced than the one represented by the slips of paper that Simha Bunim of Pzhysha advises us to keep in each pocket—one reading "For me the earth was made," the other "I am but dust and ashes"—since here there is the possibility that we are lower than the flea. The slips of paper assume our intermediate position on the ladder, then look at it from above and below. But in this darker view of Genesis we lose our footing altogether. Or we rise and fall like the angels at Padan Aram. Punning on the verse "And have dominion [*uredu*] over the fish of the sea," Rabbi Hanina said: "If he merits it, [God says,] '*uredu*' (have dominion);

while if he does not merit, [God says,] *'yerdu'* (let them descend)" (*Bereshith Rabbah* 1.28).

For all the nuances the rabbis find, it is true that the shadow of the Tower falls across this stretch of parchment. There are two creation stories in Genesis, that of the six days and that of Eden. The first is attributed to the source known as P, or the Priestly author; the second to J, or the Yahwist. Throughout the Pentateuch, P is greatly concerned to legitimize the power of the central cult and the central government in Jerusalem. Not surprisingly, his words bear something of a Mesopotamian stamp. For better or worse, it is that stamp which has most impressed the West.

"profound sentiment of love for nature . . .": *Cosmos,* vol. 2, pp. 411–13.

132 **"the plan of Zeus is accomplished":** *Iliad* 1. 5. Actually, it is just a shade ironic that the Gaia hypothesis should be taken up by neopagans, for it restores the faith in nature's harmony that moved those great Christian ecologists. Indeed, in its pop form the Gaian faith is as monotheistic as Moses'. It merely shifts metaphysical gears (from transcendence to immanence) and changes gender.

"And, behold, the Lord passed by . . .": 1 Kings 19:11–12.

133 **nature is not diminished:** The great thing about having gods in nature is that nature can fight back. Misuse of the soil earns the wrath of earth spirits. A wild mountain has its guardian; though that guardian may, like Huwawa, get beaten, at least he will put up a fight. But monotheism provides such a mechanism, too: if we disobey God's command to respect the land, he will kick us out. It's the difference between knowing that if you hit your sister she'll hit you back, and knowing that if you hit her she'll tell Father and he'll hit you. Since Father hits harder, the latter may be more effective. On the other hand, that effectiveness is compromised if, as in most monotheisms, Father likes you (the human) best.

Yet in the Bible it is often the land itself that fights back, as in Leviticus 18:25: "And the land was defiled . . . and the land vomited out her inhabitants." As the great scholar Johannes Pedersen argued, a good deal of animism persists in the Bible. It is hard to pick up in English translations, which neuter the male and female stones, trees, earth, and waters of the Hebrew text.

God would escape his cage and burst back into nature: Much of the process is traced in the revolutionary works of Gershom Scholem and in the still more revolutionary, if far less readable, works of Moshe Idel. Whereas Scholem finds the wellspring of kabbalistic ideas in Gnosticism, Idel's divining rod finds indigenous, often underground Hebrew sources at least as old as the Bible.

regarded the earth itself as a living being: See, for instance, *Abot de Rabbi Nathan* (Schechter) 46a; *Midrash Tehillim* 19.2; and even the dyed-in-the-wool rationalist Maimonides, pt. I, chap. 72. See also Patai.

two empires, the Greek and the Roman: On their role in shaping the Western view (and use) of nature, see Passmore—who, however, is more concerned with philosphers than with a general temper of mind. For general references on Greece and Rome, see the notes to chapters 13, 14, and 16.

134 **"Many things are wonderful . . .":** Sophocles *Antigone* 332–75.

filling the earth and subduing it: See above, p. 484.

Sextus Julius Frontinus boasted: *De aquis urbis Romae*, 16.

135 **Centuriation:** For evidence from aerial photography that centuriation was widely applied, see Bradford. It should be noted that, in the application of the grid, some allowance was made for natural topography.

new Roman cities were built on the same rectangular plan: See Mumford, *The City in History.*

so many roads came from Rome: Plutarch says the roads "were carried straight across the countryside. . . . Hollows were filled in, and torrents or ravines that cut across the route were bridged." Quoted in Pritchard, *Harper Concise Atlas*, p. 124.

136 **later waterworks:** On Tivoli, see chapter 18. On dams, see Reisner, *Cadillac Desert;* Worster, *Rivers of Empire;* Goldsmith. On levees and on the rebellious urges of rivers, see McPhee, who quotes the hydrologic engineer Raphael Kazmann (p. 55): "This is an extremely complicated rivers system altered by works of man. . . . Floods across the country are getting higher, low stages lower. The Corps of Engineers—they're scared as hell. . . . The more planning they do, the more chaotic it gets."

chaos as the expression of shapes in imaginary space: See Gleick, *Chaos*, pp. 213–40.

artery of wildness in a biological sense: See chapter 29.

physical law applies identically in all places: Modern physics has put a dent in this assertion, but for most purposes, on the surface of the earth, it holds.

137 **He demands that we account:** For the idea of a community as an ecological accounting device, I am indebted to Wes Jackson.

Jews never tried to impose their culture on others: A very few exceptions, such as the reign of John Hyrcanus, might be noted.

138 **crush them to the dugs of Gaia:** For a striking case study of meddling by Western environmentalists in the affairs of an Amazonian people, see Kane.

13. ARCADIA

143 **the person from Porlock:** See p. xxi.

145 **Even those of us who have never read a line of pastoral verse:** For the English reader a good place to start would be the anthology by Kermode and the tendentiously Marxist, but as such well done, anthology by Barrell and Bull. Valuable secondary works include Poggioli; Rosenmeyer; Highet, *The Classical Tradition;* and the famous work of Empson, who—by spreading his net to include any work "about the people but not by or for them"—manages to snag such things as *Troilus and Cressida, The Beggar's Opera,* and *Alice in Wonderland.*

the tablets of Sumer: There are pastoral tropes in the Akkadian Epic of Gilgamesh, as well as in Sumerian poems about the shepherd god Dumuzi. On the most ancient wellsprings of pastoral, see Halperin.

"easy, vulgar, and therefore disgusting . . .": Johnson, "Life of Pope," in *Lives of the Poets.*

146 **the savanna in which we were dandled:** See Dubos, *The Wooing*, chap. 5.

147   *Et in Arcadia ego:* Often taken in modern times to express sweet nostalgia—"I,
      too, have been in Arcadia"—the motto once meant something more sinister:
      "Even in Arcadia, I [death] am present." See Panofsky; for a more recent variant,
      see Boynton, in which the cows' chant *Et in Arcadia sumus* is translated, "Coun-
      try life is overrated."
      actual ecology of the world that pastoral depicts: On the geography and
      ecology of the ancient Mediterranean, see Zimmern, pt. 1; Semple; Pounds;
      J. Hughes; and the more general work of Grant. The groundbreaking work of
      Marsh (one of the first major books about groundbreaking) is still useful. A well-
      illustrated overview is offered by Attenborough. For a rare attempt to take some
      Latin poets, including the pastoral Virgil, out of the library and set them down
      on native ground, see Highet, *Poets*—though his view is not what we would call
      ecological.
148   **"Next morning, I was minding my flock . . .":** Menander *The Arbitrants* 256–59
      (ed. Sandbach).
      "By comparison with the original territory . . .": Plato *Critias* 111a–d. By some
      odd slipup, the word for "bees" is rendered as "trees" in the Princeton edition,
      making nonsense of the passage.
      much of that erosion was man-made: See Runnels; J. Hughes.
149   **"The gods, who live on Mount Olympus . . .":** Hesiod *Works and Days* 109–18.
150   **change is what the pastoral poet spends much of his ingenuity covering up:**
      Except in the elegiac mode.
      it has to stay spring or summer: It is true that in some types of pastoral cele-
      brating a settled agricultural life, what is celebrated most of all is the seasons and
      their soothing, bracing recurrence. There are poems about the seasons, georgics,
      shepherd's calendars, farmer's almanacs. But Hesiod, the first of that line, dis-
      liked the harsher seasons, and his golden men were golden not least in never hav-
      ing to round their backs against the cold.
      a leisure as thick and palpable as honey: The Greek term is *hasychia.*
151   **"Goat chases moon-clover . . .":** Theocritus 10. 30–31.
      people will always suffer for love: Tasso tried to get around this by claiming that
      the touchstone of the Golden Age was free love. But when in modern times free
      love has been tried, paradise has stayed out of reach—perhaps because jealousy
      could not be banished, perhaps because there was no way to guarantee what,
      according to Theocritus, was guaranteed in the Golden Age—*antephiles' ho
      philetheis,* "the beloved loved in return" (12. 16).
            The sad fact is that the near goldenness of the pastoral world, far from mak-
      ing love less painful, makes it more so. Most other problems having melted away,
      people are free to make their sex life as problematic as humanly possible. The
      same thing happens in a genre that at first blush seems the opposite of pastoral:
      the court romance, which in modern times becomes the romance of the hyper-
      sophisticated city-dweller or the idle rich. Not exactly the simple life. But here,
      too, idleness breeds romance.
            The weakness of rich foxes for sheep's clothing has not abated. For instance,
      the romance of the hypersophisticated city-dweller takes a low-rent form in tales

of bohemia, where rats and bedbugs are the main breeds of livestock. From Murger to Jarmusch, the amorous antics and tragedies of the artistic poor have been an object of fascination for the bourgeoisie. The bohemian tale might be seen as an urban form of pastoral, for pastoral has always had an element of slumming. In fact, the characters of bohemia, being free of nine-to-five jobs and other middle-class obligations, are free as pastoral swains to spend their energies on wine, women, and song. But they don't have the security and ease of those shepherds; even as they sing and smooch they dog-paddle frantically to stay afloat. And their struggles to survive add drama to their love lives. The aphrodisiac of danger, the temptations of gold-digging, the novelty of strange beds (however lousy and unlooked-for) force upon the poor excitements that the rich must seek out.

Even death and disease can be glamorous, like Mimi's: if the rich can afford to dally longer and more exquisitely with illness, the poor make up for this in sheer availability. All of this is evident in Henry Miller. (June is a good example of gold-digging, done expressly for the sake of the artist husband, which leads to the furious jealousy of which great love stories are made.) In part this is just an example of the way the extremes of society meet, the rich and poor sharing a certain immunity to middle-class constrictions. Society wears a girdle and bursts out at the top and bottom. The same kind of physics is at work in the convergence of pastoral and court romance.

From a geographic point of view, bohemia is that section of a great city which will be fashionable in ten years and flooded with wealth in twenty. As New Yorkers well know, artists are the shock troops of gentrification. If they do their job well, they soon find themselves priced out of the neighborhood they have reclaimed: first Greenwich Village, then Chelsea, SoHo, NoHo, TriBeCa, the East Village, the Lower East Side . . . Finally they are forced Aeneas-like to colonize distant shores, removing the seacoast of bohemia to Long Island City and Hoboken. In short, the scimitar's blade is as active here as in grassier Arcadias.

152   **why he sees woman as temptress and agent of his Fall:** As Enkidu no doubt does after his seduction.

Both Plato and Aristotle approved of infanticide: Plato *Republic* 459e, sq.; Aristotle *Politics* 1335^b 20–3. See Passmore, pp. 158–60.

Greek or Roman woman would use giant fennel: Riddle et al.

Exposure makes a good plot device for tragedy: As in *Oedipus the King.*

paradise is a teat: Psychologist Paul Bindrim, originator of Nude Marathon Therapy: "The major idea is to combine the warm water of the pool with nudity and massage to enable the person to reexperience the very early periods of his life. . . . The whole group concentrates on one person, and we also give him a bottle and rock him. . . . A lactating cotherapist would be better, but since we don't have one we use a bottle. . . . [It is] what some have called a Garden of Eden experience, a feeling of closeness with all mankind."

a good deal more squawling for it: This is less true in primitive societies, in some of which a child can commandeer any passing dug and is not weaned until six or seven. There is a hilarious bit in E. M. Thomas's book on the Bushmen in which a child at the breast calmly and formally curses his mother for suggesting he might be old enough to quit.

153 **Nero displaced thousands of Romans:** See chapter 16.

middle-class, more or less American-style suburbs: Of their rise in America, a canny history is K. Jackson's *Crabgrass Frontier*. A classic study (of the Boston area in the latter part of the nineteenth century) is Warner. For a longer and wider view, see Kostof; Mumford, *The City in History*. Those who think in terms of shapes, patterns, and maps may like the concise if somewhat academic survey by H. Carter, chap. 7. On the spread of American-style suburbs in Europe and elsewhere, see Heinritz and Lichtenberger.

working-class industrial suburbs: With roots in the ancient world, these suburbs, however squalid, have rarely been boring. They are rich in the one thing middle-class suburbs most conspicuously lack—edges. A taste of the way Parisian suburbs of the 1940s jumbled smokestacks and truck farms, dump heaps and picnics can be obtained from the photographs of Robert Doisneau and the text of Blaise Cendrars, the one-armed literary adventurer (and idol of Henry Miller) whose usual beat was the Amazon or Siberia. For Cendrars, the *banlieue* is straight out of the Middle Ages: raunchy, Rabelaisian, grotesque.

155 **big winners in the animal kingdom:** Garrott et al.; Spirn, p. 210; W. K. Stevens, "Opportunistic Species."

156 **Migratory songbirds need something like deep forest:** Robinson; Askins.

In the suburbs of Tucson: Bormann et al., pp. 115–16.

soils had been physically changed: Loucks.

157 **Trout and whitefish:** Spirn, p. 210.

158 **As I said, there are exceptions:** Some older suburbs, especially of the more affluent sort, may have mature tree cover and biological diversity that suburbs fresh from the bulldozer lack; some European suburbs may also be exceptions (Goudie, *The Human Impact*, chap. 3).

average American exhales: Flavin, p. 26.

row house uses 30 percent less: K. Jackson, p. 299.

14. LOST ILLUSIONS

160 **On a visit to Lucullus's villas:** Plutarch *Life of Lucullus* 39. 4.

161 **Migration has been called the breathing of the planet:** Lopez, *Arctic Dreams*, chap. 5 (titled "Migration. The Corridors of Breath"). He says, "I came to think of the migrations as breath, as the land breathing. In spring a great inhalation of light and animals. The long-bated breath of summer. And an exhalation that propelled them all south in the fall." His perspective is the poles; I have reversed the polarity.

Rome was a city of almost a million persons: On the grubby glory that was Rome, see Mumford, *The City*, chap. 8. On the escape to the countryside, see Littlewood.

"held up by slender props": Juvenal 3. 193–6.

Elder Scipio retired to his estate: Livy 38. 52–53.

villa rustica **became the** *villa pseudourbana:* Vitruvius *De Architectura* 6. 5. 3.

as Ovid did while in exile: Technically it was not an exile, as his goods were not seized nor his citizenship revoked; but to a cosmopolitan poet languishing (and finally dying) in a chill frontier town beset by fierce nomadic tribes, this was small consolation. Ovid's neglected poems of exile, the *Tristia, Epistulae ex Ponto,* and *Ibis*—neglected first of all by Augustus, who never relented—have been reclaimed as the masterpieces of kvetching they are by the moving, funny translations of Slavitt.

A country store with French-roast beans: A possibly apocryphal story from the Berkshires has a stylishly dressed woman standing on line in a grocery store one Saturday and complaining to a friend: "You'd think the locals could do their shopping during the week."

162 **"The city-dweller . . .":** Levin, pp. 43, 2. Compare the quotation from Eliade above, note to p. xxi.

"If only I had really been one of you . . .": Virgil *Eclogues* 10. 35–36.

to strike out—or imagine striking out—for wilder places: If the city person is a poet, these facts may express themselves in a mingling of pastoral with other genres. The place where the Golden Age still reigns might be accessible to a real adventurer; hence the need to embed the pastoral within romance: the sexy sort of adventure story that was invented in Hellenistic times and now is known as pulp fiction. Hence the appeal of the "Western" and other tales of frontier life to Europeans and Easterners whose own nearby pastures were hemmed by railroad tracks. The Western is a form of romance set among pastoral people and savages. The pastoral side of cowboy life is the idyllic side—already an escape into "nature"—with the wilder nature beyond represented by the Indians. The frontier has the character of in-betweenness, but instead of an equilibrium that is in danger of collapsing, there is a constant pressing forward. The pastoral is a dewdrop in the lap of a leaf; the Western is the edge of an ax. But this is really a matter of degree, or, if you like, of tempo.

In real life the cowboy's idyll is doomed, whatever the color of his hat. Before he has finished slaughtering the Indians, he finds he must protect his rear against farmers and their fences, to say nothing of sheriffs, schoolmarms, and railroad executives. All he can do is ride into the sunset, which is where his real paradise lies.

163 **as the farming disintegrates:** A little farther south, in the Berkshire foothills, in the mid-nineteenth century, the Harvard literary scholar Charles Eliot Norton and his New York friend, the journalist and editor George William Curtis, used to summer in the small town of Ashfield. F. G. Howes writes (pp. 378–79):

> They attended the meetings of the Farmers' Club and entered freely into the discussion with the members. On one such occasion when the subject was "Rural Betterments," Mr. Norton expressed his disapproval of barbed wire fences very strongly, saying he had much more respect for one of the old stone walls our fathers built, covered with beautiful vines, than for

these fences. Mr. Alvan Cross, a hard headed practical farmer, speaking soon after, said he was sorry to disagree with Professor Norton but he had noticed when cattle were turned out to pasture they had little respect for an old stone wall, even if it was partly covered with vines, and were pretty sure to jump over it somewhere, but put up a good, strong barbed wire fence, and after they had tried it two or three times they respected it, and it was for the farmer's interest to have a fence the cattle would respect, whether Professor Norton did or not.

As long as city people seek "real country"—or think they do—they carry the city ever deeper into the hinterland. In 1855, Emerson wrote in his journal (vol. XIII, pp. 415–16):

> The young people do not like the town, do not like the seashore, they will go inland, find a dear cottage deep in the mountains, secret as their hearts. They set forth on their travels in search of a home: they reach Berkshire, they reach Vermont, they look at the farms, good farms, high [on] mountain sides, but where is the seclusion? The farm 'tis near this, 'tis near that. They have got far from Boston, but 'tis near Albany, or near Burlington, or near Montreal. They explore this farm, but the house is small, old, thin: discontented people lived there, & are gone:—there's too much sky, too much outdoor, too public: this is not solitude.

When they find the perfect place, others will find it, too. Soon they will find that they have brought the city with them.

ancient world wavered between degeneration and progress: See Lovejoy and Boas; also the rather different view taken by Bury, especially pp. 15–17.

164 **good Epicurean grounds:** On much the same grounds, modern science has to do some fancy footwork to reconcile entropy and evolution.

Lucretius opts for a commonsense compromise: *De Rerum Natura* 5. 1379–1435.

Golden Age myth may seek an in-between time: On the other hand, it may be purely escapist—"soft" primitivism, in the words of Lovejoy and Boas—in which case there is no attempt at chronological realism, no need to place the Golden Age anywhere but at the very beginning.

165 **poets and critics who say that pastoral is a state of mind:** Leo Marx, for example, writes (p. 264): "For Thoreau the realization of the golden age is, finally, a matter of private and, in fact, literary experience. Since it has nothing to do with the environment, with social institutions or material reality (any facts will melt if the heat of imaginative passion is sufficient), then the writer's physical location is of no great moment." With the view that he ascribes (only half rightly) to Thoreau, Marx seems to be in full agreement.

Theocritus set most of his idylls in Sicily: Although he sometimes nods in the direction of Arcadia: remarking for instance (2. 48) that *hippomanes* is an Arcadian plant, which seems to have some bearing on its aphrodisiac effect.

The poet's nostalgia for Sicily would have been especially keen in Egypt, his adult home, which he must have found a tedious landscape (a strip of green hemmed in by sand), its formal gardens scant consolation. A Sicilian setting made good sense for other reasons, too. Daphnis, the patron saint of singing shepherds, was supposed to be Sicilian. Also, Sicily had anciently been colonized by Greeks for agricultural purposes. For centuries it had filled the bellies of Athenians and other urban Greeks with grain. Grain's goddess, Demeter, and her daughter Persephone had ties to the island. It had a reputation for abundance and for a rural, if not exactly pastoral, character.

The odd thing is that, when we look at the botanical details of the poet's settings, we find that they are not Sicilian at all, but Greek. The likely reason is that he got them out of books—probably books he studied while in medical school in Greece. If this is so, it means that the first pastoral poet had a scholarly rather than a first-hand knowledge of nature. But that is no surprise, for pastoral has always been a city boy's game.

On Theocritus, his background, and the background of his poems, see Gow.

Sicily, though still a breadbasket, was largely deforested: Also, as Snell writes, her shepherds "had entered the service of the big Roman landlords. In this new capacity they had also made their way into Roman literature; witness Lucilius' satire on his trip to Sicily. But they could no longer be mistaken for the shepherds of song and love."

distant and unknown and 'unspoilt' ": Highet, *The Classical Tradition*, p. 163. Note that Virgil sets some of his *Eclogues* in his own northern Italy, some in Sicily, some in Arcadia; but it was Arcadia that stuck.

It had other qualifications: See Rosenmeyer.

kept alive many rituals and cults: Not all of the stories about Arcadia were quaint. Plato tells us that human sacrifice was practiced at the shrine of Zeus on Mount Lycaeon, and that whoever tasted the flesh of the victim turned into a wolf (*Republic* 565d–e). If the shepherd occupies a zone between nature and culture, so does the werewolf.

The way lycanthropy slinks in the shadows of pastoral has been deftly pointed out by Silvano Arieti in his novel *The Parnas*, which has at its midpoint a kind of pastoral idyll involving the gentile peasants, and at its end a collective hallucination in which the Nazi soldiers, who have come to devour the Jews of the town, become wolves. So the two extremes, the sweetness and the danger, of the gentile or pagan closeness to nature are explored. The Italian peasants had managed to cling to their paganism unconsciously; the Nazis were recovering theirs willfully and with a vengeance.

site of the oldest cult of Artemis [and] Hermes: On Hermes especially, see Guthrie, chap. 2, sec. 5. Besides these full-strength gods, the Arcadians had a godling called Myagros or "Flycatcher," a useful sort of divinity for a pastoral people.

166 **chief God of Arcadia was Pan:** See Rosenmeyer; Hillman, "An Essay on Pan."

"Wishing to smooth and soften . . .": Polybius 4. 20–21. Highet (*The Classical Tradition*, pp. 611–12) is surely right to reject Kapp's suggestion, endorsed by Snell, that this passage was Virgil's main reason for moving pastoral to Arcadia.

He goes too far, though, when he calls the passage irrelevant, on the grounds that the music described is not primitive but highly refined. For Virgil, as for many of his followers, Arcadia bridges that spurious antithesis. Besides, as I argue in the text, it is the *function* of the music that is crucial.

167 **interest in the moral function of music that is deeply Platonic:** See Plato *Republic* 398e–401e; see also my *Recording Angel*, chap. 10.

168 **Striking a balance is a trick:** More on how this trick might be managed will be found in chapter 26.

15. THE WALLED GARDEN

The most comprehensive history of the garden in the West remains Gothein; scholars will prefer the German original, as the English translation lacks notes. Concerned with landscape architecture in a broader sense are Jellicoe and Jellicoe (a good place to start, being sparing of text) and Newton (heftier). There is much about gardens and landscape in Scully, *Architecture*, though the fine text is let down by grainy reproductions of lecture slides. Gorgeous coffee-table books are legion, some more scholarly than others; one that blends beauty with a hint of irreverence is Schinz.

170 *pairidaeza:* Just possibly, this may have meant a mound supposed to represent the World Mountain, like the tree-topped artificial hills of Assyria (see Jellicoe and Jellicoe, chap. 2, fig. 20).

171 **poets and painters have felt the need to put a wall around Eden:** In Milton (*Paradise Lost*, 4. 131–45) it is a "verdurous" wall, surrounded by thick forest; in Tertullian, Lactantius, St. John Chrysostom, and Isidore a wall of fire; in Philo (*Alleg. Interp.* 1.65–6), Barinthus (*Voyage of St. Brendan*), Augustine, and the journals of Columbus, a wall of water (making Eden an island in a river or ocean); while Dante, for good measure, combines the last two. See Grandgent's annotations to Dante, *Purgatorio*, canti 27, 28.
"at one slight bound": Milton, *Paradise Lost*, 4. 181.

172 THE PARADISE PARK: On Persian parks and gardens, see Moynihan; Sackville-West; Mcdougall and Ettinghausen. For social and ecological context see Arberry, especially vol. 1; Frye; Dandamaev and Lukonin.
And then we need a second, inner wall: Ephraim the Syrian (fourth century A.D.) says that Eden is divided into an outer part and an inner, more sacred sanctum (Grandgent, annotations to Dante, *Purgatorio,* canto 28).
Iranian highlands were once well wooded: See Whyte; Bobek.
quincunx pattern: For the advantages, both practical and mystical, of this figure, see the endlessly ramified book by Sir Thomas Browne.

173 **"I have found it to be a general Rule . . .":** Chardin, vol. 2, chap. 6. After noting that the eastern part of the country is "nothing but one continu'd Parterre, from *September*, to the End of *April*," Chardin goes on to remark that Persian gardens are "confusedly set with Flowers, and planted with *Fruit-Trees*, and *Rose-Bushes*;

and these are all the Decorations they have. They don't know what *Parterres* and *Green-Houses*, what *Wildernesses* and *Terraces*, and the other Ornaments of our Gardens are."

174   **Persian carpet:** This woven spring may have been perennial but it was not eternal. Like flowers, carpets fade and wither, and none of earlier date than the sixteenth century has survived (though a carpet and some carpet fragments from Scythian tombs in Siberia [Hillmann, chap. 2], dated from the fifth to seventh centuries B.C., reveal a remarkable continuity of design and technique). In carpets of later years marvels of botanical and zoological detail can be seen. Yet the greatest marvel may be the depiction of the single thing that was the source of all this multifarious life. In some carpets the central pool is woven from several dye lots, so that it ripples and shimmers in delicately varying blue. One can sit and watch it for a long time, just as a Persian poet would have watched the pool.

      demands of irrigation: Such is the primacy of irrigation in the Near East that, rather than speak of a square of land divided by channels of water, we might better speak of crossed channels of water whose banks form a garden. Indeed, the phrase *chahar bagh* often referred to the channels rather than the land.

175   **"Since even in Paradise itself . . .":** Browne, chap. 1.

      motif of the square or circle sliced: Jung, "Dream Symbols."

177   **"a psychic center of the personality . . .":** Ibid.

178   **throne room is perched at the junction of the waters:** Moynihan, p. 71. The mythic impulse behind this setup reappears in the oddest places. When Thomas Moore, the best-selling dispenser of James Hillman's ideas, gave a lecture at John Houseman's house in Malibu, his seat was on a narrow plank stretched across the center of the swimming pool.

      to connect that need with monotheism: See, for example, Hillman's *Re-Visioning Psychology*. But Hillman admits that "Momism," or worship of the Great Mother, is also a form of monism.

16. PATTING NATURE ON THE HEAD

179   **tales of the paradises they had seen:** For example, Xenophon *Anabasis*. On Greek gardens or the lack thereof, see Gothein; D. Thompson; Ridgway.

      no immediate rush to copy them: Xenophon himself did build a paradise park in the Persian style, but it was not a private pleasure garden; rather, it made up the grounds of a temple he dedicated to Artemis, where worshippers pastured their animals and neighbors joined in hunting game for the goddess's festival (*Anabasis* 3. 5. 7–13).

      garden of Alcinous: Homer *Odyssey* 7. 112–32.

180   **it hasn't yet become scenery:** Although we may doubt that Homer was a "naïve" poet even by Schiller's special definition, we can agree that he did not cling to nature "with fervor, with sentimentality, with sweet melancholy, as we moderns do."

181   **To say they lacked the modern notion of the unitary self:** As Bruno Snell has

done. Extinct among scholars, this notion survives in popular writing about the ancient world.

182 **Greeks made real maps:** Though the first Greek world maps, drawn by Anaximander and Hecataeus in the sixth century B.C., showed a round land mass surrounding the Mediterranean and so had a rather mythic tone, they were scoffed at by Herodotus a century later. Maps of smaller regions were a good deal more naturalistic.

183 ROMAN GARDENS: See Gothein; Macdougal and Jashemski; Grimal.
fish-sauce shop: Those hard put to imagine what Italian cuisine was like before pasta and tomatoes need only imagine the pervasive taste and smell of *garum*, a sauce made from rotten fish.
Greeks were seafarers first, farmers second: For a dissenting view, see Hanson, *The Other Greeks.*

184 **gardens of the urban bourgeoisie in Pompeii:** See Jashemski.

185 **new villas:** On Roman villas and their gardens, see the essays in Macdougall, ed., especially Littleton. The fullest contemporary accounts are in the letters of Pliny the Younger, 2. 17, 5. 6.
Quintus Hortensius walled in thirty acres: Varro *Rerum rusticarum* 3. 13.
aviaries: Ibid. 3. 4–5, which also describes an even more elaborate aviary-cum–dining room on one of Varro's own estates.
Talking birds: Pliny the Elder, *Natural History* 59, 60, 72.

186 **they traced their descent:** For a discussion of some of these stories, see Harrison.
Romans were materialists: Hillman, *The Dream,* p. 69. The great classicist Grimal (whose book I did not get my hands on until mine was almost done: a misfortune, as it stands alone in subtlety of interpretation) means something similar when he speaks (pp. 471–74) of Roman *naturalisme.* In the shade of a plane tree, we find "a Mediterranean society, a bit untidy, a bit vulgar, whose ideal is not always very lofty, and which contents itself, after having conquered the world, with sleeping in a cool place, in fine weather." While this practicality may at times have vulgarized the Hellenistic pastoral ideal, it also saved it from becoming a mere "game of the literati." In brief: "The Roman garden gave flesh to a Greek dream."
self-indulgent splashing about: On water in Roman gardens, see Gothein.

187 **shapes hewn from profusions of boxwood:** On Roman topiary, see Pliny the Elder, *Natural History* 12. 13; Pliny the Younger, 5. 6.
Caligula had built: Pliny the Elder, *Natural History* 12. 5. 10.
Elagabalus is said: Regrettably, this story comes from the disreputable, and probably apocryphal, *Historia Augusta* (*Heliogabalus* 21. 5).
"to confound the order of seasons and climates . . .": Gibbon (chap. 6), who himself is drawing on the *Historia Augusta* 17. 23.

188 **Nero in his Golden House:** The quotations are from Tacitus *Annals* 15. 42; Suetonius *Nero* 31. 2. On the Golden House, and generally on the ideals of "town in country" and "country in town," see Purcell.

189 **Cicero had a portico:** See Gothein; Littlewood. The Tusculan villa of his friend Crassus, where he sets most of *De Oratore,* is equally Greek.

17. THE CLOISTER AND THE PLOW

191 **preserved under the Vandals and Visigoths and Ostrogoths:** Littlewood, citing Venantius Fortunatus, Cassiodorus, and Luxorius. Cf. Gothein, who gives a good general account of medieval gardens. She provides plenty of period illustrations, but those wanting more should look up Crisp. On the literary and philosophical streams that fed both medieval and Renaissance gardens, see the dazzling work of Comito.

In Byzantium: Gothein.

Augustine: See Comito; Gothein; Augustine, *Confessions.* The quotation is from *De Ordine* 1. 2. 4.

192 **Abbey of St. Gall:** See Comito; Jellicoe and Jellicoe, chap. 13.

Garden of Delights: Gothein, p. 180.

193 **Christian thinkers seeking cosmic significance in numbers:** E.g., Augustine, Boethius, the School of Chartres. See Comito; Hugh of St. Victor, *De Claustro Animae* 4. 31.

city "that lieth foursquare": Revelations 21:16.

St. Gall plan is the first medieval example of design *ad quadratum*: It was a trick the Roman architect Vitruvius liked to use, and it may be no accident that a manuscript of Vitruvius was found at St. Gall, too. But the St. Gall architect is so number-crazed that form overruns function. Some evidence even suggests that the site for which the design was meant was triangular.

We have come a long way from the World Mountain. By way of Ezekiel and St. John, the wild peak of the Canaanites has come to look a lot like the Tower of the Mesopotamians.

convulsed by social, technological, and ecological changes: See White, *Medieval Technology;* Cipolla, vol. 1; Pounds; Mumford, *The City,* chaps. 9, 10.

194 **heavy wheeled plow:** See White, "The Expansion," *Medieval Technology.*

195 **borrowed from China:** See White, *Medieval Technology;* and, more particularly, Needham, *Science and Civilization in China.* Widely thought to be the greatest monument of scholarship in this century, the latter is worth visiting at a library as one might visit the Forbidden City, even if one only dips at random into one or two of the numerous volumes. More welcoming, perhaps, is the one-volume, well-illustrated popularization by Temple, which itself inspires humility (in Westerners) with its endpaper time-lines of Chinese inventions and Western also-rans.

population of Northern and Western Europe nearly trebled: J. Russell.

"In the course of three centuries...": Mumford, *The City,* p. 260; he also makes the comparison to the nineteenth-century rate of growth.

monasteries played a central role: See ibid.; White, "The Expansion," where the quote from the archbishop of Mainz is found.

196 **troubadours:** See Briffault; Lindsay; the scholarly book by Hueffer; and the delightfully unscholarly book by his son, Ford Madox Ford. For a taste of the

verses—a treat that should not be missed—see the bilingual anthology by Goldin, whose translations have a fine vernacular swing.

197 **"Go, tell him, be here in the morning…":** Marcabru, "Estornel, cueill ta volada" (No. 12 of this poet's works in Goldin's anthology).

"by the fountain…": Marcabru, "A la fontana del vergier" (No. 10).

things you are cooped up with will start to sing: She was talking about the literature of England.

"to which no low-born man had ever come": Guillaume de Lorris, 470. The Old French has *"ou onc n'avoit entré bergiers"*—literally, "no shepherds."

"…a better place than Eden for delight": Ibid., 640–3.

"a perfect measured square…"; "Why mention more?": Ibid., 1323–5, 1361.

198 **"one might lie beside his sweetheart as upon a couch":** Ibid., 1394–5.

"Not since King Pepin's time…": Ibid., 1428–9.

"Here Fair Narcissus wept himself to death": Ibid., 1437–8.

199 **"foursquare in shape…":** Ibid., 3815–6.

The lovely poem of Guillaume breaks off here: Its clear, childlike watercolors give way to the casuistic murk of Jean de Meun's much longer continuation, which—crammed with learned repartee that could be used in amatory and other situations—helped make the book a best-seller for three centuries.

By the mid-thirteenth century: See Duby.

200 **collapse into famine:** See Duby; NcNeill, *Plagues,* chap. 4; Pounds, p. 11.

18. BRINGING A STATUE TO LIFE

As an introduction to the Italian Renaissance, the great work of Burckhardt remains invaluable. Hale "temerariously" and gracefully extends Burckhardt's concerns to Europe as a whole. A scholarly and well-designed coffee-table book is Plumb; Aston is more up-to-date in both content and form, being pegged to a video sensibility. For a deeper exploration of the Italian city-states, Martines would be a good next step. Social and economic background may be found in the works of Braudel.

201 **"The saying 'Know thyself'…":** Last comma mine. Pico's *Oration* (the rest of the title was added after his death) is translated, along with works of Petrarch, Valla, Ficino, Pomponazzi, and Vives, in the benchmark work of Cassirer et al., which remains a good introduction to Renaissance philosophy.

For the sake of simplicity I have stitched the humanists, a varied lot in time and space, into a single fabric. But a distinction should be made, at least, between the Florentine humanists of the late fourteenth and early- to midfifteenth century and those of the late fifteenth century. The former, including Niccolò Niccoli, Leonardo Bruni, and Leon Battista Alberti, were often men of independent standing, counselors to the oligarchy, and active participants in Florence's civic life. The latter, including Ficino and Poliziano, were in effect courtiers to a prince (such as Lorenzo) who "banishes everyone from true political life, transforming the culture from the expression, instrument, and program of a class risen to wealth and power, into an elegant ornament of court, or into a desperate flight from the world" (Garin, chap. III, sec. I). Not surprisingly, it is among the later humanists that both pastoral and the myth of the Golden Age become obsessions.

202 **felt it grow warm and tremulous under their hands:** Late in the Renaissance, John Evelyn would find the connoisseur Hippolito Vitellesco embracing and talking to his classical statues.

Petrarch, perhaps the West's first amateur gardener of note: See Comito. In *Familiares* 11. 20. 10, Petrarch speaks of "my little gardens, planted with these hands."

ascent of Mount Ventoux: Recounted by Petrarch in a letter to his friend and teacher Francesco Dionigi de' Roberti (*Familiares* 4. 1, translated in Cassirer et al.). For an early example of its overplaying, see Burckhardt, pt. 4, chap. 3.

203 **rare and exotic plants in his garden:** Lazzaro. Another favorite of Lorenzo's circle of poets was the "Orti Oricellari" of his brother-in-law, Bernardo Rucellai, in which the attempt was made to grow every plant mentioned in classical literature.

equal attention given to stone and leaf: Even the naturalistic woods that often flanked formal gardens (and were given as much space) were spiked with mythological statues and fountains. Artists of the first rank busied themselves with the design of topiary—including Leonardo, who may have invented a way of tying branches together to form an arboreal ceiling (Battisti). By the sixteenth century, the age of the Roman High Renaissance, the same great artists were lavishing equal talent on paintings, sculpture, buildings, and gardens.

As for the buildings, the appetite for the ancient style was often fed with materials looted from the ancient ruins themselves. A report to Pope Leo X in 1519 (Hale, p. 202) noted: "All this new Rome which we see today, great as it is, so embellished with palaces, churches and other buildings, is built with mortar made from ancient marbles."

Hypnerotomachia Poliphili: The author was Francesco Colonna. A streamlined version, with much of the entablature scraped away and replaced with Jungian captions, is offered by Fierz-David.

ramming one world-pole down the throat of the other: If the greatest yearning of the garden-lover is for wholeness, then the Rinascimento meets this yearning (as it meets so many others) with a special passion and flair. A host of opposites are joined: ancient and modern, pagan and Christian, wild and civilized. The two world-poles—Mountain and Tower—are married in a way at once sensuous and cerebral. At the same time, the ancient function of the world-pole—to join heaven and earth—takes on a new meaning. Pico della Mirandola speaks of the role of "white magic"—meaning roughly what we mean by science—which begins in wonder at the harmony of nature and leads to mastery of nature. "As the farmer weds his elms to vines, even so does the magus wed earth to heaven, that is, he weds lower things to the endowments and powers of higher things."

For a glimpse of the way imagery shaded into magic, see the works of Frances Yates—or, for a real frisson, their fictionalization by John Crowley: though both deal mainly with a later period of the Renaissance.

204 **"Such was the cruelty of Heaven . . .":** Boccaccio, *Decameron*, Introduction.

A look at the plague's history: Based largely on McNeill, *Plagues and Peoples.*

205 **first large industrial proletariat:** On the economy and society of Renaissance Florence, see R. Lopez; Carus-Wilson; Brucker; Martines; Ferguson. A good popular account is Mee.

208 **original El Dorado:** Levin, p. 64.

parade held in Florence in 1513: Vasari, "Life of Jacopo da Puntormo."

Lorenzo praised the mythical Golden Age: See Levin.

209 **In 1252 the city had begun minting the florin:** Spufford.

"This is the Golden Age, all right": Ovid *Ars Amatoria* 2. 227.

"Preach what you will . . .": Quoted in Martines, p. 79.

210 **"Banish every sad thought . . .":** *"Ogni tristo pensier casci: facciam festa tuttavia."*

would shortly make pastoral the rage: Though some forms of pastoral were common already in the France and Flanders of the High Middle Ages (see Huizinga, chap. 9), it was in the Renaissance that the craze really took off.

relation of the Florentine elite to the "pastoral" landscape: See Martines, pp. 162–67; Duby; McNeill.

211 **Modern landscape painting has its origin in Venice:** On the pastoral economy of the Vicenza region, see Braudel, *The Mediterranean,* p. 91; on Venetian landscape painting, see Rosand.

In Spain pastoralism had become big business: See Braudel, *The Mediterranean,* pp. 91–92 (from which the quotation is taken); Duby; Ponting, p. 78. Since the flocks' owners lived in the hill towns, geographers call this a case of "reverse transhumance."

On a smaller scale, the same tragedy was played out in southern Italy, where in fifteenth-century Apulia alone a half-million sheep would be on the move each season, and where much of the land was soon too scarred to support the raising of crops.

212 **Behind the house, the gardens climbed up the hillside:** The basic plan went back to the ancient Romans, who liked to have the working part of the estate (the *villa rustica*) in the valley, the residence on the hillside above.

On Italian Renaissance gardens, see Shepherd and Jellicoe; Lazzaro; Coffin; the relevant sections of Newton and Gothein. For the High Renaissance and Baroque, the travel journals of Montaigne and Evelyn give valuable first-hand accounts. Lush photographs of extant gardens can be found in Agnelli.

213 **Perspective takes the place of the magic rectangles:** Comito writes (p. 158): "Renaissance perspective is an attempt to . . . claim for the optical experience itself the unity and coherence the medieval artist had discovered in the merely ideal cosmos of his art and architecture. What is involved is the redemption of the space of our fallen world, a rediscovery of its potential sacredness." Berenson (bk. 3, sec. 12) puts it still more succinctly: "Space-composition is the art which humanizes the void, making it an enclosed Eden. . . ."

Grand Duke Cosimo took special pride: See Lazzaro, pp. 191, 208–10. The central fountain of the Boboli, the Fountain of Neptune with its seventy jets, even has a rocky mass in its center, though the god who bestrides it is a sea god rather than a sky god. Ferdinando's aqueduct was completed in 1635, in honor of his marriage: as in ancient times, the god-king's union with the goddess brought heaven's bounty down to earth. So the garden's highest pool was crowned with a statue of Dovizia—"Abundance"—holding a cornucopia and a sheaf of wheat.

214   **"While the ladies are busy watching the fish . . .":** Montaigne, *Travel Journal,* entry for Oct. 15–19, 1580 (the Fuggers' summer houses in Augsburg).
longest cascades: Wylson, pp. 14, 169.
He faults the Villa d'Este: Montaigne, *Travel Journal,* entries for April 3, 1581; Nov. 22–24, 1580.
"as big and long as a pine nut . . .": Ibid., entry for Aug. 24, 1581 (at La Villa).

215   **"We have so much by our inventions surcharged . . .":** Montaigne, *Essays* I. 30. Similarly, the fountains of the Boboli Garden, once showcases for the pure water Cosimo showered on Florence, now have a greenish murk in their basins, which admittedly serves well enough for carp; and the fat dwarf who bestrides a tortoise in a small drinking fountain near the palace appears to bear the Grimmly apt name Kein Trinkwasser ("No Potable Water").
"In many of their chief merits . . .": Burckhardt, pt. 1, chap. 7.

19. LEAPING THE FENCE

216   **"You would not believe . . .":** Quoted in Comito, p. 1.
On Baroque French gardens, see Woodbridge, *Princely Gardens;* W. H. Adams; Scully, *Architecture,* chaps. 9–10. For background on the age of Louis XIV, see Hatton; Carr; Methivier; W. H. Lewis; Packard; Goubert; Petitfils. Versailles, and the figures strolling its walks, come to life in the memoirs of the Duc de Saint-Simon, which nearly match the royal grounds in extent.

217   **collecting great artists and sparkling women:** He even had the hardihood to pay court to the young king's mistress, Louise de Vallière. But it was probably Fouquet's growing power (embodied in a private army at his estate at Belle-Ile) that really sealed his fate. Indeed, his fate was sealed before the fête at Vaux. Colbert had already planned his arrest in every detail, right down to the arresting officer: the underlieutenant of musketeers, Charles de Batz-Castlemore, sieur d'Artagnan, a figure well known to readers of Dumas *père.*
Historians have wondered how a man as bright as Fouquet so signally failed to see the trap that was being set for him. Petitfils (p. 208) explains that Fouquet was in fact "a tired man, tortured periodically by malaria, a manic-depressive who alternated between enthusiasm, which overcame his too-lively imagination, and depression that reduced him to nothingness."
the rest of Fouquet's team: Notably Louis Le Vau and Charles Le Brun.
ballets de cour: "The harmony of the ballet is an invitation to compose a new political space in which absolute monarchy will be deployed. . . . Thus, in the new heliocentric order, the nobles and courtiers were asked to be no more than satellites of the royal star, turning about him with obedience and regularity and reflecting only his glory. . . . The nobility ceased to be an autonomous and chaotic world of independent asteroids and became an orderly, regulated cosmos." Petitfils, pp. 289, 294.

218   **slapping down new parterres like dominoes:** Cf. Adams, p. 86.
      power expressed itself horizontally: See Mumford, *The City,* chap. 12; Scully, *Architecture,* chap. 10; Adams, p. 86.
      France had possessions in Canada: On France's colonies and trade, see Braudel, *The Structures,* p. 335. Besides its own trade in pelts and sugar, it took part at second hand in the much greater trade of the Spanish colonies, providing textiles and other manufactures and getting gold and silver in return. More directly, it served up African slaves to the South American market. Louis's support for Colbert's maritime and colonial schemes was at best lukewarm; as Goubert notes (p. 91), "A stronghold in Flanders or the Palatinate always excited him more than India, Canada, and Louisiana put together."
      "America was lost . . .": Quoted by Braudel (*The Structures,* chap. 5), who mislaid the reference in the years between the publication of his book without notes and its republication as part of an annotated trilogy; he says it may have come from a lecture by Pirenne.

219   **a centralized nation-state:** It was Tocqueville who pointed out that this work had been largely completed long before the Revolution took it up; see *The Old Regime.* For each district of France, an *intendant* was appointed who answered directly to the king. The varied regions of this great land mass—Provence, Burgundy, Brittany, and so on—which until now had had their own cultures and in some cases their own languages, took a great leap forward in the march toward a common Frenchness. Under Colbert, most of the industries of France were likewise placed under central control or at least influence. They would now serve the national interest, as defined by mercantilism: getting as big a slice as possible of the closed-crusted pie of world trade, and thereby filling the state's treasury with precious metals. Huge state-owned factories made tapestries at Gobelin, glass at Saint-Gobain, porcelain at Sèvres. In Abbeville, a textile factory employing some three thousand workers, with hundreds under a single roof, helped meet the worldwide demand for *les nouvelles draperies*—silks and fustians and other light, flattering fabrics—a demand that was itself piqued by the influence of Versailles.
      jardin de l'intelligence: Corpechot, who dedicates his book to the nationalistic, blood-and-soil writer Maurice Barrès: unsurprisingly, given Corpechot's view that only the French genius could embody intelligence in a garden.
      failure of reason: First of all, the new nation-state displayed, as it has in spades many times since, its knack for wasting and destroying in the national interest. Throughout his reign, Louis fought wars. Some were for territory or trade overseas, some for territory in Europe, some to smother his own people when they cried out for bread. When he took the crown, France's peacetime army was a loose aggregation of perhaps 15,000 men; when the crown had settled on his head, Louis's standing army, the first in Europe, numbered some 100,000, and could swell to 350,000 in time of war. At the start of Colbert's administration, the navy did not have twenty seaworthy ships; at his death less than three decades later it had 250. To build them, timber was floated down nearly every navigable waterway in France. All these troops and ships needed cannons; the cannons

needed iron; the iron needed blast furnaces; the blast furnaces needed fuel. To meet this need, the Alpine region called the Dauphiné, where iron was mined and smelted, was stripped of its trees.

The precious metals that these wars were supposed to bring into French coffers had to be spent to fight them; but that was the least of the paradoxes involved. Precious metals were precious partly because they resisted change, but getting them required change of the most violent sort, imposed not only on people but on other species. In Canada, Louis made the fur trade a government monopoly, and proceeded to rationalize it by guaranteeing a fixed price for every beaver pelt received. The result was twofold: beavers were massacred on a spectacular scale, and the luxury hat market in Europe was glutted. "One cannot repress a feeling of indignation," writes Parkman (vol. 4, chap. 20), "at the fate of the interesting and unfortunate animals uselessly sacrificed to a false economic system."

Little Ice Age: In the present context, it would be a lovely irony if the reduced sunspot activity (the "Maunder minimum") that coincided with the Little Ice Age were somehow its cause; unfortunately, this remains unproven. See Schneider and Londer, chaps. 3 and 7.

Major famines: The authority on this subject is Lachiver, who observes (chap. V) that in the famine of 1693–94 the population of France fell by 1.7 million—as many deaths as in the First World War, but in a fraction of the time and in a population half the size. All the wars of the Revolution and the Empire produced only 1.35 million deaths, spread over twenty-three years.

Saint-Simon is eloquent on the winter of 1708–9 and the ensuing famine: "It had been so cold that bottles of Queen of Hungary water broke in the cupboards of rooms with fires and surrounded by chimneys, in several apartments of the château of Versailles. . . . The fruit trees were killed; no walnuts, olives, apples, nor vines survived, none worth mentioning. Other kinds of trees died in great numbers, the gardens were ruined, and so was the seed planted in the ground. Impossible to form any idea of the magnitude of that national disaster. Everyone hoarded the old grain, and the price of bread rose as people despaired of a harvest. The more prudent re-sowed their cornfields with barley; many others followed suit, and they were the lucky ones, for it saved their lives. The police, however, took it into their heads to forbid that activity and changed their minds when it was too late. Various decrees were published concerning wheat, searches were instituted for hoarded grain, and commissaries were sent into the provinces three months after their arrival was announced. And all that this policy achieved was to increase the poverty and the prices to a disastrous level at a time when calculation showed there to have been enough wheat in the country to feed the entire nation for two years, without reckoning on a harvest." (*Historical Memoirs*, pp. 416–17.)

The famines were only dips in an already rock-bottom standard of living. "One sees certain wild animals," wrote La Bruyère ("Of Man," *Characters*), "male and female, scattered about the countryside, black, livid, and wholly sunburned, attached to the soil they dig and turn over with an invincible stubbornness; they have a sort of articulate voice and when they stand up they show a human face, and in fact they are human; they retire at night into lairs where they live on black bread, water, and roots . . ." Those who fled to the city did not fare much better. In Paris alone there were forty thousand beggars, nearly a tenth of the city's population. In an age fixated on luxury, five-ninths of all Frenchmen, according to Vauban, lived "in utter destitution" (*Projet d'une dîme royale*, 1707).

In fairness to Louis and especially to Colbert, it should be said that many of their problems were inherited and many of their reforms really did improve things. Colbert's forestry policy, for instance, kept the woods of France from vanishing utterly in the face of the demand his other policies helped create. As recent Western history shows, a central government guided by reason can sometimes do good things for the landscape it governs. Yet the overall verdict stands. The rays of the Sun King could not thaw the Little Ice Age; the rays of human reason could not reduce nature's wildness to Apollonian order. In some ways, they made matters worse.

way it was made and maintained: See W. H. Adams; Woodbridge.

220   **"What has not been spent . . .":** Abbé Marc-Antoine Laugier, *Essai sur l'architecture*, 1755, quoted in Woodbridge, p. 223.

THE ENGLISH LANDSCAPE GARDEN: Hussey offers a good history, Hunt and Willis a superb anthology of primary sources, to which Barrell and Bull makes a good supplement. Beyond these, the travel journals of Defoe and Cobbett are invaluable; for the former, the illustrated (though abridged) edition of Furbank and Owens is a delight. For a history of the English landscape that is both politically and ecologically informed, see Williamson and Bellamy; for a more traditional antiquarian approach, see Baker and Harley. In his classic *The Country and the City*, the late leftist critic Raymond Williams speaks of the way English literature has dealt with some of the same issues that fertilized and plagued English gardens. K. Thomas provides a wealth of material on attitudes toward nature; Trevelyan's more general social history is unusually attuned to these issues.

"the flower of all the private gentlemen's palaces . . .": Defoe, Letter VI.

221   **landscape at Stourhead:** See Woodbridge, *Landscape and Antiquity; The Stourhead Landscape*.

the Dutch *landschap*: Initially, this meant simply a parcel of land.

222   **He had to prove that "this was the Taste of the Ancients in their Gardens":** This he did, rather tendentiously, by pointing to the garden of Alcinous in Homer (see text, p. 179).

"ADAM and *Eve* in Yew . . .": Essay from *The Guardian*, 1713 (excerpted in Hunt and Willis, pt. 2), from which the preceding quotes also are taken.

"first leaped the fence . . .": Horace Walpole, *History of the Modern Taste in Gardening* (written 1770, published 1780), reprinted in Hunt and Willis, pt. 3.

Kent was not really the first, of course. Juvenal (*Satires* 6. 18) had said of the subjects of Saturn that they "lived in an open garden" (*aperto viveret horto*); Ovid (*Fasti* 5. 209) spoke of Flora's "garden in the fields" (*hortus in agris*). Many British writers freely acknowledged, like Pope, their gardens' debt to Greece, Rome, and more recent Italy. In fact, Italian Renaissance gardens did have their naturalistic areas, which by the 1580s were taking up as much space as the formal gardens themselves; the Villa Mattei, Villa Lante, Castello, Boboli, and Pratolino all had sizable *boschetti* (see Macdougall, *"Ars Hortulorum"*).

China was another conscious model, at first as vague and abstract as Greece. William Temple wrote that the Chinese scorn an ordered, symmetrical way of planting, "and say a Boy, that can tell an Hundred, may plant Walks of Trees in straight Lines. . . . But their greatest Reach of Imagination is contriving Figures, where the Beauty shall be great, and strike the Eye, but without any Order or Disposition of Parts, that shall be commonly or easily observ'd." (*Upon the Garden of Epicurus* [1692], excerpted in Hunt and Willis, pt. 1.) His comments were repeated almost verbatim by Addison and so passed into common wisdom, though it was not until the time of William Chambers, almost a century later, that specific elements of Chinese design were imitated.

But if the new landscape gardens had forebears abroad, they also had a parent closer to home in the deer park. Hunting preserves had for centuries been more extensive in England than anywhere else in Europe. They were, as Walpole noted, "contracted forests, and extended gardens." Landscape gardens were the new havens of this ancient privilege, which was limited by the Game Laws to persons with incomes of at least £100 from a freehold estate (Hunt and Willis, intro.).

223   **expropriation:** Expropriation has a long pedigree in pastoral; in Virgil's very first Eclogue, Meliboeus has lost his land to a soldier (see note to p. 387). There was, in fact, a comparable process in England by which retired officers and civil servants became part of a new aristocracy of land and capital. The more useful comparison, though, is between enclosure in England and the Roman displacement of small farmers by great slave-worked plantations, or latifundia.

enclosure: Most of the sources listed for the English landscape garden have something to say about enclosure, but Williamson and Bellamy is particularly useful. See also De Maddalena; C. Wilson; E. P. Thompson, chap. 7. To experience at first hand the battle over enclosure in a particular village—or come as close to doing so as a modern person can—read *The Yellow Admiral,* the eighteenth novel in Patrick O'Brian's magnificent Aubrey-Maturin series.

"scientific" agriculture: The classic Norfolk system, common in the latter half of the eighteenth century, had two variations: a four-year rotation of grain, turnips, barley, and clover, and a six-year rotation of wheat; barley or oats; turnips; oats or barley mixed with clover; grazed clover until June 21, followed by winter wheat; and finally winter wheat. Turnips—promoted above all by Lord ("Turnip") Townshend, whose talk was "of turnips, turnips, and nothing but turnips"—fattened the cattle whose dung was used to fatten the soil. With Tull's seed drill, crops could be evenly spaced in straight lines, so that a horse-drawn hoe could till between the plants. Marling—digging up clay and lime from deeper soil layers to mix with sandy topsoil—helped keep erosion in check. Not least important was the lengthening of leases to give tenants a stake in improvement. See De Maddalena; C. Wilson; Fussell; Hyams, *Soil and Civilization.*

224   **more integrated system:** It was also a more congenial system, from the animals' point of view, since the raising of winter feed made unnecessary the massacre of flocks at autumn's end.

"One of the most scientific botanists in Europe": Quoted in K. Thomas, p. 283.
"call in the country": Pope, *An Epistle to Lord Burlington* (1731), excerpted in Hunt and Willis, pp. 211–14.

> . . . In all, let *Nature* never be forgot.
> Consult the *Genius* of the *Place* in all,
> That tells the Waters or to rise, or fall,
> Or helps th'ambitious Hill the Heav'ns to scale,
> Or scoops in circling Theatres the Vale,
> Calls in the Country, catches opening Glades,
> Joins willing Woods, and varies Shades from Shades,
> Now breaks, or now directs, th'intending Lines;
> *Paints* as you plant, and as you work, *Designs.*

Calling in the country was not a new idea, of course: Roman, Italian, and Mogul gardens had done it, in their various ways.

225 hedgerows of medieval enclosures: On the difference between medieval and modern hedgerows, see Williamson and Bellamy.
survey of parks and gardens in Hampshire: Ibid., p. 138.
Castle Howard: See Hussey, p. 115. Even before the landscape garden came along, this sort of thing was common. It was common in the sixteenth and seventeenth centuries, when pastoral verse was all the rage in England. "It is not easy to forget"—not for Raymond Williams, anyway—"that Sidney's *Arcadia*, which gives a continuing title to English neopastoral, was written in a park which had been made by enclosing a whole village and evicting the tenants" (p. 22).

But these things *were* easy for poets and landowners to forget. When Ben Jonson, sucking up to the lord of Penshurst, claims that

> The painted partrich lyes in every field
> And, for thy messe, is willing to be kill'd. . . .
> Thou hast thy ponds, that pay thee tribute fish,
> Fat aged carps that run into thy net. . . .
> The blushing apricot, and woolly peach
> Hang on thy walls, that every child may reach

—invoking the convention of the Golden Age in which nature offers up its bounty spontaneously—the claim is outrageous but not, I think, insincere. From the point of view of the poet and his patron, the laboring classes *are* nature.

226 "No Fountains murmur here . . .": Stephen Duck, "The Thresher's Labor" (1736), in Barrell and Bull, pp. 385–90.
anti-pastoral was born: Barrell and Bull use this term; Raymond Williams prefers "counter-pastoral." There is, of course, another and less political form, common since Roman times, which debunks the myth of country life without showing any particular interest in the countryman's plight. My favorite example is S. J. Perelman's *Acres and Pains*.

"I used to go around a little common . . .": Cobbett, *Political Register,* vol. 39 (May 26, 1821), pp. 518–21, quoted in Williamson and Bellamy, p. 115.

227   **"I had rather . . .":** Quoted in Williams, p. 67.

As history it is "hogwash": Donald Worster, lecture at the Land Institute, Salina, Kansas, May 30, 1987. Some critics of Hardin, such as Vandana Shiva, go too far in the other direction, idealizing the common and its stewardship, especially in non-Western and indigenous societies.

thousands of acres of wild moor, heath, and fen: Indeed, Trevelyan (vol. 3, chap. 3) estimates that, in the century from 1696 to 1795, two million acres were added to the agricultural land of England and Wales.

228   **England was one of the most densely forested countries:** The forests of the south, ravaged by Roman iron-smelters, had returned with such vigor that timber was now exported in vast quantities to Holland, Flanders, and northern France. On British forests and the transition to coal, see Perlin, *A Forest Journey,* chap. 10. On coal, iron, and industrialization, see C. Wilson; Lilley; Ashton; M. J. T. Lewis; Mumford, *Technics and Civilization.*

229   **London's air had drawn unfavorable comment:** See K. Thomas, chap. 6, sec. i.

"the suburbs of Hell": John Evelyn, *Fumifugium* (1661), quoted in Perlin, *A Forest Journey,* p. 225.

"Wood and Charcoale . . .": Dud Dudley is quoted by M. J. T. Lewis (p. 584), who says Dudley's claim has been buttressed by the discovery of his furnace at Himley and the analysis of its slag. On the Dudley family, see Raybould.

"All the activity . . .": Quoted in Prince, p. 162.

230   **use of iron for everything from railroad cars to men's collars:** See Mumford, *Technics and Civilization,* chap. 4, sec. 5, who notes, "Much of the iron that the period boasted was dead weight."

"from whence the Lord Musgrave . . .": Defoe, Letter IX.

"paying or making Satisfaction . . .": Quoted in Raybould, p. 43.

"the smoothness of the Bark and Leaves . . .": Estate trustee James Loch, quoted in Raybould, p. 102.

231   **innovations were technically piddling:** See Lilley.

pioneered by bankers like Henry Hoare: See Wilson; Gille; Woodbridge, *Landscape and Antiquity.*

232   **"Like Aeneas . . .":** Woodbridge, *The Stourhead Landscape,* p. 200. It is notable that Gilgamesh's quest begins after he builds a city, Odysseus's after he sacks a city, and Aeneas's after his city is sacked. Aeneas has a pastoral interlude by the water in which, far from rejecting the city, he is confirmed in his quest to build it. And his allies will be the Arcadians, whose king he finds worshipping Hercules, the hero of heroes, the spawn of Gilgamesh. In the same way, Hoare and his family will find refreshment and guidance by the lakeside before plunging back into the roil of Fleet Street. Arcadia will be their ally.

233   **"fruits of industry . . .":** Ibid., p. 17.

Pitt the elder: See Hussey, pp. 100–6.

plant long associated with the Indian subcontinent: In the European mind, that is: the New World had its own indigenous varieties. See Hobhouse.

"cast off Bacon, Locke & Newton . . .": *Milton,* plate 41, line 5.

"**Loom of Locke . . .**": *Jerusalem,* plate 15, lines 15–20.

234 "**mountains green . . . dark Satanic mills**": *Milton,* plate 1, lines 2, 8.

235 **Locke couldn't stand the smog:** K. Thomas, chap. VI, sec. i.

"**a town of red brick . . .**": Dickens, *Hard Times,* bk. 1, chap. v.

"**the whole of Lancashire . . .**": *Wordsworth's Guide to the Lakes* (5th ed., 1835), quoted in K. Thomas, p. 267.

236 "**For adults . . .**": K. Thomas, p. 301. The son of a South Wales farmer, Thomas betrays, perhaps, the amused disdain that farmers (and ex-farmers) often feel for those who romanticize nature.

"**The creation of the mental realm of phantasy . . .**" Sigmund Freud, *Introductory Lectures on Psycho-Analysis,* pt. iii, cited in K. Thomas, p. 242.

237 **old-fashioned . . . walled garden at Rousham:** In a contemporary plan (Jellicoe and Jellicoe, chap. 20, no. 404), this is represented by a blank rectangle labeled "The Kitchen Garden."

238 **and the word was made greenery:** That gardeners as unlike each other as Vignola, Le Nôtre, and Capability Brown should all have had in mind more or less the same classical models is less astonishing when we remember that the same was true of Dante, Racine, and Pope. One trait that has kept the classics and the Bible alive is that they are endlessly amenable (if not quite supine) to new interpretation. Both Homer and Moses have obligingly doffed their robes and allowed themselves to be dressed like dolls in cassocks, periwigs, spats, or black leather as the occasion required. Adam's garden and that of Alcinous have worn the tonsure of the cloister, the baubles of Versailles, and the flowing velvet of Blenheim and Stowe.

20. WESTWARD IN EDEN

The perceptive reader will find this chapter sketchier and breezier than most others. I find I have been influenced by the style, if not the substance, of D. H. Lawrence's *Studies in Classic American Literature,* as well as his studies of places in which he sojourned. His method was to move to a new country, take a quick look around, and start writing before he could be confused by the facts. Lawrence had the advantage of genius; I, on the other hand, have lived longer in America than unlucky Lawrence lived anywhere.

241 **farming showed signs of becoming sustainable:** If their descendants cannot claim, as some farmers of Asia can, millennia of successfully working the same soil, they can at least claim centuries.

America was plastered with every label: Among the many secondary works exploring this theme, see for example P. Miller; L. Marx; Nash; Sanford; F. Turner, *Beyond Geography.*

"**blooming, humming, fertile paradise**": McKibben, *The End of Nature,* p. 50. But McKibben, though an important writer, should not be singled out; the image of pre-Columbian America as Eden is a commonplace among modern environmentalists. For two almost random examples, one major and one minor, see Sale, *The Conquest of Paradise;* Shetler.

242 "**Adirondack**" **means** "**bark-eaters**": DiNunzio, pt. 1.

"Woodlice . . . abound here . . .": Kalm, entry for June 28, 1749. On the same date, the mosquitoes "were so eager for our blood that we could not rest all night, though we had surrounded ourselves with fire."

"dreary wilderness": Bartram, pt. I, chap. iii. The encounter with the alligator, part of a much larger struggle with vast numbers of them, occurs in pt. II, chap. v. I mention this section in particular because McKibben cites the trout-in-orange-juice menu in his praise of the pristine Eden that Bartram found.

243   a smarter Adam: On the American as Adam, and "the case against the past," see R. W. B. Lewis.

"Shall we never . . .": Hawthorne, *The House of the Seven Gables,* chap. XII. Even so diligent an ironist as Hawthorne could hardly have known what he had wrought in making Holgrave a daguerreotypist. Once a keen wind blowing away the detritus of vision, photography now buries us in an avalanche of images—a perpetual present at least as burdensome as the "giant's dead body" that was the past. (See Sontag; and below, note to p. 259.)

Puritan anticulture of America: True, this was supposed to be the culture of Christianity scraped down to its original purity. As Perry Miller has shown, the Puritans liked to think they were carrying the religious wars of Europe to a new and undefended front. Their retreat was only strategic. Then, too, the thinness of their culture has been exaggerated. At its best, Puritan culture had a Shaker simplicity and soulfulness—but Hawthorne, visiting the Shaker village in Hancock, said it was so tidy that it pained the eye. (In the same way, Henry Miller felt suffocated by America's Protestant-ethic culture, but inhaled the culture of Europe in great gulps.)

244   "a pagan suckled in a creed outworn": Wordsworth, "The World Is Too Much With Us" (1807).

"sprang from the ground when he touched it": Brooks, p. 287, referring to an incident described by Emerson ("Thoreau").

Muir had his dog-eared Emerson: Nash, chap. 8.

245   common since Ovid: *Ars Amatoria* 2. 277. See text, p. 209.

lately brought against Columbus: See, for example, Sale, *The Conquest of Paradise.*

that "Wilderness should turn a mart": Edward Johnson, *Johnson's Wonder-Working Providence* (1654), quoted in Cronon, *Changes in the Land,* p. 167: "Nor could it be imagined that this Wilderness should turn a mart for Merchants in so short a space, Holland, France, Spain, and Portugal coming hither for trade."

Columbus himself thought: See his letter to Ferdinand and Isabella, dispatched October 18, 1498, describing his third voyage to the New World (Jane, ed., vol. 2). Columbus said he had sailed some distance up this breast, and that he had tasted the fresh waters of Eden where the Orinoco and the Guarapiche met the sea; but added that no man might enter the *parayso terrenal* itself except by the will of God.

246   "inventing money in the rainy woods . . .": Tisdale, p. 102.

Emperor Khosrow made spring eternal: Moynihan.

247   "should crumble to ruin once in twenty years . . .": Hawthorne, *The House of the Seven Gables,* chap. XII.

rare for a family to live at the same address: K. Jackson, chap. 3.

"an American will build a house . . .": Tocqueville continues, "He will plant a garden and rent it just as the trees are coming into bearing; he will clear a field and leave others to reap the harvest; he will take up a profession and leave it; settle in one place and soon go off elsewhere with his changing desires" (*Democracy in America*, vol. two, pt. II, chap. 13).

248 **gridiron plan:** See H. Carter; Stilgoe.

249 **the wilderness preserve and the front lawn:** Pollan, *Second Nature*. By the time I encountered that book, I had pretty much changed my mind. But there is nothing like seeing your old views in someone else's book to enforce a clean break.

250 **As the English park gobbled up both garden and wilderness:** Richard Payne Knight (*The Landscape: A Didactic Poem* [1794], excerpted in Hunt and Willis, pp. 342–48) contrasted the old formal gardens with the work of Brown and his school:

> Though the old system against nature stood
> At least in this, 'twas negatively good:—
> Inclos'd by walls, and terraces, and mounds,
> Its mischiefs were confined to narrow bounds;
> Just round the house, in formal angles trac'd,
> It mov'd responsive to the builder's taste . . .
> While uncorrupted still, on every side,
> The ancient forest rose in savage pride. . . .
> Our modern taste, alas! no limit knows:—
> O'er hill, o'er dale, through woods and fields it flows;
> Spreading o'er all its unprolific spawn,
> In never-ending sheets of vapid lawn.

What Knight would have made of present-day America is not hard to guess. **parklike—indeed, lawnlike—vistas:** Even the wild man John Muir, in his pitch to make Yosemite a national park, spoke of its "broad velvet lawns" (see "Features"). Similarly, to the Scottish botanist David Douglas, the grand vistas of the upper Columbia "looked in spring 'like English lawns' " (Tisdale, p. 11).

251 **they could retain their direct spiritual relation to certain parts:** Another way of looking at it is this: The reason Americans feel the need to divide things up is that they are afraid of their own relentless mixing and leveling. Wilderness preserves are one expression of this fear; zoning regulations are another. American zoning is by no means the natural expression of a communal sense of fitness, such as shaped much of the landscape of Europe. Though such a sense governed some of the early settlements (especially in New England), it did not last long. What shaped the zoning bylaws was the fear that no sense of fitness remained and that the landscape was fast becoming a squalid hash. In practice, many bylaws ended up contributing to urban decay and suburban sprawl, mainly because the planners, though they had a vague sense that the landscape needed structure, had in their minds' eyes the same ideals of Eden and Arcadia that had destroyed the structure in the first place.

Students of American literature, such as Fiedler, find that for most of its career it has drawn women either as madonnas or as whores, but rarely as anything in between. Maybe here, as elsewhere, the image of woman mirrors the image of nature. And maybe here, too, strict division is a way of staving off the urge to indiscriminate mixing—an urge finally released, in our day, in the cult of Madonna.

"The only thing we have . . .": Wendell Berry, "Preserving Wildness," which remains the best brief statement on the subject that I know of.

America's biggest crop is "turfgrass": On lawns and the "turfgrass" industry, see Bormann et al.; Pollan, "Why Mow?"

252   **large net loss of oxygen:** Bormann et al., p. 93.

In an hour of pushing or riding your power mower: Ibid., p. 96.

up to 30 percent of all urban water: Ibid., p. 75.

67 million pounds of pesticide: Pollan, "Why Mow?"

sudden rush of free energy can destroy structure: See p. 54 and note.

253   **pseudo-urban forms called "conurbations":** See Mumford (Geddes's disciple), *The City,* chap. 15, sec. 7.

"build a motor car for the great multitude . . .": Henry Ford in 1909, cited in K. Jackson, p. 160.

254   **streets and cars were heavily subsidized:** To learn how America's public-transportation system was deliberately scrapped and the way paved for suburbia, see the judicious and very readable account in K. Jackson. As my text was being revised, the book by Kay appeared, offering a popular treatment of the same subject.

streetcar companies started going under: In 1890, streetcar ridership per capita had been four times as great in the U.S. as in Europe; by 1910, U.S. ridership had slumped and Europe had pulled even.

"car welfare": Douglas Foy, director of the Conservation Law Foundation, quoted in Kay, p. 117.

subsidy of over $20 billion: Roelofs and Komanoff.

255   **It also invented redlining:** Though the HOLC itself made loans to all categories, the banks that adopted its appraisal system were not so even-handed.

"We shall solve the city problem by leaving the city": Quoted in Jackson, p. 175.

256   **Usonian vision of Frank Lloyd Wright:** See Riley's catalogue of the vast exhibition at the Museum of Modern Art, New York. "Usonian" was Wright's own meaningless coinage, having something vaguely to do with the U.S.A. The slogan for Broadacre City accompanied the twelve-by-twelve-foot scale model exhibited at Rockefeller Center in 1935.

Philip Johnson's gibe: Cited in Cronon, "Inconstant Unity."

"Things are in the saddle,/And ride mankind": Emerson, "Ode (Inscribed to W.H. Channing)" (1847).

257   **Americans rarely stopped to see:** Wendell Berry and Wes Jackson are both eloquent on this point.

259   **sit in the living room and have nature delivered:** As the using up of nature loses the thrill of immediacy, Americans resort to using up culture. A steady rain of

words, images, and products falls on them and around them runs into the gutter. Since culture does not really matter, nothing is made to last. To sustain this flow of culture, nature is used up even faster.

At first glance, Americans seem to treat their cultural heritage much the way they treat their natural heritage. There is no attempt to make what is new attend responsibly to what is old. Instead a great infatuation with the idea of the old—with Colonial, and Early American, and Western, and Victorian—finds its outlet in villages and districts frozen at some arbitrary point in time. Why frozen? Because Americans know that, unless they are fenced out of a place, they will not be able to stop themselves from wrecking it. Their sun runs so fast that the only hope of controlling him is to make him stand still altogether. So forests and deserts, towns and farms—but only a few of each—are locked up in the curator's icebox.

On closer inspection, American efforts to preserve nature and culture turn out to be subtly at odds. The reason is simple. Other cultures—European ones, for instance—like to imagine themselves tucked away in a frame of timelessness. Their ideal self-image is of a quiet village in the lap of lovingly cultivated land, with a church steeple whose bell chimes an indefinite hour. Though Americans are not proof against the charms of such a picture, it is hard for them to fool themselves in just this way. In other countries it is easy to believe that a way of life just now vanishing had gone on more or less unchanged for two or three thousand years. In some cases it is even true. In America, hardly any white way of life has had a run of more than a few decades.

In other countries, nature and culture could meld in the same autumnal glow. The landscape had been domesticated for so long that one could preserve both nature and culture at the same time, or at least pretend to do so. For Americans that was much harder. How could they preserve both their nature and their culture, when their culture was based on using nature up at breakneck speed?

Here is one reason for the peculiar silliness of America's cultural preservation, its Williamsburgs and Plimoth Plantations, its Disney Worlds and Disneyfied battlefields. For nature there are national parks, for culture theme parks. Theme parks are hardly an American invention—just think of Versailles, or Hadrian's villa at Tivoli—but in America they have found a mass audience. They have become places of pilgrimage, to which every true American must journey once in his life. Some theme parks preserve the past, some the present, and some the future; what is celebrated, finally, is the speed with which both nature and culture are consumed. Even the mall (there is only one mall, a timeless, placeless forerunner of cyberspace) becomes a theme park. The Mall of America, which occupies seventy-eight acres of erstwhile prairie near Minneapolis, includes nightclubs, a post office, and an amusement park. Its name suggests that it would, if it could, enclose the whole country in an air-conditioned shell, with Canada and Mexico leased as parking lots. Then the whole country would be its own theme park, like the map imagined by Borges that is the same size as the place it depicts.

America is the oldest country in the world: Stein, "Epilogue," in *Wars I Have Seen.*

21. A GODDESS QUANTIFIED

262 **Lovelock was invited by NASA:** See Lovelock, *Gaia: A New Look at Life on Earth,*
chap. 1.

263 **We already knew:** This would not exclude the possibility, now strengthened by
bacterialike imprints in a Martian meteorite, that the red planet once did have
life. The assumption would be that life on Mars never achieved Gaian liftoff—
that it never became rich and diverse enough to form a resilient system. (Strictly
speaking, Lovelock's argument would not exclude the possibility that there is at
present some very small quantity of life on Mars, which in time will either
achieve Gaian liftoff or die out.)

    If you took the elements: See Lovelock, *The Ages of Gaia,* chap. 9.

    it did point "in the right direction": Lovelock, *Gaia,* p. 5.

264 **dropped below 15 percent:** For this and other examples of Gaian equilibrium,
see Margulis and Lovelock, which, though written for scientists, may be the
clearest brief exposition of the hypothesis; it also contains a brief bibliography
of publications on Gaia, both pro and con, through 1988.

    "faint young sun paradox": First posed by Sagan and Mullen in 1972.

266 **Thousands or millions of Gaias or near Gaias:** Ironically, the scenario that
Dawkins and Doolittle thought self-evidently absurd is now being propounded
by a respected cosmologist, Lee Smolin, in an even more bizarre form, with Darwinian selection of universes leading to the proliferation of those in which life
can exist.

    Lovelock created Daisyworld: A. Watson and Lovelock; Lovelock, *The Ages,*
chap. 3.

267 **John Harte was curt:** Harte's critique is included in Schneider and Boston, ed.,
Scientists on Gaia, the record of an American Geophysical Union meeting in
March 1988 which marked the start of serious scrutiny of the Gaia hypothesis by
mainstream science.

268 **"butterfly effect":** Gleick; see also above, p. 277.

270 **New Age mystagogues find a new kind of science:** For relatively sober and
scholarly versions of this thesis, see W. I. Thompson, ed., *Gaia;* Abram, "The
Mechanical." Even Lovelock and Margulis sometimes allow themselves to be
borne along on this warm current—though not when they are actually doing the
science.

    carbon dioxide reacts with calcium silicate: Lovelock, *The Ages,* chap. 6.

    According to one scenario: Ibid., chap. 5; D. Anderson.

    phytoplankton blanket the ocean: Charlson et al.

273 **cast of characters:** They are vividly sketched by Gleick.

274 **she has not been able to prevent major changes in climate:** Indeed, it now
appears that on at least two occasions—about 2.26 billion years ago, and about
600 to 800 million years ago—the earth came very close to being locked into
deep freeze. With glaciers reaching nearly to the equator, earth's albedo was so

high that scientists are hard put to understand how the planet could ever have warmed up again—unless volcanic eruptions or the impact of a giant meteor, usually blamed for extinctions, in this instance saved the day. See Evans et al.

symbiosis, immunological Darwinism, neural Darwinism: The first of these was discussed in chapter 2; the latter two are expounded by Edelman.

tendency toward self-organized complexity: See text, pp. 276–7, and note.

shuffle the genes: Mainly by means of transposons, the "jumping genes" discovered by Barbara McClintock.

lichen: See chapter 2.

275   **Thoreau had a special fondness for lichen:** *Journal,* entry for Dec. 15, 1841: "I seem to see somewhat more of my own kith and kin in the lichens on the rocks than in any books."

"symbiosis viewed from space": To my knowledge, the phrase comes from Gary Beluzo, a student of Margulis.

For Lovelock, Gaia and chaos: Lovelock, *The Ages,* chaps. 3 and 9. On stability, instability, and chaos in ecosystems, and on the vexed question of the relation between species diversity and stability, see Hutchinson, "Homage"; Elton; May; Botkin; Pimm, *The Balance;* Tilman.

276   **experiment with monocultured plots of perennial grass:** Tilman and Wedin.

computer model of populations of Dungeness crabs: Hastings and Higgins.

"Researchers modeled a world . . .": Yoon, "Boom and Bust."

"a demon or a dragon": Lovelock, *The Ages,* p. 220.

new science of complexity: The two popular surveys of the subject, by Lewin and Waldrop, fail to do what Gleick did for chaos: they are workmanlike, no more. Of the recent spate of popular books by the complexity theorists themselves, I can recommend the one by Kauffman, whose enthusiasm is catching even if his prose is sometimes over the top; and the one by Gell-Mann, whose cranky, obsessive style is engaging in its way.

"Despotism tempered by sloppiness": Dyson, p. 76.

278   **"If the red slayer think he slays":** Emerson, "Brahma" (1857).

279   **survival of the fittest is a factor in evolution:** Kauffman (in Waldrop, chap. 8) tells of walking the Sussex Downs with the evolutionist John Maynard Smith, who "pointed out that we weren't far from Darwin's home. Then he opined that, by and large, those who would take natural selection seriously were English country gentlemen—like Darwin. And then he looked at me and gave me his little smile, and he said, 'Those who thought that natural selection didn't have much to do with biological evolution have been urban Jews!' " If Maynard Smith's dig was aimed at his old antagonists Stephen Jay Gould and Richard Lewontin, then his "didn't have much to do" is comically overstated; nevertheless, he has a point. Perhaps it has something to do with the words of Zechariah (4:6), experimentally confirmed by three thousand years of history: "Not by might and not by strength, but by my spirit, saith the Lord of Hosts." (On the other hand, that doesn't quite explain Karl Marx.)

280   **"How much can come . . .":** Emily Dickinson, "There Came a Wind Like a Bugle" (c. 1883).

22. MANAGERS AND FETISHERS

283 **"oh yes, I've learned from my mistakes . . .":** Sir Arthur Stebe-Greebling is a creation of the comedian Peter Cook.

284 **"shows its West Coast origins at times":** McKibben, *The End of Nature,* p. 92. Besides Devall and Sessions, see the later and somewhat better anthology by Sessions and the more rigorous and engaging work of Naess. Among the more interesting writers linked, to varying degrees, to the movement are Edward Abbey, Bill McKibben, Kirkpatrick Sale, Paul Shepard, Gary Snyder, and David Abram, whose *Spell of the Sensuous* has led his colleagues into new and fragrant fields.

"thoroughly Jewish": Schopenhauer quoted in Devall and Sessions, p. 240. By contrast, they find in Christianity some redeeming qualities (chap. 6, sec. IV), citing as evidence that well-known New Testament document, the Book of Job. The converse of this is found in app. G, authored by Gary Snyder, which speaks derisively of the "Judeo-Capitalist-Marxist West."

In a later essay, Sessions cleans up his act somewhat, admitting that Spinoza drew upon "ancient Jewish pantheistic roots" and describing his utilitarian view of animals as "typically seventeenth-century" rather than typically Jewish. He even throws a bone to Maimonides for declaring that the world was good before man was created; but he ignores the fact that this view is explicit in Genesis—as even Naess (p. 184) admits—and his overall assessment of Judaism seems unchanged.

As to Heidegger, I am not one of those who think the work can be parted from the life. While the arcana of that controversy are beyond me, a passage from Claudio Magris (chap. 1, sec. 11) seems to me suggestive:

> . . . Heidegger, who more than once repeated his claim to be a Black Forest peasant, in fact profaned this very feeling of loyalty and humility. In that over-emphatic identification with a familiar and immediate community—its woods, its hearth, its dialect—there was an implicit claim to a monopoly of authenticity, almost to an exclusive, patented trademark, as if his sincere attachment to his own soil allowed no room for the loyalties of other men towards other soils and other lands—to their log cabins, or their blocked-rent tenements, or their skyscrapers.

Earth First!: See the memoir of the group's founder, Dave Foreman.

285 **notorious maiming in 1987:** See Foreman, chap. 13.

AIDS is a good thing: Miss Ann Thropy. Milder views are expressed by Christopher Manes, whose tendentious history of the movement manifests most of its faults.

286 **Planet Management has become the dominant worldview:** So dominant, in fact, that listing representative books or authors would be pointless. Go to your

bookstore's environment shelf, pick up any book with a scientific or policy slant, and chances are good that its premises will be those of Planet Management.

"Managing Planet Earth": *Scientific American,* Sept. 1989. An earlier statement of the idea of Planet Management can be found in Myers.

287 **"if, at some intolerable population density ...":** Lovelock, *Gaia,* p. 132.

288 **"a world of abundant wildlife ...":** W. T. Anderson, p. 360. Compare *The Tender Carnivore,* by the Deep Ecology hero Paul Shepard.

tools of ecosystem restoration: See note, p. 535–40.

plan that would envelop North America: See text, p. 407.

289 **things like good workmanship and thrift:** "We are going to have to see that, if we want our forests to last, then we have to make wood products that last, for our forests are more threatened by shoddy workmanship than by clear-cutting or by fire" (W. Berry, "Preserving Wildness").

291 **"All France is a garden ...":** H. Miller, p. 33.

23. BEBOP

293 **Both our perception and our action might be improved:** Sound may in fact give a better rough-and-ready estimate of an ecosystem's health than sight can. When Paul Brooks came up with the name *Silent Spring* for Rachel Carson's book, he was tugging on heartstrings, but he was also touching on a deep ecological truth. Unseen birds may just be hidden; unheard birds are probably gone.

Spectrograms of recordings made in a healthy tropical rain forest find the audible spectrum densely packed with chirps, screeches, hoots, bellows, and other purposeful sounds. The reason is plain: since the social and sexual lives of many birds, insects, and other animals depend on getting their message across, they have to divide the spectrum up between them, just as radio stations in a big city divide up the airwaves. Increasingly, though, both spectrograms and practiced ears are finding alarming gaps. In the same way, recordings made in the temperate rain forest of the Olympic Peninsula show a stunning difference between stands of healthy old-growth forest and new stands planted by Georgia-Pacific. Monoculture makes for monotone. See Krause; and, for a general discussion of "soundscapes," the pioneering work of Schafer.

294 **mongrel of splendid pedigree:** Jazz is the fruit of a troubled marriage between two cultures that have been taken, rightly or wrongly, to represent two poles in humankind's dealings with nature. Africa yields to nature's dark suction, Europe breaks free. Or, to look at it from the other side: Africa celebrates the complex rhythms of nature, Europe imposes its own rigidities. And this marriage took place on a third continent which was seen as a new Eden, a place where humans and nature would live in harmony.

Nor were Africa and Europe the only contributors. The native musics of South America and the Caribbean had already pollinated black music. And it may be that something of the keening, jubilant spirit of the Near East, the site of so much desperate and impassioned wrestling between man and nature, was conveyed by the Jewish composers whose tunes jazz used.

In jazz you can hear the meeting of the New World and the Old World—or, rather, of the newest world and the oldest world. The place where the first human-led waves began very gently to swell outward, and the place where the latest waves crashed through at breakneck speed. The place where nature had a chance to adapt gradually to human-led change, and the place where it was bowled over.

If I had to choose one fact to prove that the modern West can yet be reconciled with its ancient and non-Western self, I would choose jazz. Nor is this a mere polite reconciliation, an agreement to disagree. It is a rejoining of disjoined selves, a remarriage. And jazz is its child.

Another useful thing about jazz: most of it is mediocre. Of course, that is true of most concert music, too. But in the realm of concert music mediocrity is failure. Since Beethoven, concert music is written for the ages. Jazz is made for the moment. If it survives, so much the better.

The great jazz singer Abbey Lincoln recalls getting this advice, early in her career, from Thelonious Monk: "Don't be so perfect." Most Earth Jazz is, and always will be, mediocre. It is made by ordinary people wrestling with ordinary, messy problems. When it tries to be perfect, it generally ends up doing violence to people and planet alike. Moreover, it ceases to be jazz.

Pedants have gone gray trying to define it: I myself am partial to Peter Schickele's gloss: "It don't mean a thing if it ain't got that certain *je ne sais quoi*."

295   **"You're never quite on the beat . . .":** Ellison, chap. 1.
In the Sonoran Desert of Arizona: Nabhan, *The Desert Smells like Rain*, chap. 7.
296   **"When people live and work":** Ibid., p. 96.
Central Valley of California: Emory.
298   **it almost never works:** Ludwig et al.
299   **Each of these has been somebody's pick:** See, respectively, Henry Adams; Wendell Berry; Bookchin; Shepard, Johnson, Calder; Bookchin; Gimbutas, Eisler; Eisenberg; Sauer; Todd and Todd; Mumford. Bookchin appears twice because he uses the pueblo as a model of the Neolithic village—a serious error, for the pueblo is the end product of many millennia's fine-tuning. The Eisenberg in question is the present author at the time he started work on the present book.

These are, of course, only a few of many Edens. The roster of thinkers who located Eden or the Golden Age in classical antiquity would take a chapter of its own, as would the list of those in classical antiquity who placed it in archaic antiquity. Some modern visionaries find Eden in the prehuman past: for example, Bateson, who finds the Fall in purposive thinking; or McMennamin, who writes (with tongue in cheek) of the Garden of Ediacara, some seven hundred million years ago, when soft-bodied creatures floated peaceably (to all appearances) in shallow seas, free of the armor that carnivory would later force upon them.

301   **egret chick:** Mock and Forbes; Creighton and Schnell.
302   **bacterial community that built a fission reactor:** Lovelock, *The Ages*, chap. 5
not for his mechanical duck that Jacques Vaucanson is remembered: Mumford, *Art and Technics*, p. 32. Mumford later came to believe that careful imitation

of organic models was where the future of technology lay, or should lie, and that what was wrong with Vaucanson's clockwork duck and flute player were that they were clockwork: that the inventor had fabricated "crude mechanical equivalents" rather than "drawing on a pre-existing biological solution" (*The Myth of the Machine*, vol. 2, p. 394). He gives as an example of such "drawing" the telephone receiver, which Bell consciously modeled on the anatomy of the human ear. In subsequent chapters I will give other examples; but the tension between abstraction and imitation is not so easily resolved.

303 **clay tablet from Babylon:** Dunham, chap. 1; Jourdain, chap. 1; D. Smith, vol. II, chap. V, sec. 5; Needham; vol. 3, pp. 19–24, 257. Boyer, however, does not credit the Egyptians with a full understanding of the theorem.

Both Proclus and Herodotus say the Egyptians were forced to invent geometry by the annual flood of the Nile, which kept wiping out their landmarks and property lines. In that case, the impulse to impose geometry on nature was a reaction not so much to her irregularity in space as to her inconstancy in time.

304 **pesticides work their way through the food chain:** Carson.

24. THE WILD GARDEN

306 **Indian was no Adam:** On native-American management of the American landscape, see Cronon, *Changes in the Land;* MacLeish, *The Day Before America.*
Bartram's rambles in the Southeast have been cited: Notably by McKibben, whose account in *The End of Nature,* pp. 49–51, was discussed, from another perspective, in my text, chapter 20. The phrases in quotes are his, not Bartram's.
"wild oranges" were not wild at all: See Crosby, *The Columbian Exchange,* pp. 66–7: "In the latter part of the sixteenth century José de Acosta asked who had planted the 'whole woods and forests of orange trees' through which he walked and rode, and was told 'that oranges being fallen to the ground, and rotten, their seeds did spring, and of those which the water carried away into diverse parts, these woods grew so thicke. . . .' "
Sturtevant says a small, bitter orange may possibly be indigenous to South and Central America and the Caribbean; but Bartram's descriptions (for example, in pt. II, chap. III) suggest the larger, sweeter Old World type. See also Palmer.
land the white settlers espoused was not virgin: See Cronon, *Changes,* chap. 1.
307 **"A curious thing about the Spirit of Place . . .":** Lawrence, chap. 4.
"One of our chiefs said it best . . .": Richard Hill, in *Niagara* (Florentine Films).
everything that lives is holy: Blake, *America,* pl. 8, l. 13.
308 **Its rituals, its "magic":** Among many possible examples, Malinowski's *The Coral Gardens and Their Magic* might be singled out.
Gorotire Kayapo: Besides the articles authored or co-authored by Posey, see the book by Werner. For archeological evidence that the Amazon's "virgin" rain forest (like rain forests throughout the tropics) has been swidden-farmed and otherwise managed for thousands of years, see McDade et al. The fact that the

forest has recovered from these delicate disturbances should not be taken as license, though, for the kind of wholesale clearing that is now taking place. Nor should the fact that wilderness is a relative concept lead us to the convenient conclusion that it is a meaningless one. The Mountain has a history, but that should not lessen our awe.

311 **In western Java:** On *kebun-talun* and Javanese home gardens, see Christianty et al.; Freeman and Fricke.

312 **"It was covered . . .":** E. Anderson, pp. 136–7.

313 **In the Sonoran Desert of Mexico and Arizona:** See Castetter and Bell; Nabhan, *The Desert Smells Like Rain* and *Gathering the Desert.*

    "In all this land of the Papaguería . . .": Bolton, entry for Oct. 30, 1775 (vol. IV, p. 33). Font took part in the expeditions of Captain Juan Bautísta de Anza, who explored the route to California and organized its colonization.

314 **ethnobotanist Gary Paul Nabhan:** Nabhan, who has spent many years traveling among the Indians of the Southwest, observing their agriculture and disseminating the scarce seeds of their traditional crops, once studied a flash-flood field in one of the hottest and driest parts of the Sonoran Desert. At the beginning of August, teparies and pintos has been sown side by side (the family in question had not had enough teparies on hand to plant the whole field). They were soon joined by wild desert beans (*Phaseolus filiformis*), which had volunteered. By late October, the rains—a total of sixty-eight milimeters, or less than three inches—were long gone, and all the plants were close to death from thirst. This was a matter of indifference to the teparies and the wild beans, which had already produced a full complement of fruit. By contrast, only 3 percent of the pinto pods had matured, and even those were shriveling up. In dry weight of beans, the teparies outyielded the pintos a hundred to one. The contest between zucchini and native Sonoran squash gave a similar result. (*Gathering,* pp. 107–21.)

316 **cities of the Nabateans:** See Hillel, *Negev,* chap. 13. Between 330 A.D. and the Muslim conquest, the Nabateans were subjects of Byzantium.

    Slash-and-burn is not always done so well: Geertz, chap. 2; Weisenfeld, who discusses the link to sickle cell anemia.

    langur monkey: Hrdy. Such self-serving infanticide is not uncommon among animals.

317 **"moral luck":** I have borrowed the concept from Nussbaum, who applies it mainly to individuals.

    even more subject to accident: True, the concentration of power is less, as is the inherent dynamism; but the effect of size outweighs these factors.

318 **collapse of the Maya:** R. Hansen. Prolonged warfare, luxurious elites, overpopulation, and climate change may also have been factors; their relation to ecological malfeasance is complex and controversial.

    On the ecological folly of the "noble savage" and non-Westerners generally, see Diamond, *The Rise,* chap. 17; E. O. Wilson, *The Diversity,* chap. 12; Dubos, *The Wooing,* chap. 5; Tuan.

    volcanic highlands of central Mexico: See O'Hara et al.; and the commentary by Butzer, "No Eden." Based on detailed study of sediments in Lake Pátzcuaro, erosion seems to have started by 1900 B.C., with surges about 600 B.C. and again after 1200 A.D.

Maoris deforesting much of New Zealand: Diamond, *The Rise,* chap. 17; E. O. Wilson, *The Diversity,* chap. 12. Wilson cites the Kiwi ditty:

> No moa, no moa,
> In old Ao-tea-roa.
> Can't get 'em.
> They've et 'em;
> They've gone and there ain't no moa!

North American Indians overhunting the buffalo: See P. Matthiessen (*Wildlife,* pp. 148–49), who writes that even before the white men came the Indians were "fearfully wasteful of the bison.... They killed frequently for the buffalo tongues alone, leaving whole plateaus of rotting carcasses to the swarming plains wolves and coyotes." The white man's orgy of buffalo slaughter, intended partly to starve the Plains Indians into surrender, was joined in by defeated tribes, who "worked toward their own destruction willingly and well." On the Indians' implication in the killing off of big animals at the end of the Ice Age, see text, p. 37, and notes.

Chinese monks shaving the mountains: Tuan.

Asian medicine: See Reisner, *Game Wars;* and Burns, who writes that "a tiger penis can sell for $1,700 in Taiwan, and powdered tiger bone for as much as $500 a gram." An investigator for the World Wide Fund for Nature, notes Burns, "has given the Indian police the names of several men he believes are at the heart of the poaching network, operating from headquarters in Bombay and Delhi, but none of the leaders have been arrested." One is reminded of a passage from Eric Newby (p. 68) in which an old man, squatting in the sand by the Ganges, is posted to scare poachers from the jungle, which happens to be on the other side of the river. Asked why he doesn't report poachers to the authorities, he replies: "How can you report a poacher to himself?"

importing almost four times as much timber: World Resources Institute. Meanwhile, the price of whale meat in Japan—about $300 a pound—has led to the widespread sale of flesh from species protected by international law (Angier, "DNA Tests"). Chanterelles, matsutakes, and other edible mushrooms, which as mycorrhizae are essential to the health of trees such as the Douglas fir, are being stripped from the forests of the Pacific Northwest to feed the Japanese market ("The more phallic they look, the more they'll pay"). See Yoon, "Lucrative Harvesting."

we have much to learn: For other examples of traditional farming and gardening from which we have much to learn, see Altieri, chap. 6; Alcorn; Brush; Brush et al.; Cox and Atkins; Gade; Janzen; King; Marten; Todd. Most of these are fairly technical, but a good essay by Wendell Berry, "An Agricultural Journey in Peru," describes Brush's work on potato culture in the Andes.

25. THE TREE OF LIFE

322 **"Why would we have any more reverence . . .":** McKibben, *The End of Nature*, p. 211.

"They're a human product like rayon . . .": Dillard, p. 4.

"I used to have a cat . . .": Ibid., p. 1. To some minds, the wildness in cats makes their domestication that much more objectionable. There is an element of smugness involved, even of revenge. Big cats tossed our hominid ancestors like dolls; now death curls in our lap, purring. To me this turnabout seems fair play.

Conjugation slangily pairs members: See Sonea and Panisset; Beardsley.

323 **soil microbe that causes crown galls:** See J. Watson et al., chap. 13; Ahmadjian and Paracer, p. 45. The fit between the plasmid and the plant chromosome is uncannily neat; as with retroviruses and their mammalian victims, there are spooky homologies between the DNA of the parasite and that of the *uninfected* host. In view of this, it seems likely that the bacterium long ago borrowed genes from host plants, which it gradually retooled into weapons that it would turn against them.

324 **When biotechnologists want to introduce a new gene:** J. Watson et al.

325 **"an untold amount of breeding work . . .":** Wes Jackson, *Altars*, p. 61. His views on biotechnology inform this chapter.

company funding the research also makes the herbicide: A 1988 report ("Herbicides") by Jack Doyle of the Environmental Policy Institute found at least thirty-three companies at work engineering crops for resistance to various herbicides, including atrazine, Bromoxynil, cinmethylin (Cinch), glufosinate, glyphosate (Roundup), imidazolinones (Scepter), paraquat, penoxalin (Prowl), picloram (Tordon), 2,4-D, phenmedipham (Betanal), sulfonlurea herbicides (Glean, Ally, and Londox), and trifluralin (Treflan). Several of these are suspected carcinogens or teratogens which have shown up in groundwater. Doyle notes: "Thirteen of the 21 field tests for genetically engineered crops approved by the U.S. Department of Agriculture since late 1987 have been conducted by three chemical and pharmaceutical companies—Monsanto, Sandoz, and Du Pont— and more than half of these were for herbicide-tolerant plants." See also Schmidt.

T-cytoplasm corn: Ibid., p. 61.

326 **If black pineleaf scale insects are moved:** Yoon, "Insects Adapted." Though all this is truest of sedentary bugs—white peach scale on mulberry trees, thrips that prey on the seaside daisy—it has also been found true of moth caterpillars that feed on sand live-oaks in northern Florida. Free to relocate when in their winged stage, they nonetheless return to their home oak to breed. Even sumac flea beetles that can hop-fly from one sumac to another seem to settle on one plant as home (indeed, as universe) for themselves and their lineage.

scientists have not been able to create any genes worth mentioning: The minor exceptions have involved taking existing genes and modifying them.

327 "accumulating on the shelf . . .": Wes Jackson, *Altars*, p. 8.
"nature is hard to beat . . .": Ibid., p. 57.
"The ecosystem level of biological organization . . .": Ibid., p. 61.

328 Jackson and his associates at the Land Institute: See his *New Roots*, and my
"Back to Eden" (a title that was not mine, and that rings a bit too tinklingly in
the context of the present book).
wheat field and prairie are unlike: On the natural history of the tallgrass prairie,
see Reichman.
"Wrong side up": W. Jackson, *Altars*, p. 77. He attributes the story to Joseph Kin-
sey Howard; the farmer's name is John Christiansen, and the encounter takes
place in the spring of 1883.

329 *r* strategy, K strategy: See Odum, *Fundamentals*, chap. 6; Wilson and Bossert,
chap. 3.

330 eastern gama grass: *Tripsacum dactyloides*. The mutant is known as the pistillate
or gynomonoecious form. See L. Jackson and Dewald.
polycultures combining these promising plants: In Jackson's ideal mix, a
warm-season grass, a cool-season grass, and a legume would provide starch and
protein, while a composite would yield oil. The legume would fix nitrogen, and
chemicals released by the composite—"allelopathic" chemicals—would keep
weeds at bay. Among the plants under study are giant wild rye (*Leymus racemo-
sus*), Illinois bundleflower (*Desmanthus illinoensis*), and Maximilian sunflower
(*Helianthus maximilianii*).
      The idea of a domestic prairie is tailored to the prairie states. In time, specific
mixes of perennials might be bred for specific regions, and for soil types and
microclimates within regions. But getting too specific would be a pain in the
neck for breeders and farmers alike. A farmer can't plant, and a seed house is not
about to market, a different mix for every square foot of his land. Yet this is what
nature does. Our image of native prairie is of endless sameness, but in truth the
flora shifts kaleidoscopically, inch by inch, like the better sort of minimalist
music.
      A natural prairie adapts not only in space but in time. It adapts to changes in
rainfall and temperature, to invasions of insects and microbes. It can do so
because each population of plants has a deep pool of genes to draw on. But the
gene pool of a human-selected population is almost always going to be shal-
lower. Jackson describes the dilemma: "Successful plant breeding tends to narrow the genetic base of any crop roughly in proportion to its success. The better
adapted to the human purpose a population is, the narrower its genetic base and,
therefore, the fewer options it has available to meet future threats of insects and
pathogens." Ideally, while breeding his optimal mix for high yield, high protein,
easy harvesting, and so on, he would leave in enough diversity to give some of
the adaptive properties of a natural ecosystem. Modern agriculture will resist
this, of course. But a food-bearing prairie that retains some of the heat-and-
drought resistance of the native prairie—and some of its diversity—is likely to
adapt better to the imponderables of the next century than a field of corn plants
with genes for heat-and-drought resistance spliced in.

As Jackson imitates the prairie, so Bill Mollison is paying a similar compliment to the forests of his native Tasmania. Unlike the seed-rich herbaceous perennials that Jackson wants to breed, fruit-rich woody perennials (i.e., trees) already exist, so Mollison has a head start. In fact, his "permaculture" is already being practiced by homesteaders in Australia and elsewhere, and in theory can be adapted to almost any region whose native vegetation is forest. On the other hand, it is unlikely to lend itself to large-scale or even medium-scale production, as the edible prairie would, or to satisfy the ancient human need to eat grains.

Even less of a concession to modern agriculture is made by Masanobu Fukuoka, who six decades ago gave up a career in microbiology to come home and work his father's farm on the island of Shikoku. A penitent return to tradition was not what he had in mind. He thought farmers worked too hard—that they fussed and interfered too much with the workings of nature, instead of letting nature work for them. To test his theory, he abstained from pruning his father's orchard. The trees died. Undeterred, he continued to experiment, and learned that, if fruit trees grew up without ever being pruned, with weeds and grasses and wild vegetables growing freely among them, they would flourish and fruit prolifically without the aid of pesticides. He learned that he could broadcast his rice seed in the standing winter barley just before it was harvested, and his barley seed in the standing rice, and never have to plow at all. He learned that he could get yields as high as any in the prefecture with half the work traditional Japanese farmers put in, and none of the chemicals industrial farmers use. While industrial farmers deplete the soil and traditional farmers more or less maintain it, Fukuoka's soil gets richer each year. Young people come to his mountain, where they dwell in huts, wear loincloths, and learn how to be self-sufficient "quarter-acre farmers." To them, and to thousands in the West who have read his books, he is a sort of latter-day Lao Tse.

By the standards of mainstream technology, all of these people engage in relatively little abstraction. Rather than start with man-made systems and ask what natural principles can be grafted onto them, they start with natural systems and ask how they can be adapted, with minimal tampering, to human needs. The less tampering, the more likely it is that a natural system will retain whatever it was that made it work.

This is not a new idea. It is the idea that, consciously or not, guided the earliest proto-farmers and gardeners. It is the idea behind old-fashioned plant and animal breeding—which, however, has too often looked on the organism as the only relevant natural system. For the best traditional gardeners, from the Kayapo of the Amazon to the market growers of the Ile de France, the relation between species in communities and ecosystems has been of equal concern.

Another thing that unites these people, and sets their work apart from much of modern technology, is that they try to learn from nature not just in general, but in a particular place. Though they do not disdain the satellite's-eye view, they know that they must spend most of their time on the ground, paying close and patient heed to what nature is doing right in the neighborhood. Each of the workers I have mentioned is closely tied to place: Fukuoka to southern Japan, Jackson to the prairie, Mollison to the temperate forests of southern Australia.

A further advantage of working on a local scale is that, when you do some-thing, it is easier to listen for nature's response. The process of trial and error is faster and less disastrous when the trials are small. In the spirit of Earth Jazz, you "trade fours" with nature, each player responding and adjusting minutely to the other.

Of course, you cannot learn from nature locally if your locality no longer has any wild nature to learn from. Aldo Leopold noted that the first rule of intelli-gent tinkering is to save all the parts. From this rule the effort to protect endan-gered species draws much of its force. In the case of living systems, however, there is a second rule which is less often mentioned. That is to keep some sizable examples of the type of system in question intact, whole, so that their workings may be observed. In other words, we need Mountains on every scale.

Besides the works of Jackson, Mollison, and Fukuoka, some useful works on alternative agriculture (or, in some cases, alternatives to agriculture) are Altieri; A. Howard; McFadyen; J. R. Smith; Todd and Todd.

"Would it not be well to consult with Nature . . .": Thoreau, "The Succession of Forest Trees."

paper clip: Invented by Johann Vaaler in 1900.

331 *Bacillus thuringiensis:* See W. K. Stevens, "New Microbial Agents"; "Power of Natural Pest Killer Wanes"; Rennie; Carr et al.; Perkins. Resistance has been found in populations of Colorado potato beetles, Indian meal moths, diamondback moths, and tobacco budworms. Early in 1995, the U.S. Environmental Protection Agency began granting limited approval to crops that had been genetically engi-neered to produce a Bt toxin, and in 1997 more than 3 million acres were planted with such crops. On September 16, 1997, a petition was filed with the E.P.A. by a number of groups and individuals, including Greenpeace International, the Sierra Club, and the International Federation of Organic Agriculture Movements, charging the E.P.A. with the "wanton destruction" of the world's foremost bio-logical pesticide and demanding that the approvals be rescinded.

It would no doubt be impertinent to ask Monsanto (the prime maker of Bt plants) to put its efforts into designing subtler crops that would express the toxin only in the part of the plant that is most economically valuable (such as the fruit), or only at certain crucial points in the plant's lifetime, or only when trig-gered by a chemical signal the farmer might spray on the field. According to Monsanto, it will be a decade before plant genetics are understood well enough to make such crops possible; in the meantime they have no choice but to go ahead and market the crude form of Bt-toxin-laced crop they have in hand. They seem unfazed by the prospect that their subtler crops may be greeted, a decade hence, with derisive hoots from races of bugs wholly resistant to Bt.

In a rational world, Monsanto would restrain itself, or be restrained, from releasing a product that is likely to speed the development of resistance to a val-ued biological control. In the actual world we should at least demand (or demand that government demand) that Monsanto mix a certain proportion of old-fashioned toxin-free seeds with each batch of engineered seeds.

Additional note, 1999: New studies (see Mackenzie) suggest that Bt spores can cause lung inflammation, internal bleeding, and death in laboratory mice that inhale them. Exactly which strains of Bt do this, and which toxin or toxins might be responsible, is unclear. What is clear is that the question of engineering Bt toxins into crops, especially food crops, has now been further complicated.

332 **"limitations inherent in biological control ...":** W. K. Stevens, "New Microbial Agents."

333 **how largely it has already been realized in traditional farming:** In fact, in some parts of the world traditional farming and gardening are being remembered, and revived, by those who once reviled them. Thirty years ago, the International Rice Research Institute, in the Philippines, was in the vanguard of the Green Revolution, promoting with equal zeal new hybrid varieties and the chemicals they were designed to require. Today, the institute is recanting. Its own studies found that farmers who used pesticides—the majority of Filipino farmers—suffered five diseases at a rate two to five times higher than that of nonsprayers, so that they lost more money to illness than they gained from higher yields. For that matter, the "higher yields" were often lower than they had been before the revolution. The new strains lacked the diversity and natural resistance of the old; the chemical abuse encouraged by corporate pushers had destroyed the ancient balance of power in the rice paddies. Formerly, the brown leafhopper had been held in check by at least one hundred natural enemies. With most of these killed off by scattershot spraying, the leafhopper's fearsome fertility—the female laying a thousand eggs at a go, the race able to multiply as much as a thousandfold every three weeks—was unleashed. A minor pest became a plague. A series of such disasters prompted the Philippine government, in the mid-1970s, to begin a program to recover native varieties of rice and other crops. See Doyle, *Altered Harvest,* chap. 13; Zwerdling.

26. THE TREE OF KNOWLEDGE

335 **"Cease from grinding, ye women ...":** *Greek Anthology,* 4. 418.

336 **"Find the shortest, simplest way ...":** Lanza del Vasto, quoted in Wendell Berry, *The Unsettling of America,* p. 96.
"Locke sank into a swoon ...": Yeats, "Fragments" (added to *The Tower* in *Collected Poems,* 1933). In an earlier draft I ascribed the poem to Blake and I still can hardly believe it is *not* Blake's. (In the second fragment Yeats asks, "Where got I that truth?" and answers, "Out of a medium's mouth. . . .")
"... Instead of engineering for all America ...": Emerson, "Thoreau." He goes on, "Pounding beans is good to the end of pounding empires one of these days; but if, at the end of years, it is still only beans!"
he could sit happily at the foot of a telegraph pole: Thoreau, *Journal,* entries for Sept. 3 and 12, 1851; Jan. 9, 1853 ("It stings my ear with everlasting truth").

337 **first music to depend on records:** See my book *The Recording Angel,* chap. 8.

338 **mechanizing of matter preceded:** Mumford, *Technics and Civilization,* chap. II; *Art and Technics,* pp. 64ff.
George Inness's *Lackawanna Valley:* See Novak, chap. VIII. Compare Emerson

("The Young American"): "Railroad iron is a magician's rod, in its power to evoke the sleeping energies of land and water."

339 **recycling in a city called Phoenix:** My account of the Phoenix project is based on Muschamp; K. Stein; and conversations with Michael Singer. See Rub for a brief account of some other Singer projects, the most visible, perhaps, being his interior garden for Concourse C of the Denver International Airport: this, with its overgrown, ruined stonework, is so wickedly subversive of the usual aseptic, blandly reassuring airport ethos that one wonders how he got away with it. (All that's missing is Amelia Earhart's wreck.)

340 **"in a Gothic light":** Muschamp.
a kind of chapel: In truth, recycling centers ought to be our churches. In some rural communities, they already are: the place where people come on Saturday and Sunday mornings to meet their neighbors, to purge themselves of filth, and to feel virtuous.

341 **amounts of nitrogen and sulfur:** W. Clark.

343 **small city of Kalundborg:** Tibbs.

How can the growth of industrial ecosystems be encouraged? They are unlikely to arise by fiat. Although regulation can sometimes help, as in the German example above, for the most part industry is allergic to regulation. Industry likes to think of itself as forward-looking, adventurous, active, and independent. Regulation cramps its style. Chafing under red tape for the past three or four decades, industry has come to regard environmentalists—and, by extension, the environment—as the enemy.

In fact, some kinds of regulation may now be suppressing the first shoots of a future industrial ecosystem. In the U.S., a plant whose waste includes cyanide is not free to reclaim that cyanide and sell it; the waste must be handled and disposed of in prescribed ways. Yet a company that needs cyanide for its processes is free to buy it in freshly made form. The same goes for heavy metals and toxic hydrocarbons. When steel for cars is treated to prevent corrosion, the wastewater that runs off makes a sludge rich in zinc. In the past, the sludge was smelted down, the zinc recovered and reused. But in 1985, such wastewater-treatment sludges were classified as hazardous. Wrapped in forbidding regulations, the zinc-rich sludge was no longer accepted by smelters. Instead, it ended up in landfills.

Even if we set such paradoxes aside, it is clear that industrial ecosystems will arise not because they are mandated, but because industry comes to see them as the next step in its own development; as the natural response of its own inner logic to a world no longer boundless; and as a way to make more money.

In Kalundborg, industry has made more money. So has the Minnesota Mining and Manufacturing Company, better known as 3M, since starting its "Pollution Prevention Pays" program, better known as 3P, in 1975. Many of 3M's products, such as magnetic tape, are made by coating something with something else, usually with the help of a solvent. Then heat is applied and the solvent disperses in air, which would be fine if many of the solvents were not toxins such as toluene, xylene, and methyl ethyl ketone. If you try to filter these out of a factory's exhaust, you will spend a great deal of money, and at least a tenth part of the pollutants will still sneak through. Instead, 3M decided to solve the problem at the source by reformulating its products and redesigning its processes and equipment. In its first fifteen years, the program reduced 3M's annual worldwide release of air, water, sludge, and solid-waste pollutants by half a million tons; cut pollution per unit of production in half; and saved the company half a billion dollars.

Of course, it was environmental regulation that goaded 3M into moving along this path. And in fact the free market cannot be counted on to make companies do the right thing. In the long run, the free market might do the job, just as scarcer resources force an ecosystem to mature; but it would be a very long run, during which many of the earth's natural systems would be battered underfoot beyond all hope of recovery.

But that is because the free market is not really free at all. It forces most of the earth's people (to say nothing of its nonpeople) to pay the hidden costs of practices they don't engage in and products they don't buy. Or—in the case of practices they do engage in and products they do buy—it forces them to ignore those hidden costs and hence to behave irresponsibly.

For instance, if you factor in the social costs of air and water pollution, oil spills, environmental damage at drilling sites, the likely effects of global warming and ozone depletion, and the military effort to protect the oil wells of the Near East (most visibly and recently those of Kuwait), you can readily see that the real cost of oil is many times what you and I pay at the pump, and even more times what industry pays its suppliers. You and I make up the difference in our taxes and in the decline of our quality of life. But since the real price is not charged for the oil or gasoline itself, there is little incentive to be thrifty in its use.

Nor is there much incentive to try alternative forms of energy, feedstock, or industrial process, which necessarily seem expensive (at least at first) next to their secretly subsidized rivals. Indeed, tax breaks, research support, and other "subsidies" for alternative energy and technology would have to reach many times their present piddling levels before they could begin to match the hidden subsidies to fossil fuel. Meanwhile, a new process like the biocatalysis of catechol from cornstarch, mentioned in the next note, will have trouble setting up shop as long as catechol made from petroleum, using heat from petroleum, remains so absurdly cheap.

The same goes for any number of resources, products, and processes. The "externalities" that make up much of the real cost are not imposed on the makers and users, but are borne by society at large. Baneful in its outcomes, this "free market" is also plain unfair, since it rewards plunder, punishes care, and allows a few to get rich at the expense of many.

A number of ways have been proposed of making the free market more free, more fair, and more able to lead industry to maturity before its youthful high spirits have turned the planet gray. "Green" or "Pigovian" taxes (proposed in 1920 by the British economist A. C. Pigou) could be imposed on the release of various pollutants in proportion to the damage they do or the cost of cleaning them up. Using up various nonrenewable resources, or failing to renew renewable ones, could be taxed in the same way. Those who warn that these costs would just be passed along to the consumer are missing the point; the costs have already been passed along to the consumer in the form of income taxes that help pay for government cleanup programs; health-care expenses; and other, less measurable losses. The point of Pigovian taxes is to put those costs back where they belong and to label them properly, so that both corporations and consumers will have an incentive to reduce them.

Alternatively, a government could issue pollution permits—certificates entitling the bearer to release a certain amount of a certain substance—that could be traded on the open market. A company that cleaned up its act could then clean up financially by selling its permits. If the permits were annual, the government could issue fewer each year, whittling away at emissions until some were phased out altogether and others reduced to what nature might painlessly take in. The EPA used this method to phase out leaded gasoline and has lately extended it to some industrial emissions. The main drawback is that certain industries in certain regions may find it cheaper to keep emitting, in which case the people of that region are out of luck.

When it comes to toxic wastes, the market may work best if their creators are not allowed to "release" them at all, but must keep them safely confined on their own property, at their own expense, until they figure out what to do with them. With the demand that such pent-up wastes would create, it would not be long before new technologies emerged for recyling them or transmuting them into some harmless form.

At bottom, this "parking-lot" concept is a way of making wastes stick to the hands that would toss them away. The same sort of glue can be applied to manufactured goods. When a corporation sells a product to a consumer, it cleverly passes along the headache of getting rid of that product after it is used. By what right? After all, what you really want to buy is not the object, but the use of the object for as long as it stays useful. Germany's Environmental Protection Encouragement Agency has recognized this in its "intelligent product system." Durable goods such as cars and large appliances revert to the manufacturer when they die: Krupp giveth and Krupp taketh away. See Frosch; Tibbs; Hawken.

344 **technology has much to learn about metabolism:** Take the chemical industry, which in its broad sense produces not only "chemicals" but much of the substance of our clothing, tools, fuels, furniture, vehicles, and (alas) food. Broadly speaking, the recipe is simple: start with some petrochemicals or other cheap "feedstocks," assault them with a series of catalysts, and you end up with a magical stuff that can be shaped into clock-radios, stockings, or fat-free frozen yogurt. The drawbacks to the recipe are also simple. The feedstocks are generally nonrenewable; the catalysts are often concentrated acids and other toxins; both can cause various kinds of environmental harm.

For decades, "solid acids," porous minerals with a lot of protons loosely clinging to their pores, have been replacing liquid acids as catalysts of industrial reactions (J. M. Thomas). Though potent—in some cases, ten million times more potent than concentrated sulfuric acid—they can safely be held in the palm of your hand. A kind of solid acid called a zeolite, a crystal so porous that all its atoms are surface atoms, can offer as many as ten million trillion active sites per gram, and the shape of its pores can determine which of several possible products of the reaction will emerge. In recent years chemists have come up with new kinds of solid acids, including clays and microscopically sandwiched metals, that can broker industrially vital reactions more efficiently, at lower temperatures, and with less environmental damage than old-fashioned catalysts.

Using the shapes of molecules, rather than the brute forces of heat and valence, to promote a desired reaction brings us closer to the methods of nature's chemical matchmakers, the enzymes. So why not go the whole hog and use the enzymes themselves, which can make chemical reactions happen thousands, perhaps millions of times faster than in the nonliving world? (See Todd and Josephson.)

Consider a typical industrial process: benzene derived from petroleum is made into catechol, which in turn can be either made into vanilla flavoring or used in the manufacture of nylon. Versatile though the process may be (for the chemist, the idea of edible lingerie can only draw a yawn), it does have its foibles. Benzene is a nasty carcinogen whose release into the atmosphere is now illegal in many countries. Its transmutation into catechol takes place at hellish temperatures and gives off nitrous oxide, which thins the ozone layer even as it thickens the greenhouse layer.

Encouraged by an EPA program, chemists at Purdue have started out instead with glucose derived from cornstarch. The simple sugar is set upon by a genetically altered strain of *Escherichia coli*, a bacterium that is even more common in our intestines than in biotechnology labs. Ordinary, gut-variety *E. coli* would use most of the sugar for energy and the leftovers for making amino acids, but in this strain more of the sugar is pushed into the amino-acid pathway. When it is only halfway down the road to proteinhood, it is hijacked by an alien enzyme (the genes for which have been transplanted from other microbes) and turned into catechol. Not only does this process get rid of the need for benzene; it gives off not a whiff of nitrous oxide, and it takes place at a mere 37 degrees C., or 98.6 F.—the temperature of your body, where *E. coli* customarily lives.

Another reason for using biological processes to make things is that (in many cases) what nature makes nature can unmake. Bacteria can make a plastic called polyhydroxybutyrate (PHB), which other bacteria can break down (see Nawrath et al.). Apart from being biodegradable, PHB is pretty much like polypropylene—though, bacterial work rules and pay scale being what they are, PHB at present costs more to make. To bring the cost down, scientists at Stanford have taken the bacterial genes responsible for PHB-making and sandwiched them into the genome of mustard plants. Synthesized in the chloroplasts, the plastic makes up one-fifth of the harvested plants' dry weight.

Letting things decompose is often less of a bother than recycling them, and in many cases the lighter use of energy tips the scale—both economic and ecological—in its favor. As a bonus, the methane released when a plastic like PHB breaks down can be collected and used as fuel.

Admittedly, PHB is not the world's most useful plastic. It is prone to break down when melted for molding and is brittle after it sets. But researchers at

Monsanto are trying to use plants to make other, better plastics. Ultimately, soybean fields may produce a crop of plastics as versatile as tofu.

345   **two million organic compounds:** Todd and Josephson; Corcoran.

Silk leaves materials scientists agape: See Corcoran; Vollrath; Lin et al.; Leary.

346 . **Renzo Piano:** See Dini.

"No one understands protein folding . . .": Corcoran. Other researchers are studying the nacre of abalone, in which calcium carbonate—the stuff that is always breaking against blackboards—is arranged in microscopic layers of overlapping hexagonal bricks, thus becoming thirty times stronger than lab-made calcium carbonate and as strong as the best synthetic ceramics (Leary). Learning at the knee of mother-of-pearl, these scientists have approximated its structure in carbon, boron, and aluminum and come up with substances 30 to 50 percent tougher than other ceramics made from the same ingredients. The new stuff would be tougher still if technology could make the layers as thin as mollusks make them. (For surveys of biomimensis, see Willis; Benyus.)

347   **solved by a swarm of nucleotides:** Adleman; Kolata. Even more impressive than its speed is the energy efficiency of a DNA computer, which could in principle perform 2×10^{19} operations per joule (the amount of energy needed to light a hundred-watt lightbulb for one second). Not only is this ten trillion times more efficient than any supercomputer; it is also astonishingly close to the theoretical limit of 34×10^{19} irreversible operations per joule set down by the second law of thermodynamics. Spurred by Adleman's experiment, scientists have proposed doing in four months of molecular lab work a computation that would take more operations than have been performed by all the computers ever built, combined. They have proposed floating a pound of DNA in a tank of liquid about a yard square, forming a memory bank roomier than those of all the computers ever built, combined.

Size is perhaps the main factor driving the interest in biological computers, whether the biological stuff in question is nucleic acid or protein. For decades, computers have been shrinking and concentrating like a good sauce, but silicon chips are starting to hit an economic limit: each halving of size now entails a quintupling of cost. Typically, the etched transistor that now serves as a logic gate—the basic unit of computing, an on-off switch holding one bit of information—is one-millionth of a meter across. But it still contains thousands of molecules. As you try to whittle it down to something closer to molecular size, the random motions of molecules start to mess things up. A protein molecule, on the other hand, can change from one state to another in a perfectly predictable way; after all, that is the scale it was designed to work on.

While materials scientists play with spiders and seashells, computer scientists are splashing about in the purple muck of certain salt marshes (Birge). The purple stuff is bacteriorhodopsin, a protein found in the membrane of *Halobacterium salinarium*. When struck by light, the protein changes its structure and pops a proton across the membrane, providing energy that keeps the cell going when oxygen is scarce. A cousin of the rhodopsin in your retina, which also changes shape in response to light, the protein drew the interest of Soviet scientists in the 1970s. Though still enmurked in secrecy, their work is thought to have produced microfiche film, optical data processors, and a processor for military radar. Now American scientists have taken up the scent. In theory, a computer using protein molecules as its logic gates could be one-fiftieth the size of a comparable box of silicon. Since smaller gate size usually translates into greater speed, a protein computer could be one thousand times faster than present models. Even an intermediate step—a hybrid of silicon and organic molecules, which is what one already finds in the liquid-crystal screens of laptops—could offer a hundredfold jump in speed.

As exciting as brute speed is the likely affinity of protein computers for the favored "architectures" of the future, such as three-dimensional memory, parallel processing, and neural networks. All of these mimic the workings of the brain. Not surprisingly, nature's processes seem easier to copy when something like nature's materials is used.

348   **work of John Todd and his associates:** See the books and articles authored and co-authored by Todd, especially Todd and Josephson.

"an immense bin of parts . . .": Odum, quoted in Todd and Josephson, pt. I.

350   **"simpler systems . . .":** Ibid., pt. I.

all five kingdoms: But see above, note to p. 12.

living machine is a "mesocosm": The same might be said of some of the man-made ecosystems that the Todds studied in Indonesia under the tutelage of Margaret Mead. In "The Practice of Stewardship," John Todd describes a farm in central Java where trees, grains, forage grasses, vegetables, livestock, and fish had coexisted for centuries to great mutual advantage. Rainwater falling on hills planted with useful trees ran through an aqueduct under the livestock sheds and the household latrine, where it gathered nutrients; over a small waterfall, which aerated it; deep below raised-bed crops, which it fed and irrigated but did not contaminate; into fishponds, where it received a second helping of nutrients; over rice paddies, where it dropped the nutrients off; and finally into a communal pond, from which sediments were periodically dredged and carted uphill to fertilize the trees.

Todd was impressed. Yet it seems obvious that the whole arrangement was "unnatural"—at least, there was little outward resemblance between it and the virgin forest that had preceded it. It sounds like a prime example of Geertz's "agricultural involution": the desperately ingenious overelaboration to which Javanese farmers were driven by the greed of their Portuguese overlords. But if there was little outward resemblance between the farm and the primeval forest, there was a deep inner resemblance, for both were complex ecosystems that honored the principle of flow. And since the claims once made by the Portuguese are now made, throughout the world, by clamoring populations on thinning turf, it may be that involution—or at least ingenuity—is just what we need.

351   **inferior in most ways to the hand scythe:** In an essay ("A Good Scythe") that is his closest approach to *Consumer Reports*, Wendell Berry makes the comparison between a Marugg hand scythe and a Sears "power scythe," which seems to have been a primitive form of weed-eater, using a rotary blade instead of a monofila-

ment string. A real weed-eater is safer than either of those tools, but otherwise his ratings stand.

352   **"Canst thou send forth lightnings . . .":** Job 38:35.
       "biophilia": See E. O. Wilson's book of that name; also Kellert and Wilson.
       Ezekiel's "Eden, the garden of God": See text, pp. 92–3, and notes.
       downstream from Eden, in Havilah: Genesis 2:11.

354   **vision of a "soft technology":** Besides the "industrial-ecology" and "biorealism" writers mentioned above, see, for example, Schumacher; Lovins; Callenbach, *Ecotopia*.
       water hyacinth in Africa: See above, p. 44, and note. The problem of invasive exotics is likewise slighted by Mollison's permaculture.

356   **Board of Technology invites:** Sclove, *Democracy and Technology*, chap. 12, sec. 7. In 1997, the first U.S. citizens' technology jury, on the Danish model, was held in Amherst, Massachusetts, under the auspices of the Loka Institute.
       populist watchdog was put to sleep: The lameness of the present system of technology review in the United States could be illustrated with dozens of examples. A recent case that stands out for sheer idiocy is that of rBST, or recombinant bovine somatotropin. A genetically engineered form of the growth hormone naturally present in cows, rBST is injected to increase milk production. Since it is chemically indistinguishable from the natural form, its manufacturer, Monsanto, argued that its presence in milk could have no effect on human health. After a review of various studies, the Food and Drug Administration agreed, and the drug was approved.

       In the classic manner of Western science, the panel focused on the narrow question of whether the substance itself could be harmful to people. As a result, a host of other questions were left by the wayside. Would boosting lactation above levels that already are biologically obscene cause the health of cows to be further strained, leading to even greater use of antibiotics? Would this lead to the birth of yet more strains of antibiotic-resistant bacteria, with unguessable consequences for human health? Would increased milk production entail more feed, and more protein in the feed? Would that, in turn, encourage the practice of mixing meat by-products into the feed—that is, of feeding chopped-up cows (and sheep) to other cows? Would that practice lead in America, as it seems to have done in Britain, to epidemics of prion-mediated diseases that affect the brain tissue: "mad-cow disease" in cattle, Creutzfeldt-Jakob disease and suchlike neurological horrors in people?

       These questions were scanted by the FDA, a fact that is not surprising when one considers that three top officials involved in the matter had ties to Monsanto. But the most important question of all was not even asked: why on earth do we need this stuff?

       America is a land flowing with milk. Wholesale prices are so low that small dairy farmers are unable to survive. In many parts of the country the traditional rural landscape is being destroyed, and increased production will increase the glut and hasten the destruction. The only conceivable social use of rBST is to fill the coffers of Monsanto.

Americans are not greatly interested in cheaper milk. Surveys show that they do not want rBST in the milk they drink, and will avoid it if given the choice (a choice which the FDA, in its labeling guidelines, has done its best to deny them). But the American people have not been asked.

It is not enough for society to choose among technologies: Judging technologies after they have emerged, you might say, is like shutting the barn door after the horses have bolted. Though government may be able to ban or restrain a new technology for a while, sooner or later it will break out—if not here, then somewhere else. The most famous exception to this rule (see Perrin) actually proves it: though guns were phased out of use in Japan, starting in the early seventeenth century, for fear they would destroy the noble sword culture of the samurai, their use by Western powers eventually forced Japan, in the late nineteenth century, to follow suit.

Society must help set the agenda of research, both basic and applied. By this I do not mean that scientists should labor under the eye of an all-powerful bureaucracy, as they have done in some communist countries. As I said, technology from which all wildness and self-will have been wrung will be fairly limp technology. But anyone who thinks scientists in the West just soar and swoop, free as eagles, wherever the currents of nature lead them is either birdbrained or ill-informed. What scientists mostly follow is money—not from venality, but because cyclotrons and chromatographs don't come cheap. And money has come from two main sources: government, whose interest has mainly been military, and business, whose interest is mercenary.

Set aside in-house corporate research, and think of the money corporations pump into universities. Though the volume they pump in is great, the influence they command is far greater, as is the benefit they suck out. A great deal of research that is done largely at public expense, in public universities (and even in public research facilities like the NIH), is later turned over to corporations for private profit. In recent years, the courting of corporate patronage by universities has become so toadying, and their activities have been so bespoke-tailored to corporate needs, that education suffers. Indeed, the case has been made that the main reason tuition fees keep rising is that they are subsidizing corporate-fitted research. (Admittedly, those corporate needs have some relation to "the market," which in turn has some relation to larger social needs—but the only people who think that last relation is an exact one are people who have been soaking so long in right-wing think tanks that their minds have gone pruney.) At present, the main counterweight to the pull of these highly specialized needs is the scientist's vague sense of "progress" or "the public good."

In the Netherlands: See Sclove, "Putting Science to Work"; *Democracy and Technology,* chap. 12, sec. 8. At bottom, the science shop is a marriage of convenience between two sets of needs. On the one side, ordinary people—and such clumps of ordinary people as public-interest and community groups—need help with scientific problems that bear on their daily lives and dearest goals. On the other side, students and faculty need research topics. In most of the world, these two sets of needs meet only furtively or by chance. Students and faculty dream up research topics that are of no earthly interest to anyone and of slight intellectual interest even to themselves; or they pick up the kindly hints of big government and big business. Community groups, meanwhile, unless they can gain the ear of the rare friendly scientist, must dig up the money to hire consultants and pay for studies.

In the Netherlands, this farce of cross-purposes is at last being untangled. To use a science shop, a group must show that it cannot afford to pay for the research it wants done, and that it will put the research to good use—that use not to include making money. (The means test is not hard and fast; some shops make Dutch-treat arrangements with groups that have some money, such as local governments or national environmental organizations.) Staff members and student interns screen questions. If a question is easily answered, they answer it or tell the questioner where the information can be found. If a question is tough, they pass it along to faculty and students in the relevant departments.

Workers have gotten help in judging the likely effects of new production methods. Social workers have deepened their understanding of troubled teenagers. Environmentalists have learned the exact ingredients of industrial stews. One shop tested the market appeal of a planned women's radio station; another did a study for Amnesty International to assess whether graphic photographs of torture victims would help or hurt its fund-raising; a third dealt with the problem of air pollution in the black-lunged city of Dorog, Hungary.

The goods purveyed by the Dutch shops are mostly facts, but there is no reason why tools and technologies could not also be shaped in response to people's needs—needs that may get lost in the thrashings of the "free market." For instance, a study of the effects of a new production method might lead to the development of a better one.

The costs to the universities are minimal and the benefits great. Scientists, both seasoned and green, are prompted to think about the effects of their work in the real world. Often, their work takes off in whole new directions. As a bonus, publicity of the shops' efforts brings greater public support (and funds) for higher education.

Though America already has a set of institutions chartered, in part, to answer the needs of ordinary people for scientific knowledge and technologies—to wit, the land-grant universities and the related Cooperative Extension system of agricultural agents—the heavy tread of business and military funding has trampled that charter in the dust. (See Wendell Berry's critique in The Unsettling, chap. 8.) The science-shop concept could restore it to its proper place, and then some.

In the meantime, various enterprises—some of them fairly informal, less shops than pushcarts—are bringing scientists into contact with ordinary people's concerns: for instance, Chicago's Center for Neighborhood Technology, the School for Workers of the University of Wisconsin, and the involvement of scientists from the Brookhaven National Laboratory in reclaiming Brooklyn's badly fouled Gowanus Canal. In western Massachusetts, the Institute for Science and Interdisciplinary Studies (ISIS) is harnessing some of the plentiful academic horsepower of that region to address a wide range of issues, among them the economic and ecological problems of farmers in the fertile but suburbanizing Connecticut River Valley; the plight of people with multiple chemical sensitivity in a synthetic world; and the struggles of the Amazonian Secoya people against the poisoning of their homeland by oil drillers. Perhaps the most remarkable thing ISIS has done so far is to mediate between community activists and the engineers and officers in charge of the cleanup of various toxic sites at Westover Air Force Base. A flat, barren scene that might easily have become as soaked in bile and spleen as it already is in jet fuel and wing de-icer, Westover has instead become the locus of fruitful, if sometimes testy, cooperation. ISIS has brought in fresh eyes and fresh ideas. Geology classes at local colleges are mapping the base's hydrology; an expert on biological remediation (loosely, the use of bacteria for cleanup) has suggested improvements on the methods of the base's engineers. (An account of ISIS's work will appear in the forthcoming book by Fortun and Bernstein.)

   While the dispensation of public funds can help turn science and technology toward humanly helpful (or at least nonharmful) goals, too often the "democratic process" dilutes and distorts people's real needs. By contrast, science shops and their shopless like encourage scientists to slog about in the muck of real human concerns.

357   **moratorium on the deployment of new biotechnologies:** Wes Jackson is quoted in my "Back to Eden." A much narrower moratorium—on research into human cloning—was proposed by President Clinton in 1997, in response to the creation of the sheep known as Dolly.

   Let every seventh year: Waskow, *Down-to-Earth Judaism*, chap. xix. Back in 1975, Waskow proposed a modern adaptation of the Jubilee, with radical economic implications (see *Godwrestling*, chap. XI). More recently, a similar proposal by the Dutch rabbi Awraham Soetendorp (voiced as well by the Dutch cleric Father Dominik Grania) has received serious attention from the Global Forum, the International Council of Churches, and the Vatican (Rosenthal). One of its key points, the write-off by creditor nations of developing nations' debts, could help the planet as well as the poor by slowing the mad scramble to exploit natural resources and export cash crops.

   On that day we meddle with nature as little as we can: See A. J. Heschel.

358   **while the need to build the Tower is acknowledged:** I am improvising here on remarks by Arthur Green at the Jewish Community of Amherst, February 25, 1995.

359   **rate of invention seems to have been as high in the Old Stone Age:** Shepard, *The Tender Carnivore.*

27. THE URBAN ANIMAL

361   *tsimtsum:* See Scholem, *Major Trends,* lecture 7; *On the Kabbalah,* chap. 3; *On the Mystical Shape of the Godhead,* lecture 2.

362   **"In an abstract sense, there is a global 'limit to growth' . . .":** Commoner, *Making Peace,* pp. 147–8.

"megazoos": Dr. William Conway, president of the New York Botanical Garden, quoted in Stevens, "Zoos Find a New Role." See also Sullivan and Shaffer.

363   *r* **and K:** See text, p. 329, and note. A less biological, more conventional explanation goes like this: In agricultural and early industrial societies, an extra pair of hands outweighs the cost of an extra mouth, and is the best hedge against destitution in old age; in mature industrial societies, this is no longer true.

in India, high-yield grains are reckoned to have spared: Stevens, "Feeding."

Sometimes nature heals itself: Consider New England, and in particular a part I know well, western Massachusetts. The frenzy of logging and farming that picked the hills clean within two centuries of their first sighting has subsided. The hills are woolly again, like sheep two months after shearing, and the houses and cars lodged in them seem of no more consequence than fleas. Climb Mount Holyoke and see the Connecticut Valley spread out before you like an endless fleece. You will see more forest than Timothy Dwight saw in 1796, when he stood on the same peak and admired "the lofty forests, wildly contrasted with the rich scene of cultivation"—a scene that fully satisfied his "relish for landscape." Fields (you will need a telescope to pick out corn, tobacco, strawberries, asparagus) skirt the river, and patches of pasture splotch the Berkshire foothills. You can see the spires of Northampton and Holyoke, the towers of Springfield, and on a clear day the gray glint of Hartford. What you cannot see are the dozens of smaller towns, for they are drowned in the flood of green. (In addition to Dwight, see the essay by J. B. Jackson.)

From the ground, admittedly, things can look different. There are subdivisions, malls, and commercial strips that might be interchanged with their counterparts in New Jersey or Ohio without causing much more upset than a brief flurry of jammed keys, puzzled spouses, and exchanges of franchise papers. Yet the sea of green is no illusion. Swimming in it, we take it for granted, like fish. Chlorophyll is our element.

The forest, like a giant turning in its sleep, has stretched out and recovered its vast bed. It is not the "forest primeval"—but then, the forest the settlers found was not really primeval either. One might even claim that the new forest is more "natural" than the old, which was heavily managed by Indian torches. (Cf. McKibben, *Hope.*)

A similar *tsimtsum* is taking place in many other parts of the northeastern United States, whose humid temperate climate is among the most forgiving on earth. In the developing world, there are as yet few regions where nature has been cut such large-scale slack. There are, however, narrower places where nature has managed to restore itself with no human help other than the cessation of human harm. When the Gogol Valley of Papua New Guinea was clearcut in the 1970s, the loss of this great lowland rain forest was accounted a great catastrophe. A year and a half later, a visiting biologist (Hamilton) could hardly make his way through the tangled regrowth. By the end of ten years, only 7 percent of the trees were small pioneers; 23 percent belonged to large pioneer species, and 70 percent to species typical of the mature forest before it was felled.

Unfortunately, rain forests often fail to bounce back so bouncily, especially if their soil has been turned to farming or grazing. Many dry or mesic (medium dry) forests will not bounce back at all—not for centuries, anyway—and the same goes for many kinds of grassland and wetland. Even with such success stories as the Gogol Valley and the American Northeast, generations may pass before we know just how successful they are. Already, in this crescendo of life, sour notes and missing notes can be detected. The woods of the Northeast are lovely, dark, and deep, but some of its former denizens have passed into the darker realm of extinction. The exiled wolf, wolverine, cougar, lynx, and caribou have so far had only modest success in reclaiming their realms. The chestnut languishes under the spell of an alien blight that condemns it to sprout endlessly from its mighty stumps. Sugar maples have come back, but their niche is being usurped by red and Norway maples. Hemlocks have come back but are now being smothered by the woolly adelgid. And the forest's return is being reversed, on much private land in Maine, New Hampshire, and Vermont, by large-scale clearcutting, and by rampant development in the Adirondacks. (See Kunstler.)

Another case of a natural recovery that, while cause enough for breaking out the bubbly, may on closer inspection deserve a split rather than a magnum is that of the prairie potholes of midcontinental North America (mainly the Dakotas, Montana, and Saskatchewan). Once the great maternity ward of the continent's ducks, most of the potholes were—like most other wetlands—drained during the course of white settlement. The great fluffy eiderdowns of southward-sweeping ducks that blanketed the skies of America every fall were reduced to shreds and tatters. Things changed in 1985 with the creation of the Conservation Reserve Program. Designed to fight erosion on fallow land, the program swept along on its coattails more than ten million acres of farmland in the pothole region. The method of "restoration" was simple: just break some drainage tiles or block a ditch. In other words, end human interference and count on nature to do the rest.

To the duck fancier's eye and ear, the program has been a spectacular success. The rapture of ecologists, though, has been modified. Many potholes had been dry for so long, the native seeds biding their time in the soil had given up the ghost and could no longer germinate. Instead, exotic species rushed in: brome and alfalfa in place of big bluestem and cord grass, scattered tufts of barnyard grass in place of the former dense sedge. Wrens, yellowthroats, bitterns, rails, and other birds needing dense cover saw no percentage in a sedgeless pothole, and did not come back. (See Galatowitsch and van der Valk, pp. 3–9; Holloway.)

As these instances show, we cannot hope to bring about a real *tsimtsum* just by getting out of nature's face. This is a good start, but it is only a start. In many cases, in many places, the harm we have done can be undone only by ourselves.

This is not exactly a new idea. Reforestation has been practiced since ancient times. Only in the last decade or so, however, have the varied skills and experiences of foresters, engineers, hydrologists, soil scientists, ecologists, and others begun to coalesce in a single discipline known (not so singly) as environmental restoration or restoration ecology. (See Berger, both references; Holloway.)

Prairie potholes are just beginning to benefit from this new science. Instead of just letting the water back in, restorers are now trying such methods as seeding with native species, inoculating with fresh pothole soil, spreading wild hay, burning, maintaining a grassland buffer, eradicating weedy species, and even screening inlets and outlets to keep carp out (Galatowitsch and Valk, pp. 151–3). Restorers of other, more common kinds of ecosystems have bigger, more sophisticated bags of tricks at their disposal. For decades, German bioengineers have been refining the art of restoring the banks of lakes, rivers, and streams. Often erosion, strong currents or waves, and other nasty features of degraded or channelized bodies of water keep vegetation from getting a toehold. The Germans deal with this problem by anchoring rolls of coir fiber in the subsoil. These wieners of shredded coconut husk, either preplanted or postplanted with wetland species, calm the water and retain sediment. Behind them, carpets of native plants are unrolled on a sloping bank, or rafts of plants are floated in a would-be wetland. In either case, the matrix is coir fiber. The coir biodegrades; the plants grow up, attracting other native plants, animals, and microbes and forming something like a native wetland community. (See W. Goldsmith.)

In some of these cases, the land or water to be restored is in a state that the innocent eye would find perfectly nice; in others, even the innocent eye would turn away in disgust. The deserts formed worldwide by human action may be returned by human action to their former sparse but vital greenness. Desertification seals the soil like asphalt; a process called imprinting opens it up again. Drawn behind a seed-scattering tractor, a giant steel wheel makes diamond-shaped indentations in the soil, rather like a pastry wheel crimping the edge of a pie crust. When rain comes—whenever it does come—water, litter, and topsoil, instead of running off and away, run into these hollows and help the seeds germinate. As the seedlings grow, they are sheltered in their foxholes from drying sun and wind. Finally, mature communities of plants, animals, and microbes take over the work of keeping the soil textured, porous, and hospitable to themselves.

Mining sites can look like deserts even in wet climates. For many years they have been targets of restoration efforts, mostly token or cosmetic. Recently the efforts have gotten more serious. In 1985, the American Cyanamid Company was ordered to "mitigate" the damage it had done to a 3.8-acre forested wetland near Tampa, Florida, in the course of mining for phosphorus. The forest had been wholly obliterated; now it was to be wholly restored. Guided by nearby, undisturbed forest, the botanist in charge saw that if he could get the topography right, the hydrology would follow; if the hydrology was right, the ecology would follow. Nine years later, the once-barren, sore-thumb site had melted almost without a trace into the surrounding forest.

A less hospitable matrix surrounds the quondam salt marsh of the Arthur Kill. One of the busiest waterways in the world, the Kill is glowered over by the smokestacks of northeastern New Jersey, whose chemical plants and oil refineries it serves. On the Staten Island side, the remnant of a pre-Columbian salt marsh struggled gamely on, elbowing its way like a subway commuter through each new layer of spilled oil and other gunk, until 1990: in that year, over half a million gallons of heating oil burst from an Exxon pipeline, coating several hundred acres and apparently killing the marsh once and for all. Though it drew less notice than Exxon's other indiscretion in the photogenic Gulf of Alaska, it did draw a $15 million settlement, a part of which is funding the efforts of the New York City parks department to bring the Kill back to life.

Many experts said they shouldn't bother. Concentrations of oil were higher than in a Kuwaiti oil field. Even conservationists wanted to write the blighted area off, urging that the money be used instead to buy other land for a wading-bird preserve. But the parks department had a new strategy and insisted on trying it out.

Over the course of three years, some four hundred volunteers (ghetto kids, retirees, an off-track betting manager taking a week off) collected more than six million seeds of cordgrass—*Spartina alterniflora,* a plucky native of eastern tidal marshes—sowed them in nurseries, and transplanted the shoots to the Kill.

Three years later, the cordgrass, fortified with ten kinds of fertilizer, was waving gaily, dense and serried, in almost the whole of the six-acre area in which it had so far been planted. Nor was it merely decorative. It was drawing oxygen through its stems into the sediment, jolting certain bacteria into a frenzy of oil-eating that had already cut saturation by 70 percent. It was also the keystone of a rebuilt trophic pyramid: microbes, mussels, fiddler crabs, blue crabs, mummi-chogs, killifish, muskrats, ducks, geese, herons, snowy egrets, and even a pair of peregrine falcons had all returned to the marsh.

Impressive as this recovery is, it should be seen as a steep rise following a steep fall (the Exxon spill) in a gentler but much larger upward trend. The health of the Arthur Kill, like that of the Hudson and of many other waterways across the country, has improved impressively since the passage of the Clean Water Act in 1972. In 1967, scientists had lowered cages of clams, blue crabs, and killifish into the Kill; when they drew them up a few days later, not one was left alive. Most of the improvement since then can be chalked up to nature's self-repair, once the flow of human insults had been curtailed. (See D. Martin.)

Restoration ecology has some small successes under its belt, but its ambitions are ungirded. Probably the biggest restoration project now under way anywhere is the one that would return to the Everglades—with the widest and shallowest river on earth and one of earth's richest and strangest ecosystems—some semblance of its former glory. The main action being taken is the undoing of former action: unchanneling streams, breaching levies, and generally dismantling some of the structures that, over the past century, have showered largesse on sugar barons and real-estate speculators while leaving nature athirst. But other actions are being taken, too, such as hunting down and destroying some of the exotic plants that are choking out natives.

Restoration is still more art than science. It may always be. How does one define success? What is the prior state to which the land or water is supposed to be restored? (Is it the state that existed before European settlement? That concept is tricky enough in America and a good deal trickier in Europe. Or is it the state that existed before human settlement? But how does that work in Africa?) Where do you find an extant example of such an ecosystem? Even if its final goal is clear, the success of a restoration project may still be up in the air when its designers are mouldering underground. Even the simplest ecosystem has to survive long-term wriggles of climate and all manner of cyclical booms and busts. More complex ecosystems may have reached their mature forms only by grace of a series of climate changes, migrations, fires, and other quirks that happened arbitrarily over the course of millennia and may never happen again in the same sequence (Cairns; cf. Pimm, *The Balance*).

Some ecologists argue that the idea of restoration—at least in the form of "mitigation"—is becoming a license for the trashing of natural ecosystems. To them, this is a bit like chopping off your fingers on the assumption that modern medicine can sew them back on or make new ones. At best, they say, restoration diverts desperately scarce resources from the more vital—and more reliable—work of saving the wild places we still have.

In fact, restoration in one part of the world—whether spontaneous or helped along by human efforts—often seems to depend on devastation in another part. Return for a moment to the case of western New England. When the forest began its comeback, halfway through the nineteenth century, this was seen not as a seal of economic maturity but as an omen of decline. The trees came back because the fields were abandoned. As the steel plow unzipped the plush flesh of the West and the Erie Canal gave its fruits passage, the bony flanks of New England no longer seemed worth cosseting. Farmers went west or left for the mill towns and cities.

Western Massachusetts has become a Potemkin paradise. To all appearances, it is a place where humans and nature have made peace. But the peace they have made is like the peace lately enjoyed by the superpowers while their surrogates in the Third World made bloody war. In this case, the war goes on not only in the Third World but in every part of the world where food is grown for market, oil is drilled, or bauxite is mined.

Maybe the idea of turning back the waves of human-led change really is a fatuous one, worthy only of Canute's courtiers. Maybe the best we can hope is that, as the waves of Western-style agriculture and industry sweep across the globe, there may be lulls or troughs behind them in which nature can recover; there may even be another wave, a wave of healing and right living. Maybe that is the way things work and have always worked, in human and even prehuman history. The breaking of the prairie let the hills of Massachusetts recover; now the breaking of the Australian outback and the Argentine pampas may let the prairie recover. The logging of the Pacific Northwest let the forests of Maine and New York recover; now the logging of Siberia and Malaysia may let the Northwest recover. And so on.

But can we afford to wait for the trough, or for the next wave? There are many places on earth where the damage now being done to nature is unlikely to be righted even with the best human help. Rather than keep moving the zone of exploitation outward, perhaps it would be wiser to do our exploiting more wisely. We might, for example, make sure that a good part of New England's farmland stays farmland—that we keep a working landscape, but one that works side by side with nature. We might find ways to keep some of the prairie for raising food, whether bison or some set of crops compatible with the ecology of the place. In so doing, we might take some of the pressure off the places that are now feeling the full force of the waves of change. If those waves cannot be stopped, at least their force may be broken in some places, transformed in others.

For all these qualms, restoration is something we must do. And we must do it now, while natural ecosystems are still around to draw models and materials from.

In Genesis we are commanded to "work and protect" the Garden: to work the earth, and to protect it against our work. But despite our best efforts at protection, the earth will be damaged by our work. So a third commandment comes into play: that of *tikkun olam*, the repairing of the world. "Be careful that you do not damage or destroy my world," Adam is told in the midrash, "for after you there will be no one to repair it"(Kohelet Rabbah 7. 13). Damage it we do, however, and the task of repairing the damage falls, by default, to us.

364   ARE CITIES UNNATURAL: My thinking about cities has been shaped by three books above all: Mumford's *The City in History*, Spirn's *The Granite Garden*, and Jacobs's *The Death and Life of Great American Cities*.
       cities have been around for six or seven thousand years: The bold proposal of Jane Jacobs (*The Economy of Cities*, chap. 1) that cities may have preceded farming—arising first as trading posts—has not won many converts, but it has drawn attention to the city's central role in the human economy.
       marriage between the two previous "stages" of human living: Mumford, *The City*, chap. 1, sec. 8.

365   first city is founded by a wanderer, Cain: Genesis 4:17. On the other hand, Cain was a farmer, Abel a herdsman, and their story belongs in the line of stories, common from Sumerian times to the present, about the conflict between those groups.
       "urban animal": Aristotle *Politics* 1253^{a}3. To be sure, Aristotle meant by "city" both the walled city and its associated countryside: in other words, the city-state. This may be a healthier unit, both ecologically and socially, than the modern nation-state.
       "preadaptation": See Gould, "The Problem of Perfection."

367   in America in the 1940s: Jackson, *New Roots*, chap. 7.
       A young man can have few more paradisiacal experiences: Updike (p. 117) writes of like sensations:

> There, the seasons spoke less in the flora of the hard-working parks than in the costumes of the human fauna . . . of the young women who rose up from the surfaces of stone as tirelessly as flowers out of mud. New York was so very sexy, in memory: the indoorness of it all, amid circumambient peril, and the odd good health imposed upon everyone by the necessity of hiking great distances in the search for taxis. . . . On this island of primitive living copulation occurred as casually as among Polynesians, while Scarlatti pealed from the stereo and the garbage truck whined its early-morning song two blocks away.

       "the cave paintings of Cro-Magnon man . . .": Reference mislaid.

368   the big city is where young men and women go to prove themselves: A few still head for wilderness; but if the challenge and the danger there are real enough in

a physical sense, they are not real in the sense of being necessary. Daring the wilderness is a dangerous game, but as far as society is concerned it is only a game.

Young people go to the city to "find themselves," and with good reason. From Mesopotamian times to the present, the ideal shape of the city has commonly been that of a circle or square divided into four quarters—the mandala shape of the world and of the self. Like the garden, the city is a place where wholeness is forged from disparity. But there is a crucial difference.

If the garden really does join Mountain and Tower, it does so by excluding what is wild and unruly in nature and culture alike. The city's proper claim is less ambitious—"Nothing human is alien to me"—but this means it has more room for what is wild and messy in the human realm.

In dreams, the city's "quarters" often seem to represent different parts of the self, with the rougher districts hinting at dark, primitive needs. So wildness finds a place—human wildness, mainly, which is a part of nature's wildness. By limiting its scope to the human, the city not only offers a fuller image of the human self; paradoxically, it may offer a deeper, truer, less pedicured image of the world as a whole.

In a sense, then, the Mesopotamians were right to make the city the center and image of the world. For certain human purposes, it is. And for civilized human beings—even those who do not actually live in a city—it may be the best mirror of the self.

369   **making weights, painting trade names:** From a comic strip by Ben Katchor.
"the contrast between town and country . . .": Engels quoted in R. Williams, p. 304.

370   **cities began to have walls:** See Mumford, *The City;* H. Carter; and Kostoff, chap. 1, who notes the etymological link of "town" to "hedge" or "enclosure," a link matched in other languages (like the garden, the city is verbally engirt). The marking of a city's limits was often a sacred act, accomplished by plowing a furrow (Indian and Etruscan traditions), spreading wheat and stationing elephants (Antioch), or perhaps tracing a circle in ash (Baghdad). In Rome, goatskin-clad priests ran a circuit of the walls each February, striking women to make them fruitful.

371   **instead they were parts of the king's domain:** True, a strong state has at times protected wilderness, as in the case of English or French royal forests and American national parks. More often, it has systematized plunder. Before the Amazon rain forest was brought under the control of the governments of Brazil and its neighbors, settlers scratched at its edges like lice. Only when governments built roads, gave tax breaks to ranchers, and used DDT and quinine to suppress the malaria parasite and troops to suppress the Indians did the rain forest start to shrink at anything like its present rate. On the resurgence of the city-state in theory and in practice, see Kemmis; Morris.

372   **proponents of the "ecocity":** For example, Register; Hough; Berg et al.; Calthorpe; Todd and Todd; many of the writers in Walter et al.
In the drawings of one ecocity theorist: Register.

no visible nonhuman life except for cats and pigeons: Indeed, some of the wildlife we prize in cities is there only because other wildlife is not: for instance, the peregrine falcon, which not only finds ideal cliff-face habitat on skyscrapers, but is safe from chick predation by great horned owls. See Garrott et al.

373   **Biomass accounts for one part in ten billion:** E. O. Wilson, *The Diversity*, p. 35. Even if we confine ourselves to what we ordinarily mean by nature—and what I ordinarily mean by that word in this book—namely, the living and nonliving components of the biosphere, or that part of the earth's crust, soil, water, and air in which life resides—excluding human material culture—even so, the living mass is far less than the nonliving.

"If we want to find sustainable landscapes . . .": Hough, "Design with City Nature."

flora of that "place": *Acer platanoides, Robinia pseudoacacia, Populus tremuloides, Ailanthus altissima.* See Spirn, p. 27.

374   **deep nature of cities is denied:** What about the idea (see L. Brown) that cities should be "ecologically transparent"—that they should be designed in such a way as to make apparent the natural processes that go on within them? At first blush, this idea would seem to be right up my alley. If the city's cardinal sin is to forget the fact of its dependence on wildness, then surely the first step toward repentance is to have that fact brought before our eyes.

We have to watch our step, though. Part of the joy of city life is precisely the play of illusion—the sense of a magical, man-made world that follows its own rules. In a great city, anything is possible. You can nest like an eagle a thousand feet in the air. You can work or play all night and sleep all day, making light of sun and moon. You can gastronomically, if not geographically, have lunch in Paris, supper in Rome, and a midnight snack in Beijing. You can move in odd directions and at strange speeds through spaces of light and sound unmatched in dreams.

For the designer or planner, the trick is to remind people of their dependence on wildness without rubbing their noses in it. This can be done by playing cheekily with natural forms, rather than aping them. By pointing to the facts of a city's natural setting—its climate, its topography, its soils and bedrock, its native flora and fauna, its orientation to sun, moon, and stars—but not insisting that these determine everything that goes on within the city limits. By using the limits set by nature not as a cage, but as a trellis or arbor about which the tendrils of art can twine.

In much of this chapter, one should perhaps read the word "city" as "big city." It is the big city that is really the Tower, the seat and playground of civilization. Of course there are, or ought to be, Mountains and Towers on every scale. What is a big city for people in lower New York State, and what is a big city for people in the Gobi Desert, are two very different things. But much of what I have said about cities applies best to really big cities. In particular, as one thinks about smaller and smaller cities, and cities closer and closer to wild places, the idea of the ecocity makes more and more sense. If I were asked to map my ideal world, I would probably draw a dozen big cities on every continent in which culture would be allowed (within certain bounds) to run wild. With great sweeps of my green pencil, I would draw vast tracts of wilderness. In between, an interwoven skein of farmland and wild places would be spangled with small cities of, say, ten thousand to a hundred thousand persons. And many of these small cities would be, in varying degrees, ecocities.

some cities have solved these problems: Besides the watery examples mentioned in the text, a few more might be noted.

Though tricks for maintaining a livable climate in a city are surely as old as the oldest cities, the first exposition of them we know is the Hippocratic treatise "Of Air, Water, and Places," dated to the fourth century B.C. This tells how cities should be sited to avoid the unhealthy effects of marshy ground; how pure water should be secured, from a distance if need be; how streets should be widened to allow the circulation of air; how buildings and streets should be aligned to catch the sun in winter and the breeze in summer.

As early as the fifth century B.C., new towns and neighborhoods were routinely planned with parallel streets that gave all homes a southern exposure. Generally, the main rooms of each house looked south across a portico into an open courtyard. As Xenophon has Socrates explain, "The sun penetrates the portico in winter, while in summer the path of the sun is right over our heads and above the roof so that there is shade." (*Memorabilia* 3. 8. See Butti and Perlin, chap. 1; Mumford, *The City*, chap. 5.)

In Hellenistic and Roman times, these ideas were applied to the planning of colonies, garrison towns, and other new foundations. They were systematized in the works of Vitruvius, the great Latin architectural writer of the first century A.D. Though a penchant for grids and other regularities made them less supple and less useful than they might have been, many of the basic notions were sound. Whether gleaned from Vitruvius or just from common sense, they have been applied fairly often, if irregularly, over the last two millennia. In modern times, though, they have generally been ignored both by developers looking for a quick profit and by architects and engineers eager to show off their virtuosity with grand expanses of glass and ductwork.

When things get bad enough, however, common sense has a way of reasserting itself, and that is what seems to have happened in Stuttgart (see Spirn). An industrial city that sits at the bottom of an inland valley, Stuttgart also spent an average of 247 days a year sitting under inversion layers. With few strong winds to stir things up, the air was often unfit to breathe. In summer it was often unfit to touch, being unbearably warm. In the early 1950s, a phalanx of planners, architects, landscape architects, and climatologists attacked the problem. Their solutions have been revolutionary—a breath of fresh air that is also a blast from the past.

Using smoke signals at first, then infrared aerial photos, they charted the pattern of air flow in and around town. What little fresh air Stuttgart was getting on the doldrum days was flowing from wooded hillsides, down ravines, and into the city. The bad news was that those same hillsides were quickly turning into suburbs. And this news was even worse than it sounded, for the vital flow was so feeble that even a few houses could choke it off.

The first step, then, was to restrict development of the hillsides. The next was to make them part of a network of open space starting in the hinterland and converging on the city center. Each channel would be at least one hundred meters wide, and would be kept in grass and trees. As hot, dirty air rose from the downtown, cool, clean air would stream down from the hills to take its place.

Of course, this stream (or, more often, trickle) of fresh air would not by itself make Stuttgart a pleasant place to live and breathe. There was too much hot, dirty air to start with. The city now addressed this underlying—or, rather, overlying—problem by taking its incinerators and power plants, moving them downwind of the city, and fitting them out with smokestacks at least 160 meters tall: tall enough to break through the top of the inversion layer. (Many private factories have since followed suit.) In congested, airless parts of the downtown, motor vehicles and the burning of oil and coal were outlawed. Now stoves in these neighborhoods run on natural gas; and heat comes from water warmed by waste heat from municipal incinerators and electric generators.

Shady parks have been built, some aligned with air-flow corridors. Streets have been widened and planted with large trees. Roof gardens and wet roofs, which hold a few inches of water, reduce the absorption of heat. So do parking lots paved with turf blocks instead of asphalt. (Both wet roofs and turf blocks also help to curb the problem of stormwater runoff.) Plans for new buildings are vetted for their likely effects on the surrounding air.

By all accounts, Stuttgarters now breathe easier and sweat less than they did four decades ago. Their success has inspired like efforts not only in Germany, but in places as far afield as Dayton, Ohio. Even without imitators, a city that helps itself ecologically can help the outside world as well. By reducing the use of fossil fuels for heating and cooling, Stuttgart not only sweetens its own air, but lessens its complicity in acid rain, global warming, and other ills. By holding stormwater in wet roofs and allowing its absorption by parks and turf, the city lessens the overflow of sewage into the Neckar River and thence into the Rhine.

Of all the measures Stuttgart has taken, maybe the most obvious is the planting of trees. But while city fathers and mothers everywhere pay lip service to the virtues of trees—their ability to cool the air, clean the air and water, restrain flooding, and mute noise—few really think them worth the trouble and cost of a vigorous planting-and-maintenance program. They ought to look at a recent study of metropolitan Chicago, which found that planting and maintaining ninety-five thousand trees, at a cost of $21 million, would yield benefits worth $59 million, for a net gain of $38 million over the course of thirty years (W. K. Stevens, "Money Grow on Trees?").

In most departments, the greatest benefits come when a tree is mature. Unfortunately, many trees in Chicago, as in most cities, are invasive exotics that spring up in disturbed soil, grow fast, and die young. In the study, the prescribed species was the green ash, a common Chicagoan, but in practice other species would be planted as well. Norway maple and poplar share the ash's considerate habit of sprouting its leaves late in spring and dropping them early in fall, so that the sun gets in when and only when it is wanted. By planting two deciduous trees of this sort and an evergreen to block winter winds, the average urban homeowner can cut his fuel bill by 10 percent.

In hotter climates the savings can be even greater. In one simulation, three trees slashed air-conditioning costs for a house in Sacramento by 53 percent, with much of the cooling effect coming from transpiration (in effect, the sweating of leaves, which cools not only the tree but the surrounding air). But the savings, though a good carrot for the homeowner, are dwarfed by the benefits that accrue to the city and the planet at large. To maximize these, "peak-load landscaping" supplies maximum shade a few hours before local utilities feel the maximum demand for air-conditioning watts—the lag allowing for the time it takes heat absorbed by exterior walls to find its way indoors. Thus, in South Florida, where peak demand typically hits between 5:00 and 6:00 p.m. in early August, trees would be placed so as to cast their best shade between 3:00 and 4:00 (Macpherson).

Stuttgart, I said, has banned the internal-combustion engine from parts of its downtown. Many other European cities have done the same—not only to clear the air, but to make the streets safer, quieter, and withal livelier. Amsterdam and Copenhagen, two of the cities that have gone furthest in this direction, are widely conceded to be two of the pleasantest cities on earth, and it is unlikely that this is wholly coincidental.

If the city is the place where culture and technology should be given free rein, by what right is this particular technology kicked out? For one thing, culture and technology should get free rein only within the limits nature sets, and cars—by virtue of the resources they use, the pollutants they put out, and the shapeless shape they give the city—swerve dangerously beyond those limits. For another thing, cars are a fossilizing of technology, not a free play. As Mumford has said, they are paleotechnic machines in futurist disguise.

Few American cities, of course, have dared to bar cars from more than a few streets or demistreets. The idea is dismissed as alien to our culture. Interestingly, Danes made the same objection when the notion of banning cars from much of downtown Copenhagen was first proposed in the 1960s (see Walljasper). As the urban designer Jan Gehl later recalled, the headlines were scathing: " 'We are Danes, not Italians. A public space will never be used. It is contrary to Scandinavian tradition. Danes will never leave their cars and their homes. The city is bound to die out if you take any cars out.' . . . However, it became a success right away." Public transit has been expanded, as have bikeways: half of all downtown commuters now pedal their way to work. Despite removing parking spots—one by one and two by two, like seats in a game of musical chairs—the city has seen a business-and-shopping boom. Crowded streets keep crime infrequent. Rundown neighborhoods have been revived. There has been a threefold rise in leisure use of the downtown, which is as busy on weekends and evenings as on workdays. "A good city is like a good party," Gehl observed. "People don't want to leave early."

Other Danish cities are throwing their own parties. In Aalborg, at the tip of the Jutland Peninsula, a public brainstorming process led to a plan that would ban private cars from the downtown, expand bikeways and public transit, clean up a local fjord, and wreathe it with beaches. Even low-rent, high-rise districts full of immigrants, old people, drug addicts, and the mentally ill are transforming themselves. "People don't want to live outside the city anymore," said the city architect. "It's now hard to sell houses out there."

Like the Danes, the Dutch have a pastoral self-image and a fondness for suburban life. Having made their own land, they like to get a private piece of it. Yet the Amsterdam experiment has been equally successful. The Dutch have added another twist: the *woonerf,* or "residential yard," a compromise between the foot and the wheel that clearly favors the foot (see Spirn). In a *woonerf,* there are no sidewalks. Cars and people share the same pavement. A speed limit in the neighborhood of ten miles an hour is imposed not only by law, but by a slalom course of tree islands, irregular house fronts, raised planters, benches, speed bumps, and other obstacles. Pioneered in Delft, *woonerven* have spread from the Netherlands (where there are now over a thousand in several hundred cities) to Germany (where they are called *Wohnberichten*), and even to the United States. Thanks to a *woonerf,* Appleton Street in Boston's South End is no longer a shortcut to the Southeast Expressway, but a place where children play and grown-ups chat. The *woonerf* has adapted to downtown, nonresidential uses, too. In Springfield, Massachusetts, the plaza about which the courthouse, symphony hall, and other public buildings sit is a *woonerf.* It is friendly and calm, but not too calm—as it might have been if autos were banned outright in the struggling center of a city addicted to wheels.

Another Dutch treat: Amsterdam has made bike travel more convenient by leaving bikes around for anyone to use. A fleet of specially marked public bicycles is scattered about the city. You can grab one, ride it to where you are going, and leave it there for the next comer. Portland, Oregon, now has its own fleet of yellow bikes, brightly spray-painted, unmistakable, so far unstolen.

Hopeful signs can be found even in Athens, long the most mephitic of European capitals (see Simons; Kongshaug). Having doubled in size in three decades, Athens now holds almost four million of Greece's ten million people. In Third World style, perhaps a third of its buildings went up illegally. Over 97 percent of its land is built or paved over—more than any other European capital. Much of the bottomland that once was Attica is now a tangle of television antennas and snarled cars; hemmed in by mythic crags, the *nephos,* or smog, can quickly turn lethal, as it did during a heat wave in 1987, when two thousand Athenians died. Yet the catharsis of cars has begun. First the Plaka, the ancient, seedy district at the foot of the Acropolis, was renovated and closed to traffic. It soon became the favored place for tourists to sleep, stroll, and eat—unsurprisingly, since it was almost the only place where they could breathe. For the rest of the city, emergency measures were imposed when the *nephos* was at its worst: on some days of the week, only cars with even-numbered license plates were allowed to drive, on others only odd. In the summer of 1995, however, the car-free zone of the Plaka was expanded into the key downtown triangle whose apexes are Sindagma Square, Omonia Square, and the Agora. Shortly, two other neighborhoods joined the program. Work began on an electric-trolley system (Athens's dirt being so densely packed with archeological strata that a major subway was out of the question). The dream of the city's late mayor Antonis Tritsis—to join its major historical sites, from Plato's Academy to the Temple of Zeus, in a single pedestrian district liberally splashed with green—no longer seems the textbook case of hubris it did a few years ago.

In the long view, in fact, it seems a rectification of hubris. The myth of Athena's birth—that she sprang full-grown and fully armed from the head of Zeus—makes vivid one strain in the self-image of Athens and many other cities. The city owes nothing to Mother Earth. It is purely a child of the thrusting, questing brain. The trivia with which living systems must concern themselves—growth, change, breath, ingestion, excretion—are beneath its notice. Its pattern is eternal.

Anyone who has tried to walk through downtown Athens on a hot summer's day knows where such hubris leads. To snap out of that tragic blindness is no mean feat. If Athens can do it, there is hope for almost any city you can name.

For further examples of ecological problem solving in cities, see Nicholson-Lord; Platt; and the case study of Curitiba, Brazil, in McKibben, *Hope*.

374 **Boston:** On Boston's natural and unnatural history, see Spirn, which remains the best single book on the ecological prospects of cities.

375 **"would be novel, certainly . . .":** Olmsted, quoted in Spirn.

376 **low-income housing project in Minneapolis:** Catherine Brown, lecture at Harvard Graduate School of Design, March 18, 1995. Among many other important projects, Morrish and Brown were involved in the Phoenix Arts Commission's urban-design plan, which resulted (somewhat fortuitously, perhaps) in Singer and Glatt's recycling center (see above, pp. 339–41).
Strawberry Creek: Lyman.
Rome's Trevi Fountain: See Morrish and Brown; the splendidly illustrated book by C. Moore; and, more generally, the interesting chapter entitled "Fountain-blues" in Rudofsky.

377 **"The Sea is, so to speak . . .":** Salvi, quoted in Moore, pp. 22ff.
"monstruous devices in marble . . .": Hawthorne, *The French and Italian Notebooks* (1876), quoted in Rudofsky.
local play of Mountain and Tower: In places where water is scarce, a less splashy statement may be appropriate—perhaps playing up the image of the urn that stores water, or relying on modest trickles and shimmers as some Near Eastern fountains do. In any case, something more than the modern equivalent of Salem's town pump is called for—though, in a modern context, Salem's town pump itself might be just the thing.

28. RECLAIMING ARCADIA

379 **excised like a tumor:** See, for example, Manes. The image of humankind as a tumor or gangrene on the body planetary, like that of humankind as parasite, is ancient, dating back at least to Ovid (*Metamorphoses* 1.190).
"decoupling": M.W. Lewis. Interestingly, at least one Deep Ecologist has proposed a version of this: see Shepard, *The Tender Carnivore*.

380 **"I am a child of the suburbs . . .":** McKibben, *The End of Nature*, p. 69.

381 **Just this is the problem with modern nature-love:** See W. Berry, "Getting Along with Nature."

Here the attitudes of Manager and Fetisher oddly converge. Each in his own way tries to achieve the American dream of Eden: that is, of a direct relation to nature. On one side, the direct spiritual relation of the backpacker; on the other, the direct pragmatic relation of the technocrat. Even if the backpacker and the technocrat happen to live in the same body, you have to wonder if they can think with the same mind.

A purely aesthetic relation to nature is not natural for man or any other creature. Humans learn—and learn to love—by doing, touching, making, breaking. We learned to love the antelope by hunting, butchering, and eating it. We learned to love stone by chipping hand-axes out of it.

A Deep Ecologist might respond: All this is true of early humans and surviving tribal peoples. But modern exploiters of nature—the whole crowd of trappers, *voyageurs*, woodsmen, pioneers, loggers, ranchers, prospectors, miners, and oil drillers from the fifteenth century to the present—are a breed apart. The only love that moves them is the love of mastery and gain. The only green that catches their eye is that of greenbacks.

This is a false dichotomy, unfair to primitives and moderns alike. Tribal peoples are unsentimental about the world around them and, in their way, keen to master it. And modern men who contend with nature are often full of respect for the beauty and power of their adversary. If anything, it is an excess of respect that causes trouble, when it takes the form of the belief that nature is a magic barrel with no bottom at all. (No modern writer better conveys the love for nature of pioneers and other exploiters of nature than Edward Hoagland, both in his fiction, such as *Seven Rivers West*, and in his nonfiction, such as *Notes from the Century Before*.)

382 **community-supported agriculture:** Bio-Dynamic Farming and Gardening Association, *Introduction*. According to Jean Yeager of the association (personal communication), there were 561 farms on their list as of November 1996. Similar farms in Japan are hard to classify since they are associated with large food cooperatives.

A careful three-year study of what members spent on and got from one CSA farm found that buying the same produce at the supermarket would have cost each family 37 percent more. Admittedly, in one key respect the study was flawed: the supermarket produce would not have been "the same" at all, since it would have been chemically grown, less fresh, and by all odds less tasty.

383 **"What's the use of a house . . .":** Thoreau, letter to H. G. O. Blake, May 20, 1860.
different kind of suburb: See Langdon; Rowe.

384 **New Urbanists:** See Langdon; Katz; Lindgren; Newman; Calthorpe; and the earlier, profoundly influential books by Alexander and Jacobs. Though the automobile suburb has always claimed the raising of children as its main mission on earth, anyone who has been a child can see that the small-town arrangement is far better. Instead of being stranded in front of a screen or carted, like a blindfolded hostage, from one placeless place to the next, a child is free to increase the radius of his world, block by block, as he feels ready. People of various ages, con-

ditions, and walks of life can be encountered. Various kinds of adult work can be seen and (in some cases) participated in. Wildness, both natural and cultural, is available in bite-size increments.

Safe public transportation makes the span of a child's world still greater. Writing in the 1960s, Albert Eide Parr recalled his childhood sixty years before in a Norwegian seaport of seventy-five thousand inhabitants.

> Not as a chore, but as an eagerly desired pleasure, I was fairly regularly entrusted with the task of buying fish and bringing it home alone. This involved the following: walking to the station in five to ten minutes; buying ticket; watching train with coal-burning steam locomotive pull in; boarding train; riding across long bridge over shallows separating small-boat harbor (on the right) from ship's harbor (on the left), including small naval base with torpedo boats; continuing through a tunnel; leaving train at terminal, sometimes dawdling to look at railroad equipment; passing central town park where military band played during midday break; strolling by central shopping and business district, or alternatively, passing fire station with horses at ease under suspended harnesses, ready to go, and continuing past fish market and fishing fleet; selection of fish; haggling about price; purchase and return home.

At the time, Albert was four years old. Short of subscribing to Bertie Wooster's view that regular fish-eating strengthens the brain, we have to assume that Albert's abilities were a function of his surroundings. In fact, the architect and planner Elizabeth Plater-Zyberk remembers that, in suburban Philadelphia in the 1950s and '60s, "tiny kids would commute on the train" (Langdon, p. 127).

Not coincidentally, Plater-Zyberk (who is the child of Polish émigrés) and her husband, Andres Duany (who is a Cuban émigré), have become the most prominent leaders of the New Urbanism. The chief interest of DPZ (as their firm is called) is in helping people of all ages live civilized lives. The second most prominent leader of the movement, the San Franciscan Peter Calthorpe, shares that concern but is even more interested, perhaps, in letting nature live without too much interference from people. DPZ finds inspiration in architecture and historical preservation, Calthorpe in ecology and left-wing politics. DPZ wants to tame the automobile, making it a constant but unthreatening presence, like a horse and buggy; Calthorpe wants to reduce its use to a bare minimum. (Calthorpe's Laguna West, a thousand-acre development near Sacramento, has a concentrated grid-plan center, but long blocks and many culs-de-sac in the outer residential neighborhoods which, though meant to discourage tires, may end up foiling feet.) DPZ's many projects, of which the most celebrated are Seaside in Florida and Kentlands in Maryland, are known for using subtle guidelines to make the work of numerous builders and architects cohere in a single glimmering, nostalgia-quenching whole. Calthorpe's designs incorporate solar energy, restored wild places, organic farming, and other touchstones of sustainability, which may explain why only two of his big projects have been built so far.

In Los Angeles, DPZ worked with the landscape architect Laurie Olin to design Playa Vista, a 1,087-acre development on the site of a former Howard Hughes airport. The largest infill project yet attempted in any American city, it presses a subtly varied grid against three miles of restored coastal wetlands. Strictly, "restored" is the wrong word: the wetlands cannot be restored to what they were—the delta of the Los Angeles River before that body was channelized and rerouted twenty miles south—so they are being re-created in a more or less "natural" shape. Regraded and replanted with thirty-seven thousand native plants, the freshwater and saltwater marshes will hold and filter storm runoff as well as offer hospitality to wildlife. Parks, a ballfield, a playground, and other landscapes with mostly native flora will move the eye and body gently from the "scruffy" marshes to the "highly structured" streets.

Calthorpe, too, has an infill project under way, which, though far more modest than Playa Vista, may be a better example of what the suburbs most urgently need. In Mountain View, a suburban "edge city" near Palo Alto, Calthorpe took a dying 1960s-vintage mall, wrecked it, and used the crushed detritus to make new streets and houses. Closely spaced single-family houses have already been built, with two small green spaces rather like town commons; town houses will follow, adjoining a new transit stop, for a total of 540 homes on the eighteen-acre site. Meanwhile, Mountain View's ghost-town center has been recalled to life. Sidewalk cafés, galleries, and other small shops, a new town hall, a park, a swimming pool and gymnasium, and high-density housing have all sprung up around a new regional-transit station.

Though Mountain View's resurrection was performed with the aid of a master plan (by the San Francisco firm of Freedman, Tung, and Bottomley), sometimes the mere laying on of hands by a mass-transit system is enough to do the trick. In the 1970s, the Washington, D.C., suburbs of Arlington and Alexandria—the one a typical "edge city," the other blessed with a historic waterfront—were both sinking. In the 1980s, the advent of the Washington Metro, and of high-density housing surrounding the stations, brought new blood and new money to both.

Murderer of town centers under normal circumstances, a shopping center can *become* a town center if it plays its cards right. In Mashpee, Massachusetts, on frantically suburbanizing Cape Cod, a shopping center from the 1960s—a flat fake-mansard barge afloat in a lake of blacktop—has been made over as an old-fashioned downtown, complete with narrow streets, sidewalks, public squares, and a post office. There is even a new town green trimmed with a new town library and a Catholic church and maraschinoed with a bandstand. Housing is going up on nearby streets. Whether such a planned, privately owned "town center" can play the role of a real town center—which Mashpee had historically lacked—remains to be seen. The lake of blacktop has not evaporated, nor has the odor of malldom; but a sort of street life appears in the evenings outside the ice-cream parlor. At the very least, such cases give reason to hope that strip malls and shopping centers, the outsized coprolites of the car age, need not litter our landscape for all eternity.

384 declining suburbs account for one-third of the total: According to studies based on the 1990 census; see Lindgren. In these suburbs (many of them in the older, "inner ring" of their metropolitan areas) property values and tax bases are falling, crime and poverty rising, often faster than in the parent cities themselves. Here an investor or community partnership might buy up houses and move them about, producing small-town density on one block and freeing the next for a vest-pocket park. Three- or four-story condominium or rental apartment buildings might be sandwiched between single-family homes. The sullen, monotonous skyscape of the suburb might suddenly come alive with dormers, balconies, stairways, and catwalks. Of course, all of this would require a zoning board more open-minded than most.

385 **cluster developments:** Besides the works of Arendt and Yaro, readers might try the brief account in Hiss's fine book.

biggest obstacle is not greedy developers: For years, developers have been doing well with communities built around golf courses. Oddly, most people who buy into these communities don't play golf. What attracts them is the great expanse of common land, for which they are glad to sacrifice a big home lot of their own. Slowly, developers are catching on that the expense, noise, and chemical dangers of golf-course maintenance can be dispensed with—that meadows, woods, and a trail system will draw at least as many buyers.

386 **Grayslake:** See F. E. Martin. Standing at the frontier of that city's suburban sprawl (and beset by commercial strips spun off by the Gurnee outlet mall), Grayslake has seen one century-old homestead after another conquered and subdivided. One 677-acre site, though, has escaped that fate, thanks largely to the vision of a Chicago printer named Gaylord Donnelley. Boyishly in love with farms and farm equipment, Donnelley wanted to see housing that would let a farm stay a farm.

At first, focus groups threw cold water on his notion. People just didn't like the idea of living amid farmland. The reason, it turned out, was their (very sensible) sense that, on the urban fringe, a field of beans today would be a field of houses tomorrow, of unpredictable density and design. Assured that the beanfield would stay a beanfield—that the farm would be protected from development—people quickly changed their tune.

Donnelley died in 1992, but his project has been carried on by the landscape architect William Johnson and the development firm Shaw Homes. Now under construction, Prairie Crossing will include 317 houses, an organic farm, and restored wetland and prairie. Homeowning families will be able to get a bushel of produce and flowers each week for a yearly payment of $400. They will also be able to rent community-garden plots to do some farming of their own.

Particulars of the project are worth noting. Old farmhouses on the site are being restored for the farmers to live in; a big old barn will serve as community center and regional farmers' market. Old windbreaks of maple and osage orange have been preserved and melded with the new plantings—almost all of them native, except for traditionally Midwestern lilacs, crabapples, and the like around the village green. House lots take up only one-fifth of the site—the reverse of the usual ratio—with fields, prairie, wetland, and a twenty-two-acre lake accounting for the rest. Since no synthetic chemicals may be applied to the fields, none may be applied to lawns or shrubs, either. Planted swales, prairie, and wetland will all help filter runoff from homes and farm. Most important, perhaps, are the connections to Mountain and Tower: contiguity to the twenty-two-hundred-acre Liberty Prairie Reserve on the one hand, and a direct light-rail link to Chicago on the other.

suburban trend may be nearing its peak: See K. Jackson, chap. 16.

387 **bioregionalists:** See Sale, *Dwellers;* Snyder. There is some overlap with Deep Ecology, Green politics, etc. After reading my manuscript, Mitch Thomashow sent me a forthcoming essay of his that boldly moves toward a "cosmopolitan bioregionalism" (a term owed to Gary Snyder).

All the world is watered with the dregs of Eden: *Taanit* 10a.

"a swaying nipple and a shower of gold": Chatwin, *The Songlines*, p. 182. Chatwin, of course, makes the perfect foil to Berry. Both give weight to the notion that eponymy is destiny: Chatwin derives his name from "winding path," while a berry grows in place. On the paradoxes of Chatwin's art, see my essay "The Voyage Out."

We are exiled from a state of wandering: In Virgil's First Eclogue, the herdsman and farmer Meliboeus is an exile. He has lost his land to an unknown soldier, presumably a veteran of Philippi. (The rulers of Rome had come to realize that the cheapest way to pay their partisans was to give them the land of people who had taken the wrong side or no side.) His friend Tityrus has narrowly escaped the same fate, having had his land restored to him by a great man's favor. (It is usually assumed that Tityrus is Virgil and the great man Octavius, but that is by no means clear.)

Exile here is exile from the pastoral life—a life that contains already a good deal of wandering, or ruggedness, of exposure to the wind and rain. Meliboeus is forced into a kind of parody of nomadic pastoralism ("heartsick, I drive my goats along"), aimless and open-ended. God knows where he will end up. Among the Africans, maybe, or the Scythians, or even the Britons, "wholly cut off from the world"—savages all, but by some accounts noble ones, keeping alive the rough virtues of the Golden Age. More likely, though unmentioned, is the possibility that he will end up on the dole in Rome.

However it will finally end, his exile from the pastoral life begins in Tityrus's unblemished pastoral domain. The last lines of the poem are poised heartbreakingly between the pleasures of home and those, strangely similar, of the wandering life:

> Yet tonight you might have rested here with me
> On these green boughs. We have ripe apples, chestnuts,
> Plenty of good pressed cheese. Already smoke is rising
> From the houses, from the mountains longer shadows fall.

388 **Mogul rulers of India:** Moynihan.
"almost a perpetual journey . . .": Gibbon, vol. I, chap. 1.

Ornamental gardens are the hallmark: "And a Man shall euer see, that when Ages grow to Ciuility and Elegancie, Men come to *Build Stately*, sooner then to *Garden Finely*; As if *Gardening* were the Greater Perfection." Francis Bacon, *Essays*, "Of Gardens."

prophet would arise among the Tupi-Guarani: See Clastres; Eliade, "Paradise and Utopia." In 1539, for example, some fourteen thousand Tupi Indians set out from Brazil. Ten years later, three hundred arrived in Peru. Warfare, famine, and disease accounted for the rest.

Why did they go? They were not so badly off in Brazil. They enjoyed a large agricultural surplus; half of what they produced went into drinking festivals. Nor were they politically oppressed. Their migrations were recorded by missionaries long before Spanish power penetrated the forests in which they lived. In fact, they had recently taken these lands by conquest from another people. (Whether they did this while in quest of the Land Without Evil is not known.)

The migration, it has been suggested, was its own destination. The Land Without Evil offered freedom from the strictures of society. The migration, as long as it lasted, delivered exactly that. We can take this a bit further. One aspect of the freedom from society is the freedom from work. Among primitive peoples, as elsewhere, agricultural work is usually thought the most onerous kind. A major feature of the great migrations was the reversion to hunting and gathering, planting being resorted to only if famine loomed. As much time as possible was spent in sacred dancing, as little as possible in getting food. So, though the bounty of the Land Without Evil was not attained, the lifestyle very nearly was.

Clastres is concerned to distinguish this *prophetisme* from messianism. The first, she says, is a negation of traditional society, the second an attempt to restore a society damaged or menaced by outside forces. But it would be a mistake to deny any similarity at all. Each is an attempt to recapture a lost paradise—a paradise perhaps equally imaginary in each case. For the Tupi-Guarani, it is the perfect world that was destroyed before the present imperfect world was created.

389 **"unaccommodated man":** Shakespeare, *King Lear* III. 4. 111.

391 **part of what makes them special:** On the value of moving between fully equipped city and stripped-down country life, see Hoagland, "Home."

393 **"The summer exodus of the East Londoners . . .":** Mumford, *The City*, chap. 9, sec. 3, from which the preceding examples also are taken.

In Mongolia, many city-dwellers still return: Possehl.

Both young and old visitors: Two fragrant bales preserved from this process, from two different generations, are the books by Klinkenborg and Donald Hall.

394 **city greenmarkets:** See MacFadyen, chap. 4.

"There is undoubtedly a deep affinity . . .": Powys, chap. IX.

395 **Aranda people:** Strehlow.

29. TWO NETWORKS

398 **"Make a hedge around the Torah":** *Pirkei Avot* 1. 1. "Hedged with roses" is a rabbinic play on Canticles 7.3 whose source I have mislaid. The word *seyag* can also
be translated as "fence."

399 **regional plan for the Baikal watershed:** A mile-deep lens holding one-fifth of
the earth's fresh water, Lake Baikal entertains some eighteen hundred species
known nowhere else on earth. Its fabled purity is at risk from industrial effluents
and acid rain, which could easily get worse once free enterprise gets rolling.

The aim of Davis's team was not to nobble free enterprise, but to spur it
onward while keeping hold of the reins. Places and uses would be matched in a
way the region's land and water could sustain. After a series of public hearings (a
concept to which Siberians had to get acclimated), the team came up with a map
that set aside roughly a third of the total area—52 million acres, including the
lake itself—as the heart and major arteries of wildness: national parks, scientific
reserves, scenic rivers, greenbelts, and the like. The rest of the watershed was
divided into twenty-five kinds of zones, each with "preferred" and "conditional"
uses. Besides meeting local environmental standards, foreign-owned companies
would have to meet those of their own country; Davis's firm set about scouting
for American companies willing to do so. The U.S. Agency for International
Development promised to sweeten the pot with $3.4 million for projects in for-
est management, ecotourism, and the like.

So far, the Buryat Republic and the Chita Oblast have signed on to the plan,
Irkutsk is dragging its feet, and Mongolia is considering a plan of its own along
similar lines. Meanwhile, Davis is working with American, Russian, and Chinese
scientists to fashion a master plan for the Ussuri River basin, a 60-million-acre
region straddling Russia and China and encompassing some of China's last wetlands, as well as the last 250 Siberian tigers, the last thirty Amur leopards, and two
imperiled species of cranes. Davis has also been invited to help in the creation of
regional plans by the Haisla Indians of British Columbia, by the Miskito Indians
of Nicaragua, and by authorities in Bolivia and Chile. See Gibbs.

400 **trying to protect endangered species too late:** Beatly.
landscape linkages or corridors: The study of such linkages has become one of
the principal concerns of the new field of landscape ecology; Forman and
Godron provide a good introductory text. See also Hudson.

402 **they do so at peril of their lives:** On a higher level, the same logic applies to
birds. In the fall, blue jays with the whole sky of Wisconsin at their disposal fly
over fencerows, where they can dive for cover if attacked by a migrating hawk.

403 **role of roads in bringing settlers into the Amazon:** Recently, plans to close the
last gap in the Pan-American Highway, conceived in 1923 as an unbroken,
sixteen-thousand-mile route from Alaska to Argentina, have run into road-
blocks. Environmentalists balk at the route of the missing seventy-mile stretch,
which cuts through national parks in Panama and Colombia. They point to a

study of the most recent segment, finished in 1983 and ending twenty miles shy of the Colombian park. By 1992, the study found, 60 percent of the forest on both sides of the highway had vanished. ("Parkland Route.")

city greenways hug the sides: See Hudson; Little; Spirn.

404   **railroads that radiate from London:** Spirn, p. 223.

we flow globally and choke locally: In the text I talk about correcting the latter, but the reader who has been hoping for answers to the questions posed by the New Pangaea may be left hanging. In the biological realm, the most useful model is offered by New Zealand, which, like most remote islands, has suffered disproportionately from invasion. Whereas U.S. policy worries mainly about agricultural pests, New Zealand regards any exotic organism as a potential threat to its ecological health (Office of Technology Assessment, *Harmful Non-indigenous Species*). Arriving passengers, baggage, and goods are intensively screened and aircraft are treated with insecticide. Ecosystems are monitored for the inevitable escapees, with detailed contingency plans in place for dealing with them. Those who wish to import alien species, for whatever purpose, must bear most of the costs of inspection, surveillance, computerized tracking, scientific analysis, and enforcement, and often must post a bond to cover damage that might be caused if their alien escapes. Recently, New Zealand made plans to consolidate the work of various agencies in a single Hazards Control Commission which, in adversarial hearings, would balance the benefits that might be obtained from new organisms, including genetically engineered organisms, against the risks of damage "to the environment and to the health, safety and economic, social and cultural well-being of people and communities." People and communities are, of course, crucial to the success of all these efforts, and their support has been widely enlisted. (For a case study of grassroots involvement in Hawaii, see Loope.)

When it comes to economic and cultural invasion, answers are harder to come by. The end of the Cold War has brought with it an upsurge of ethnic, religious, and tribal violence. Conventional wisdom chalks this up to the thawing of passions unnaturally frozen by communism. Let the tribal mind have its last spasm, we say, before it settles into the warm bath of reason and prosperity.

Maybe the tribal mind is smarter than we think. Maybe it knows that it is about to move from one kind of enforced uniformity to another. The rage of Eastern-bloc peoples against a world that offers them wealth at the price of self-determination—that would have them sell their birthright for a mess of over-packaged pottage—may not be so irrational after all, especially when they see most of the pottage doled out in large lumps to local elites, gangsters, and born-again apparatchiks. (There is no contradiction in the fact that it is often those same apparatchiks who stir up the rage; remember that National Socialism, with its rage against the world capitalist conspiracy, was funded by German capitalists.) Nor is it any surprise that people defend their own cultures by attacking other, smaller cultures—that Serbs fend off the "Judeo-Masonic New Worlds Order" (a phrase used by a Belgrade law student interviewed by a reporter for National Public Radio) by committing genocide against Bosnian Muslims, while Indonesia asserts its national sovereignty by killing or uprooting the tribesmen of West Papua and East Timor. By a perfectly normal logic, the sameness pressing from without gets mixed up with difference that exists within.

Am I in danger of fostering this mixup? After all, I have claimed that a burst of local diversity may be the prelude to global uniformity. And sometimes it is; but sometimes it is not. By no stretch of the unfevered imagination can Bosnian Muslims or Arawaks be seen as agents of a homogenizing world culture. Even if they could, it would not justify even the mildest violation of their human rights.

In ecology, economics, and culture alike, a pure nativism, or an attempt to resist all change, can only end in failure, or worse. But a pure universalism is also misguided, and worse than misguided when it tries to impose itself. Threatened communities have a right to draw a line—a line that is properly sharpest in matters ecological, less sharp in matters economic, and least sharp in matters cultural. (On the "seed savers" who are addressing a problem straddling these realms—the loss of crop and livestock varieties—see Nabhan; Eisenberg, "Save.")

Much depends on which species: One of the first controlled experiments in the use of corridors by animals points up some of the problems. Salamanders traveling between one-by-three-meter plots of moist woodland habitat were twice as likely to use moist, near-natural corridors than corridors of dry, scraped earth. But that left about a third of the salamanders wandering into the dry corridors, where many expired. The researchers drew the lesson that, if the landscape as a whole is inhospitable, corridors may not do the trick. (See Ezzeli.)

In gauging the likely effectiveness of such bridges, we first have to distinguish between "movement corridors" and "habitat linkages." Movement corridors are ways of getting from one patch of habitat to another; they are not complete habitats in themselves. They are like highways: you might be able to grab a snack or a snooze, but you can't live a life. For the most part, they are meant for animals that can move fairly fast. Habitat linkages, on the other hand, are strips of complete habitat, wide enough and complex enough to offer all the amenities: food, water, shelter, breeding places. These can be used by crawling insects and creeping animals, as well as by plants, fungi, and microbes, all of which can inch across them, generation by generation, whether by way of normal dispersal and gene flow or in response to something like climate change.

Which of these functions will be served by a linkage, and for which species, depends in large part on dimensions. A long, narrow, lightly wooded strip may be a dandy movement corridor for deer but a death trap for salamanders, draining population from one habitat patch without adding any to the next. A narrow linkage, such as a typical hedgerow, may be all outside and no inside: that is, it may be hospitable to creatures that like edge habitat, but not to those that need interior habitat. Few forest herbs, for example, will grow in hedgerows less than eight meters wide. But a narrow linkage that widens in places—at intersections, say, or at other occasional swellings—may support and transport some interior species. If birds and mammals using a hedgerow drop seeds in the right places, forest plants may take root hundreds of feet from the nearest patch of forest; in time, their descendants may work their way to the next patch.

405 **"Outside of the wall of the moat...":** Bonvesin da la Riva, quoted in Mumford, *The City,* chap. 10, sec. 4.

406 **Leonardo da Vinci conceived:** Mumford, "The Garden City Idea."
Town and Country Planning Act: On London's greenbelt and England's New Towns, see Peter Hall.

407 **Portland, Oregon:** In 1903, shortly after the death of Frederick Law Olmsted, his son and stepson, who were keeping the family business alive and well, were asked to propose a park system for the city of Portland. In the gleaming tradition of their father's Emerald Necklace, they proposed to hang a Forty-Mile Loop of parks around the city's still-petite neck, using streams and rivers, where possible, as the string. In 1995, the loop was closed and the clasp fastened: except that now, with Portland's neck engorged by the years, the length of the necklace was 140 miles.

What keeps the greenbelt from being embedded in the city's growth, like a ring of barbed wire among the growth rings of a tree, is an Urban Growth Boundary, beyond which new development is taboo. The effect of this and other regional-planning efforts has been to funnel growth inward, giving Portland one of the most livable, walkable, and generally desirable downtowns in the nation, and letting Portlanders rub shoulders with wild things in a way few other urbanites can do—finding elk, black bear, and pileated woodpeckers in Forest Park; great blue herons, sandhill cranes, Canada geese, and (in winter) bald eagles in the wetlands of Sauvie Island and Vancouver Lake; and beaver, cutthroat trout, and chinook salmon in nearby rivers and streams. See Platt; Houck; Little.
Wildlands Project: Pennisi.
In an earlier part of the book: See text, p. 407.

409 **a fairly spectacular *tsimtsum* could be achieved:** Unfortunately, some of the very people who ought to be floating this idea are busy shooting it down. In a stunningly ill-considered piece in *The New York Times Magazine*, the usually considerate garden writer Michael Pollan attacks the advocates of "wild gardening"—in particular, Stein and Ken Druse—as wild-eyed ideologues and, in effect, racists. Wiping his brush in the muck of Nazi Germany, where native German plants were in vogue, he proceeds to tar Stein and Druse with it. Nativism is nativism, he says: if we are to be cosmopolitan about people, we ought to be cosmopolitan about plants, too. Aliens and immigrants should be welcomed with open arms. Most exotic plants are harmless, he says, indeed are happy additions to the landscape. Nativists talk about kudzu, always kudzu—but what other examples are there?

A phone call to any competent ecologist or a visit to any decent library would have furnished Pollan with dozens of examples in America, and hundreds worldwide, of exotic plants that have done grievous harm to native ecosystems. (For a few such examples, refer back to chapter 4.) As far as I can tell, the ecologists who study these matters are as concerned about the effect of their own region's plants on foreign ecosystems as about the effect of foreign plants on their own ecosystems. What moves them (and gardeners like Stein) is not xenophobia, but a love of the many landscapes and waterscapes of this planet and a wish that they not be boiled down to one.

Parallels can be drawn between natural and human diversity, but they should not be pushed too far. Humans are one species. A Jew in Germany, or a Korean in America, is not an invasive exotic, but a member of the same species. Even in nature, gene flow and cultural exchange between populations of the same species are usually a good thing.

Pollan is tickled by the burst of local diversity he can bring about in his own gardens with the help of exotics. Ecologists know that this burst will be brief, and that it will help bring about a global loss of diversity that may last for millennia.

410   **They may escape into the wild:** Unfortunately, this sort of "naturalizing" is just what Pollan likes to do—as do a slew of other "naturalistic" gardeners, including many of those written up by Druse in his book *The Natural Garden*. Rather than blast the book, Pollan should have saluted a kindred spirit.

30. HOT AND COOL

412   **No one has told my cat about global warming:** Since this part of the book was written, my cat has died of natural causes. As it would be awkward, and painful, to rewrite the section in the past tense, I ask the reader to indulge me in the historical present.

    flurry of new inequities and recriminations: Under the Rio Convention of 1992, thirty-five industrialized nations committed themselves to holding their greenhouse gas emissions to 1990 levels in the year 2000. At present, only half appear likely to meet that goal. See Flavin and Tunali.

    Briefly, the facts are these: See Houghton et al.; Watson et al.; Karl et al.; The first part of McKibben's *The End of Nature* is still one of the best expositions of the science involved. Revkin provides a concise picture-book alternative.

413   **temperatures will rise 1.8 to 6.3 degrees F.:** Houghton et al.

    temperatures have already risen: Hanson et al.

    "the balance of evidence suggests": Houghton et al., p. 5.

417   **renewable energy:** See Flavin and Tunali; Flavin and Lenssen. Natural gas, though not renewable, releases less carbon than coal or oil, and for technical reasons makes a good bridge to hydrogen power.

    replacement of gasoline with ethanol: Care must be taken, however, not to stress an agriculture that already uses too many chemicals and loses too much soil.

419   **"Fresh-firecoal chestnut-falls . . .":** Hopkins, "Pied Beauty" (1877).

    we are losing dozens each day: See text, p. 36, and note.

420   **"Humans may have chosen . . .":** Lovelock, *The Ages*, p. 157.

421   **"Generations have trod . . .":** Hopkins, "God's Grandeur" (1877).

BIBLIOGRAPHY

I have generally listed the final or latest edition of a given book, adding the date of the first edition in brackets. For works in foreign languages, I have listed a translation; the title and date of the original publication follow in brackets. An asterisk pinned to the translator's name means that quotations from the work in question are (unless the note says otherwise) taken from that translation. (For most quotations from the Bible I have used the translation listed under the name of its editor in chief, Max L. Margolis.) Where no credit is given, the translation is my own.

Although the bibliography is much too fat to be called "select," neither does it aim at comprehensiveness: only works and authors mentioned in the text or notes are included.

Abbey, Edward. *Desert Solitaire: A Season in the Wilderness*. New York: McGraw-Hill, 1968.

———. *The Monkey Wrench Gang*. New York: Avon, 1976.

———. *Slumgullion Stew: An Edward Abbey Reader*. New York: Dutton, 1984.

Abram, David. "The Mechanical and the Organic: The Impact of Metaphor in Science." See Schneider and Boston.

———. *The Spell of the Sensuous: Perception and Language in a More-Than-Human World*. New York: Pantheon, 1996.

Adams, Henry. *Mont Saint-Michel and Chartres*. Boston: Houghton Mifflin, 1904.

Adams, Lowell W., and Daniel L. Leedy, eds. *Integrating Man and Nature in the Metropolitan Environment*. Columbia, Md.: National Institute for Urban Wildlife, 1987.

———, eds. *Wildlife Conservation in Metropolitan Environments*. Columbia, Md.: National Institute for Urban Wildlife, 1991.

Adams, Robert McCormick. *Heartland of Cities: Surveys of Ancient Settlement and Land Use on the Central Flood Plain of the Euphrates*. Chicago: University of Chicago, 1981.

———. *Land Behind Baghdad: A History of Settlement on the Diyala Plain*. Chicago: University of Chicago, 1965.

Adams, Robert McCormick, and Hans J. Nissen. *The Uruk Countryside: The Natural Setting of Urban Societies*. Chicago: University of Chicago, 1972.

Adams, William Howard. *The French Garden, 1500–1800*. New York: Braziller, 1979.

Adleman, Leonard M. "Molecular Computation of Solutions to Combinatorial Problems." *Science* 266 (1994): 1021–4.

Agnelli, Marella. *Gardens of the Italian Villas*. New York: Rizzoli, 1987.

Aharoni, Yohanan. *The Land of the Bible: A Historical Geography.* 2d ed. Translated and edited by A. F. Rainey. Philadelphia: Westminster, 1979. [*Erets-Yisrael bi-tekufat ha-Mikra*]

Ahmadjian, Vernon. *The Lichen Symbiosis.* Waltham, Mass.: Blaisdell, 1967.

Ahmadjian, Vernon, and Surindar Paracer. *Symbiosis: An Introduction to Biological Associations.* Hanover, N.H.: University Press of New England, 1986.

Albright, William Foxwell. *The Archaeology of Palestine.* Rev. ed. Harmondsworth: Penguin, 1960. [1949]

———. *From the Stone Age to Christianity: Monotheism and the Historical Process.* 2d ed. Philadelphia: Johns Hopkins, 1957. [1940]

Alcorn, Janice B. "Huastec Noncrop Resource Management: Implications for Prehistoric Rain Forest Management." *Human Ecology* 9, no. 4 (1981): 395–417.

Alexander, Christopher. *A Timeless Way of Building.* New York: Oxford, 1979.

Alexander, Christopher, et al. *A Pattern Language: Towns, Buildings, Construction.* New York: Oxford, 1977.

Alster, Bendt. "Dilmun, Bahrain, and the Alleged Paradise in Sumerian Myth and Literature." See Potts.

Altieri, Miguel A. *Agroecology: The Scientific Basis of Alternative Agriculture.* Boulder, Colo.: Westview, 1987.

Ammerman, Albert J., and L. L. Cavalli-Sforza. *The Neolithic Transition and the Genetics of Populations in Europe.* Princeton: Princeton University Press, 1984.

Anati, Emmanuel. *The Mountain of God: Har Karkom.* New York: Rizzoli, 1986. [*Montagna di Dio*]

———. *Palestine Before the Hebrews.* New York: Knopf, 1963.

Anderson, D. L. "The Earth as a Planet: Paradigm and Paradoxes." *Science* 223 (1984): 347–55.

Anderson, Edgar. *Plants, Man and Life.* Berkeley: University of California Press, 1952.

Anderson, Philip. "More is Different." *Science* 177 (1972): 393–6.

Anderson, Walter Truett. *To Govern Evolution.* New York: Harcourt Brace Jovanovich, 1987.

Angier, Natalie. "DNA Tests Find Meat of Endangered Whales for Sale in Japan." *New York Times,* September 13, 1994.

Angotti, Thomas. *Metropolis 2000: Planning, Poverty, and Politics.* London, N.Y.: Routledge, 1993.

Anthony, David W., et al. "Archaeology and Language." *Current Anthropology* 29, no. 3 (1988): 437–68.

Arberry, A. J., ed. *The Cambridge History of Iran.* 7 vols. in 8. Cambridge: Cambridge University Press, 1968–9.

Arendt, Randall G. *Conservation Design for Subdivision: A Practical Guide to Creating Open Space Networks.* Washington, D.C.: Island Press, 1996.

———. *Rural by Design: Maintaining Small Town Character.* Chicago: Planners Press, 1994.

Arieti, Sylvano. *The Parnas.* New York: Basic Books, 1979.

Armstrong, John. *The Paradise Myth.* London: Oxford University Press, 1969.

Artzy, Michal, and Daniel Hillel. "A Defense of the Theory of Progressive Soil Salinization in Ancient Southern Mesopotamia." *Geoarchaeology* 3, no. 3 (1988): 235–8.

Ashton, T. S. *The Industrial Revolution 1760–1830.* Rev. ed. London: Oxford University Press, 1964. [1948]

Ashton, T. S., and Joseph Sykes. *The Coal Industry of the Eighteenth Century.* 2d ed. New York: A. M. Kelley, 1967. [1964]

Askins, Robert. "Hostile Landscapes and the Decline of Migratory Songbirds." *Science* 267 (1995): 1956–7.

Aston, Margaret, ed. *The Panorama of the Renaissance.* New York: Abrams, 1996.

Atkins, P. W. *The Second Law: Energy, Chaos, and Form.* Rev. ed. New York: Scientific American, 1994. [1984]

Atsatt, Peter R. "Are Vascular Plants 'Inside-Out' Lichens?" *Ecology* 69, no. 1 (1988): 17–23.

Attenborough, David. *The First Eden: The Mediterranean World and Man.* Boston: Little, Brown, 1987.

Bacon, Francis. *Essays.* London: Oxford University Press, 1937. [1597]

Baker, Alan R. H., and J. B. Harley, eds. *Man Made the Land: Essays in English Historical Geography.* Totowa, N. J.: Rowman and Littlefield, 1973.

Barlow, Connie, and Tyler Volk. "Open Systems Living in a Closed Biosphere: A New Paradox for the Gaia Debate." *BioSystems* 23 (1990): 371–84.

Barrell, John, and John Bull, eds. *The Penguin Book of English Pastoral Verse.* Harmondsworth: Penguin, 1982. [*A Book of English Pastoral Verse,* 1974]

Barricelli, Nils Aall. "Symbiogenetic Evolution Processes Realized by Artificial Methods." *Methodos* 9, nos. 35, 36 (1957): 143–82.

Bartram, William. *Travels through North & South Carolina, Georgia, East & West Florida.* Philadelphia: James & Johnson, 1791.

Bar-Yosef, Ofer, and Anatoly Khazanov, eds. *Pastoralism in the Levant: Archaeological Materials in an Anthropological Perspective.* Madison, Wis.: Prehistory Press, 1992.

Bar-Yosef, Ofer, and Bernard Vandermeersch. "Modern Humans in the Levant." *Scientific American* 268, no. 4 (1993): 94–100.

Bateson, Gregory. *Steps to an Ecology of Mind.* San Francisco: Chandler, 1972.

Bateson, Gregory, and Margaret Mead. *Balinese Character: A Photographic Analysis.* New York: New York Academy of Sciences, 1942.

Battisti, Eugenio. *"Natura Artificiosa* to *Natura Artificialis."* See Coffin, ed.

Bayliss-Smith, T. P. *The Ecology of Agricultural Systems.* Cambridge: Cambridge University Press, 1982.

Beardsley, Tim. "La Ronde." *Scientific American* 270, no. 6 (1994): 26–9.

Beatly, Timothy. "Reconciling Urban Growth and Endangered Species: The Coachella Valley Habitat Conservation Plan." See Platt.

Beauchamp, Benoit, et al. "Cretaceous Cold-Seep Communities and Methane-Derived Carbonates in the Canadian Arctic." *Science* 244 (1989): 53–6.

Beck, Evelyn Torton, ed. *Nice Jewish Girls.* Rev. ed. Boston: Beacon, 1989. [1982]

Benyus, Janine M. *Biomimicry: Innovation Inspired by Nature.* New York: Morrow, 1997.

Berenson, Bernard. *Italian Painters of the Renaissance.* Cleveland: World, 1957. [1894–1907]

Berg, Peter, et al. *A Green City Program for the San Francisco Area and Beyond.* San Francisco: Planet Drum Books, 1989.

Berger, John J., ed. *Environmental Restoration: Science and Strategies for Restoring the Earth.* Washington, D.C.: Island Press, 1990.

———. *Restoring the Earth: How Americans Are Working to Renew Our Damaged Environment.* New York: Knopf, 1986.

Bermudes, David, and Richard C. Back. "Symbiosis Inferred from the Fossil Record." In *Symbiosis as a Source of Evolutionary Innovation,* edited by Lynn Margulis and René Fester. Cambridge, Mass.: MIT Press, 1991.

Bernbaum, Edwin. *Sacred Mountains of the World.* San Francisco: Sierra Club, 1990.

Berry, Thomas. *The Dream of the Earth.* San Francisco: Sierra Club, 1988.

Berry, Wendell. "An Agricultural Journey in Peru." See *The Gift of Good Land.*

———. "Getting Along With Nature." See *Home Economics.*

———. *The Gift of Good Land: Further Essays Cultural and Agricultural.* San Francisco: North Point, 1981.

———. "A Good Scythe." See *The Gift of Good Land.*

———. *Home Economics.* San Francisco: North Point, 1987.

———. "Preserving Wildness." See *Home Economics.*

———. *The Unsettling of America: Culture and Agriculture.* 2d ed. San Francisco: Sierra Club Books, 1986. [1977]

Bibby, Geoffrey. *Looking for Dilmun.* New York: Knopf, 1970.

Bindrim, Paul. "Nude Marathon Therapy." In *Inside Psychotherapy,* edited by Adelaide Bry. New York: Basic Books, 1972.

Bio-Dynamic Farming and Gardening Association. *Introduction to Community Supported Farms/Gardens, Farm/Garden Supported Communities.* Kimberton, Penn.: The Association, n.d.

Birge, Robert R. "Protein-Based Computers." *Scientific American* 272, no. 3 (1995): 90–5.

Bleek, W. H. I., and L. C. Lloyd. *Specimens of Bushman Folkore.* London: George Allen, 1911.

Bobek, H. "Vegetation." See Arberry, vol. 1.

Boccaccio, Giovanni. *Decameron.* Translated by Richard Aldington.* Garden City, N.Y.: Doubleday, 1949.

Bolton, Herbert Eugene, trans. and ed. *Anza's California Expeditions.* 5 vols. New York: Russell and Russell, 1966. [1930]

Bonner, John Tyler. *The Evolution of Culture in Animals.* Princeton.: Princeton University Press, 1980.

Bookchin, Murray. *The Ecology of Freedom.* Palo Alto, Calif.: Cheshire Books, 1982.

Bormann, F. Herbert, et al. *Redesigning the American Lawn: A Search for Environmental Harmony.* New Haven: Yale, 1993.

Borowski, Oded. *Agriculture in Iron Age Israel.* Winona Lake, Ind.: Eisenbrauns, 1987.

Botkin, Daniel B. *Discordant Harmonies: A New Ecology for the Twenty-first Century.* New York: Oxford University Press, 1990.

Boucher, Douglas H., ed. *The Biology of Mutualism.* New York: Oxford, 1985.

Bowersock, G. W. "Tylos and Tyre: Bahrain in the Graeco-Roman World." See Khalifa and Rice.

Boyce, Mary, ed. and trans. *Textual Sources for the Study of Zoroastrianism.* Totowa, N.J.: Barnes & Noble, 1984.

Boyer, Carl B. *A History of Mathematics.* 2d ed. New York: Wiley, 1989. [1968]

Boynton, Sandra. *Grunt: Pigorian Chant from Snouto Domoinko de Silo.* New York: Workman, 1996.

Bradford, John. *Ancient Landscapes: Studies in Field Archaeology.* London: Bell, 1957.

Brady, Nyle C. *The Nature and Properties of Soils.* 10th ed. New York: Macmillan, 1990.

Braudel, Fernand. *Civilization & Capitalism, 15th–18th Century.* 3 vols. Translated by Siân Reynolds. New York: Harper & Row, 1981. [*Civilisation materielle et capitalisme, XVe–XVIIIe siècle,* 1967–79]

———. *The Mediterranean and the Mediterranean World in the Age of Philip II.* Translated by Siân Reynolds.* New York: Harper & Row, 1989. [*La Mediteranée et le monde mediterranéen à l'époque de Philippe II,* 1949, 1966]

Brice, William C., ed. *The Environmental History of the Near and Middle East Since the Last Ice Age.* New York: Academic Press, 1978.

Briffault, Robert. *The Troubadours.* Translated by the author. Bloomington: Indiana University Press, 1965. [*Les troubadours et le sentiment romanesque*]

Brooks, Van Wyck. *The Flowering of New England 1815–1865.* New York: E. P. Dutton, 1936.

Brown, Larissa. "The Ecologically Transparent City: Making Nature Manifest Through a New Aesthetic." Unpublished paper.

Brown, Peter. "A More Glorious House." *New York Review of Books,* May 29, 1997, 19–24.

Browne, [Sir] Thomas. *The Garden of Cyrus.* In *The Major Works,* edited by C. A. Patrides. Harmondsworth: Penguin, 1977. [1658]

Brucker, Gene. *Renaissance Florence.* Berkeley: University of California, 1969.

Brush, Stephen B. *Mountain, Field, and Family: The Economy and Human Ecology of an Andean Valley.* Philadelphia: University of Pennsylvania, 1977.

Brush, Stephen B., et al. "Dynamics of Andean Potato Agriculture." *Economic Botany* 35, no. 1 (1981): 70–88.

Buber, Martin. *I and Thou.* Translated by Walter Kaufmann. New York: Scribner, 1970. [*Ich und du,* 1923]

———. *Tales of the Hasidim.* 2 vols. Translated by Olga Marx. New York: Schocken, 1948. [*Die chassidischen Bucher,* 1928]

Buchmann, Stephen L., and Gary Paul Nabhan. *The Forgotten Pollinators.* Washington, D.C.: Island Press, 1996.

Bult, C. J., et al. "Complete Genome Sequence of the Methanogenic Archaeon, *Methanococcus jannaschii.*" *Science* 273 (1996): 1058–73.

Burckhardt, Jacob. *The Civilization of the Renaissance in Italy.* 2 vols. Translated by S. G. C. Middlemore. [*Die Cultur der Renaissance in Italien: Eine Versuch,* 1860]

Burke, Peter. *The Fabrication of Louis XIV.* New Haven: Yale University Press, 1992.

Burns, John F. "Medicinal Potions May Doom Tiger to Extinction." *New York Times,* March 15, 1994.

Bury, J. B. *The Idea of Progress.* New York: Macmillan, 1932.

Butterworth, E. A. S. *The Tree at the Navel of the Earth.* Berlin: De Gruyter, 1970.

Butti, Ken, and John Perlin. *A Golden Thread: 2,500 Years of Solar Architecture and Technology.* Palo Alto: Cheshire Books, 1980.

Butzer, Karl W. "No Eden in the New World." *Nature* 362 (1993): 15–17.

Byatt, A. S. *Possession.* New York: Random House, 1990.

Cairns-Smith, A. G. *Genetic Takeover and the Mineral Origins of Life.* Cambridge: Cambridge University Press, 1982.

Cairns, John, Jr. "Some Factors Affecting Management Strategies for Restoring the Earth." See Berger, ed.

Calder, Nigel. *Eden Was No Garden: An Inquiry into the Environment of Man.* New York: Holt, Rinehart and Winston, 1967.

Callenbach, Ernest. *Ecology for Everyone.* Berkeley: University of California. In press.

————. *Ecotopia.* Berkeley: Banyan Tree Books, 1975.

Calthorpe, Peter. *The Next American Metropolis: Ecology, Community, and the American Dream.* New York: Princeton Architectural Press, 1993.

Campbell, Joseph. *Historical Atlas of World Mythology.* 2 vols. in 5. New York: Harper & Row, 1988–89.

————. *The Masks of God.* 4 vols. New York: Viking, 1959–69.

Carlton, James T., and Jonathan B. Geller. "Ecological Roulette: The Global Transport of Nonindigenous Organisms." *Science* 261 (1993): 78–82.

Carneiro, Robert L. "A Theory of the Origin of the State." *Science* 169 (1970): 733–8.

Carr, Anna, et al. *Rodale's Chemical-Free Yard & Garden.* Emmaus, Pa.: Rodale, 1991.

Carr, John Laurence. *Life in France Under Louis XIV.* New York: Putnam, 1970.

Carson, Rachel. *Silent Spring.* Boston: Houghton Mifflin, 1962.

Carter, Harold. *An Introduction to Urban Historical Geography.* London: Edward Arnold, 1983.

Carter, Vernon Gill, and Tom Dale. *Topsoil and Civilization.* Rev. ed. Norman, Okla.: University of Oklahoma Press, 1974. [1955]

Cassirer, Ernest, et al. *The Renaissance Philosophy of Man.* Chicago: University of Chicago, 1948.

Cassuto, Umberto. *A Commentary on the Book of Genesis.* Translated by Israel Abrahams. 2 vols. Jerusalem: Magnes Press, 1961.

Castetter, Edward F., and Willis H. Bell. *Pima and Papago Indian Agriculture.* Albuquerque: University of New Mexico, 1942.

Cavalli-Sforza, Luigi L., and Francesco Cavalli-Sforza. *The Great Human Diasporas: The History of Diversity and Evolution.* Translated by Sarah Thorne. Reading, Mass.: Addison-Wesley, 1995. [*Chi siamo*]

Cavalli-Sforza, Luigi L., et al. "Demic Expansions and Human Evolution." *Science* 259 (1993): 639–46.

————. *The History and Geography of Human Genes.* Princeton: Princeton University Press, 1994.

Cendrars, Blaise, and Robert Doisneau. *La Banlieue de Paris.* Paris: Seghers, 1949.

Chapela, Ignacio H., et al. "Evolutionary History of the Symbiosis Between Fungus-Growing Ants and Their Fungi." *Science* 266 (1994): 1691–4.

Chardin, Sir John. *Travels in Persia.* Translated by Edmund Lloyd. London: J. Smith, 1720. [*Journal du Voyage . . . de Chardin en Perse et aux Indes Orientales,* 1686]

Charles-Picard, Gilbert, and Colette Charles-Picard. *Daily Life in Carthage at the Time of Hannibal.* Translated by A. E. Foster. New York: Macmillan, 1961. [*Vie quotidienne à Carthage au temps d'Hannibal*]

Charlson, Robert J., et al. "Oceanic Phytoplankton, Atmospheric Sulphur, Cloud Albedo and Climate." *Nature* 236 (1987): 655–66.

Chatwin, Bruce. *In Patagonia.* New York: Summit Books, 1977.

———. *The Songlines.* New York: Viking, 1987.

Childress, James J., et al. "A Methanotrophic Marine Molluscan (Bivalvia, Mytilidae) Symbiosis: Mussels Fueled by Gas." *Science* 233 (1989): 1306–8.

Christianty, Linda, et al. "Traditional Agroforestry in West Java: The *Pekarangan* (Homegarden) and *Kebun-Talun* (Annual-Perennial Rotation) Cropping Systems." See Marten.

Cipolla, Carlo M., ed. *The Fontana Economic History of Europe.* 6 vols. in 9. New York: Harvester Press / Barnes & Noble, 1976–77.

Clark, Grahame. *Mesolithic Prelude: The Paleolithic-Neolithic Transition in Old World Prehistory.* Edinburgh: Edinburgh University Press, 1988.

Clark, William C. "Managing Planet Earth." *Scientific American* 261, no. 3 (1989): 46–54.

Clastres, Hélène. *La terre sans mal: le prophetisme tupi-guarani.* Paris: Editions du Seuil, 1975.

Clifford, Richard J. *The Cosmic Mountain in the Old Testament.* Cambridge: Harvard University Press, 1972.

Cloud, Preston. *Oasis in Space: Earth History from the Beginning.* New York: Norton, 1988.

Cobbett, William. *Rural Rides.* 2 vols. London, 1853. [1830]

Coffin, David R., *Gardens and Gardening in Papal Rome.* Princeton: Princeton University Press, 1991.

———, ed. *The Italian Garden.* Washington, D.C.: Dumbarton Oaks, 1972.

Cohen, Jeremy. *Be Fertile and Increase, Fill the Earth and Master It: The Ancient and Medieval Career of a Biblical Text.* Ithaca: Cornell, 1989.

Colonna, Francesco. *Hypnerotomachia Poliphili.* 2 vols. Edited by Giovanni Pozzi and Lucia A. Ciapponi. Padova: Antenore, 1980. [1499]

Columbus, Christopher. *The Four Voyages of Columbus.* 2 vols. in 1. Edited and translated by Cecil Jane. New York: Dover, 1988.

Comito, Terry. *The Idea of the Garden in the Renaissance.* New Brunswick, N.J.: Rutgers University Press, 1978.

Commoner, Barry. *The Closing Circle: Nature, Man, and Technology.* New York: Knopf, 1971.

———. *Making Peace with the Planet.* New York: Pantheon, 1990.

Conklin, Harold C. "An Ethnoecological Approach to Shifting Agriculture." See Vayda.

Conniff, Richard. "When the Music in Our Parlors Brought Death to Darkest Africa." *Audubon,* July 1987, 76–93.

Coote, Robert B. *Early Israel: A New Horizon.* Minneapolis: Fortress Press, 1990.

Coote, Robert B., and Keith W. Whitelam. *The Emergence of Early Israel in Historical Perspective.* Sheffield, England: Almond, 1987.

Corcoran, Elizabeth. "Charlotte's Patent." *Scientific American* 266, no. 4 (1992): 138–40.

Corpechot, Lucien. *Les Jardins de l'intelligence.* Paris: Emile-Paul, 1912.

Costanza, Robert, et al. "The Value of the World's Ecosystem Services and Natural Capital." *Nature* 387 (1997): 253–60.

Cowen, Richard. "Algal Symbiosis and Its Recognition in the Fossil Record." See Tevensz and McCall.

Cox, George W., and Michael D. Atkins. *Agricultural Ecology: An Analysis of World Food Production Systems*. San Francisco: Freeman, 1979.

Crawford, Harriet E. W. *Sumer and the Sumerians*. Cambridge: Cambridge University Press, 1991.

Creighton, J. C., and G. D. Schnell. "Proximate Control of Siblicide in Cattle Egrets: A Test of the Food-Amount Hypothesis." *Behavioral Ecology and Sociobiology* 38, no. 6 (1996): 371–7.

Crèvecoeur, J. Hector St. John de. *Letters from an American Farmer*. New York: Dutton, 1957. [1782]

Crisp, [Sir] Frank. *Mediaeval Gardens*. 2 vols. London: John Lane, 1924.

Cronon, William. *Changes in the Land: Indians, Colonists, and the Ecology of New England*. New York: Hill and Wang, 1983.

———. "Inconstant Unity: The Passion of Frank Lloyd Wright." See Riley.

———. *Nature's Metropolis: Chicago and the Great West*. New York: Norton, 1991.

Cronon, William, ed. *Uncommon Ground: Toward Reinventing Nature*. New York: Norton, 1995.

Crosby, Alfred W. *The Columbian Exchange: Biological and Cultural Consequences of 1492*. Westport, Conn.: Greenwood, 1972.

———. *Ecological Imperialism: The Biological Expansion of Europe, 900–1900*. Cambridge: Cambridge University Press, 1986.

Cross, Frank Moore. *Canaanite Myth and Hebrew Epic*. Cambridge, Mass.: Harvard University Press, 1973.

Crowley, John. *Aegypt*. New York: Bantam, 1987.

———. *Love and Sleep*. New York: Bantam, 1994.

Daily, Gretchen C., ed. *Nature's Services: Societal Dependence on Natural Ecosystems*. Washington, D.C.: Island Press, 1997.

Dandamaev, Muhammad, and Vladimir Lukonin. *The Culture and Social Institutions of Ancient Iran*. Cambridge: Cambridge University Press, 1989.

Dante Alighieri. *La Divina Commedia*. Edited and annotated by C. H. Grandgent. Revised by Charles S. Singleton. Cambridge, Mass.: Harvard, 1972.

Darwin, Charles. *The Autobiography of Charles Darwin, 1809–1882*. Edited by Nora Barlow. New York: Norton, 1958. [1892]

———. *The Formation of Vegetable Mould, Through the Action of Worms*. London: John Murray, 1881.

Daum, Annette. "Blaming the Jews for the Death of the Goddess." See Beck.

Davies, Julian. "Inactivation of Antibiotics and the Dissemination of Resistance Genes." *Science* 264 (1985): 375–82.

Dawkins, Richard. *The Extended Phenotype*. New York: Oxford University Press, 1982.

Defoe, Daniel. *A Tour Through the Whole Island of Great Britain*. Edited by P. N. Furbank and W. R. Owens. New Haven: Yale, 1991. [1724–27]

De Maddalena, Aldo. "Rural Europe 1500–1750." Translated by Muriel Grindrod. See Cipolla, vol. 2.

Devall, Bill, and George Sessions. *Deep Ecology: Living as if Nature Mattered.* Salt Lake City: Gibbs M. Smith, 1985.

Diamond, Jared. *The Rise and Fall of the Third Chimpanzee.* London: Radius, 1991.

Dillard, Annie. *Pilgrim at Tinker Creek.* New York: Harper's Magazine Press, 1974.

Dini, Massimo. *Renzo Piano: Projects and Buildings 1964–1983.* Translated by Richard Sadleir. New York: Electa/Rizzoli, 1984. [*Renzo Piano, progetti e architteture, 1964–1983*]

DiNunzio, Michael G. *Adirondack Wildguide: A Natural History of the Adirondack Park.* Elizabethtown, N.Y.: The Adirondack Conservancy Committee and the Adirondack Council, 1984.

Dolittle, W. Ford. "Is Nature Really Motherly?" *CoEvolution Quarterly* 29 (1981): 58–65.

Doyle, Jack. *Altered Harvest: Agriculture, Genetics, and the Fate of the World's Food Supply.* New York: Viking, 1985.

———. *Herbicides & Biotechnology: Extending the Pesticide Era.* Washington, D.C.: Environmental Policy Institute, 1988.

Driver, G. R. *Canaanite Myths and Legends.* Edited by J. C. L. Gibson, Edinburgh: Clark, 1978. [1956]

Druse, Ken. *The Natural Garden.* New York: Potter, 1989.

Dubos, René. *The Wooing of Earth.* New York: Scribner's, 1980.

Duby, Georges. "Mcdieval Agriculture 900–1500." See Cipolla, vol. 1.

Duerr, Hans Peter. *Dreamtime: Concerning the Boundary Between Wilderness and Civilization.* Translated by Felicitas Goodman. Oxford: Blackwell, 1985. [*Traumzeit: Über die Grenze zwischen Wildnis und Zivilisation,* 1978]

Duff, Michael J. "The Theory Formerly Known as Strings." *Scientific American* 278, no. 2 (1998): 64–9.

Dunham, William. *Journey Through Genius: The Great Theorems of Mathematics.* New York: Wiley, 1990.

Dwight, Timothy. *Travels in New-England and New-York.* 4 vols. New Haven: T. Dwight, 1821–22.

Dyson, Freeman. *Origins of Life.* Cambridge: Cambridge University Press, 1985.

Edelman, Gerald M. *Bright Air, Brilliant Fire: On the Matter of the Mind.* New York: Basic Books, 1992.

———. *Neural Darwinism: The Theory of Neuronal Group Selection.* New York: Basic Books, 1987.

Ehrlich, Paul. "Co-Evolution and its Application to the Gaia Hypothesis." See Schneider and Boston.

Ehrlich, Paul, and Anne Ehrlich. *Extinction: The Causes and Consequences of the Disappearance of Species.* New York: Random House, 1981.

Eigen, Manfred et al. "The Origin of Genetic Information." *Scientific American* 244, no. 4 (1981): 88–118.

Eisenberg, Evan. "Back to Eden." *The Atlantic Monthly,* November 1989, 57–89.

———. "The Call of the Wild." *The New Republic,* April 30, 1990, 30–8.

———. *The Recording Angel.* New York: McGraw-Hill, 1987.

———. "Save the Veggies." *Connoisseur,* March 1988: 136–43.

———. "The Voyage Out: Bruce Chatwin's Long and Winding Road." *Voice Literary Supplement*, March 1989, 25–7.

Eisler, Rianne. *The Chalice and the Blade.* San Francisco: Harper & Row, 1987.

Eldredge, N., and S. N. Salthe. "Hierarchy and Evolution." *Oxford Surveys in Evolutionary Biology* 1 (1984): 184–208.

Eliade, Mircea. *Cosmos and History: The Myth of the Eternal Return.* Translated by Willard R. Trask.* New York: Harper & Row, 1959. [*Le Mythe de l'éternel retour: archétypes et répétition*, 1949]

Eliade, Mircea, ed. *The Encyclopedia of Religion.* 16 vols. New York: Macmillan, 1987.

———. "The Myth of the Nobel Savage or, The Prestige of the Beginning." See *Myths, Dreams and Mysteries.*

———. *Myths, Dreams and Mysteries.* Translated by Philip Mairet.* New York: Harper & Row, 1961. [*Mythes, rêves, et mystères*]

———. "Nostalgia for Paradise in the Primitive Traditions." See *Myths, Dreams and Mysteries.*

———. "Paradise and Utopia." See *Myths, Dreams and Mysteries.*

Elliot, J. H. *The Old World and the New 1492–1650.* Cambridge: Cambridge University Press, 1970.

Elton, Charles S. *The Ecology of Invasions by Animals and Plants.* London: Methuen, 1958.

Emerson, Ralph Waldo. *Journals and Miscellaneous Notebooks.* 16 vols. Edited by William H. Gilman et al. Cambridge, Mass.: Harvard, 1960–82.

———. "Thoreau." In *Complete Works*, vol. X. Boston: Houghton Mifflin, 1883.

———. "The Young American." In *Nature: Essays and Addresses.* Boston: James Munroe, 1849.

Emory, Jerry. "Just Add Water." *Nature Conservancy* 44, no. 6 (1994): 10–15.

Empson, William. *Some Versions of Pastoral.* Rev ed. New York: New Directions, 1974. [1935]

Evelyn, John. *Diary.* Edited by E. S. de Beer. London: Oxford University Press, 1959.

Ewald, Paul W. *Evolution of Infectious Disease.* Oxford: Oxford University Press, 1994.

———. "The Evolution of Virulence." *Scientific American* 268, no. 4 (1993): 86–93.

Ezzeli, C. "Wilderness Corridors May Not Benefit All." *Science News* 142 (1992): 135.

Falk, Marcia. *The Book of Blessings: New Jewish Prayers for Daily Life, the Sabbath, and the New Moon Festival.* San Francisco: HarperSanFrancisco, 1996.

Farb, Peter. *Living Earth.* New York: Harper & Row, 1959.

Farlow, James O. "Speculations about the Diet and Digestive Physiology of Herbivorous Dinosaurs." *Paleobiology* 13, no. 1 (1987): 60–72.

Ferguson, Wallace K. *The Renaissance.* New York: Holt, 1940.

Fieldler, Leslie. *Love and Death in the American Novel.* 2d ed. New York: Stein and Day, 1966. [1960]

Fierz-David, Linda. *The Dream of Poliphilo.* Translated by M. Hottinger. New York: Pantheon, 1950. [*Liebestraum des Poliphilo*]

Fine, Lawrence, trans. and ed. *Safed Spirituality: Rules of Mystical Piety, the Beginning of Wisdom.* New York: Paulist Press, 1984.

Fitt, William K., and Robert K. Trench. "Spawning, Development, and Acquisition of

Zooxanthellae by *Tridacna Squamosa* (Mollusca, Bivalvia)." *Biological Bulletin* 161 (1981): 213–35.

Flavin, Christopher. *Slowing Global Warming: A Worldwide Strategy.* Worldwatch Paper 91. Washington, D.C.: Worldwatch Institute, 1989.

Flavin, Christopher, and Nicholas Lenssen. "Global Warming: The Energy Policy Challenge." In Richard A. Geyer, ed., *A Global Warming Forum.* Boca Raton, Fla.: CRC Press, 1993.

Flavin, Christopher, and Odil Tunali. *Climate of Hope: New Strategies for Stabilizing the World's Atmosphere.* Worldwatch Paper 130. Washington, D.C.: Worldwatch Institute, 1996.

Flora of North America Editorial Committee. *Flora of North America North of Mexico.* 2 vols., ongoing. New York: Oxford, 1993–.

Ford, Ford Madox. *Provence: From Minstrels to the Machine.* Philadelphia: Lippincott, 1935.

Foreman, Dave. *Confessions of an Eco-Warrior.* New York: Harmony, 1991.

Forman, Richard T. T., and Michel Godron. *Landscape Ecology.* New York: Wiley, 1986.

Fortun, Mike, and Herbert J. Bernstein. *Muddling Through: Pursuing Science and Truths in the 21st Century.* Washington, D.C.: Counterpoint. 1998.

Frazer, [Sir] James George. *The Golden Bough: A Study in Magic and Religion.* 3d ed. 9 vols in 13. London: Macmillan, 1911–36. [1890]

———. *The New Golden Bough.* Edited by Theodor H. Gaster. New York: Criterion Books, 1959.

Freeman, Peter, and Tomas Fricke. "The Success of Javanese Multi-Storied Gardens." *The Ecologist* 14 (1984): 150–2.

Freud, Sigmund. *The Interpretation of Dreams.* Rev. ed. Translated by A. A. Brill. New York: Macmillan, 1933. [*Die Traumdeutung*]

Friedman, Richard Elliot. *Who Wrote the Bible?* New York: Summit Books, 1987.

Frosch, Robert A. "The Industrial Ecology of the 21st Century." *Scientific American* 273, no. 3 (1995): 178–81.

Frye, Richard N. *The History of Ancient Iran. Handbuch der Altertumswissenschaft,* Abt. 3, Teil 7. München: C. H. Beck, 1984.

Frymer-Kensky, Tikva. *In the Wake of the Goddesses: Women, Culture, and the Biblical Transformation of Pagan Myth.* New York: Free Press, 1992.

Fukuoka, Masanobu. *The Natural Way of Farming: The Theory and Practice of Green Philosophy.* Translated by Frederic P. Metreaud. Tokyo: Japan Publications, 1985.

———. *The One-Straw Revolution: An Introduction to Natural Farming.* Translated by Chris Pearce et al. Emmaus, Pa.: Rodale, 1978. [*Shizen Noho Wara Ippon No Kakumei,* 1975]

Fussell, G. E. *Farming Technique from Prehistoric to Modern Times.* Oxford: Pergamon Press, 1966.

Gade, Daniel W. *Plants, Man and the Land in the Vilcanota Valley of Peru.* The Hague: W. Junk, 1975.

Galatowitsch, Susan M., and Arnold G. van der Valk. *Restoring Prairie Wetlands: An Ecological Approach.* Ames, Iowa: Iowa State University Press, 1994.

Gardner, Gary. *Shrinking Fields: Cropland Loss in a World of Eight Billion.* Worldwatch Paper 131. Washington, D.C.: Worldwatch Institute, 1996.

Garin, Eugenio. *L'Umanismo Italiano: Filosofia e vita civile nel rinascimento.* 2d ed. Bari: Laterza, 1958. [*Der italienische Humanismus,* 1947]

Garrott, Robert A., et al. "Overabundance: An Issue for Conservation Biologists?" *Conservation Biology* 7, no. 4 (1993): 946–9.

Gascoigne, Bamber. *The Treasures and Dynasties of China.* London: Jonathan Cape, 1973.

Geertz, Clifford. *Agricultural Involution: The Processes of Ecological Change in Indonesia.* Berkeley: University of California, 1966.

Gell-Mann, Murray. *The Quark and the Jaguar: Adventures in the Simple and the Complex.* New York: Freeman, 1994.

Gendler, Everett. "On the Judaism of Nature." In *The New Jews,* edited by James A. Sleeper and Alan L. Mintz. New York: Random House, 1971.

Gibbon, Edward. *The Decline and Fall of the Roman Empire.* Edited by J. B. Bury, 7 vols. London: Methuen, 1909–14. [1776–88]

Gibbons, Ann. "How the Akkadian Empire Was Hung Out to Dry." *Science* 261 (1993): 985.

Gibbs, W. Wayt. "No-Polluting Zone." *Scientific American* 271, no. 6 (1994): 14–16.

Gille, Bertrand. "Banking and Industrialization in Europe 1730–1914." Translated by Roger Greaves. See Cipolla, vol. 3.

Gimbutas, Marija. *The Civilization of the Goddess: The World of Old Europe.* San Francisco: HarperSanFrancisco, 1991.

———. *The Language of the Goddess.* San Francisco: Harper & Row, 1989.

Ginzberg, Louis. *The Legends of the Jews.* Translated by Henrietta Szold and Paul Radin. 7 vols. Philadelphia: Jewish Publication Society, 1909–38.

Glacken, Clarence J. *Traces on the Rhodian Shore: Nature and Culture in Western Thought from Ancient Times to the End of the Eighteenth Century.* Berkeley: University of California, 1967.

Gleick, James. *Chaos: Making a New Science.* New York: Viking, 1987.

Goldin, Frederick. *Lyrics of the Troubadours and Trouvères: An Anthology and a History.* Garden City, N.Y.: Anchor Books, 1973.

Goldsmith, Edward, and Nicholas Hildyard. *The Social and Environmental Effects of Large Dams.* San Francisco: Sierra Club, 1984.

Goldsmith, Wendi. "Working with Nature to Stabilize Shorelines." *Land and Water,* Nov. 11, 1991.

Goodall, Jane. *In the Shadow of Man.* Boston: Houghton Mifflin, 1971.

Goodenough, Erwin Ramsdell. *Jewish Symbols in the Greco-Roman Period.* 13 vols. New York: Pantheon, 1953–68.

Gore, Al. *Earth in the Balance: Ecology and the Human Spirit.* Boston: Houghton Mifflin, 1992.

Gothein, Marie Luise. *A History of Garden Art.* Translated by Mrs. Archer-Hind. 2 vols. New York: Dutton, 1928. [*Geschichte der Gartenkunst,* 1926]

Gottwald, Norman K. *The Tribes of Yahweh: A Sociology of the Religion of Liberated Israel, 1250–1050 B.C.E.* Maryknoll, N.Y.: Orbis, 1979.

Goubert, Pierre. *Louis XIV et vingt millions de français.* Enl. ed. Paris: Fayard, 1965.

Goudie, Andrew. *Environmental Change.* 2d. ed. New York: Oxford, 1983. [1977]

———. *The Human Impact on the Natural Environment.* 2d ed. Cambridge, Mass.: MIT, 1986. [1982]

Gould, Stephen Jay. "Human Equality Is a Contigent Fact of History." In *The Flamingo's Smile*. New York: Norton, 1985.

————. "The Problem of Perfection, or How Can a Clam Mount a Fish on Its Rear End?" In *Ever Since Darwin: Reflections in Natural History*. New York: Norton, 1977.

Grant, Michael. *The Ancient Mediterranean*. New York: Scribner, 1969.

Graves, Robert. *The Greek Myths*. 2 vols. Baltimore, Md.: Penguin, 1955.

Gray, John. *The Legacy of Canaan: The Ras Shamra Texts and Their Relevance to the Old Testament*. 2d ed. Leiden: E. J. Brill, 1965.

————. *The Canaanites*. London: Thames and Hudson, 1964.

The Greek Anthology. 5 vols. Translated by W. R. Paton.* Cambridge, Mass.: Harvard, 1916–18.

Green, Arthur. *Seek My Face, Speak My Name: A Contemporary Jewish Theology*. Northvale, N.J.: Aronson, 1992.

Grimal, Pierre. *Les Jardins romains; à la fin de la république et aux deux premiers siècles de l'empire. Essai sur le naturalisme romain*. Paris: E. de Boccard, 1943.

Grollenberg, Luc. H. *The Penguin Shorter Atlas of the Bible*. Translated by Mary F. Hedlund. Harmondsworth: Penguin, 1978. [*Kleine Atlaas de Bijbel*, 1959]

Groves, R. H., and J. J. Burdon. *Ecology of Biological Invasions*. Cambridge: Cambridge University Press, 1986.

Guillaume de Lorris, and Jean de Meun. *The Romance of the Rose*. Translated by Harry W. Robbins.* New York: Dutton, 1962. [*Le Roman de la rose*]

Hale, John. *The Civilization of Europe in the Renaissance*. New York: Atheneum, 1994.

Hall, Donald. *String Too Short to Be Saved*. New York: Viking, 1961.

Hall, Peter. *The World Cities*. 3d ed. New York: St. Martin's Press, 1984. [1966]

Halperin, David M. *Before Pastoral: Theocritus and the Ancient Tradition of Bucolic Poetry*. New Haven: Yale, 1983.

Hamilton, Lawrence S. "Restoration of Degraded Tropical Forests." See Berger.

Hansman, J. "Gilgamesh, Humbaba and the Land of the Erin-Trees." *Iraq* 38 (1976): 23–35.

Hanson, James, et al. "Table of Global-Mean Monthly, Annual and Seasonal Land-Ocean Temperature Index, 1950–Present." Goddard Institute for Space Studies. Available online at www.giss.nasa.gov/Data/GISTEMP.

Hareuveni, Nogah. *Nature in Our Biblical Heritage*. Translated and adapted by Helen Frenkley. Kiryat Ono, Israel: Neot Kedumim, 1980.

Hanson, Victor Davis. *The Other Greeks: The Family Farm and the Agrarian Roots of Western Civilization*. New York: Free Press, 1995.

Harden, Donald B. *The Phoenicians*. Rev. ed. Harmondsworth: Penguin, 1980.

Hardin, Garrett. "The Tragedy of the Commons." *Science* 162 (1968): 1243–48. Reprinted in *Managing the Commons*, edited by Garrett Hardin and John Baden. San Francisco: Freeman, 1977.

Harley, J. L., and S. E. Smith. *Mycorrhizal Symbiosis*. London: Academic Press, 1983.

Harris, David R., and Gordon C. Hillman, eds. *Foraging and Farming: The Evolution of Plant Exploitation*. London: Unwin Hyman, 1989.

Harrison, Robert Pogue. *Forests: The Shadow of Civilization*. Chicago: University of Chicago, 1992.

Harte, John. "Ecosystem Stability and Diversity." See Schneider and Boston.

Hastings, Alan, and Kevin Higgins. "Persistence of Transients in Spatially Structured Ecological Models." *Science* 263 (1994): 1133–6.

Hatton, Ragnhild. *Louis XIV and His World.* London: Thames and Hudson, 1972.

Hawken, Paul. *The Ecology of Commerce.* New York: Harper Business, 1993.

Hawthorne, Nathaniel. "Earth's Holocaust." In *Mosses from an Old Manse.* New ed. Boston: Ticknor and Fields, 1854. [1846]

———. *The House of the Seven Gables.* Boston: Ticknor, Reed and Fields, 1851.

Hearne, Vicki. *Adam's Task: Calling Animals by Name.* New York: Knopf, 1986.

Heidel, Alexander. *The Babylonian Genesis.* 2d ed. Chicago: University of Chicago, 1951. [1942]

———. *The Gilgamesh Epic and Old Testament Parallels.* 2d ed. Chicago: University of Chicago, 1949. [1946]

Heithaus, E. Raymond, et al. "Models of Some Ant–Plant Mutualisms." *American Naturalist* 116 (1980): 347–61.

Heinritz, Günter, and Elisabeth Lichtenberger, eds. *The Take-Off of Suburbia and the Crisis of the Central City.* Stuttgart: Frank Steiner, 1986.

Heiser, Charles B., Jr. *Seed to Civilization: The Story of Food.* Rev. ed. Cambridge, Mass.: Harvard University Press, 1990. [1973]

Henderson, Lawrence Joseph. *The Fitness of the Environment: An Enquiry into the Biological Significance of the Properties of Matter.* New York: Macmillan, 1913.

Henry, Donald O. *From Foraging to Agriculture: The Levant at the End of the Ice Age.* Philadelphia: University of Pennsylvania, 1989.

Heschel, Abraham Joshua. *The Sabbath.* New York: Farrar, Straus, and Young, 1951.

Hesiod and Theognis. *Theogony and Works and Days.* Translated by Dorothea Wender.* Harmondsworth: Penguin, 1973.

Hiebert, Theodore. *The Yahwist's Landscape: Nature and Religion in Early Israel.* Oxford: Oxford University Press, 1996.

Highet, Gilbert. *The Classical Tradition.* Oxford: Oxford University Press, 1949.

———. *Poets in a Landscape.* New York: Knopf, 1957.

Hillel, Daniel. *Negev: Land, Water, and Life in a Desert Environment.* New York: Praeger, 1982.

———. *Out of the Earth: Civilization and the Life of the Soil.* New York: Free Press, 1991.

———. *Rivers of Eden: The Struggle for Water and the Quest for Peace in the Middle East.* New York: Oxford, 1994.

Hillman, James. *The Dream and the Underworld.* New York: Harper & Row, 1979.

———. "An Essay on Pan." In *Pan and the Nightmare,* edited by W. H. Roscher and J. Hillman. New York: Spring Publications, 1972.

———. *Re-Visioning Psychology.* New York: Harper & Row, 1975.

Hillmann, Michael Craig. *Persian Carpets.* Austin: University of Texas Press, 1984.

Hinkle, Gregory, et al. "Phylogeny of the Attine Fungi Based on Analyses of Small Subunit Ribosomal RNA Sequences." *Science* 266 (1994): 1695–7.

Hinkle, Maureen K. "Problems with Conservation Tillage." *Journal of Soil and Water Conservation* 38, no. 3 (1983): 207–11.

Hiss, Tony. *The Experience of Place.* New York: Knopf, 1990.

Hoagland, Edward. "Home Is Two Places." In *The Courage of Turtles.* New York: Random House, 1971.

———. *Notes from the Century Before: A Journal from British Columbia.* New York: Random House, 1969.

———. *Seven Rivers West.* New York: Summit, 1986.

Hoar, Sheila K., et al. "Agricultural Herbicide Use and Risk of Lymphoma and Soft-Tissue Sarcoma." *Journal of the American Medical Association* 256 (1986): 1141–7.

Hobhouse, Henry. *Seeds of Change: Five Plants That Transformed Mankind.* London: Sidgwick & Jackson, 1985.

Hofstadter, Douglas. *Metamagical Themas.* New York: Basic Books, 1985.

Hölldobler, Bert, and Edward O. Wilson. *The Ants.* Cambridge, Mass.: Harvard, 1990.

———. *Journey to the Ants.* Cambridge, Mass.: Harvard University Press, 1994.

Holloway, Marguerite. "Nurturing Nature." *Scientific American* 270, no. 4 (1994): 98–108.

Holmberg, Scott D., et al. "Animal-to-Man Transmission of Antimicrobial-Resistant *Salmonella*: Investigations of U.S. Outbreaks, 1971–1983." *Science* 229 (1984): 833–5.

Hopkins, David C. *The Highlands of Canaan: Agricultural Life in the Early Iron Age.* Sheffield, England: Almond, 1985.

Horgan, John. "In the Beginning." *Scientific American* 264, no. 2 (1991): 117–25.

Hoskins, W. G. *The Making of the English Landscape.* 9th ed. London: Hodder and Stoughton, 1970.

Houck, Michael C. "Metropolitan Wildlife Refuge System: A Strategy for Regional Resource Planning." In Adams and Leedy, eds., *Wildlife Conservation in Metropolitan Environments.*

Hough, Michael. *Cities and Natural Process.* London: Routledge, 1995. [Revised edition of *City Form and Natural Processes,* 1984]

———. "Design with City Nature: An Overview of Some Issues." See Platt et al.

Houghton, J. T., et al., eds. *Climate Change 1995: The Science of Climate Change.* Cambridge: Cambridge University Press, 1996.

Howard, Ebenezer. *Garden Cities of Tomorrow.* Cambridge, Mass.: MIT Press, 1965. [*To-Morrow: A Peaceful Path to Real Reform,* 1898]

Howard, [Sir] Albert. *An Agricultural Testament.* London: Oxford, 1940.

Howes, Frederick G. *History of the Town of Ashfield, 1742–1910.* Ashfield, Mass., 1910.

Hrdy, Sarah Blaffer. *The Langurs of Abu: Female and Male Strategies of Reproduction.* Cambridge, Mass.: Harvard University Press, 1977.

Hruska, Blahoslav. "Dilmun in den vorsargonischen Wirtschaftstexten aus Suruppak und Lagas." See Potts.

Huber, Claudia, and Günter Wächtershäuser. "Activated Acetic Acid by Carbon Fixation on (Fe,Ni)S Under Primordial Conditions." *Science* 276 (1997): 245–47.

Hudson, Wendy E., ed. *Landscape Linkages and Biodiversity.* Washington, D.C.: Island Press, 1991.

Hueffer, Francis. *The Troubadours.* London: Chatto & Windus, 1878.

Hughes, J. Donald. *Ecology in Ancient Civilizations.* Albuquerque: University of New Mexico, 1975.

———. *Pan's Travail: Environmental Problems of the Ancient Greeks and Romans.* Baltimore: Johns Hopkins, 1994.

Hughes, Thomas P. *American Genesis: A Century of Invention and Technological Enthu-siasm, 1870–1970.* New York, Viking, 1989.

Huizinga, Johan. *The Autumn of the Middle Ages.* Translated by Rodney J. Payton and Ulrich Mommitzsch. Chicago: University of Chicago, 1996. [*Herfsttij der Middeleeuwen*]

Humboldt, Alexander von. *Cosmos: A Sketch of a Physical Description of the Universe.* Translated by E. C. Otte.* 2 vols. London: George Bell, 1886.

Hunt, John Dixon, and Peter Willis. *The Genius of the Place. The English Landscape Gar-den 1620–1820.* 2d ed. Cambridge, Mass.: MIT Press, 1988. [1975]

Hunter, Malcolm L., Jr. *Fundamentals of Conservation Biology.* Cambridge, Mass.: Blackwell Science, 1996.

Hussey, Christopher. *English Gardens and Landscapes 1700–1750.* London: Country Life, 1967.

Hutson, Vivian, et al. "Dynamics of Ecologically Obligate Mutualisms—Effects of Spa-tial Diffusion on Resilience of the Interacting Species." *American Naturalist* 126 (1985): 445–9.

Hutchinson, G. Evelyn. *The Ecological Theatre and the Evolutionary Play.* New Haven: Yale, 1965.

———. "Homage to Santa Rosalia, or Why Are There So Many Kinds of Animals?" *American Naturalist* 93 (1959): 145–59.

Hyams, Edward. *A History of Gardens and Gardening.* New York: Praeger, 1971.

———. *Soil and Civilization.* London: Thames and Hudson, 1952.

Idel, Moshe. *Kabbalah: New Perspectives.* New Haven: Yale, 1988.

Isager, Signe, and Jens Erik Skydsgaard. *Ancient Greek Agriculture.* London: Routledge, 1992.

Jackson, John Brinckerhoff. *Discovering the Vernacular Landscape.* New Haven: Yale, 1984.

Jackson, Kenneth T. *Crabgrass Frontier: The Suburbanization of the United States.* New York: Oxford, 1985.

Jackson, L., and C. Dewald. "Predicting Evolutionary Consequences of Greater Repro-ductive Effort in Monocultures and Mixtures." *Ecology* 75 (1994): 627–41.

Jackson, Wes. *Altars of Unhewn Stone: Science and the Earth.* San Francisco: North Point, 1987.

———. *New Roots for Agriculture.* Rev. ed. Lincoln: University of Nebraska Press, 1985. [1980]

Jackson, Wes, et al., eds. *Meeting the Expectations of the Land: Essays in Sustainable Agri-culture and Stewardship.* San Francisco: North Point, 1984.

Jacobs, Jane. *The Death and Life of Great American Cities.* New York: Random House, 1961.

———. *The Economy of Cities.* New York: Random House, 1969.

Jacobsen, Thorkild. "Mesopotamia." In *Before Philosophy,* edited by Henri Frankfort et al. Baltimore, Md.: Penguin, 1949. [*The Intellectual Adventure of Ancient Man,* 1946]

———. *The Treasures of Darkness.* New Haven: Yale, 1976.

Jacobsen, Thorkild, and Robert McC. Adams. "Salt and Silt in Ancient Mesopotamian Agriculture." *Science* 128 (1958): 1251–8.

Jansen, D. H. "The Natural History of Mutualisms." See Boucher.

———. "Tropical Agroecosystems." *Science* 182 (1973): 1212–9.

Jashemski, Wilhelmina F. "The Campanian Peristyle Garden." See Macdougall and Jashemski.

———. *The Gardens of Pompeii, Herculaneum, and the Villas Destroyed by Vesuvius.* New Rochelle, N.Y.: Caratzas Bros., 1979.

Jellicoe, Geoffrey, and Susan Jellicoe. *The Landscape of Man: Shaping the Environment from Prehistory to the Present Day.* Rev. ed. New York: Thames and Hudson, 1987. [1975]

Jeon, Kwang W., ed. *Intracellular Symbiosis: International Review of Cytology.* Supplement 14. New York: Academic Press, 1983.

Jantsch, Erich. *The Self-Organizing Universe.* Oxford: Pergamon Press, 1980.

Johnson, Samuel. *The Lives of the Most Eminent English Poets.* New ed., corr. 4 vols. London: T. Longman, 1794. [1781]

Johnson, Warren. *Muddling Toward Frugality.* San Francisco: Sierra Club, 1978.

Jourdain, Philip E. B. *The Nature of Mathematics.* London: T. C. and E. C. Jack, 1913.

Jung, Carl. "Dream Symbols in the Process of Individuation." In *The Integration of the Personality,* translated by S. M. Drell. London: Kegan Paul, 1940.

———. *Memories, Dreams, Reflections.* Edited by Aniela Jaffe and translated by Richard and Clara Winston.* New York: Pantheon, 1963. [*Erinnerungen, Träume, Gedanken*]

Kalm, Peter [Pehr]. *Travels in North America.* Translated by John Reinhold Forster.* 2 vols. New York: Dover, 1964. [*En Resa til Norra America,* 1753–61]

Kane, Joe. *Savages.* New York: Knopf, 1995.

Kappen, Ludger, and Otto L. Lange. "Kälteresistenz von Flechten aus verschiedenen Klimagebieten." In *Vorträge aus dem Gesamtgebiet der Botanik.* Neue Folge, no. 4. Stuttgart: Fischer, 1970.

Karl, Thomas R., et al. "The Coming Climate." *Scientific American* 276, no. 5 (1997): 78–83.

Katz, Peter. *The New Urbanism: Toward an Architecture of Community.* New York: McGraw-Hill, 1994.

Kauffman, Stuart. *At Home in the Universe: The Search for the Laws of Self-Organization and Complexity.* New York: Oxford, 1995.

———. *The Origins of Order: Self-Organization and Selection in Evolution.* New York: Oxford, 1993.

Kaufmann, Yehezkel. *The Religion of Israel.* Translated and abridged by Moshe Greenberg. Chicago: University of Chicago, 1960. [*Toldot ha-emunah ha-Yisreelit*]

Kay, Jane Holtz. *Asphalt Nation: How the Automobile Took Over America, and How We Can Take It Back.* New York: Crown, 1997.

Kellert, Stephen R., and Edward O. Wilson, eds. *The Biophilia Hypothesis.* Washington, D.C.: Island Press, 1993.

Kemmis, Daniel. *The Good City and the Good Life.* Boston: Houghton Mifflin, 1995.

Kermode, Frank, ed. *English Pastoral Poetry from the Beginnings to Marvell.* London: G. G. Harrap, 1952.

Khalifa, Shaikha Haya Ali al, and Michael Rice. *Bahrain Through the Ages: The Archae-ology.* London, N.Y.: KPI, 1986.

Khazanov, Anatoly M. *Nomads and the Outside World.* Translated by Julia Crookenden. Cambridge: Cambridge University Press, 1984.

King, F. H. *Farmers of Forty Centuries; or, Permanent Agriculture in China, Korea, and Japan.* Madison, Wis.: Mrs. F. H. King, 1911.

Kirk, G. S. *Myth: Its Meaning and Functions in Ancient and Other Cultures.* Cambridge: Cambridge University Press, 1970.

Klinkenborg, Verlyn. *Making Hay.* New York: N. Lyons, 1986.

Knight, Douglas A., and Gene M. Tucker. *The Hebrew Bible and its Modern Interpreters.* Chico, Calif.: Scholars Press, 1985.

Knoll, Andrew H. "Biological Interactions and Precambrian Eukaryotes." See Tevensz and McCall.

Kolata, Gina. "A Vat of DNA May Become Fast Computer of the Future." *New York Times,* April 11, 1995.

Kongshaug, Nils. "Traffic-Snarled and Smoggy, Athens Cleans Up." *Christian Science Monitor,* June 12, 1995.

Kook, Abraham Isaac. *The Lights of Penitence, The Moral Principles, Lights of Holiness, Essays, Letters, and Poems.* Translated by Ben Zion Bokser. New York: Paulist Press, 1978.

Kostof, Spiro. *The City Assembled: The Elements of Urban Form Through History.* Boston: Little, Brown, 1992.

Kramer, Mark. *Three Farms: Making Milk, Meat, and Money from the American Soil.* Rev. ed. Cambridge, Mass.: Harvard University Press, 1987. [1980]

Kramer, Samuel Noah. "Dilmun: Quest for Paradise." *Antiquity* 37 (1963): 111–15.

———. *History Begins at Sumer.* New York: Doubleday, 1959. [*From the Tablets of Sumer,* 1956]

———. *In the World of Sumer.* Detroit: Wayne State, 1986.

———. *The Sumerians.* Chicago: University of Chicago Press, 1963.

Kramer, Samuel Noah, and John Maier. *Myths of Enki, the Crafty God.* New York: Oxford, 1989.

Krause, Bernard L. 'The Habitat Niche Hypothesis: A Hidden Symphony of Animal Sounds." *The Literary Review* 36, no. 1 (1992): 40–3.

Kunstler, James Howard. "For Sale." *New York Times Magazine,* June 18, 1989.

Labandeira, Conrad C., and J. John Sepkoski, Jr. "Insect Diversity in the Fossil Record." *Science* 261 (1993): 310–15.

Lachiver, Marcel. *Les Anées de misère: La Famine au temps du Grand Roi 1680–1720.* Paris: Fayard, 1991.

Lamberg-Karlovsky, C.C. "Death in Dilmun." See Khalifa and Rice.

———. "Dilmun: Gateway to Immortality." *Journal of the Near Eastern Society* 41 (1982): 45–50.

Langdon, Philip. *A Better Place to Live: Reshaping the American Suburb.* Amherst: University of Massachusetts, 1994.

Lawrence, D. H. *Studies in Classic American Literature.* New York: T. Seltzer, 1922.

Lazzaro, Claudia. *The Italian Renaissance Garden.* New Haven: Yale, 1990.

Leakey, Richard. *The Origin of Humankind.* New York: Basic Books, 1994.

Leary, Warren E. "Science Takes a Lesson from Nature, Imitating Abalone and Spider Silk." *New York Times,* August 31, 1993.

Lee, Richard Borshay. *The !Kung San: Men, Women and Work in a Foraging Society.* Cambridge: Cambridge University Press, 1979.

Lee, Richard B., and Irven DeVore, eds. *Kalahari Hunter-Gatherers.* Cambridge, Mass: Harvard University Press, 1976.

Leopold, Aldo. *A Sand County Almanac, and Sketches Here and There.* New York: Oxford, 1949.

Levenson, Jon. *Sinai and Zion: An Entry into the Jewish Bible.* San Francisco: Harper and Row, 1985.

———. *Theology of the Program of Restoration of Ezekiel 40–48.* Cambridge, Mass.: Scholars Press, 1976.

Levin, Harry. *The Myth of the Golden Age in the Renaissance.* Bloomington: Indiana University Press, 1969.

Lewin, Roger. *Complexity: Life at the Edge of Chaos.* New York: Macmillan, 1992.

Lewis, D. H. "Symbiosis and Mutualism: Crisp Concepts, Soggy Semanitics." See Boucher.

Lewis, M. J. T. "Industrial Archaeology." See Cipolla, vol. 3.

Lewis, Martin W. *Green Delusions: An Environmentalist Critique of Radical Environmentalism.* Durham, N.C.: Duke University Press, 1992.

Lewis, R. W. B. *The American Adam: Innocence, Tragedy, and Tradition in the Nineteenth Century.* Chicago: University of Chicago, 1955.

Lewis, W. H. *The Splendid Century.* New York: Sloane, 1954.

Liddell, Henry George, and Robert Scott. *A Greek-English Lexicon.* Revised by Henry Stuart Jones. Oxford: Oxford University Press, 1968.

Lilley, Samuel. "Technological Progress and the Industrial Revolution 1700–1914." See Cipolla, vol. 3.

Lin, Lorraine H., et al. "Structural Engineering of an Orb-Spider's Web." *Nature* 373 (1995): 146–8.

Lindsay, Jack. *The Troubadours & Their World of the Twelfth and Thirteenth Centuries.* London: F. Muller, 1976.

Lindgren, Hugo. "Bourgeois Dystopias." *Landscape Architecture* 85, no. 8 (1995): 50–5.

Little, Charles E. *The Dying of the Trees: The Pandemic in America's Forests.* New York: Viking, 1995.

———. *Greenways for America.* Baltimore: Johns Hopkins, 1990.

Littlewood, A. R. "Ancient Literary Evidence for the Pleasure Gardens of Roman Country Villas." See Macdougall, ed.

Loope, Lloyd. "Community Outreach and Interagency Involvement: Examples from Maui, Hawaii." See Vitousek, et al., eds.

Lopez, Barry. *Arctic Dreams: Imagination and Desire in a Northern Landscape.* New York: Scribner, 1986.

Lopez, Robert S. "Trade of Medieval Europe: The South." See Postan and Habakkuk, vol. 2.

Loucks, Orie L. "Sustainability in Urban Ecosystems: Beyond an Object of Study." See Platt et al.

Lovejoy, Arthur O., and George Boas. *Primitivism and Related Ideas in Antiquity.* Baltimore: Johns Hopkins, 1935.

Lovelock, James. *The Ages of Gaia: A Biography of Our Living Earth.* New York: Norton, 1988.

———. *Gaia: A New Look at Life on Earth.* Oxford: Oxford University Press, 1979.

Lovins, Amory. *Soft Energy Paths: Toward a Durable Peace.* New York: Harper & Row, 1979. [1977]

Lovins, Amory B., et al. "Energy and Agriculture." See Jackson et al.

Lutz, Richard A., et al. "Rapid Growth at Deep Sea Vents." *Nature* 371 (1994): 663–4.

Lyman, Francesca. "Ginkgoes, Lagoons, and Lanterns: Once and Future Cities." *Orion* 13, no. 4 (1994): 7–13.

MacArthur, Robert H., and Edward O. Wilson. *The Theory of Island Biogeography.* Princeton: Princeton University Press, 1967.

Macdougall, Elisabeth B. *"Ars Hortulorum:* Sixteenth-Century Garden Iconography and Literary Theory in Italy." See Coffin, ed.

———. ed. *Ancient Roman Villa Gardens.* Washington, D.C.: Dumbarton Oaks, 1987.

Macdougall, Elisabeth B., and Richard Ettinghausen. *The Islamic Garden.* Washington, D.C.: Dumbarton Oaks, 1976.

Macdougall, Elisabeth B., and Wilhelmina F. Jashemski, eds. *Ancient Roman Gardens.* Washington, D.C.: Dumbarton Oaks, 1981.

MacFadyen, J. Tevere. *Gaining Ground: The Renewal of America's Small Farms.* New York: Holt, Rinehart, and Winston, 1984.

Mackenzie, Debora. "Red Flag for Green Spray." *New Scientist,* May 29, 1999.

MacLeish, William H. *The Day Before America: Changing the Nature of a Continent.* Boston: Houghton Mifflin, 1994.

Madigan, Michael T., et al. *Brock Biology of Microorganisms.* 8th ed. New York: Prentice-Hall, 1997.

Magris, Claudio. *Danube.* Translated by Patrick Creagh.* New York: Farrar, Straus, Giroux, 1989. [*Danubio,* 1986]

Maimonides, Moses. *The Guide of the Perplexed.* 2 vols. Translated by Shlomo Pines. Chicago: University of Chicago, 1963. [*Dalalat al-hairin*]

Maize, Kennedy P. "The great Kern County mouse war." *Audubon* November 1977, 158–60.

Malloch, D. W., et al. "Ecological and Evolutionary Significance of Mycorrhizal Symbioses in Vascular Plants (A Review)." *Proceedings of the National Academy of Sciences* 77, no. 4 (1980): 2113–18.

Mallory, J. P. In *Search of the Indo-Europeans: Language, Archaeology, and Myth.* London: Thames and Hudson, 1989.

"Managing Planet Earth." *Scientific American* 261, no. 3 (1989).

Manuel, Frank E., and Fritzie P. Manuel. *Utopian Thought in the Western World.* Cambridge, Mass.: Harvard University Press, 1979.

Manes, Christopher. *Green Rage: Radical Environmentalism and the Unmaking of Civilization.* Boston: Little, Brown, 1990.

Margolis, Max L., ed. *The Holy Scriptures According to the Masoretic Text.* Translated by Marcus Jastrow et al. Rev. ed. Philadelphia: Jewish Publication Society, 1955. [1917]

Margulis, Lynn. *Early Life*. Boston: Jones and Bartlett, 1984.

———. "Spirochetes and the Origin of Undulipodia." See Margulis and Olendzenski.

———. *Symbiosis in Cell Evolution*. 2d ed. New York: Freeman, 1993. [1981]

Margulis, Lynn, and J. E. Lovelock. "Gaia and Geognosy." In *Global Ecology: Towards a Science of the Biosphere*, edited by Mitchell B. Rambler et al. Boston: Academic Press, 1989.

Margulis, Lynn, and Lorraine Olendzenski, eds. *Environmental Evolution*. Cambridge, Mass.: MIT Press, 1992.

Margulis, Lynn, and Dorion Sagan. *Microcosmos: Four Billion Years of Microbial Evolution*. New York: Summit Books, 1986.

———. *What is Life?* New York: Simon & Schuster, 1995.

Margulis, Lynn, and Karlene V. Schwartz. *Five Kingdoms: An Illustrated Guide to the Phyla of Life on Earth*. 2d ed. San Francisco: Freeman, 1988.

Marsh, George Perkins. *Man and Nature*. Edited by David Lowenthal. Cambridge, Mass.: Harvard University Press, 1965. [1864]

Marten, Gerald G., ed. *Traditional Agriculture in Southeast Asia*. Boulder, Colo.: Westview Press, 1986.

Martin, Douglas. "Helping Nature Restore Life to a Shoreline Left for Dead." *New York Times*, September 30, 1994.

Martin, Frank Edgerton. "Riverside Revisited?" *Landscape Architecture* 85, no. 8 (1995): 56–9.

Martin, Michael M. "The Biochemical Basis of the Fungus-Attine Ant Symbiosis." *Science* 169 (1970): 16–20.

Martin, Paul, and Richard Klein. *Quaternary Extinctions*. Tucson: University of Arizona, 1984.

Martines, Lauro. *Power and Imagination: City-States in Renaissance Italy*. Baltimore: Johns Hopkins, 1988.

Marx, Leo. *The Machine in the Garden: Technology and the Pastoral Ideal in America*. New York: Oxford, 1964.

Masson, Jeffrey Moussaieff, and Susan McCarthy. *When Elephants Weep: The Emotional Lives of Animals*. New York: Delacorte, 1995.

Matthiessen, Peter. *Wildlife in America*. Rev. ed. New York: Viking, 1987. [1959]

Mattson, D. J., and M. M. Reid. "Conservation of the Yellowstone Grizzly Bear." *Conservation Biology* 5 (1991): 364–72.

Maturana, Humberto R., and Francisco J. Varela. *The Tree of Knowledge: The Biological Roots of Human Understanding*. Boston: Shambhala, 1988.

Mauer, Gerlinde. "Agriculture of the Old Babylonian Period." *Journal of the Ancient Near Eastern Society* 15 (1983): 63–78.

May, Robert, M. *Stability and Complexity in Model Ecosystems*. Princeton: Princeton University Press, 1973.

May, Robert M., ed. *Theoretical Ecology*. 2d ed. Sunderland, Mass.: Sinauer, 1981. [1976]

McDade, Lucinda A., et al., eds. *La Selva: Ecology and Natural History of a Neotropical Rainforest*. Chicago: University of Chicago, 1994.

McEvedy, Colin. *The Penguin Atlas of Ancient History*. Harmondsworth: Penguin, 1967.

McHarg, Ian L. *Design with Nature*. Garden City: Natural History Press, 1969.

McPhee, John. *The Control of Nature*. New York: Farrar Straus Giroux, 1989.

McPherson, E. Gregory. "Cooling Urban Heat Islands with Sustainable Landscapes." See Platt et al.

McKibben, Bill. *The Comforting Whirlwind: God, Job, and the Scale of Creation*. Grand Rapids, Mich.: Eerdmans, 1994.

———. *The End of Nature*. New York: Random House, 1989.

———. *Hope, Human and Wild: True Stories of Living Lightly on the Earth*. Boston: Little, Brown, 1995.

McNeill, William H. *Plagues and Peoples*. Garden City, N.Y.: Anchor Press/Doubleday, 1976.

McKinley, James C., Jr. "In Lake Victoria, Amazonian Weed Wreaks Havoc." *New York Times*, August 5, 1996.

McMenamin, Mark A., and Dianna L. Schulte McMenamin. *The Emergence of Animals: The Cambrian Breakthrough*. New York: Columbia University Press, 1990.

Mee, Charles L. *Daily Life in Renaissance Italy*. New York: American Heritage, 1975.

Meiggs, Russell. *Trees and Timber in the Ancient Mediterranean World*. Oxford: Clarendon Press, 1982.

Mendenhall, George E. *The Tenth Generation*. Baltimore: Johns Hopkins, 1973.

Methivier, Hubert. *Le siècle de Louis XIV*. 9th ed. Paris: Presses Universitaires de France, 1983. [1950]

Millard, A. R., and P. Bordeuil. "A Statue from Syria with Assyrian and Aramaic Inscriptions." *Biblical Archaeology* 45 (1982): 135–41.

Miller, Henry. *The Colossus of Maroussi*. New York: New Directions, 1941.

Miller, Perry. *Errand Into the Wilderness*. Cambridge, Mass.: Harvard, 1956.

Mitchinson, Avrion. "Will We Survive?" *Scientific American* 269, no. 3 (1993): 136–44.

Mock, D. W., and L. S. Forbes. "The Evolution of Parental Optimism." *Trends in Ecology and Evolution* 10, no. 3 (1995): 130–4.

Mollison, Bill. *Permaculture Two: Practical Design for Town and Country in Permanent Agriculture*. Stanley, Tasmania: Tagari, 1979.

Mollison, Bill, and David Holmgren. *Permaculture One: A Perennial Agriculture for Human Settlements*. 2d ed. Tyalgum, New South Wales: Tagari, 1982. [1978]

Montaigne, Michel de. *Essays*. Translated by John Florio.* New York: Modern Library, 1933. [*Essais*, 1580–88]

———. *Travel Journals*. Translated by Donald M. Frame.* San Francisco: North Point, 1983. [*Journal de voyage en italie*, 1774]

Mooney, Pat Roy. "The Law of the Seed: Another Development and Plant Genetic Resources." *Development Dialogue* 1983, nos. 1–2.

Moore, Charles W. *Water and Architecture*. New York: Abrams, 1994.

Morris, David. *The New City-States*. Washington, D.C.: Institute for Local Self-Reliance, 1982.

Morrison, Philip. "A Mayan in St. Petersburg." *Scientific American* 268, no. 4 (1993): 126–7.

Morrish, William, and Catherine Brown. "Infrastructure for the New Social Covenant." In *The Productive Park: New Waterworks as Neighborhood Resources*, edited by Architectural League of New York. New York: Princeton Architectural Press, 1994.

Morse, Stephen S. "Factors in the Emergence of Infectious Diseases." *Emerging Infectious Diseases* 1, no. 1 (1995).

Moscati, Sabatino. *Ancient Semitic Civilizations.* New York: Putnam, 1960. [*Storia e civiltà dei Semiti*]

———. *The Phoenicians.* New York: Abbeville, 1988. [*I Fenici*]

———. *The World of the Phoenicians.* Translated by Alastair Hamilton. London: Weidenfeld & Nicolson, 1968. [*Il mundo dei Fenici*]

Moynihan, Elizabeth B. *Paradise as a Garden: In Persia and Mughal India.* New York: Braziller, 1979.

Muir, John. "Features of the Proposed Yosemite National Park." In *Nature Writings.* New York: Library of America, 1997. [1890]

———. *My First Summer in the Sierra.* In *Nature Writings.* [1911]

Mumford, Lewis. *Art and Technics.* New York: Columbia, 1952.

———. *The City in History.* New York: Harcourt, Brace & World, 1961.

———. "The Garden City Idea and Modern Planning." See Howard.

———. *The Myth of the Machine.* 2 vols. New York: Harcourt, Brace & World, 1967–70.

———. *Technics and Civilization.* New York: Harcourt, Brace & World, 1963. [1934]

Muschamp, Herbert. "When Art Is a Public Spectacle." *New York Times,* August 29, 1993.

Myers, Norman, ed. *Gaia: An Atlas of Planet Management.* New York: Doubleday, 1984.

Nabhan, Gary Paul. *The Desert Smells Like Rain: A Naturalist in Papago Indian Country.* San Francisco: North Point, 1982.

———. *Gathering the Desert.* Tucson: University of Arizona, 1985.

———. *Songbirds, Truffles, and Wolves: An American Naturalist in Italy.* New York: Pantheon, 1993.

Naess, Arne. *Ecology, Community, and Lifestyle: Outline of an Ecosophy.* Revised and translated by David Rothenberg. Cambridge: Cambridge University Press, 1989. [Adapted from *Økologi, Samfunn, og Livsstil,* 1976]

Nash, Roderick. *Wilderness and the American Mind.* 3d ed. New Haven: Yale, 1982. [1967]

National Academy of Sciences. *Genetic Vulnerability of Major Crops.* Washington, D.C.: Government Printing Office, 1972.

Nawrath, Christiane, et al. "Targeting of the Polyhydroxybutyrate Biosynthesis Pathway to the Plastids of Arabidopsis thaliana Results in High Levels of Polymer Accumulation." *Proceedings of the National Academy of Sciences* 91 (1994): 12760–4.

Naylor, Rosamond. "Socio-Economic Aspects of Biological Invasion, A Case Study: The Golden Apple Snail." See Vitousek et al.

Needham, Joseph. *Science and Civilization in China.* 6 vol. in 16, ongoing. Cambridge: Cambridge University Press, 1954–.

Newby, Eric. *Slowly Down the Ganges.* New York: Scribner, 1967.

Newman, Morris. "The New Urbanism Meets Nature." *Landscape Architecture* 85, no. 8 (1995): 60–3.

Newton, Norman T. *Design on the Land: The Development of Landscape Architecture.* Cambridge, Mass.: Harvard University Press, 1971.

Nicholson-Lord, David. *The Greening of the Cities.* London: Routledge & Kegan Paul, 1987.

Nietzsche, Friedrich. "The Genealogy of Morals." In *Basic Writings of Nietzsche*. Edited and translated by Walter Kaufmann. New York: Modern Library, 1966. [*Zur Genealogie der Moral, Eine Streitschrift,* 1887]

Nissen, H.J. "The Occurrence of Dilmun in the Oldest Texts of Mesopotamia." See Khalifa and Rice.

Novak, Barbara. *Nature and Culture: American Landscape and Painting, 1825–1875.* Rev. ed. New York: Oxford, 1995. [1980]

Nussbaum, Martha C. *The Fragility of Goodness: Luck and Ethics in Greek Tragedy and Philosophy.* Cambridge: Cambridge University Press, 1986.

O'Brian, Patrick. *The Yellow Admiral.* New York: Norton, 1996.

O'Brien, Patrick, and Roland Quinault. *The Industrial Revolution and British Society.* Cambridge: Cambridge University Press, 1993.

Odum, Eugene P. *Fundamentals of Ecology.* 3d ed. Philadelphia: W. B. Saunders, 1971. [1953]

Office of Technology Assessment, U.S. Congress. *Harmful Non-Indigenous Species in the United States.* OTA F565. Washington, D.C.: Government Printing Office, 1993.

O'Hara, Sarah L., et al. "Accelerated Soil Erosion Around Mexican Highland Lake Caused by Prehispanic Agriculture." *Nature* 362 (1993): 48–51.

Oldeman, L. R., et al. *World Map of the Status of Human-Induced Soil Degradation: An Explanatory Note.* 2d ed. Wageningen, Netherlands: International Soil Reference and United Nations Environment Programme, 1991.

Olsen, Gary J., and Carl R. Woese. "Lessons from an Archaeal Genome: What Are We Learning from *Methanococcus jannaschii?*" *Trends in Genetics* 12, no. 10 (1996): 377–9.

Oparin, A. I. *The Origin of Life on the Earth.* 3rd ed. Translated by Ann Synge. Edinburgh: Oliver and Boyd. [*Proiskhozhdenie Zhizni,* 1924]

Oppenheim, A. Leo. *Ancient Mesopotamia: Portrait of a Dead Civilization.* Rev. ed. Edited by Erica Reiner. Chicago: University of Chicago, 1977. [1964]

Organization for Economic Co-operation and Development. *Environmental Policies for Cities in the 1990s.* Paris: OECD, 1990.

Otto, Rudolf. *The Idea of the Holy.* 2d ed. Translated by John W. Harvey. London: Oxford, 1957. [*Das heilige*]

Packard, Laurence Bradford, *The Age of Louis XIV.* New York: Holt, Rhinehart, and Winston, 1929 [1966].

Paglia, Camille. *Sexual Personae: Art and Decadence from Nefertiti to Emily Dickinson.* New Haven: Yale, 1990.

Palmer, E. Laurence. *Fieldbook of Natural History.* 2d ed. Revised by H. Seymour Fowler. New York: McGraw-Hill, 1975. [1949]

Panofsky, Erwin. "*Et in Arcadia Ego*: Poussin and the Elegiac Tradition." In *Meaning in the Visual Arts.* Garden City, N.Y.: Doubleday, 1955.

"Parkland Route of Pan-American Highway Sets Off Storm." *New York Times,* March 21, 1995.

Parkman, Francis. *France and England in North America.* 7 vols. Boston: Little, Brown, 1865–92.

Parr, Albert Eide. "Urbanity and the Urban Scene." *Landscape,* Spring 1967, 1.

Passmore, John. *Man's Responsibility for Nature: Ecological Problems and Western Tradition.* New York: Scribner, 1974.

Patai, Raphael. *The Hebrew Goddess.* 3d ed. enl. Detroit: Wayne State University Press, 1990. [1967]

———. *Man and Temple in Ancient Jewish Myth and Ritual.* 2d ed., enl. New York: Ktav, 1967. [1947]

Paul, E. A., and F. E. Clark. *Soil Microbiology and Biochemistry.* New York: Academic Press, 1989.

Pedersen, Johannes. *Israel, Its Life and Culture.* 4 vols. in 2. Translated by Mrs. Aslaug Moller. London: Oxford, 1964. [1920–34]

Peirce, Neal R. *Citistates: How Urban America Can Prosper in a Competitive World.* Washington, D.C.: Seven Locks Press, 1993.

Pennisi, Elizabeth. "Conservation's Ecocentrics." *Science News* 144 (1993): 168–70.

Perkins, S. "Transgenic plants provoke petition." *Science News* 152 (1997): 199.

Perlin, John. *A Forest Journey: The Role of Wood in the Development of Civilization.* New York: Norton, 1989.

Perrin, Noel. *Giving Up the Gun: Japan's Reversion to the Sword, 1543–1879.* Boston: Godine, 1979.

Petitfils, Jean-Christian. *Louis XIV.* Paris: Perrin, 1995.

Pico della Mirandola, Giovanni. "Oration on the Dignity of Man." Translated by Elizabeth Livermore Forbes.* See Cassirer et al. [*Opera,* 1495–96]

Pigou, A. C. *The Economics of Welfare.* 4th ed. London: Macmillan, 1932. [1920]

Pimentel, David, et al. "Environmental and Economic Costs of Soil Erosion and Conservation Benefits." *Science* 267 (1995): 1117–23.

Pimm, Stuart. *The Balance of Nature? Ecological Issues in the Conservation of Species and Communities.* Chicago: University of Chicago, 1991.

———. "In Search of Perennial Solutions." *Nature* 389 (1997): 126–7.

———. "The Value of Everything." *Nature* 387 (1997): 231–2.

Pimm, Stuart, et al. "Times to Extinction for Small Populations of Large Birds." *Proceedings of the National Academy of Sciences* 90 (1993): 10871–5.

Pirozynski, K. A., and D. W. Malloch. "The Origin of Land Plants: A Matter of Mycotrophism." *BioSystems* 6 (1975): 153–64.

Plaskow, Judith. "Blaming the Jews for the Birth of Patriarchy." See Beck.

———. *Standing Again at Sinai: Judaism from a Feminist Perspective.* San Francisco: Harper & Row, 1990.

Plato. *Collected Dialogues.* Corr. ed. Edited by Edith Hamilton and Huntington Cairns. Princeton: Princeton University Press, 1963. [1961]

Platt, Rutherford H., et al. *The Ecological City: Preserving and Restoring Urban Biodiversity.* Amherst: University of Massachusetts, 1994.

Plimpton, George. "Death in the Family." *New York Review of Books,* March 4, 1993, 28–33.

Pliny the Elder. *Natural History.* 10 vols. Translated by H. Rackham. Cambridge, Mass.: Harvard, 1938–63.

Plumb, J. H., ed. *The Horizon Book of the Renaissance.* New York: American Heritage, 1961.

Poggioli, Renato. *The Oaten Flute: Essays on Pastoral Poetry and the Pastoral Ideal.* Cambridge, Mass.: Harvard University Press, 1975.

Pollan, Michael. "Against Nativism." *New York Times Magazine*, May 15, 1994.

———. *Second Nature: A Gardener's Education.* New York: Atlantic Monthly Press, 1991.

———. "Why Mow? The Case Against Lawns." *New York Times Magazine*, May 28, 1989.

Possehl, Suzanne. "In Steppe and Desert, U.N. Seeks to Protect Mongolia Environment." *New York Times,* August 22, 1995.

Ponting, Clive. *A Green History of the World.* New York: St. Martin's, 1991.

Posey, Darrell A. "Keepers of the Campo." *Garden,* November/December 1984, 8–32.

———. "Keepers of the Forest." *Garden,* January/February 1982, 18–24.

Posey, Darrell, et al. "Ethnoecology as Applied Anthropology in Amazonian Development." *Human Organization* 43, no. 2 (1984): 95–107.

Postan, M. M., and H. J. Habakkuk, eds. *The Cambridge Economic History of Europe.* 2d ed. Cambridge: Cambridge University Press, 1966–. [1941–]

Potts, Daniel T., ed. *Dilmun: New Studies in the Archaeology and Early History of Bahrain.* Berlin: D. Reimer, 1983.

Pounds, N. J. G. *An Historical Geography of Europe.* Cambridge: Cambridge University Press, 1990.

Powell, M. A. "Salt, Seed and Yields in Sumerian Agriculture: A Critique of the Theory of Progressive Salinization." *Zeitschrift der Assyrologie* 75 (1985): 7–38.

Powys, John Cowper. *The Meaning of Culture.* New York: Norton, 1929.

Prince, Hugh C. "Georgian Landscapes." See Baker and Harley.

Pritchard, James B., ed. *Ancient Near Eastern Texts Relating to the Old Testament.* 3d ed., with Supplement. Princeton: Princeton University Press, 1969. [1950]

———. ed. *The Ancient Near East in Pictures Relating to the Old Testament.* 2d ed. Princeton: Princeton University Press, 1969. [1954]

———. ed. *The Harper Concise Atlas of the Bible.* New York: HarperCollins, 1991.

Purcell, Nicholas. "Town in Country and Country in Town." See Macdougall, ed.

Quammen, David. *The Song of the Dodo: Island Biogeography in an Age of Extinctions.* New York: Scribner, 1996.

Raloff, Janet. "Dying Breeds." *Science News* 152 (1997): 216–18.

Rank, Otto, and Hans Sachs. "The Significance of Psychoanalysis for the Humanities." In *Psychoanalysis as an Art and a Science.* Detroit: Wayne State University Press, 1968.

Rappoport, A. S. *Myth and Legend of Ancient Israel.* 3 vols. New York: Ktav, 1966.

Rappaport, Roy A. *Pigs for the Ancestors: Ritual in the Ecology of a New Guinea People.* Enl. ed. New Haven: Yale, 1984. [1968]

Raup, David M. *Extinction: Bad Genes or Bad Luck?* New York: Norton, 1991.

Raven, Peter H., et al. *Biology of Plants.* 4th ed. New York: Worth, 1986. [1970]

Raver, Anne. "What's a Flying Female Gypsy Moth? It's Trouble." *New York Times,* September 25, 1994.

Raybould, T. J. *The Economic Emergence of the Black Country: A Study of the Dudley Estate*. Newton Abbot, Devon: David & Charles, 1973.

Rebek, Julius. "Synthetic Self-Replicating Molecules." *Scientific American* 271, no. 1 (1994): 48–55.

Reed, Charles A., ed. *Origins of Agriculture*. The Hague: Mouton, 1977.

Register, Richard. *Ecocity Berkeley: Building Cities for a Healthy Future*. Berkeley: North Atlantic Books, 1987.

Reichman. O. J. *Konza Prairie: A Tallgrass Natural History*. Lawrence: University Press of Kansas, 1987.

Reifenberg, A. *The Struggle Between the Desert and the Sown: The Rise and Fall of Agriculture in the Levant*. Jerusalem: The Jewish Agency, 1955.

Reisner, Marc. *Cadillac Desert: The American West and Its Disappearing Water*. New York: Viking, 1986.

———. *Game Wars: The Undercover Pursuit of Wildlife Poachers*. New York: Viking, 1991.

Renfrew, Colin. *Archaeology and Language: The Puzzle of Indo-European Origins*. London: Jonathan Cape, 1987.

Rennie, John. "Getting Out the Bugs." *Scientific American* 266, no. 3 (1992): 109.

Revkin, Andrew. *Global Warming: Understanding the Forecast*. New York: Abbeville, 1992.

Rice, Michael. *Search for the Paradise Land: An Introduction to the Archaeology of Bahrain and the Arabian Gulf*. London; N.Y.: Longman, 1985.

Rice, Richard E., et al. "Can Sustainable Management Save Tropical Forests?" *Scientific American* 276, no. 4 (1997): 44–9.

Rich, Adam C., et al. "Defining Forest Fragmentation by Corridor Width: The Influence of Narrow Forest-Dividing Corridors on Forest-Nesting Birds in Southern New Jersey." *Conservation Biology* 8, no. 4 (1994): 1109–21.

Richards, B. N. *Introduction to the Soil Ecosystem*. New York: Longman, 1974.

Riddle, John M., et al. "Ever Since Eve . . . Birth Control in the Ancient World." *Archaeology* 47, no. 2 (1994): 29–35.

Ridgway, Brunilde Sismondo. "Greek Antecedents of Roman Garden Sculpture." See Macdougall and Jashemski.

Ries, Julien. "The Fall." See Eliade, ed.

Riley, Terence, ed. *Frank Lloyd Wright: Architect*. New York: The Museum of Modern Art, 1994.

Rindos, David. *The Evolution of Agriculture: An Evolutionary Perspective*. New York: Academic Press, 1984.

Ringel, Michael S., et al. "The Stability and Persistence of Mutualisms Embedded in Community Interactions." *Theoretical Population Biology* 50 (1996): 281–97.

Ringgren, Helmer. *Religions of the Ancient Near East*. Translated by John Sturdy. London: S.P.C.K., 1973. [*Framre Orientens religioner i gammal tid*]

Robinson, Scott K., et al. "Regional Forest Fragmentation and the Nesting Success of Migratory Birds." *Science* 267 (1995): 1987–90.

Rolls, Eric C. *They All Ran Wild: The Story of Pests on the Land in Australia*. Sydney: Angus and Robertson, 1969.

Roelofs, Cora, and Charles Komanoff. *Subsidies for Traffic: How Taxpayer Dollars Underwrite Driving in New York State.* New York: Komanoff Energy Associates, 1994.

Rosand, David. "Giorgione, Venice, and the Pastoral Vision." In *Places of Delight: The Pastoral Landscape,* edited by Robert C. Cafritz et al. New York: Clarkson Potter, 1988.

Rosenberg, David, and Harold Bloom. *The Book of J.* New York: Grove Weidenfeld, 1990.

Rosenmeyer, Thomas G. *The Green Cabinet: Theocritus and the European Pastoral Lyric.* Berkeley: University of California, 1969.

Rosenthal, Ruvik. "Jubilee for All." *The Jerusalem Report* 8, no. 11 (1997): 40–1.

Rothblatt, Donald M., and Daniel J. Garr. *Suburbia: An International Assessment.* New York: St. Martin's, 1986.

Rowe, Peter G. *Making a Middle Landscape.* Cambridge, Mass.: MIT, 1991.

Rub, Timothy. *Michael Singer, Artist in Residence: Recent Projects.* Hanover, N. H.: Dartmouth College, 1996.

Rudofsky, Bernard. *Streets for People: A Primer for Americans.* Garden City, N.Y.: Doubleday, 1969.

Runnels, Curtis N. "Environmental Degradation in Ancient Greece." *Scientific American* 272, no. 3 (1995): 96–9.

Rushdie, Salman. *The Moor's Last Sigh.* New York: Pantheon, 1995.

Russell, J. C. "Population in Europe 500–1500." See Cipolla, vol. 1.

Sackville-West, Vita. "The Persian Garden." In *The Legacy of Persia,* edited by A. J. Arberry. Oxford: Oxford University Press, 1952.

Sagan, Carl, and George Mullen. "Earth and Mars: Evolution of Atmospheres and Surface Temperatures." *Science* 177 (1972): 52–6.

Saint-Simon, Louis de Rouvroy Duc de. *Historical Memoirs: A Shortened Version.* 2 vols. Translated by Lucy Norton.* New York: McGraw-Hill, 1967.

———. *Saint-Simon at Versailles.* Translated by Lucy Norton. New York: Harper, 1958.

Sale, Kirkpatrick. *The Conquest of Paradise: Christopher Columbus and the Columbian Legacy.* New York: Knopf, 1990.

———. *Dwellers in the Land: The Bioregional Vision.* San Francisco: Sierra Club, 1985.

Salonen, Armas. *Agricultura Mesopotamica nach sumerisch-akkadischen Quellen.* Helsinki: Suomalaisen Tiedeakatemian Toimituksia, 1968.

Sanford, Charles L. *The Quest for Paradise: Europe and the American Moral Imagination.* Urbana: University of Illinois, 1961.

Sarna, Nahum M. *Genesis: The JPS Torah Commentary.* Philadelphia: Jewish Publication Society, 1989.

Sauer, Carl. *Agricultural Origins and Dispersals.* New York: American Geographical Society, 1952.

———. *Land and Life.* Berkeley: University of California, 1963.

———. "Man's Dominance by Use of Fire." *Geoscience and Man* 10 (1975): 1–13. Reprinted in *Selected Essays.*

———. *Selected Essays 1963–1975.* Berkeley: Turtle Island Foundation, 1981.

Schafer, R. Murray. *The Tuning of the World.* New York: Knopf, 1977.

Schama, Simon. *Landscape and Memory.* New York: Knopf, 1995.

Schapiro, Robert. *Origins: A Skeptic's Guide to the Creation of Life on Earth.* New York: Summit, 1986.

Schiller, Friedrich. *Naive and Sentimental Poetry,* and *On the Sublime.* Translated by Julius A. Elias.* New York: F. Ungar, 1967. [*Über naive und sentimentalische Dichtung.*]

Schinz, Marina. *Visions of Paradise: Themes and Variations on the Garden.* New York: Stewart, Tabori & Chang, 1985.

Schmidt, Karen. "Whatever Happened to the Gene Revolution?" *New Scientist* 145, no. 1959 (1995): 21–5.

Schneider, Stephen H. *Global Warming: Are We Entering the Greenhouse Century?* San Francisco: Sierra Club, 1989.

Schneider, Stephen H., and Penelope J. Boston. *Scientists on Gaia.* Cambridge, Mass.: MIT Press, 1991.

Schneider, Stephen H., and Randi Londer. *The Coevolution of Climate and Life.* San Francisco: Sierra Club, 1984.

Scholem, Gershom G. *Major Trends in Jewish Mysticism.* 3d. ed. New York: Schocken, 1954. [1941]

———. *On the Kabbalah and Its Symbolism.* Translated by Ralph Manheim. New York: Schocken, 1965. [*Zur Kabbala und ihrer Symbolik,* 1960]

———. *On the Mystical Shape of the Godhead.* Translated by Joachim Neugroschel. Rev. ed. New York: Schocken, 1991. [*Von der mystischen Gestalt der Gottheit,* 1962]

Schopf, J. William, ed. *Earth's Earliest Biosphere: Its Origin and Evolution.* Princeton: Princeton University Press, 1983.

Schrödinger, Erwin. *What Is Life? The Physical Aspect of the Living Cell.* Cambridge: Cambridge University Press, 1944.

Schumacher, E. F. *Small Is Beautiful: A Study of Economics as if People Mattered.* London: Blond and Briggs, 1973.

Sclove, Richard E. *Democracy and Technology.* New York: Guilford, 1995.

———. "Putting Science to Work in Communities." *Chronicle of Higher Education* XLI, no. 29 (1995): B1–2.

Scully, Vincent. *Architecture: The Natural and the Manmade.* New York: St. Martin's, 1991.

———. *The Earth, the Temple, and the Gods: Greek Sacred Architecture.* Rev. ed. New Haven: Yale, 1979. [1962]

Semple, Ellen Churchill. *The Geography of the Mediterranean Region: Its Relation to Ancient History.* New York: Holt, 1931.

Sessions, George, ed. *Deep Ecology for the 21st Century.* Boston: Shambhala, 1995.

Sharifi, Esfandiar. "Parasitic Origins of the Nitrogen-Fixing *Rhizobium*–Legume Symbioses." *BioSystems* 16 (1984): 269–89.

Shepard, Paul. *Man in the Landscape: A Historic View of the Aesthetics of Nature.* New York: Knopf, 1967.

———. *Nature and Madness.* San Francisco: Sierra Club, 1982.

———. *The Tender Carnivore and the Sacred Game.* New York: Scribner, 1973.

———. *Thinking Animals: Animals and the Development of Human Intelligence.* New York: Viking, 1978.

Shepherd, J. C., and G. A. Jellicoe. *Italian Gardens of the Renaissance.* New York: Architectural Book Pub. Co., 1966.

Shetler, Stanwyn G. "Three Faces of Eden." See Viola and Margolis.

Simons, Marlise. "At the Foot of the Acropolis, a Big Dream Unfolds." *New York Times,* February 21, 1991.

Smil, Vaclav. *Global Ecology: Environmental Change and Social Flexibility.* London: Routledge, 1993.

Smith, D. C., and Douglas, A. E. *The Biology of Symbiosis.* London: Edward Arnold, 1986.

Smith, David Eugene. *History of Mathematics.* 2 vols. New York: Dover, 1958. [1925]

Smith, J. Russell. *Tree Crops: A Permanent Agriculture.* New York: Devin-Adair, 1950.

Smith, Jonathan Z. "Golden Age." See Eliade, ed.

Smolin, Lee. *The Life of the Cosmos.* New York: Oxford University Press, 1997.

Smyth, Herbert Weir. *Greek Grammar.* Revised by Gordon M. Messing. Cambridge, Mass.: Harvard University Press, 1956. [1916]

Snell, Bruno. "Arcadia: The Discovery of a Spiritual Landscape." In *The Discovery of the Mind: The Greek Origins of European Thought.* Translated by T. G. Rosenmeyer. Cambridge, Mass.: Harvard University Press, 1953. [*Die Entdeckung des Geistes*]

Snyder, Gary. *The Practice of the Wild.* San Francisco: North Point, 1990.

———. *The Real Work: Interviews and Talks, 1964–79.* New York: New Directions, 1980.

Sonea, Sorin, and Maurice Panisset. *A New Bacteriology.* Boston: Jones and Barlett, 1983. [Rev. trans. of *Introduction à la nouvelle bactériologie,* 1980]

Sontag, Susan. *On Photography.* New York: Farrar, Straus, Giroux, 1977.

Soulé, Michael E., and Bruce A. Wilcox, eds. *Conservation Biology.* Sunderland, Mass.: Sinauer, 1980.

Speiser, E. A. "The Rivers of Paradise." In *Festschrift Johannes Friedrich,* edited by R. von Kienle et al. Heidelberg: Carl Winter, 1959.

Spirn, Ann Whiston. *The Granite Garden: Urban Nature and Human Design.* New York: Basic Books, 1984.

Sprent, J. I. *The Biology of Nitrogen-fixing Organisms.* New York: McGraw-Hill, 1979.

Sprent, J. I., and J. A. Raven. "Evolution of nitrogen-fixing symbioses." *Proceedings of the Royal Society of Edinburgh* 85B (1985): 215–37.

Spufford, Peter. "Coinage and Currency." See Postan and Habakkuk, v.2.

Strehlow, T. G. H. *Aranda Traditions.* Carlton, Australia: Melbourne University Press, 1947.

Stein, Gertrude. *Wars I Have Seen.* New York: Random House, 1945.

Stein, Karen. "Making Art of Trash." *Architectural Record,* June 1994, 98–103.

Stein, Sara. *Noah's Garden: Restoring the Ecology of Our Own Back Yards.* Boston: Houghton Mifflin, 1993.

Stevens, William K. "Feeding a Booming Population Without Destroying the Planet." *New York Times,* April 5, 1994.

———. "Money Grow on Trees? No, but Study Finds Next Best Thing." *New York Times,* April 12, 1994.

———. "New Microbial Agents May Help Kill Crop Pests." *New York Times,* July 10, 1990.

———. "Opportunistic Species Gain in Battle for the Backyard." *New York Times,* March 1, 1994.

———. "Power of Natural Pest-Killer Wanes from Overuse." *New York Times,* December 29, 1992.

——. "Zoos Find a New Role in Conserving Species." *New York Times,* September 22, 1994.

Stewart, Omer C. "Fire as the First Great Force Employed by Man." See W. L. Thomas, vol. 1.

Stilgoe, John R. *Borderland: Origins of the American Suburb, 1820–1939.* New Haven: Yale University Press, 1988.

——. *The Common Landscape of America, 1580–1845.* New Haven: Yale, 1982.

Stringer, Christopher, and Robin McKie. *African Exodus: The Origin of Modern Humans.* New York: Holt, 1996.

Struever, Stuart, ed. *Prehistoric Agriculture.* Garden City, N.Y.: The Natural History Press, 1971.

Stuckey, Ronald L., and Theodore M. Barkley. "Weeds." See Flora of North America, vol. 1.

Sturtevant, Edward Louis. *Sturtevant's Notes on Edible Plants.* Edited by U. P. Hedrick. Albany: J. B. Lyon, 1919.

Sullivan, Arthur L., and Mark Shaffer. "Biogeography of the Megazoo." *Science* 189 (1975): 13–17.

Tarn, W. W. *Hellenistic Civilization.* 3d ed. New York: St. Martin's, 1952.

Tattersall, Ian. "Out of Africa Again … and Again?" *Scientific American* 276, no. 4 (1997): 60–7.

Temple, Robert K. G. *The Genius of China: 3,000 Years of Science, Discovery, and Invention.* New York: Simon and Schuster, 1986.

Tevensz, Michael J. S., and Peter L. McCall, eds. *Biotic Interactions in Recent and Fossil Benthic Communities.* New York: Plenum, 1983.

Theocritus. 2d ed. 2 vols. Edited and translated by A. S. F. Gow. Cambridge: Cambridge University Press, 1952.

Thirgood, J. V. *Man and the Mediterranean Forest.* New York: Academic Press, 1981.

Thomas, Elizabeth Marshall. *The Harmless People.* New York: Knopf, 1959.

——. *Reindeer Moon.* Boston: Houghton Mifflin, 1987.

——. *Warrior Herdsmen.* New York: Knopf, 1965.

Thomas, Sir John Meurig. "Solid Acid Catalysts." *Scientific American* 266, no. 4 (1992): 112–18.

Thomas, Keith. *Man and the Natural World.* New York: Pantheon, 1983.

Thomas, Lewis. "An Earnest Proposal." In *The Lives of a Cell: Notes of a Biology Watcher.* New York: Viking, 1974.

Thomas, William L., Jr., ed. *Man's Role in Changing the Face of the Earth.* 2 vols. Chicago: University of Chicago, 1956.

Thomashow, Mitchell. "Cosmopolitan Bioregionalism." In Michael McGinnis, ed., *Bioregionalism.* London: Routledge, in press.

Thompson, Dorothy Burr. "Ancient Gardens in Greece and Italy." *Archaeology* 4, no. 1 (1951): 41–7.

——. *Garden Lore of Ancient Athens.* Princeton: American School of Classical Studies at Athens, 1963.

Thompson, E. P. *The Making of the English Working Class.* New York: Pantheon, 1964.

Thompson, John N. *Interaction and Coevolution.* New York: Wiley, 1982.

Thompson, William Irwin, ed. *Gaia, A Way of Knowing: Political Implications of the New Biology*. Great Barrington, Mass.: Lindisfarne Press, 1987.

Thoreau, Henry David. *The Correspondence of Henry David Thoreau*. Edited by Walter Harding and Carl Bode. New York: New York University Press, 1958.

———. *Journal*. 14 vols. Edited by Bradford Torrey and Francis H. Allen. Boston: Houghton Mifflin, 1906.

———. *The Maine Woods*. Boston: Ticknor and Fields, 1864.

———. "The Succession of Forest Trees." In *The Writings of Henry David Thoreau*. 20 vols. Boston: Houghton Mifflin, 1906, vol. 5.

———. *Walden, or, Life in the Woods*. Boston: Ticknor and Fields, 1854.

Tibbs, Hardin B. C. "Industrial Ecology: An Environmental Agenda for Industry." *Whole Earth Review*, Fall/Winter 1992.

Tigay, Jeffrey H. *The Evolution of the Gilgamesh Epic*. Philadelphia: University of Pennsylvania, 1982.

Tilman, David. "Biodiversity and Ecosystem Functioning." See Daily.

Tilman, David, and David Wedin. "Oscillations and Chaos in the Dynamics of a Perennial Grass." *Nature* 353 (1991): 653–5.

Tisdale, Sallie. *Stepping Westward: The Long Search for Home in the Pacific Northwest*. New York: Holt, 1991.

Tocqueville, Alexis de. *Democracy in America*. Translated by George Lawrence.* New York: Harper & Row, 1966. [*De la démocratie en Amérique*, 1835, 1850.]

———. *The Old Regime and the French Revolution*. Translated by Stuart Gilbert. Garden City, N.Y.: Doubleday, 1955. [*Ancien régime et la révolution*]

Todd, John. "The Practice of Stewardship." See Jackson et al.

Todd, John, and Beth Josephson. "Living Machines: Theoretical Foundations and Design Precepts." *Annals of Earth* XII, no. 1 (1994): 16–25.

———. "Living Machines: Part Two. Theory Applied: A Mesocosm for the Treatment of Sewage." *Annals of Earth* XII, no. 2 (1992): 14–21.

Todd, Nancy Jack, and John Todd. *From Eco-Cities to Living Machines: Principles of Ecological Design*. Berkeley: North Atlantic Books, 1994. [Rev. ed. of *Bioshelters, Ocean Arks, City Farming*, 1984]

Trench, R. K., et al. "Observations on the Symbiosis with Zooxanthellae Among the Tridacnidae (Mollusca, Bivalvia)." *Biological Bulletin* 161 (1981): 180–98.

Trevelyan, G. M. *Illustrated English Social History*. London: Longmans, Green, 1949–1952. [*English Social History*, 1942]

Trillin, Calvin. *Messages from My Father*. New York: Farrar, Straus and Giroux, 1996.

Tuan, Yi-Fu. "Our Treatment of the Environment in Ideal and Actuality." *American Scientist* 58 (1970): 244–9.

Turner, B. L. II, and Stephen B. Brush. *Comparative Farming Systems*. New York: Guilford Press, 1987.

Turner, Frederick. "Cultivating the American Garden." *Harper's*, August 1985, 45–52.

———. "A Field Guide to the Synthetic Landscape." *Harper's*, April 1988, 49–55.

Turner, Frederick Jackson. "The Significance of the Frontier in American History." In *The Frontier in American History*. New York: Holt, 1921.

Turner, Frederick [W.]. *Beyond Geography. The Western Spirit Against the Wilderness.* New York: Viking, 1980.

Turner, James. *The Politics of Landscape: Rural Scenery and Society in English Poetry, 1630–1660.* Oxford: Blackwell, 1979.

United States Department of Agriculture. *The Second RCA Appraisal: Soil, Water, and Related Resources on Nonfederal Land in the United States.* Washington, D.C.: 1989.

Updike, John. *Bech Is Back.* New York: Knopf, 1982.

van Seters, John. *Abraham in History and Tradition.* New Haven: Yale University Press, 1975.

Varela, F. G., et al. "Autopoiesis: The Organization of Living Systems, Its Characterization and a Model." *BioSystems* 5 (1974): 187–96.

Vasari, Giorgio. *The Lives of the Painters, Sculptors and Architects.* 4 vols. Translated by A. B. Hinds. London: J. M. Dent, 1927. [*Le vite dé più eccelenti pittori scultori e architettori italiani,* 1550, 1568]

Vavilov, N. I. *The Origin, Variation, Immunity, and Breeding of Cultivated Plants.* Translated by K. Starr Chester. Waltham, Mass.: Chronica Botanica, 1951.

Viola, Herman J., and Carolyn Margolis, eds. *Seeds of Change: A Quincentennial Commemoration.* Washington, D.C., and London: Smithsonian Institution Press, 1991.

Vitousek, Peter, et al., eds. *Biological Invasion as a Global Change.* Aspen, Colo.: Aspen Global Change Institute, 1995. Available online at www.gcrio.org/ASPEN/science/eoc94/eoc.html.

———. "Summary." See Vitousek et al.

Vollrath, Fritz. "Spider Webs and Silks." *Scientific American* 266, no. 3 (1992): 70–6.

Wagstaff, J. Malcolm. *The Evolution of Middle Eastern Landscapes.* Totowa, N.J.: Barnes & Noble, 1985.

Waldrop, M. Mitchell. *Complexity: The Emerging Science at the Edge of Order and Chaos.* New York: Simon & Schuster, 1992.

Walker, James C. G. *Earth History.* Boston: Jones and Bartlett, 1986.

Wallace, Alfred Russell. *The Geographical Distribution of Animals.* London: Macmillan, 1876.

Wallace, Howard N. *The Eden Narrative.* Atlanta: Scholars Press, 1985.

Walljasper, Jay. "Something Urban in Denmark." *Utne Reader* 65 (1994): 158–9.

Wallwork, John A. *The Distribution and Diversity of Soil Fauna.* London, N.Y.: Academic Press, 1976.

Walter, Bob, et al., eds. *Sustainable Cities: Concepts and Strategies for Eco-City Development.* Los Angeles: Eco-Home Media, 1992.

Walton, John H. *Ancient Israelite Literature in Its Cultural Context.* Grand Rapids, Mich.: Regency Reference Library, 1989.

Warner, Sam Bass, Jr. *Streetcar Suburbs: The Process of Growth in Boston, 1870–1900.* 2d ed. Cambridge, Mass.: Harvard University Press, 1978. [1962]

Waskow, Arthur. *Down-to-Earth Judaism: Food, Money, Sex, and the Rest of Life*. New York: Morrow, 1995.

———. *Godwrestling*. New York: Schocken, 1978.

Watson, Andrew, and Lovelock, James E. "Biological homoeostasis of the global environment: The parable of Daisyworld." *Tellus* 35B (1983): 284–9.

Watson, James D., et al. *Recombinant DNA: A Short Course*. New York: Scientific American, 1983.

Watson, Robert T., et al., eds. *Climate Change 1995: Impacts, Adaptations and Mitigation of Climate Change: Scientific-Technical Analyses*. Cambridge: Cambridge University Press, 1996.

Weiner, Jon. *The Next One Hundred Years: Shaping the Fate of Our Living Earth*. New York: Bantam, 1990.

Weisenfeld, Stephen L. "Sickle-Cell Trait in Human Biological and Cultural Evolution." See Vayda.

Weisman, Jonathan. "Tilting at Windmills." *Wildlife Conservation*, January/February 1994, 52–7.

Weiss, H., et al. "The Genesis and Collapse of Third Millennium North Mesopotamian Civilization." *Science* 261 (1993): 995–1004.

Wellisch, H. "Alexandrian Library." In *Encyclopedia of Library History*, edited by Wayne A. Wiegand and Donald G. Davis. New York: Garland, 1994.

Werner, Dennis. *Amazon Journey: An Anthropologist's Year Among Brazil's Mekranoti Indians*. New York: Simon & Schuster, 1984.

Westbrooks, Randy. "APHIS and Strengthening the U.S. Response to Invasions." See Vitousek et al.

White, Gilbert. *The Natural History and Antiquities of Selbourne*. 3d ed. London, 1813. [1789]

White, K. D. *Roman Farming*. Ithaca: Cornell, 1970.

White, Lynn, Jr. "The Expansion of Technology 500–1500." See Cipolla, vol. 1.

———. *Medieval Technology and Social Change*. Oxford: Oxford University Press, 1962.

———. "The Historical Roots of Our Ecological Crisis." *Science* 155 (1967): 1203–7. Reprinted in *Machina ex Deo: Essays in the Dynamism of Western Culture*. Cambridge, Mass.: MIT Press, 1968.

Whyte, R. O. "Evolution of Land Use in South-Western Asia." In *A History of Land Use in Arid Regions*, edited by L. Dudley Stamp. Paris: UNESCO, 1961.

Wilcove, David. "Biological Invasions in Fresh Water Ecosystems." See Vitousek et al.

Wilford, John Noble. "Clues Etched in Bone Debunk Theory of a Plague's Spread." *New York Times*, November 17, 1992.

Williams, Raymond. *The Country and the City*. New York: Oxford, 1973.

Williamson, Tom, and Liz Bellamy. *Property and Landscape: A Social History of Land Ownership and the English Countryside*. London: George Philip, 1987.

Willis, Delta. *The Sand Dollar and the Slide Rule: Drawing Blueprints From Nature*. Reading, Mass.: Addison-Wesley, 1995.

Wilson, Charles. *England's Apprenticeship 1603–1763*. 2d ed. London: Longman, 1984. [1965]

Wilson, Edward O. *Biophilia*. Cambridge, Mass.: Harvard University Press, 1984.

———. *The Diversity of Life*. Cambridge, Mass: Harvard University Press, 1992.

Wilson, Edward O., and William H. Bossert. *A Primer of Population Biology.* Sunderland, Mass.: Sinauer, 1971.

Wittfogel, Karl August. *Oriental Despotism.* New Haven: Yale, 1957.

Woodbridge, Kenneth. *Princely Gardens: The Origins and Development of the French Formal Style.* New York: Rizzoli, 1986.

———. *Landscape and Antiquity.* Oxford: Oxford University Press, 1970.

———. *The Stourhead Landscape.* New ed. The National Trust, 1982. [1970]

Worster, Donald. *Rivers of Empire: Water, Aridity, and the Growth of the American West.* New York: Pantheon, 1985.

Wright, H. E., Jr., ed. *Late-Quaternary Environments of the United States.* 2 vols. Minneapolis: University of Minnesota, 1983.

———. *Quaternary Geology and Climate.* Washington: National Academy of Sciences, 1969.

Wylson, Anthony. *Aquatecture: Architecture and Water.* London: The Architectural Press, 1986.

Yaro, Robert D., et. al. *Dealing with Change in the Connecticut River Valley: A Design Manual for Conservation and Development.* 2 vols. in 1. Amherst, Mass.: Massachusetts Dept. of Environmental Management, Center for Rural Massachusetts, 1988.

Yates, Frances. *Giordano Bruno and the Hermetic Tradition.* Chicago: University of Chicago, 1964.

Yonge, C. M. "Giant Clams." *Scientific American* 232, no. 4 (1975): 96–105.

Yoon, Carol Kaesuk. "Boom and Bust May Be the Norm in Nature, Study Suggests." *New York Times,* March 15, 1994.

———. "Insects Adapted to a Single Twig: Specialization in the Extreme." *New York Times,* September 27, 1994.

———. "Lucrative Harvesting of Edible Mushrooms Puts Supply in Danger." *New York Times,* July 28, 1992.

Zahm, S. H., et al. "Pesticides and Non-Hodgkin's Lymphoma." *Cancer Research* (Suppl.) 52 (1992): 5485s–8s.

Zimmern, Alfred. *The Greek Commonwealth.* 5th ed. Oxford: Oxford University Press, 1931. [1911]

Zvelebil, Marek, and Kamil V. Zvelebil. "Agricultural Transition and Indo-European Dispersals." *Antiquity* 62 (1988): 547–83.

Zwerdling, Jon. Report. *All Things Considered.* National Public Radio, July 31, 1994.

INDEX

PERMISSIONS ACKNOWLEDGMENTS

GRATEFUL ACKNOWLEDGMENT is made to the following for permission to reprint previously published material:

DIETRICH REIMER VERLAG: Excerpts by Bendt Alster from *Dilmun: New Studies in the Archaeology and Early History of Bahrain*, edited by Daniel T. Potts (Berlin: Dietrich Reimer Verlag, 1983). Reprinted by permission of Dietrich Reimer Verlag.

HARVARD UNIVERSITY PRESS: Poem #1593 from *The Poems of Emily Dickinson*, edited by Thomas H. Johnson (Cambridge, Mass.: The Belknap Press of Harvard University Press), copyright © 1951, 1955, 1979, 1983 by the President and Fellows of Harvard College. Reprinted by permission of the publishers and the Trustees of Amherst College. Excerpt from *The Cosmic Mountain in Canaan and The Old Testament* by Richard J. Clifford, copyright © 1972 by the President and Fellows of Harvard College. Reprinted by permission of Harvard University Press.

ALFRED A. KNOPF, INC.: "Anecdote of the Jar" from *Collected Poems* by Wallace Stevens, copyright © 1923, copyright renewed 1951 by Wallace Stevens; excerpt by poet Cecco Angiolieri from *Power and Imagination: City States in Renaissance Italy* by Lauro Martines, copyright © 1979 by Lauro Martines. Reprinted by permission of Alfred A. Knopf, Inc.

PENGUIN BOOKS LTD.: Ten lines from *Theogony and Works and Days; Elegies* by Hesiod and Theognis, translated by Dorothea Wender, copyright © 1973 by Dorothea Wender (London: Penguin Classics, 1973). Reprinted by permission of Penguin Books Ltd.

PRINCETON UNIVERSITY PRESS: Excerpts from translations by H. L. Ginsburg, A. K. Grayson, and E. A. Speiser from *Ancient Near Eastern Texts*, edited by James Pritchard, copyright © 1969 by Princeton University Press. Reprinted by permission of Princeton University Press.

SCRIBNER AND A. P. WATT LTD.: Excerpt from "Fragments" from *The Collected Works of W. B. Yeats, Volume I: The Poems*, revised and edited by Richard J. Finneran, copyright © 1933 by Macmillan Publishing Company, copyright renewed 1961 by Bertha Georgie Yeats. Rights outside the United States administered by A. P. Watt Ltd., London, on behalf of Michael Yeats from *The Collected Poems of W. B. Yeats*. Reprinted by permission of Scribner, a division of Simon & Schuster, and A. P. Watt Ltd.

THE UNIVERSITY OF CHICAGO PRESS: Excerpts from *The Sumerians* by Samuel Kramer, copyright © 1963 by The University of Chicago.. Reprinted by permission of The University of Chicago Press.

YALE UNIVERSITY PRESS: Excerpt from *The Treasures of Darkness* by Thorkild Jacobsen, copyright © 1976 by Yale University. Reprinted by permission of Yale University Press.

A NOTE ON THE TYPE

This book was set in Minion, a typeface designed by Robert Slimbach and released by the Adobe Corporation in 1990. Minion is an attractive text face that combines the elegance of Italian Renaissance models with modern versatility and economy. Robert Slimbach began working seriously on type and calligraphy four years prior to joining the Adobe type design staff in 1987. Slimbach has designed typefaces for International Typeface Corporation as well as other Adobe Originals, including Adobe Garamond, Minion Cyrillic, Utopia, Myriad (codesigned with Carol Twombly), Sanvito, Poetica, Caflish Script, Adobe Jenson, Cronos, and Kepler. In 1991, he received the Charles Peignot Award from Association Typographique Internationale for excellence in type design.

Composed by Creative Graphics,
Allentown, Pennsylvania

Printed and bound by Berryville Graphics,
Berryville, Virginia

Designed by M. Kristen Bearse